Florencio Zaragoza Dörwald

Organic Synthesis
on Solid Phase

Related Titles from Wiley-VCH

G. Jung (Ed)
Combinatorial Chemistry-Synthesis, Analysis, Screening
1999. XXXII; 602 pages with 212 figures and 105 tables.

Hardcover. ISBN 3-527-29899-X

H.-G. Schmalz (Ed)
Organic Synthesis Highlights IV
2000. ca. 400 pages with ca. 300 figures and 10 tables.

Softcover. ISBN 3-527-29916-5

H. Waldmann, J. Mulzer (Eds)
Organic Synthesis Highlights III
1998. XV. 444 pages with 308 figures and 10 tables.

Softcover. ISBN 3-527-29500-3

Florencio Zaragoza Dörwald

Organic Synthesis on Solid Phase

Supports, Linkers, Reactions

⊛WILEY-VCH

Weinheim · New York · Chichester · Brisbane · Singapore · Toronto

Dr. Florencio Zaragoza Dörwald
Novo Nordisk A/S MedChem Research
Novo Nordisk Park
DK-2760 Måløv
Denmark

Library of Congress Card No. applied for
A catalogue record for this book is available from the British Library

Die Deutsche Bibliothek – CIP Cataloguing-in-Publication-Data
A catalogue record for this publication is available from Die Deutsche Bibliothek

© WILEY-VCH Verlag GmbH, D-69469 Weinheim (Federal Republic of Germany), 2000

Printed on acid-free and chlorine-free paper

Composition: Kühn & Weyh, D-79111 Freiburg
Printing: Strauss Offsetdruck, D-69509 Mörlenbach
Bookbinding: Osswald & Co., D-67433 Neustadt (Weinstraße)

Printed in the Federal Republic of Germany

Preface

Although the concept of performing organic synthesis on insoluble supports is almost 40 years old, there has been rapid growth of this research area in recent years. It was probably the idea of 'parallel synthesis', proposed by several groups in the late 1980s, which spurred renewed interest in solid-phase chemistry. Parallel synthesis enables the rapid production of large numbers of compounds, and is therefore of critical importance to all organizations which depend on discovering new, patentable substances. In particular the implementation of robotic high-throughput screening, which enables the testing of thousands of compounds per day, has originated the need for comparably fast compound production. Solid-phase synthesis is well suited to perform reactions in parallel, because it readily enables the automated performance of multistep synthetic sequences. Accordingly, solid-phase synthesis is becoming an increasingly important tool for the synthetic chemist, as the trend towards automation and miniaturization of synthesis continues.

My main motivation for writing this book was that my colleagues and I required a practical guide to solid-phase chemistry, in which clear-cut answers to specific questions could be readily found. Thus, the aim of this book is to give the reader a well-structured and exhaustive collection of synthetic procedures realizable on insoluble supports. The subject has been organized according to the type of product resulting from a given synthesis. All preparations or types of linker for a specific type of product should be easy to find in this book, either by using the index or with the aid of the table of contents.

In most chapters illustrative examples are listed in tables, and the precise reaction conditions are given as reported by the authors. If closely related examples have been published, these are listed under the remark 'see also'. These references need be consulted, however, only if more detailed information about a specific reaction is required.

The composition of mixtures of solvents or reagents has been specified by *volume* ratios throughout the book (e.g. TFA/DCM 1:1). Yields are given for cleavage reactions only. These yields often refer, however, not only to the cleavage reaction but to several synthetic steps before cleavage. Yields for the conversion of one support-bound intermediate into another are not given, because such yields cannot generally be accurately determined. In many of the cleavage reactions reported, products were isolated as salts (e.g. trifluoroacetates). To keep the tables more comprehensible, counter-ions have generally been omitted, and the products have been sketched in their uncharged form.

I would like to thank my colleagues and supervisors at Novo Nordisk A/S, in particular Jesper Lau and Behrend F. Lundt, for their steadfast support and motivation. Thanks are also due to Marie Grimstrup, Ulrich Sensfuß, Kilian W. Conde-Friboes, and Bernd Peschke, who have been kind enough to read various sections of the manuscript and to give me valuable suggestions. I wish to thank also Kjeld Madsen, Nils Langeland Johansen, and Leif Christensen for helpful discussions about solid-phase peptide synthesis. Special thanks are due to Dr Anette Eckerle and to the other editors at Wiley-VCH Verlag GmbH for their assistance during the preparation of the manuscript and for the production of the finished book.

Ballerup, Denmark, October 1999   Florencio Zaragoza Dörwald

Contents

Glossary and Abbreviations

Ac	acetyl, MeCO
acac	pentane-2,4-dione
Acm	acetamidomethyl, MeCONH–CH$_2$
ADDP	azodicarboxylic acid dipiperidide
Ade	adenine
Adpoc	1-(1-adamantyl)-1-methylethoxycarbonyl
agarose	see sepharose
AIBN	azobis(isobutyronitrile)
Ala	*S*-alanine
Alloc	allyloxycarbonyl
aq	aqueous
Arg	*S*-arginine
Argogel™	PEG-grafted cross-linked polystyrene
Asn	*S*-asparagine
Asp	*S*-aspartic acid
Azoc	1-(4-phenylazophenyl)-1-methylethoxycarbonyl
(B)	heterocyclic base of nucleotides
BAL	backbone amide linker
9-BBN	9-borabicyclo[3.3.1]nonane
BEMP	2-*tert*-butylimino-2-diethylamino-1,3-dimethylperhydro-1,3-diaza-2-phosphorine
BINAP	2,2′-bis(diphenylphosphino)-1,1′-binaphthyl
Bn	benzyl
Bnpeoc	2,2-bis(4-nitrophenyl)ethoxycarbonyl
Boc	*tert*-butyloxycarbonyl
Bom	benzyloxymethyl
BOP	(1,2,3-benzotriazol-1-yloxy)-tris(dimethylamino)phosphonium hexafluorophosphate, [(Me$_2$N)$_3$P–OBt][PF$_6$]
Bop-Cl	*N,N′*-bis(2-oxo-3-oxazolidinyl)phosphinic chloride
Bpoc	1-(4-biphenylyl)-1-methylethoxycarbonyl
BSA	*N,O*-bis(trimethylsilyl)acetimidate
Bsmoc	(1,1-dioxobenzothiophen-2-yl)methoxycarbonyl
Bt	1-benzotriazolyl
BTPP	*tert*-butylimino-tris(pyrrolidino)phosphorane

Bu	butyl
Bz	benzoyl
CAN	ceric ammonium nitrate, $(NH_4)_2Ce(NO_3)_6$
Cbz	Z, benzyloxycarbonyl, $PhCH_2OCO$
CDI	carbonyldiimidazole
CIP	2-chloro-1,3-dimethyl-2-imidazolinium hexafluorophosphate (Figure 13.6)
coll	collidine, 2,4,6-trimethylpyridine
conc	concentrated
CPG	controlled pore glass
CSA	10-camphorsulfonic acid
Cy	cyclohexyl
Cys	*R*-cysteine
Cyt	cytosine
DABCO	1,4-diazabicyclo[2.2.2]octane
dba	1,5-diphenyl-1,4-pentadien-3-one
DBN	1,5-diazabicyclo[4.3.0]non-5-ene
DBU	1,8-diazabicyclo[5.4.0]undec-5-ene
DCC	*N,N'*-dicyclohexylcarbodiimide
DCE	1,2-dichloroethane
DCM	dichloromethane
DCP	1,2-dichloropropane
Dde-OH	2-acetyl-5,5-dimethyl-1,3-cyclohexanedione
DDQ	2,3-dichloro-5,6-dicyano-1,4-benzoquinone
Ddz	1-(3,5-dimethoxyphenyl)-1-methylethoxycarbonyl
de	diastereomeric excess
DEAD	diethyl azodicarboxylate, $EtO_2C–N=N–CO_2Et$
DECP	diethyl cyanophosphonate, $(EtO)_2P(O)CN$
dextran	see sephadex
(DHQD)$_2$PHAL	dihydroquinidine 1,4-phthalazinediyl diether
DIAD	diisopropyl azodicarboxylate, $iPrO_2C–N=N–CO_2iPr$
DIBAH	diisobutylaluminum hydride
DIC	diisopropylcarbodiimide
dipamp	1,2-bis[phenyl(2-methoxyphenyl)phosphino]ethane
DIPEA	diisopropylethylamine
DMA	*N,N*-dimethylacetamide
DMAD	dimethyl acetylenedicarboxylate, $MeO_2C–C≡C–CO_2Me$
DMAP	4-(dimethylamino)pyridine
DME	1,2-dimethoxyethane, glyme
DMF	*N,N*-dimethylformamide
DMI	1,3-dimethylimidazolidin-2-one
DMSO	dimethyl sulfoxide
DMT	4,4'-dimethoxytrityl
DNA	deoxyribonucleic acid
Dnp	2,4-dinitrophenyl

DPPA	diphenylphosphoryl azide, $(PhO)_2P(O)N_3$
dppe	1,2-bis(diphenylphosphino)ethane
dppf	1,1'-bis(diphenylphosphino)ferrocene
dppp	1,3-bis(diphenylphosphino)propane
DTBMP	2,6-di-*tert*-butyl-4-methylpyridine
DTBP	2,6-di-*tert*-butylpyridine
Dts	dithiasuccinoyl
DVB	divinylbenzene (mixture of regioisomers)
EDC	*N*-ethyl-*N'*-[3-(dimethylamino)propyl]carbodiimide hydrochloride
EDT	1,2-ethanedithiol
ee	enantiomeric excess
EE	1-ethoxyethyl
EEDQ	2-ethoxy-1-ethoxycarbonyl-1,2-dihydroquinoline
eq	equivalent
Et	ethyl
Expansin™	cross-linked polyacrylamide
Fmoc	9-fluorenylmethyloxycarbonyl
FT	Fourier transform
Gln	*S*-glutamine
Glu	*S*-glutamic acid
Gly	glycine
glycan	synonym of 'polysaccharide'
Gua	guanosine
HAL linker	hypersensitive acid-labile linker
HATU	7-aza-3-[(dimethyliminium)(dimethylamino)methyl]-1,2,3-benzotriazol-1-ium-1-olate hexafluorophosphate (Figure 13.6)
HBTU	3-[(dimethyliminium)(dimethylamino)methyl]-1,2,3-benzotriazol-1-ium-1-olate hexafluorophosphate (Figure 13.6)
HDTU	*O*-(4-oxo-3,4-dihydro-1,2,3-benzotriazin-3-yl)-*N,N,N',N'*-tetramethyluronium hexafluorophosphate (Figure 13.6)
Hex	hexyl
His	*S*-histidine
Hmb	2-hydroxy-4-methoxybenzyl
HMBA	4-hydroxymethylbenzoic acid linker
HMPA	hexamethylphosphoric triamide, $(Me_2N)_3PO$
HOAt	3-hydroxy-3*H*-[1,2,3]triazolo[4,5-*b*]pyridine, 4-aza-3-hydroxybenzotriazole
HOBt	1-hydroxybenzotriazole
HODhbt	3-hydroxy-3,4-dihydro-1,2,3-benzotriazin-4-one
HOSu	*N*-hydroxysuccinimide
HPLC	high pressure liquid chromatography
Ile	*S*-isoleucine, 2-amino-3-methylvaleric acid
IPA	2-propanol
*i*Pr	isopropyl

IR	infrared
ivDde-OH	2-(3-methylbutyryl)-5,5-dimethyl-1,3-cyclohexanedione
LCAA	long chain alkylamine spacer
LC–MS	liquid chromatography coupled with mass spectrometry
LDA	lithium diisopropylamide
Leu	*S*-leucine
Lys	*S*-lysine
Macrosorb™	cross-linked polyacrylamide adsorbed on to kieselguhr
MALDI–TOF MS	matrix-assisted laser desorbtion/ionization time-of-flight mass spectrometry
MAS	magic-angle spinning
MBHA	4-methylbenzhydrylamine
MCPBA	3-chloroperbenzoic acid
Me	methyl
MeOPEG	poly(ethylene glycol) monomethyl ether
Merrifield resin	partially chloromethylated, cross-linked polystyrene
MES	2-(4-morpholino)ethanesulfonic acid
Met	*S*-methionine
MMT	monomethoxytrityl
Mom	methoxymethyl
Mpc	1-(4-methylphenyl)-1-methylethoxycarbonyl
Ms	methanesulfonyl
MS	molecular sieves, mass spectrometry
MSNT	1-(mesitylene-2-sulfonyl)-3-nitro-1,2,4-triazole
Mtr	4-methoxy-2,3,6-trimethyl-1-benzenesulfonyl
Mts	mesitylene-2-sulfonyl, 2,4,6-trimethylbenzene-1-sulfonyl
Multipin™	polymer crowns grafted with various supports
nbd	norbornadiene
NBS	*N*-bromosuccinimide
NCS	*N*-chlorosuccinimide
NIS	*N*-iodosuccinimide
NMM	*N*-methylmorpholine
NMO	*N*-methylmorpholine-*N*-oxide
NMP	*N*-methyl-2-pyrrolidinone
NMR	nuclear magnetic resonance
Nos	4-nitrobenzenesulfonyl
Npeoc	2-(4-nitrophenyl)ethoxycarbonyl
Nsc	2-(4-nitrophenyl)sulfonylethoxycarbonyl
Oxone™	2 $KHSO_5 \cdot KHSO_4 \cdot K_2SO_4$, potassium peroxymonosulfate
PA	polyacrylamide, polyacrylates
(PA)	PA with linker or spacer
PAL	5-(4-aminomethyl-3,5-dimethoxyphenoxy)valeric acid linker
PAM	4-(hydroxymethyl)phenylacetic acid linker
Pbf	2,2,4,6,7-pentamethyl-2,3-dihydrobenzo[*b*]furan-5-sulfonyl
PE	polyethylene

PEG	poly(ethylene glycol)
(PEG)	PEG with linker or spacer
PEGA	polyacrylamide cross-linked with PEG
Pepsyn™	cross-linked polyacrylamide
Pepsyn K™	cross-linked polyacrylamide adsorbed on to kieselguhr
Ph	phenyl
Phe	*S*-phenylalanine
Pht	phthaloyl
Piv	pivaloyl, 2,2-dimethylpropanoyl
Pmc	2,2,5,7,8-pentamethyl-6-chromanesulfonyl
PNA	peptide nucleic acid
Pol	undefined polymeric support
Polyhipe™	polyacrylamide on macroporous polystyrene
PPTS	pyridinium tosylate
Pr	propyl
Pro	*S*-proline
PS	cross-linked polystyrene
(PS)	PS with linker or spacer
PTFE	polytetrafluoroethylene
Py	pyridine
PyAOP	(4-aza-1,2,3-benzotriazol-3-yloxy)-tris(pyrrolidino)phosphonium hexafluorophosphate (Figure 13.5)
PyBOP	(1,2,3-benzotriazol-1-yloxy)-tris(pyrrolidino)phosphonium hexafluorophosphate (Figure 13.5)
PyBrOP	bromo-tris(pyrrolidino)phosphonium hexafluorophosphate (Figure 13.5)
RAM	Rink amide linker, (2,4-dimethoxyphenyl)(4-alkoxyphenyl)-methylamine
Red-Al™	sodium bis(2-methoxyethoxy)aluminum hydride
salen	bisimine from ethylenediamine and salicylaldehyde
Sasrin™	cross-linked polystyrene with 4-alkoxy-2-methoxybenzyl alcohol linker
satd	saturated
L-Selectride™	lithium tri(2-butyl)borohydride
sephadex	dextran; a branched glycan consisting of 1,6-α-linked gluco-pyranose
sepharose	agarose; an unbranched glycan consisting of D-galactose and 3,6-anhydro-L-galactose
Ser	*S*-serine
SG	silica gel
Sieber linker	XAL linker, 3-alkoxy-9*H*-9-xanthenylamine
Su	*N*-succinimidyl
TBAF	tetrabutylammonium fluoride
TBDPS	*tert*-butyldiphenylsilyl
TBS	*tert*-butyldimethylsilyl

TBTU	3-[(dimethyliminium)(dimethylamino)methyl]-1,2,3-benzotriazol-1-ium-1-olate tetrafluoroborate (Figure 13.6)
Tentagel™	PEG-grafted cross-linked polystyrene
Teoc	2-(trimethylsilyl)ethoxycarbonyl
TES	triethylsilane
Tf	trifluoromethanesulfonyl
TFA	trifluoroacetic acid
TFFH	tetramethylfluoroformamidinium hexafluorophosphate, $[(Me_2N)_2CF][PF_6]$
TfOH	triflic acid, trifluoromethanesulfonic acid
TG	PEG-grafted polystyrene (e.g. Tentagel)
(TG)	TG with linker or spacer
THF	tetrahydrofuran
THP	2-tetrahydropyranyl
Thr	*S*-threonine
Thy	thymine
TIPS	triisopropylsilyl
TMAD	*N,N,N′,N′*-tetramethyl azodicarboxamide
TMEDA	*N,N,N′,N′*-tetramethylethylenediamine
TMG	*N,N,N′,N′*-tetramethylguanidine
TMS	trimethylsilyl
TOF–SIMS	time of flight secondary ion mass spectrometry
Tol	4-tolyl, 4-methylphenyl
TPAP	tetrapropylammonium perruthenate, $[Pr_4N][RuO_4]$
Tr	trityl, triphenylmethyl
TRIS	tris(hydroxymethyl)aminomethane, $(HOCH_2)_3CNH_2$
Triton™ B	benzyltrimethylammonium hydroxide
Trp	*S*-tryptophan
Ts	tosyl, *p*-toluenesulfonyl
TSTU	*O*-(1-succinimidyl)-*N,N,N′,N′*-tetramethyluronium hexafluorophosphate, $[(Me_2N)_2C–OSu][PF_6]$
Tyr	*S*-tyrosine
Ura	uracil
UV	ultraviolet
Val	*S*-valine
Wang resin	cross-linked polystyrene with 4-benzyloxybenzyl alcohol linker
XAL linker	Sieber linker, 3-alkoxy-9*H*-9-xanthenylamine
Z	Cbz, benzyloxycarbonyl

Experimental Procedures

1 General Techniques and Analytical Tools for Solid-Phase Organic Synthesis

In this chapter some general techniques for the performance of solid-phase reactions are presented. An overview of analytical tools suitable for the characterization of support-bound intermediates is also given, and strategies for the selection of reagents and reactions for parallel solid-phase synthesis are discussed.

1.1 General Techniques for Performing Syntheses on Insoluble Supports

Solid-phase organic synthesis seduces by its simplicity. Typically, reactions on solid phase are performed by shaking a support with a mixture of solvents and reagents for a given time, filtering the mixture, and washing the support with suitable solvents. Cleavage of the product from the support often yields products of high purity, which can either be used directly or purified further by recrystallization or chromatography.

Most supports for solid-phase synthesis are produced and sold as small spherical particles (beads, 0.04–0.15 mm; see Chapter 2). Beaded polymers can be filtered with glass or polypropylene fritts, and, to keep losses of support low, reaction sequences are generally conducted in fritted reactors (Figure 1.1). When handling beaded polymers, mechanical stress on the beads should be avoided (e.g. grinding of the beads with a magnetic stirring bar or with a spatula), because powdered supports tend to clogg fritts irreversibly.

A typical setup for manual solid-phase synthesis at room temperature is sketched in Figure 1.1. A fritted polypropylene tube (e.g. a disposable syringe) fixed on an orbital shaker can be used as a reactor. Because polypropylene reactors are light, several reactors per shaker can be used. An additional advantage of polypropylene is that this material is almost transparent, and enables the visual inspection of the reaction mixture. Polypropylene can, however, not be heated, and if heating is required, PTFE or glass reactors should be used.

After having been charged with support, each reactor is closed with a septum and connected via polyethylene tubing to a filtering flask, which serves as waste container for all reactors.

The support does not usually need to be dried between different reactions, but only washed sufficiently (2–4 times) with a suitable solvent. Only the last wash before cleavage deserves special attention, in particular if the products are to be analyzed or used without further purification. Inpurities physically adsorbed by the support (DMF, NMP, DIPEA, piperidine) will usually be desorbed during cleavage and contaminate the final products. A safe washing protocol consists in 3–10 short (10–30 s) washings alternately with MeOH and DCM, followed by a long washing cycle with DCP (5–18 h).

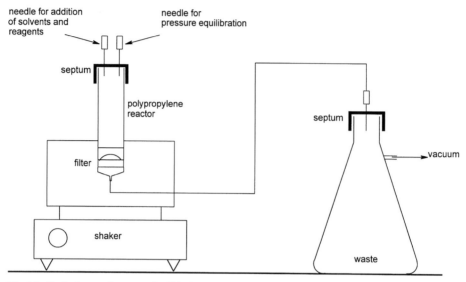

Fig. 1.1. Typical setup for manual solid-phase synthesis.

Syntheses on solid phase can in principle be performed with standard equipment for organic synthesis. This is, however, only justifiable if large amounts of support are being handled. For small-scale preparations or for the development of solid-phase synthetic methodology, the simplicity of solid-phase synthesis should be fully exploited by conducting experiments in parallel. This requires more systematic planning of the experiments, but will be amply rewarded by the number of results obtained.

Synthetic transformations on solid phase need to be driven to completion, because the purification of intermediates is not possible. For long reaction sequences (e.g. for the synthesis of oligomeric compounds) high yields in each step are essential, but for shorter syntheses lower conversions per step might still lead to acceptably pure final products (Table 1.1). If a given transformation does not proceed smoothly, yields can often be increased by using excess and high concentrations of reagent, by increasing the reaction temperature, or by repeating the reaction several times.

For the development of new reaction sequences for solid-phase synthesis, the optimum conditions for each step must be identified. This includes determination of the minimum reaction time and temperature, and the minimum amounts of reagents required. If reaction times or amounts of reagents are excessive, library production will be costly and inefficient.

Because the characterization of support-bound intermediates is difficult (see below), solid-phase reactions are most conveniently monitored by cleaving the inter-

Table 1.1. Yields of final product as a function of yield per step and number of steps.

Yield per step	Number of steps								
	2	3	4	5	6	7	8	9	10
99%	98%	97%	96%	95%	94%	93%	92%	91%	90%
95%	90%	86%	82%	77%	74%	70%	66%	63%	60%
90%	81%	73%	66%	59%	53%	48%	43%	39%	35%
85%	72%	61%	52%	44%	38%	32%	27%	23%	20%
80%	64%	51%	41%	33%	26%	21%	17%	13%	11%

mediates from the support and analyzing them in solution. Depending on the loading, 5–20 mg support will usually deliver sufficient material for analysis by HPLC, LC–MS, and NMR, and enable assessment of the outcome of a reaction. Analytical tools which are particularly suitable for the rapid analysis of small samples resulting from solid-phase synthesis include MALDI–TOF MS [1–3], ion-spray MS [4–6], and LC–MS. MALDI–TOF MS can even be used to analyze the product cleaved from a single bead [3], and is therefore well suited to the identification products synthesized by the mix-and-split method (Section 1.2).

The absolute amount of product resulting from solid-phase synthesis can often be readily determined by ^{1}H NMR with an internal standard or, less efficiently, by purification of the product followed by weighing and full characterization. Volatile samples can be cleaved with a calibrated mixture of hexamethyldisiloxane/TFA/CDCl$_3$, from which ^{1}H NMR spectra can be recorded directly [7]. In our laboratory we use with success DMSO-d_6 as solvent and DMSO-d_5 as internal standard for yield determination by ^{1}H NMR. The weight of crude products from solid-phase synthesis is generally unsuitable for estimating the real yield, because these products are often contaminated with significant amounts of salts, which lead to overestimation of the yield.

1.2 Strategies for Parallel Synthesis

Various techniques have been developed which enable the rapid preparation and screening of large numbers of different compounds by parallel solid-phase synthesis. Reactions can, for instance, be conducted in an array of reactors such as that sketched in Figure 1.1. Synthesizers with up to approximately 600 discrete reactors are commercially available; these enable the automated preparation of compound libraries containing up to 0.1 mmol of each compound. Discrete reactors are also well suited for the development and optimization of solid-phase chemistry.

Alternatively, the so-called mix-and-split method [8–13] can be used to prepare mixtures of support particles (beads, paper disks, etc.) or of small portions of support (e.g. 'tea bags') with a well-defined quantity of one discrete compound linked to each portion of support. These compound libraries can be screened either directly on the support, or in solution after partial or total cleavage of the product from the support.

A schematic representation of the mix-and-split method is sketched in Figure 1.2. In the example shown a support (e.g. Wang resin) is split into four portions and each portion is coupled with one of the four monomers A, B, C, and D. When coupling is complete the support particles are combined in one flask, mixed, and again divided into four portions. Although each portion contains all four types of support particle, each particle is bound to a single monomer only. Now each of the four portions is again coupled with one pure monomer. The first portion will contain support particles with the products A-A, A-B, A-C, and A-D, but on each particle there is still only one compound. Our library at this stage consists of a total of 16 compounds distributed into four portions. This mix-and-split protocol can now be repeated several times, and thereby, in principle, enables the preparation of all possible oligomers of the four monomers A, B, C, and D. The amount of compound available depends on the loading of the support particles chosen.

The mix-and-split method is particularly well suited to the preparation of oligonu-cleotide libraries [14–16] and peptide libraries [1,10,17–21], because even small amounts of these products enable screening and the unambiguous structural elucida-tion [22]. Alternatively, if direct structural elucidation is not possible, each support particle might be tagged during each coupling reaction, so that the final tag of a given support particle is unique to the synthesized oligomer attached to it. If tea bags or paper disks are used, tagging can be achieved simply by marking with a pencil [14]. The tagging of single polymer beads has been accomplished by chemical tagging with polyhalobenzenes [23–25], amines [26–28], carboxylic acids [29], or oligonucleotides [30]. Larger portions of support can also be linked to a chip which enables electronic tagging with a radio emitter [31–33].

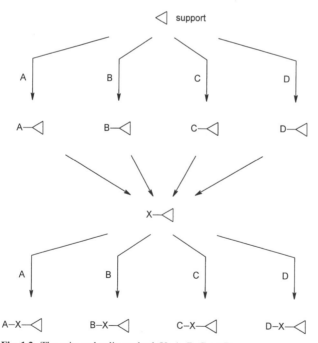

Fig. 1.2. The mix-and-split method. X: A, B, C, or D.

Some types of compound are also well suited to being screened as mixtures. This is, for instance, possible for peptides, which often either bind very strongly or not at all to a given receptor. In this instance the most potent peptide can be identified by deconvolution, and tagging of the support particles is not required. As shown in Figure 1.2, oligomers prepared by the mix-and-split method can contain positions with defined monomers and positions with undefined monomer. If each mixture of oligomers A-X, B-X, C-X, and D-X is cleaved from the support and screened directly, and only one of the 16 oligomers X-X gives a distinct positive response in an assay, then the mixture which contains this oligomer will usually also give a positive response. The active oligomer can now be identified by synthesizing and screening the four pure oligomers contained in the active mixture. Hence, 16 compounds have been screened with only eight assays. For larger oligomers the increased efficiency achieved by this methodology is much larger than in the example shown in Figure 1.2, and deconvolution of compound mixtures prepared by the mix-and-split method has been used with success to screen large peptide libraries containing several million peptides [20,21,34–38]. Mixtures of less selective ligands, however, are not always easy to deconvolute, and for the identification of new, non-peptidic lead structures, parallel synthesis and screening of single compounds is generally the preferred strategy.

1.3 Analytical Methods for Support-bound Intermediates

As alternatives to the cleavage of intermediates from the support and their characterization in solution, various methods have been developed for analyses of support-bound intermediates. The most common analytical tools include combustion analysis, colorimetric assays for specific functional groups, IR, MALDI–TOF MS, TOF–SIMS, and NMR.

1.3.1 Combustion Analysis

Combustion (elemental) analysis of polymeric supports has mainly been used to determine the amount of halogens, nitrogen, or sulfur present in samples of cross-linked polystyrene (see, e.g., [39]). This information can be used to estimate the loading of a support, and to verify if the displacement of a halide has proceeded to completion. In solid-phase peptide synthesis nitrogen determination has been used to estimate the loading of the first amino acid [40].

1.3.2 Colorimetric Assays

One of the first assays used for monitoring the solid-phase synthesis of peptides was the reaction of ninhydrin with primary amines to yield a blue dye ('Kaiser test', Experimental Procedure 1.1 [41]). This rather sensitive assay enables the detection of even small amounts of primary amines on a support, and thereby the monitoring of acylation reactions. Other reagents suitable for detecting amines include 2,4,6-trinitro-benzenesulfonic acid and *p*-chloranil/RCOMe [42,43]. The latter reagent gives blue stained beads with secondary amines also, e.g. proline.

Experimental Procedure 1.1: Detection of primary amines with the Kaiser test [42]

Reagent A: ninhydrin (5 g) in EtOH (100 mL).
Reagent B: phenol (80 g) in EtOH (20 mL).
Reagent C: KCN (2 mL, 1 mmol/L in H_2O), pyridine (98 mL).
   A few beads of the support are washed with EtOH and mixed with two drops of each of the reagents A, B, and C. The resulting mixture is heated to 120 °C for 4–6 min. If primary amino groups are present on the support the beads turn blue.

Experimental Procedure 1.2: Detection of primary or secondary amines with the chloranil test [43]
   Acetone (0.2 mL; detection of secondary amines) or acetaldehyde (0.2 mL; detection of primary amines) is added to a small sample of resin (approx. 1 mg). To this suspension is added a saturated solution of chloranil (2,3,5,6-tetrachloro-1,4-benzoquinone) in toluene (0.05 mL) and the mixture is shaken at room temperature for 5 min. Blue or green beads indicate the presence of amino groups.

There are few sensitive assays for the detection of alcohols. One assay has been described (Experimental Procedure 1.3) which is suitable for monitoring the esterification of support-bound alcohols [44]. In this assay the alcohol is first converted into a tosylate which is then used to quaternize 4-(4-nitrobenzyl)pyridine. Upon deprotonation, the pyridinium salt absorbs visible light strongly, giving rise to a deep blue or red color. Other alkylating agents, such as support-bound alkyl halides, reactive aryl halides, Mannich bases, or reactive epoxides, can also give a positive result in this assay. A negative result is obtained with serine, phenols, carboxylic acids, and amines.

Experimental Procedure 1.3: Detection of primary or secondary aliphatic alcohols [44,45]

Reagent A: TsCl (5 %) in pyridine/toluene 1:1.
Reagent B: 4-(4-nitrobenzyl)pyridine (2 %) in acetone.
Reagent C: Na_2CO_3 (1 mol/L) in water.
   A small portion of support is treated with reagent A for 1.5 h and then washed with DCM and toluene. Reagent B is added and the mixture is heated to 110 °C for at least 20 min. A blue or red color upon treatment with reagent C indicates the presence of primary or secondary aliphatic alcohols on the original support.

Thiols can be detected on insoluble supports by treatment with the symmetric disulfides of 4-mercaptonitrobenzene or 5-mercapto-2-nitrobenzoic acid [46]. These reagents are reduced by thiols to the corrsponding thiophenolates, which are intensely colored.

The amount of acylable amino, hydroxyl, or thiol groups on a support can also be determined by derivatization of these groups with a chromophore-containing reagent, followed by release of this reagent from the support and photometric quantification. Suitable reagents for the derivatization of primary or secondary amines are 3-[(dimethoxytrityloxy)methyl]-4-nitrophenyl isothiocyanate [47], picric acid [48], and *N*-Fmoc amino acids. During the deprotection of DMT-protected, support-bound alcohols and thiols the release of the DMT cation can be monitored photometrically and used to estimate the loading of alcohol.

1.3.3 Infrared Spectroscopy

IR spectroscopy is a fast and simple method for the qualitative detection of certain functional groups on insoluble supports [49,50]. Dried supports can be used directly to prepare KBr pellets for standard recording of IR spectra [39,51,52]. Newer IR-based techniques, which require much less support than a KBr pellet, include single-bead FT-IR spectroscopy [7,49,53,54], single-bead Raman spectroscopy [55], near-IR multi-spectral imaging [56], and the simultaneous analysis of several different beads by FT-IR microscopy for analysis of combinatorial libaries [57].

IR spectroscopy is not a very sensitive analytical tool and is, therefore, not well suited to the detection of small amounts of material. If, however, intermediates have intense and well-resolved IR absorptions, the progress of their chemical transformation can be followed by IR spectroscopy [53,58–60].

1.3.4 Mass Spectrometry

Photosensitive linkers (see Section 3.1.3) enable the direct analysis of support-bound intermediates by MALDI–TOF MS [61,62]. Alternatively, compounds linked to insoluble supports by non-photolabile linkers can be analyzed directly with TOF–SIMS [63].

In both MALDI–TOF MS and ion-spray MS, molecules must be positively charged to be detected. To facilitate detection of all types of support-bound compound, linkers incorporating a charged spacer (e.g. a quaternary ammonium salt), which is also released during MALDI, have proven particularly convenient [4,61].

1.3.5 Nuclear Magnetic Resonance Spectroscopy

Standard (gel-phase) NMR spectra of polymers usually show significant line broadening, mainly because of chemical-shift anisotropy and dipolar coupling [64]. Only nuclei with strong chemical-shift dispersion, e.g. ^{13}C [65–71], ^{15}N [72], ^{19}F [73–75], and ^{31}P [76] give sufficiently resolved gel-phase NMR spectra. The resolution of NMR spectra is improved when the mobility of support-bound molecules increases. Hence, gel-phase NMR of PEG–polystyrene graft supports, for instance, will generally give better spectra than if normal cross-linked polystyrene is used as support. Gel-phase ^{1}H NMR spectra, even if recorded on well-solvated and flexible supports, are, however, too poorly resolved to be of use for the characterization of support-bound intermediates.

A technique especially developed for recording NMR spectra of support-bound compounds is magic-angle spinning (MAS) NMR. This technique requires a special accessory, which keeps a sample of swollen support spinning at 1–2 kHz at the 'magic angle' relative to the magnetic field [64]. MAS NMR enables the recording of much better resolved spectra than gel-phase NMR, and ^{1}H NMR spectra of high quality can be obtained under optimum conditions [39,77–81]. C,H-Correlated and other two-dimensional NMR spectra can also be recorded with MAS NMR [82–84].

1.4 Strategies for the Selection of Reactions and Reagents for Parallel Solid-Phase Synthesis

For companies depending on the identification of novel molecular entities (drugs, herbicides, pesticides, catalysts, dyes, flavors, etc.), rapid access to large amounts of different compounds is of critical importance. Large amounts of compounds are generally required when new, patentable molecules with particular properties are being sought, or when the properties of a given lead structure must be optimized. In the evaluation of compound collections purchased or prepared for this purpose, not only the size but also the quality (i.e. purity, chemical stability, patentability, diversity, etc., of the compounds) is an important issue, which must be carefully considered.

Solid-phase chemistry can readily be automated, and is therefore well suited to automated, parallel, high-throughput compound production. Programmable synthesizers with hundreds of reactors for solid-phase synthesis have become commercially available; these enable the production of large arrays of compounds (one different compound per reactor). This is usually achieved by performing the same synthesis in each reactor, but using different reagents.

Compound libraries designed for the identification of new lead structures should not contain large numbers of closely related compounds (which are likely to have similar properties), but highly diverse compounds with widely different properties. Because compound libraries are prepared using the same reaction sequence for each member of a library, all compounds will contain a repetitive structural element. Diverse libraries will result only if the repetitive element is small and if the synthesis enables the incorporation of widely different side chains (Figure 1.3).

Which kind of solid-phase chemistry and which type of reagent will be most suitable for the production of high-quality, diverse libraries? As shown in Table 1.1 the purity of the final products quickly drops with the total number of synthetic steps, if each step does not proceed with more than 95 % yield. Because reactions cannot always be optimized to such an extent, it is advisable to keep reaction sequences for library production as short as possible.

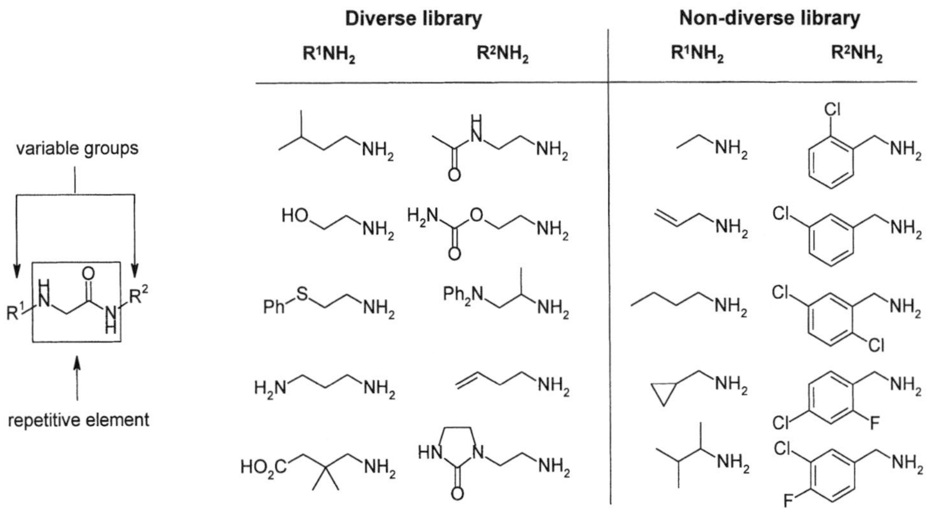

Fig. 1.3. Examples of diverse and non-diverse compound libraries.

Furthermore, reactions should be chosen which enable the use of unprotected, polyfunctional reagents available in large number. Analysis of commercially available reagents reveals that few types of reagent are available which, in addition to the required, reactive functionality, contain a broad selection of further functional groups. The reagents which contain a suitable reactive group and have the highest structural diversity are amines, carboxylic acids, alcohols, and thiols.

Suitable chemistry must be chosen to enable the use of polyfunctional, unprotected reagents. In general, acylations and other reactions with electrophiles of a support-bound substrate will require the protection of functionalized side chains, both in the substrate and in the reagent. Protected, polyfunctional reagents are, however, rare and expensive, and protections/deprotections add further synthetic steps to the synthesis. Nucleophilic transformations, on the other hand, often enable the direct use of highly functionalized, unprotected reagents. Hence, nucleophilic transformations (e.g. aliphatic or aromatic nucleophilic substitutions at support-bound electrophiles, acylations with support-bound, weak acylating agents) should play a central role in reaction sequences for parallel synthesis of compound libraries.

References for Chapter 1

[1] Dawson, P. E.; Fitzgerald, M. C.; Muir, T. W.; Kent, S. B. H. *J. Am. Chem. Soc.* **1997**, *119*, 7917–7927.
[2] Lyttle, M. H.; Hudson, D.; Cook, R. M. *Nucleic Acids Res.* **1996**, *24*, 2793–2798.
[3] Haskins, N. J.; Hunter, D. J.; Organ, A. J.; Rahman, S. S.; Thom, C. *Rapid Commun. Mass Spectrom.* **1995**, *9*, 1437–1440.
[4] McKeown, S. C.; Watson, S. P.; Carr, R. A. E.; Marshall, P. *Tetrahedron Lett.* **1999**, *40*, 2407–2410.
[5] Bray, A. M.; Chiefari, D. S.; Valerio, R. M.; Maeji, N. J. *Tetrahedron Lett.* **1995**, *36*, 5081–5084.
[6] Gao, J. M.; Cheng, X. H.; Chen, R. D.; Sigal, G. B.; Bruce, J. E.; Schwartz, B. L.; Hofstadler, S. A.; Anderson, G. A.; Smith, R. D.; Whitesides, G. M. *J. Med. Chem.* **1996**, *39*, 1949–1955.

[7] Hamper, B. C.; Kolodziej, S. A.; Scates, A. M.; Smith, R. G.; Cortez, E. *J. Org. Chem.* **1998**, *63*, 708–718.
[8] Lebl, M.; Krchnák, V.; Sepetov, N. F.; Seligmann, B.; Strop, P.; Felder, S. *Biopolymers (Peptide Science)* **1995**, *37*, 177–198.
[9] Furka, A. *Drug Develop. Res.* **1995**, *36*, 1–12.
[10] Frank, R. *J. Biotech.* **1995**, *41*, 259–272.
[11] Burgess, K.; Liaw, A. I.; Wang, N. *J. Med. Chem.* **1994**, *37*, 2985–2987.
[12] Dittrich, F.; Tegge, W.; Frank, R. *Bioorg. Med. Chem. Lett.* **1998**, *8*, 2351–2356.
[13] Zhao, P. L.; Zambias, R.; Bolognese, J. A.; Boulton, D.; Chapman, K. *Proc. Natl. Acad. Sci. USA* **1995**, *92*, 10212–10216.
[14] Frank, R.; Heikens, W.; Heisterberg-Moutsis, G.; Blöcker, H. *Nucleic Acids Res.* **1983**, *11*, 4365–4377.
[15] Frank, R.; Meyerhans, A.; Schwellnus, K.; Blöcker, H. *Methods Enzymol.* **1987**, *154*, 221–249.
[16] Markiewicz, W. T.; Markiewicz, M.; Astriab, A.; Godzina, P. *Collect. Czech. Chem. Commun.* **1996**, *61*, S315–S318.
[17] Frank, R.; Döring, R. *Tetrahedron* **1988**, *44*, 6031–6040.
[18] Blankemeyer-Menge, B.; Frank, R. *Tetrahedron Lett.* **1988**, *29*, 5871–5874.
[19] Furka, A.; Sebestyén, F.; Asgedom, M.; Dibó, G. *Int. J. Pept. Prot. Res.* **1991**, *37*, 487–493.
[20] Pinilla, C.; Appel, J.; Blondelle, S. E.; Dooley, C.; Eichler, J.; Houghten, R. A. *Biopolymers (Peptide Science)* **1995**, *37*, 221–240.
[21] Houghten, R. A.; Appel, J. R.; Blondelle, S. E.; Cuervo, J. H.; Dooley, C. T.; Pinilla, C. *Pept. Res.* **1992**, *5*, 351–358.
[22] Youngquist, R. S.; Fuentes, G. R.; Lacey, M. P.; Keough, T. *J. Am. Chem. Soc.* **1995**, *117*, 3900–3906.
[23] Baldwin, J. J.; Burbaum, J. J.; Henderson, I.; Ohlmeyer, M. H. J. *J. Am. Chem. Soc.* **1995**, *117*, 5588–5589.
[24] Nestler, H. P.; Bartlett, P. A.; Still, W. C. *J. Org. Chem.* **1994**, *59*, 4723–4724.
[25] Ohlmeyer, M. H. J.; Swanson, R. N.; Dillard, L. W.; Reader, J. C.; Asouline, G.; Kobayashi, R.; Wigler, M.; Still, W. C. *Proc. Natl. Acad. Sci. USA* **1993**, *90*, 10922–10926.
[26] Boussie, T. R.; Coutard, C.; Turner, H.; Murphy, V.; Powers, T. S. *Angew. Chem. Int. Ed. Engl.* **1998**, *37*, 3272–3275.
[27] Ni, Z. J.; Maclean, D.; Holmes, C. P.; Murphy, M. M.; Ruhland, B.; Jacobs, J. W.; Gordon, E. M.; Gallop, M. A. *J. Med. Chem.* **1996**, *39*, 1601–1608.
[28] Scott, R. H.; Barnes, C.; Gerhard, U.; Balasubramanian, S. *Chem. Commun.* **1999**, 1331–1332.
[29] Hilaire, P. M. S.; Lowary, T. L.; Meldal, M.; Bock, K. *J. Am. Chem. Soc.* **1998**, *120*, 13312–13320.
[30] Nielsen, J.; Brenner, S.; Janda, K. D. *J. Am. Chem. Soc.* **1993**, *115*, 9812–9813.
[31] Moran, E. J.; Sarshar, S.; Cargill, J. F.; Shahbaz, M. M.; Lio, A.; Mjalli, A. M. M.; Armstrong, R. W. *J. Am. Chem. Soc.* **1995**, *117*, 10787–10788.
[32] Nicolaou, K. C.; Xiao, X. Y.; Parandoosh, Z.; Senyei, A.; Nova, M. P. *Angew. Chem. Int. Ed. Engl.* **1995**, *34*, 2289–2291.
[33] Xiao, X. Y.; Parandoosh, Z.; Nova, M. P. *J. Org. Chem.* **1997**, *62*, 6029–6033.
[34] Blondelle, S. E.; Takahashi, E.; Houghten, R. A.; Pérez-Payá, E. *Biochem. J.* **1996**, *313*, 141–147.
[35] Eichler, J.; Lucka, A. W.; Pinilla, C.; Houghten, R. A. *Mol. Diversity* **1996**, *1*, 233–240.
[36] Blondelle, S. E.; Takahashi, E.; Dinh, K. T.; Houghten, R. A. *J. Appl. Bacteriol.* **1995**, *78*, 39–46.
[37] Lutzke, R. A. P.; Eppens, N. A.; Weber, P. A.; Houghten, R. A.; Plasterk, R. H. A. *Proc. Natl. Acad. Sci. USA* **1995**, *92*, 11456–11460.
[38] Pinilla, C.; Chendra, S.; Appel, J. R.; Houghten, R. A. *Pept. Res.* **1995**, *8*, 250–257.
[39] Stranix, B. R.; Gao, J. P.; Barghi, R.; Salha, J.; Darling, G. D. *J. Org. Chem.* **1997**, *62*, 8987–8993.
[40] Wang, S. *J. Org. Chem.* **1975**, *40*, 1235–1239.
[41] Kaiser, E.; Colescott, R. L.; Bossinger, C. D.; Cook, P. I. *Anal. Biochem.* **1970**, *34*, 595–598.
[42] Novabiochem Catalog and Peptide Synthesis Handbook, Läufelfingen, **1999**.
[43] Christensen, T. *Acta Chem. Scand. B* **1979**, *33*, 763–766.
[44] Kuisle, O.; Quiñoá, E.; Riguera, R. *Tetrahedron Lett.* **1999**, *40*, 1203–1206.
[45] Pomonis, J. G.; Severson, R. F.; Freeman, P. J. *J. Chromatogr.* **1969**, *40*, 78–84.
[46] Ellman, G. L. *Arch. Biochem. Biophys.* **1959**, *82*, 70–77.
[47] Chu, S. S.; Reich, S. H. *Bioorg. Med. Chem. Lett.* **1995**, *5*, 1053–1058.
[48] Gisin, B. F. *Anal. Chim. Acta* **1972**, *58*, 248–249.
[49] Luo, Y.; Ouyang, X. H.; Armstrong, R. W.; Murphy, M. M. *J. Org. Chem.* **1998**, *63*, 8719–8722.
[50] Bing, Y. *Acc. Chem. Res.* **1998**, *31*, 621–630.
[51] Fréchet, J. M.; Schuerch, C. *J. Am. Chem. Soc.* **1971**, *93*, 492–496.
[52] Chen, C.; Randall, L. A. A.; Miller, R. B.; Jones, A. D.; Kurth, M. J. *J. Am. Chem. Soc.* **1994**, *116*, 2661–2662.

[53] Yan, B.; Fell, J. B.; Kumaravel, G. *J. Org. Chem.* **1996**, *61*, 7467–7472.
[54] Yan, B.; Sun, Q.; Wareing, J. R.; Jewell, C. F. *J. Org. Chem.* **1996**, *61*, 8765–8770.
[55] Rahman, S. S.; Busby, D. J.; Lee, D. C. *J. Org. Chem.* **1998**, *63*, 6196–6199.
[56] Fischer, M.; Tran, C. D. *Anal. Chem.* **1999**, *71*, 2255–2261.
[57] Haap, W. J.; Walk, T. B.; Jung, G. *Angew. Chem. Int. Ed. Engl.* **1998**, *37*, 3311–3314.
[58] Li, W.; Yan, B. *J. Org. Chem.* **1998**, *63*, 4092–4097.
[59] Marti, R. E.; Yan, B.; Jarosinski, M. A. *J. Org. Chem.* **1997**, *62*, 5615–5618.
[60] Haap, W. J.; Kaiser, D.; Walk, T. B.; Jung, G. *Tetrahedron* **1998**, *54*, 3705–3724.
[61] Carrasco, M. R.; Fitzgerald, M. C.; Oda, Y.; Kent, S. B. H. *Tetrahedron Lett.* **1997**, *38*, 6331–6334.
[62] Fitzgerald, M. C.; Harris, K.; Shevlin, C. G.; Siuzdak, G. *Bioorg. Med. Chem. Lett.* **1996**, *6*, 979–982.
[63] Enjalbal, C.; Maux, D.; Subra, G.; Martinez, J.; Combarieu, R.; Aubagnac, J.-L. *Tetrahedron Lett.* **1999**, *40*, 6217–6220.
[64] Keifer, P. A.; Baltusis, L.; Rice, D. M.; Tymiak, A. A.; Shoolery, J. N. *J. Magn. Resonance Series A* **1996**, *119*, 65–75.
[65] Gordeev, M. F.; Patel, D. V.; Wu, J.; Gordon, E. M. *Tetrahedron Lett.* **1996**, *37*, 4643–4646.
[66] Barn, D. R.; Morphy, J. R.; Rees, D. C. *Tetrahedron Lett.* **1996**, *37*, 3213–3216.
[67] Vidal-Ferran, A.; Bampos, N.; Moyano, A.; Pericàs, M. A.; Riera, A.; Sanders, J. K. M. *J. Org. Chem.* **1998**, *63*, 6309–6318.
[68] Lee, H. B.; Balasubramanian, S. *J. Org. Chem.* **1999**, *64*, 3454–3460.
[69] Epton, R.; Wellings, D. A.; Williams, A. *Reactive Polymers* **1987**, *6*, 143–157.
[70] Look, G. C.; Murphy, M. M.; Campbell, D. A.; Gallop, M. A. *Tetrahedron Lett.* **1995**, *36*, 2937–2940.
[71] Gordeev, M. F.; Patel, D. V.; Gordon, E. M. *J. Org. Chem.* **1996**, *61*, 924–928.
[72] Swayze, E. E. *Tetrahedron Lett.* **1997**, *38*, 8643–8646.
[73] Shapiro, M. J.; Kumaravel, G.; Petter, R. C.; Beveridge, R. *Tetrahedron Lett.* **1996**, *37*, 4671–4674.
[74] Svensson, A.; Bergquist, K. E.; Fex, T.; Kihlberg, J. *Tetrahedron Lett.* **1998**, *39*, 7193–7196.
[75] Albericio, F.; Pons, M.; Pedroso, E.; Giralt, E. *J. Org. Chem.* **1989**, *54*, 360–366.
[76] Johnson, C. R.; Zhang, B. R. *Tetrahedron Lett.* **1995**, *36*, 9253–9256.
[77] Wehler, T.; Westman, J. *Tetrahedron Lett.* **1996**, *37*, 4771–4774.
[78] Pop, I. E.; Dhalluin, C. F.; Déprez, B. P.; Melnyk, P. C.; Lippens, G. M.; Tartar, A. L. *Tetrahedron* **1996**, *52*, 12209–12222.
[79] Chin, J.; Fell, B.; Shapiro, M. J.; Tomesch, J.; Wareing, J. R.; Bray, A. M. *J. Org. Chem.* **1997**, *62*, 538–539.
[80] Riedl, R.; Tappe, R.; Berkessel, A. *J. Am. Chem. Soc.* **1998**, *120*, 8994–9000.
[81] Rademann, J.; Meldal, M.; Bock, K. *Chem. Eur. J.* **1999**, *5*, 1218–1225.
[82] Anderson, R. C.; Jarema, M. A.; Shapiro, M. J.; Stokes, J. P.; Ziliox, M. *J. Org. Chem.* **1995**, *60*, 2650–2651.
[83] Anderson, R. C.; Stokes, J. P.; Shapiro, M. J. *Tetrahedron Lett.* **1995**, *36*, 5311–5314.
[84] Ruhland, T.; Andersen, K.; Pedersen, H. *J. Org. Chem.* **1998**, *63*, 9204–9211.

2 Supports for Solid-Phase Organic Synthesis

Solid-phase organic synthesis refers to syntheses in which the starting material and synthetic intermediates are linked to an insoluble material (support), which enables the facile mechanical separation of the intermediates from reactants and solvents (Figure 2.1).

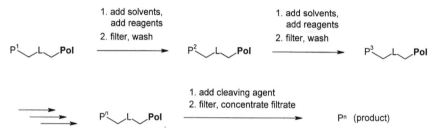

Fig. 2.1. Schematic representation of a synthesis on solid phase. Pol: support, L: linker, P: synthetic intermediate.

Supports of different macroscopic shape have been used for solid-phase synthesis. The most common are spherical particles (beads, 0.04–0.15 mm), which are readily weighed, filtered, and dried, and well-suited for most applications. Other forms of insoluble support include sheets [1], crown-shaped 'pins' [2], or small discs [3].

The general requirements for a support are mechanical stability and chemical inertness under the reaction conditions to be used. Mechanical stability is required to avoid the breaking down of the polymer into smaller particles, which could lead to the clogging of filters. Supports also need to be chemically functionalized, so that the synthetic intermediates can be covalently attached to the support via a suitable linker. Moreover, if the intermediates are located *within* the support (and not only on the surface), diffusion of reagents into the support particles will be necessary, and materials with sufficient permeability or swelling capacity need to be chosen.

Soluble polymers, such as non-cross-linked polystyrene or poly(ethylene glycol), can also be used as supports for organic synthesis [4]. These polymers generally can be precipitated from certain solvents or purified by membrane filtration or recrystallization [5,6], and in this way enable the separation of the desired intermediate from the reagents. Unlike insoluble polymers, soluble supports enable the use of insoluble reagents or catalysts and also the characterization of intermediates by NMR in a homogeneous phase. However, synthesis on soluble supports has a number of drawbacks, which make this technology less attractive for high-throughput synthesis than the use of insoluble supports. Most importantly, synthesis on soluble supports is difficult to automate [7]. Furthermore, it can easily happen that a reagent or solvent strongly binds physically to the polymer, which then precipitates as sticky mass, which can no longer be filtered [8]. It has been shown that reactions on soluble polymers do not proceed significantly faster than on insoluble, cross-linked polymers [9].

Several different support materials have proven useful for solid-phase organic synthesis, but not all materials are compatible with all types of solvents and reagents. Therefore, for each application the proper type of support has to be selected. Some review articles on supports for solid-phase synthesis have recently appeared [1,8,10,11].

2.1 Polystyrene

In the current section the use of polystyrene and copolymers of styrene with various cross-linking agents as supports for solid-phase organic synthesis is discussed. Copolymers of styrene with divinylbenzene are the most common supports for solid-phase synthesis. Depending on the kind of additives used during the polymerization and on the ratio styrene/divinylbenzene, various different types of polystyrene can be prepared. However, also non-cross-linked polystyrene has been used as support for organic synthesis [9,12–17]. Linear, non-cross-linked polystyrene is soluble in organic solvents such as toluene, pyridine, ethyl acetate, THF, chloroform, or DCM, even at low temperatures, but can be selectively precipitated by addition of methanol or water.

2.1.1 Microporous Styrene–Divinylbenzene Copolymers

One of the supports most frequently used for solid-phase organic synthesis are styrene–divinylbenzene copolymers (cross-linked polystyrene, Figure 2.2).

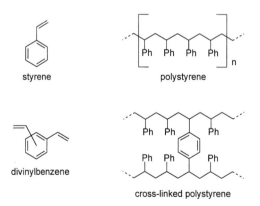

styrene

polystyrene

divinylbenzene

cross-linked polystyrene

Fig. 2.2. Structure of linear and cross-linked polystyrene.

Copolymers of styrene and divinylbenzene were initially developed for the production of ion-exchange resins [18], and are still being used for this purpose [8]. These polymers are essentially insoluble if cross-linking exceeds 0.2 %, but can swell to a variable extent in organic solvents (Table 2.1). The capacity of polystyrene to swell generally decreases with increasing cross-linking [19].

Cross-linked polystyrenes are usually prepared by radical polymerization of suspensions of styrene and divinylbenzene in such a way that polymer beads (0.04–0.15 mm) are directly obtained. The size of the beads can be controlled by addition of surfactants and by adjustment of the stirring speed [20]. Depending on the precise polymerization conditions, polymers of different porosity might result. If the polymerization is conducted in water or other solvents which neither dissolve the monomers nor swell the polymer, microporous, gel-type resins with low internal surface are formed. Microporous polystyrene cross-linked with 1–2 % divinylbenzene is the most common polymer for solid-phase synthesis, and was the support initially chosen by Merrifield for solid-phase peptide synthesis [21]. Resins with less cross-linking (e.g. 0.5 % divinylbenzene [22]) have also been used, but resins with 1–2 % cross-linking remain the most common type of polystyrene support. These polymers are commercially available as beads with a broad choice of functional groups and linkers. These groups are not only located on the surface of the beads, but are (almost) uniformly distributed throughout the polymer [23,24]. Highly cross-linked, non-porous polystyrene has been used for the solid-phase synthesis of oligonucleotides [25].

2.1.1.1 Swelling Behavior

Solvents can penetrate to different amounts into cross-linked polystyrene, causing the size of the beads to increase (Table 2.1). Swelling is strongest in solvents which can bind in a non-covalent manner to the polymer. Because polystyrene is a hydrophobic, polarizable material, swelling is generally strong in dipolar aprotic solvents but poor in alkanes, protic solvents, or water.

The volume of swollen, polystyrene-based supports in a selection of solvents is given in Table 2.1. Because swelling will vary between different batches of polymer, the values given in Table 2.1 should only be considered as approximate guideline.

The size of the beads is also affected by the amount of synthetic intermediate attached to it. For instance, when a peptide with a molecular weight of 5957 g/mol is prepared on polystyrene (1 % cross-linked; loading: 0.95 mmol/g) the volume of the beads (which consist of 81 % peptide) is five times as large as the unloaded support [23,27]. When this resin is suspended in DMF the beads swell further, attaining a volume 26-fold that of the original unloaded, non-swollen polystyrene beads. This strong swelling makes it difficult to use gel-type cross-linked polystyrene for continuous-flow synthesis (where reagents are pumped through a column containing the support).

Diffusion of lipophilic reagents into swollen, microporous polystyrene is generally fast [19,28]. For instance, N-acylation of polystyrene-bound alanine (0.05 mm diameter beads) with the symmetric anhydride of *N*-BPoc leucine reaches 99 % after

Table 2.1. Swelling of polystyrene and Tentagel in different solvents [26]. Volume of swollen, drained resin (1.0 g) in mL.

Solvent	Wang resin (PS, 2 % DVB, 0.6 mmol/g)	Tentagel S Ram[a] (0.3 mmol/g)
NMP	6.4	4.4
pyridine	6.0	4.6
THF	6.0	4.0
N,N-dimethylacetamide	5.8	4.0
quinoline [b]	5.7	5.2
DMPU [b]	5.7	4.5
CHCl$_3$	5.6	5.6
2,6-dimethylpyridine [b]	5.6	5.1
NMM [b]	5.6	4.4
dioxane	5.6	4.2
CH$_2$Cl$_2$	5.4	5.6
DMF	5.2	4.4
1,2-dichlorobenzene	4.8	5.2
MeOCH$_2$CH$_2$OMe	4.8	2.0
1,2-dichloropropane [b]	4.5	4.5
ClCH$_2$CH$_2$Cl	4.4	5.4
benzene	4.4	4.4
2-butanone	4.4	2.0
nitrobenzene [b]	4.3	4.8
DMSO	4.2	3.8
AcOEt	4.2	2.0
toluene	4.0	3.6
MeOCH$_2$CH$_2$OH	4.0	2.4
acetone	3.6	2.8
EtOCH$_2$CH$_2$OH	3.6	2.0
xylene	3.0	2.0
THF/H$_2$O 1:1	2.8	5.2
AcOH	2.8	5.2
Et$_2$O	2.8	2.0
CCl$_4$	2.4	2.8
tert-butyl methyl ether	2.4	1.6
TFA	2.0	6.4
CF$_3$CH$_2$OH [b]	2.0	5.8
MeNO$_2$	2.0	4.4
MeCN	2.0	4.0
MeCN/H$_2$O 1:1	2.0	4.0
DMF/H$_2$O 1:1	2.0	4.0
1-butanol	2.0	2.0
1-propanol	2.0	2.0
2-propanol	2.0	2.0
ethanol	2.0	1.8
DIPEA [b]	1.8	1.6
methanol	1.6	3.6
water	1.6	3.6
HOCH$_2$CH$_2$OH [b]	1.6	1.7
tert-butanol	1.6	1.6
heptane	1.6	1.6

[a] Tentagel with Rink amide linker.
[b] Determined by the author.

only 14 s [27]. On the other hand, reactions involving ionic reagents are often sluggish in polystyrene; this might be because of rate-limiting diffusion of charged particles into the hydrophobic support.

2.1.1.2 Chemical Stability

Cross-linked polystyrene tolerates well a broad range of reaction conditions, including treatment with weak oxidants (ozone, DDQ), strong bases (LDA), and acids (HBr, TfOH). Saponifications with 40 % aqueous sodium hydroxide at 180 °C for 10 h can be realized on 5 % cross-linked macroporous polystyrene without deterioration of the polymer [29]. Strong oxidants at high temperatures, and other reagents which lead to a chemical modification of alkylbenzenes, will, however, not surprisingly, also attack polystyrene.

2.1.1.3 Functionalization

Polystyrene does not enable the covalent, reversible attachment of synthetic intermediates, unless the support is derivatized with suitable functional groups. The most common groups for the reversible attachment of intermediates are chloromethyl and hydroxymethyl groups, whereas aminomethyl groups are mainly used as non-cleavable attachment point for linkers (see Chapter 3).

Functionalized polymers can be prepared either by chemical transformation of the unfunctionalized polymer, or by copolymerization of functionalized monomers. Neither of these strategies provides for perfectly homogeneous distribution of functional groups: during chemical modification of polymer beads the exterior will be more exposed to reagents than the core; this can lead to uneven distribution of functionality. On the other hand, different monomers (e.g. styrene and 4-(chloromethyl)-styrene or divinylbenzene) do not polymerize at the same rate and will be incorporated to a variable extent into the polymer as polymerization proceeds [24,30]. For this reason also the amount of cross-linking of non-functionalized cross-linked polystyrene will be different in the core from in the exterior of the beads. Accordingly, the quality of a functionalized support depends strongly on the monomers used and on the precise conditions used for its preparation.

Cross-linked polystyrene can be functionalized in many ways [31–34]. Those functionalized resins which are frequently used as supports for solid-phase synthesis are commercially available, and their preparation will be mentioned only in brief here.

Almost all electrophilic substitutions known to proceed in solution with isopropyl-benzene can also be performed with polystyrene, using solvents such as nitrobenzene, carbon disulfide, or carbon tetrachloride. These substitutions include bromination [35], nitration [36,37], sulfonylation, Friedel–Crafts acylations [38–42] and alkylations [43–45]. Chloromethyl polystyrene (Merrifield resin) has been prepared by chloromethylation of polystyrene [18,21,46,47], by copolymerization of 4-chloromethylstyrene with styrene [16,20,48,49], and by chlorination of poly(4-methylstyrene) [50,51].

Aminomethyl polystyrene is most conveniently prepared by direct amidomethylation of polystyrene with (hydroxymethyl)amides or (halomethyl)amides under acidic conditions [52–55], but has also been prepared from chloromethyl polystyrene ([56], see Section 10.1.1.1).

Metallated polystyrenes are versatile intermediates for the preparation of a number of polystyrene derivatives. Metallated polystyrene has been prepared from halogenated polystyrenes by halogen–metal exchange [34,35,57,58] and by direct metallation of polystyrene [59–61] (see Chapter 4). Electrophiles suitable for the derivatization of metallated polystyrene include carbon dioxide, carbonyl compounds, sulfur, trimethyl borate, isocyanates, chlorosilanes, alkyl bromides, chlorodiphenylphosphine, DMF, oxirane, selenium [62], dimethyldiselenide [63], organotin halides [61], oxygen [64], etc. [34,35,57–59].

The loading of a support with functional groups is usually expressed in mmol/g. This means that the loading will decrease with increasing weight of the compound attached to the support, because the number of attachment sites remains constant but the weight of the resin increases. Poly(4-chloromethyl)styrene would have a chloride loading of 6.55 mmol/g. Such high loadings can generally not be attained by direct chloromethylation of polystyrene, because interstrand cross-linking competes efficiently with chloromethylation in the later stages of the reaction and leads to a highly cross-linked polymer with poor swelling properties [18,48,65]. A similar effect is also observed in the amidomethylation of polystyrene [53].

Commercially available functionalized polystyrenes generally have loadings of 0.5–1.5 mmol/g; this corresponds roughly to 20 % derivatization of all available phenyl groups. Higher loadings can also be realized, e.g. by polymerization of functionalized styrenes [66], but these supports have not yet found broad application. The reason for this might be that for the solid-phase synthesis of large peptides high loading leads to less predictable swelling properties of the support in later stages of the synthesis. However, despite scepticism among peptide chemists, it has been shown that there is sufficient 'space' in 1–2 % cross-linked polystyrene to accommodate large peptides [23]. Coupling difficulties in peptide synthesis are generally not related to excessive loading of a support but to folding of the peptide on to itself (β-sheet formation), or to phase transitions from a polystyrene-like resin to a polyamide-like resin [67]. Peptides have, in fact, been successfully prepared on polyacrylamide supports with initial loadings as high as 5 mmol/g [68]. High loading will usually not be a problem for the solid-phase synthesis of small molecules, but rather an advantage, because smaller reactors and less solvent will be required for the synthesis of a given amount of product.

High loading can, however, be an inconvenience if resin-bound substrates are able to react with themselves (e.g. during olefin cross-metathesis or macrocyclizations), or if bi- or polyfunctional, unprotected reagents are to be used in solid-phase synthesis. It is, e.g., possible to mono-derivatize symmetric, bifunctional reagents with resin-bound acylating or alkylating agents [69,70]. However, the higher the loading of the support, the higher will be the degree of bis-derivatization of the bifunctional reagent (i.e. the degree of cross-linking). This unwanted reaction can be suppressed by choosing a support with lower loading, or by partial capping (e.g. by silylation) of attachment sites on the support [71].

2.1.2 Macroporous Styrene–Divinylbenzene Copolymers

As mentioned above, polymerization of a suspension of undissolved styrene and divinylbenzene in water leads to the formation of microporous, gel-type supports with low internal surface. Macroporous resins with high internal surface result when a suspension of *dissolved* monomers is polymerized [72,73]. As solvents can be used, for instance, toluene, xylene, or diphenylmethane [74], and the polymerization can also be conducted in the presence of higher alcohols or linear polystyrene ('porogens'). During the polymerization porogens are trapped within the cross-linked polymer, generating large pores. When polymerization is complete the porogens are removed either by washing or by evaporation under reduced pressure. The porous structure of the resulting polymers (Figure 2.3) remains stable even in the absence of solvents, leading to a large internal surface [8,75]. To provide for sufficient mechanical stability these polymers are generally highly cross-linked (> 10 % divinylbenzene); this implies limited swelling properties.

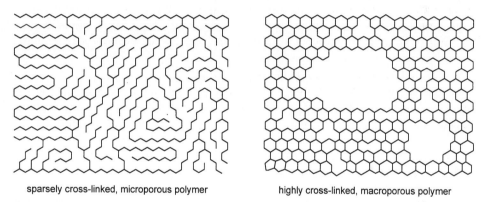

sparsely cross-linked, microporous polymer   highly cross-linked, macroporous polymer

Fig. 2.3. Schematic representation of polymer chains in microporous and macroporous polymers.

Macroporous, highly cross-linked styrene–divinylbenzene copolymers are mainly used as ion-exchange resins (after sulfonylation or aminomethylation). However, macroporous (or 'macroreticular') polystyrene [72,73] is increasingly receiving attention as support for solid-phase synthesis [10,32,76–78]. Although these polymers do not swell significantly, a large internal surface is available for functionalization and remains accessible in a variety of solvents, including alcohols and water. Hence, the choice of solvent is less critical when using macroporous polystyrene than in the case of microporous polystyrene. The capacity of macroporous polystyrene can reach up to 0.8–1.0 mmol/g [10,76].

The fact that macroporous, highly cross-linked polystyrene does not swell makes this support particularly interesting for continuous-flow synthesis in columns. This support has also been successfully used as alternative to CPG for the solid-phase synthesis of oligonucleotides [78,79]. Furthermore, because reagents do not need to penetrate into the polystyrene network, enzyme-mediated reactions should also proceed smoothly on macroporous polystyrene [73].

2.1.3 Miscellaneous Polystyrenes

With the aim of fine-tuning the physicochemical properties of polystyrene-based supports and thereby improve their suitability for solid-phase synthesis, cross-linking agents other than divinylbenzene have been investigated. These include ethylene glycol dimethacrylate [80], hexane-1,6-diol diacrylate [81,82], 1,4-(4-vinylphenoxy) butane [83], and tetraethylene glycol diacrylate [84–87], which lead to supports which swell more than Merrifield resin. Supports with higher ethylene glycol content have been prepared by polymerization of styrene with α,ω-bis(4-vinylphenoxy)oligo- or poly(ethylene glycol) [88,89]. Unlike divinylbenzene-cross-linked polystyrene, these supports swell both in dipolar aprotic solvents and in protic solvents. Supports of this type devoid of benzyl ether groups have also been prepared [90], and proven more stable towards acids than conventional PEG–polystyrene composite supports [90].

Linear polystyrene can be generated on insoluble polymers by γ-irradiation of the latter in a solution of styrene [91]. Polystyrene grafted on to polytetrafluoroethylene [92–97] or polyethylene [2,91,98] can be functionalized in the same way as cross-linked polystyrene, and loadings of up to 1.0 mmol/g can be attained. These supports, which are also available as crown-shaped pins (Multipin, 2–3 mm diameter, 8–10 μmol per crown), have been used for the synthesis of peptides [2,91,92,99], oligonucleotides [93–96,98,100], and small molecules [101–103].

Beads of cross-linked polystyrene containing magnetite crystals (Fe_3O_4, a ferromagnetic material) have been prepared and used with success for solid-phase synthesis [104,105]. The magnetic beads could be readily separated from the reaction mixture with a magnet.

2.2 Poly(ethylene glycol)–Polystyrene Graft Polymers

PEG grafted on to cross-linked polystyrene can be prepared either by linking PEG to suitably functionalized polystyrene [106,107] or by polymerization of oxirane on to a hydroxylated polystyrene support [5,44,108] (Figure 2.4). The former strategy generally gives inferior results, because PEG or symmetric derivatives thereof are bifunctional and can lead to extensive cross-linking [5,7]. An additional strategy for the production of PEG-grafted polystyrene supports is the partial derivatization of aminomethyl polystyrene with PEG monomethyl ether [55] (Figure 2.4).

Commercially available Tentagel consists of about 30 % of a porous matrix of approx. 1 % cross-linked polystyrene on to which 70 % PEG with an average molecular weight of 3000 g/mol has been grafted by oligomerization of oxirane. This support is more hydrophilic than pure polystyrene, and swells in a broad variety of solvents (Table 2.1). Loadings of commercially available Tentagel range between 0.15–0.30 mmol/g. Supports closely related to Tentagel are available commercially under the trademark Argogel [109,110] (Figure 2.4).

In swollen Tentagel and related supports the PEG chains are more mobile than the cross-linked polystyrene matrix. This enables the recording of well-resolved NMR spectra of samples linked to Tentagel [5,7,111–113]. Polystyrene or other, less mobile,

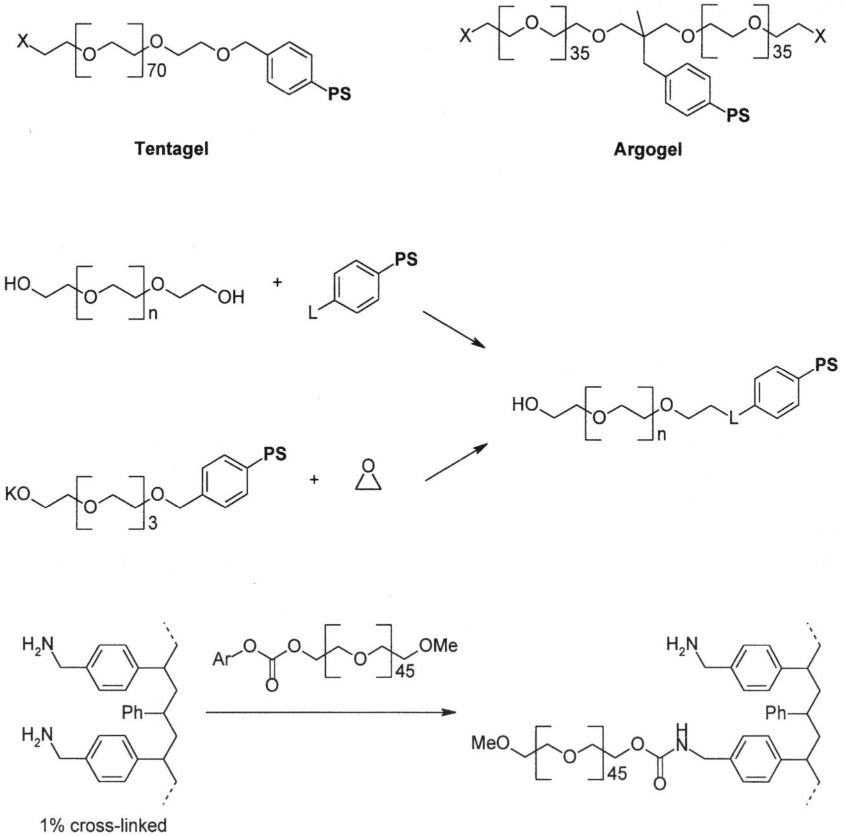

Tentagel

Argogel

1% cross-linked
aminomethyl polystyrene

Fig. 2.4. Preparation of PEG-polystyrene graft supports [5,10,44,55,109]. Ar: $4\text{-}(NO_2)C_6H_4$; L: linker; X: OH, NH_2.

supports generally give broad NMR signals, unless special techniques are used (e.g. magic-angle spinning NMR, [114–116]).

It has been proposed that the increased mobility of intermediates linked to Tentagel also provides for a more 'solution-like' environment and higher reaction rates [5,117]. Recent investigations show, however, that reaction rates on polymeric supports depend on the precise type of reaction and reagents used, and can be higher on cross-linked polystyrene than on Tentagel [28,118]. Polymeric supports generally need to be swollen if dissolved reagents are to reach the interior of the polymer particles [72]. Only reactions involving small ions and/or protic solvents will therefore usually proceed faster on Tentagel than on cross-linked polystyrene.

Unfortunately, PEG-grafted polystyrene supports have some disadvantages, the most critical being the low loading and the release of PEG upon treatment with TFA [8,55,119] or upon heating [120]. Further, because of their high PEG content, these supports become sometimes adherent and are difficult to dry [8]. Increased stability towards acid has been achieved with PEG-grafted polystyrene devoid of benzylic C–O bonds [10,44,55,110].

2.3 Poly(ethylene glycol)

The most common PEG used as support is poly(ethylene glycol) monomethyl ether with a molecular weight of about 5000 g/mol (MeOPEG$_{5000}$). Occasionally non-etherified PEG is used and both hydroxyl groups of each oligomer (e.g. PEG$_{3400}$) are coupled with the starting material [121]. These polymers are soluble in water and most organic solvents, but can be precipitated with hexane, diethyl ether or *tert*-butyl methyl ether. Applications of this class of support in organic synthesis have been reviewed [4,122].

The synthesis of peptides on soluble poly(ethylene glycol) (PEG) was described as early as 1971 [123]. The main motivation for seeking new supports at that time was that some peptides could not be prepared on cross-linked polystyrene. It was later found that this was not because of the support, but mainly the folding of the growing peptide on to itself or because of β-sheet formation [124,125]. Such problems are solved today by introducing *N*-Hmb-amino acids at certain positions of the peptide (see Section 16.1.5), or by extending the coupling times, but generally not by choosing a more hydrophilic or a soluble support.

Despite the drawbacks of soluble supports mentioned in the introduction of Chapter 2, soluble PEG has often been used as support for the synthesis both of biopolymers [6,126–131] and of small organic molecules [14,132–136]. Related supports with higher loading have been described [137], in which 3,5-bis(chloromethyl)phenoxy groups were linked to both ends of PEG$_{5000}$.

Cross-linked, insoluble PEG can be prepared by copolymerizing PEG with epichlorohydrin [89,138] or with the tosylate of 3-methyl-3-(hydroxymethyl)oxetane [139]. These supports are highly permeable and hydrophilic, and enable the use of enzymes such as subtilisin (27 kDa) to catalyze transformations of support-bound substrates [139].

2.4 Polyacrylamides

With the aim of finding more hydrophilic supports than polystyrene, suitable for solid-phase peptide synthesis, polyacrylamides have been extensively investigated [140,141]. In the same way as polystyrene, non-cross-linked polyacrylamides are soluble and hence not well suited for solid-phase synthesis, and have only occasionally been used for this purpose [4,142,143]. Cross-linked polyacrylamides, however, do not dissolve but only gelate in different solvents. Because polyacrylamide is very hydrophilic and therefore incompatible with most organic solvents, derivatives of poly-*N,N*-dimethylacrylamide (e.g. Pepsyn), poly(*N*-acryloylpyrrolidine) [144–146] (e.g. Expansin [147]), or poly(*N*-acryloylmorpholine) [143,148] have usually been chosen as supports. A selection of typical monomers used for the preparation of polyacrylamide-based supports is given in Figure 2.5. Unlike polystyrene, polyacrylamides swell strongly both in aprotic (DMF, pyridine, DCM) and in protic (methanol, water) solvents [149].

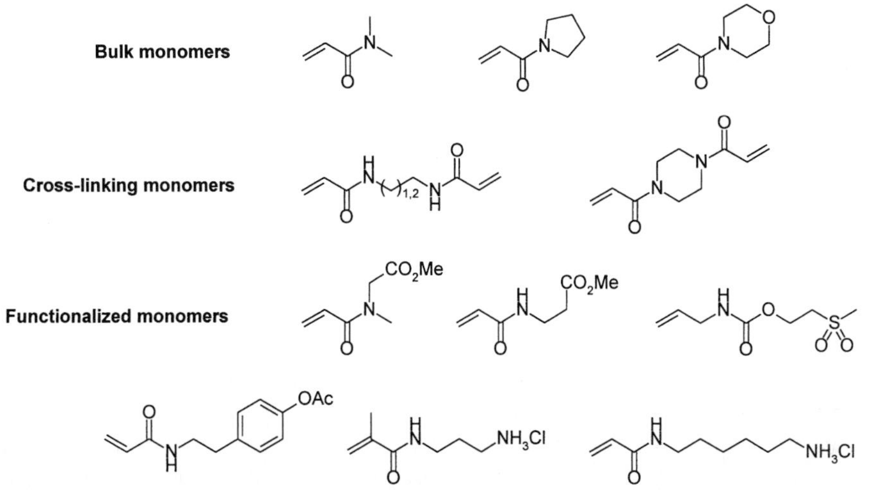

Fig. 2.5. Typical components of cross-linked polyacrylamide supports for solid-phase synthesis.

As an illustrative example, the monomers used for the preparation of a typical polyacrylamide support (Pepsyn) and its schematic representation are sketched in Figure 2.6. The ester functionality, obtained by copolymerization with N-acryloylsarcosine methyl ester, can be used as the attachment point for a suitable linker. This can be achieved by aminolysis with ethylenediamine, followed by acylation of the resulting primary amine with 3-(4-hydroxymethyl)phenylpropionic acid [140] or other linkers. Similar supports have been prepared by copolymerization of protected allylamine [30] or N-(acryloyl)-1,3-diaminopropane [150] with N,N-dimethylacrylamide and N,N'-bis(acryloyl)-1,3-diaminopropane. Amine deprotection after polymerization leads directly to the functionalized support.

Polyacrylamides in which all bulk-monomers contained an attachment site have also been prepared [68]. N-Acryloyl-N-2-(4-acetoxyphenyl)ethylamine (Figure 2.5) has been polymerized in the presence of 1,4-bis(acryloyl)piperazine to yield a support with an initial loading of 5.0 mmol/g. The successful synthesis of a decapeptide on this support demonstrated that the use of highly loaded supports is feasible and a cost-efficient alternative to standard supports [68].

Cross-linked polyacrylamides, such as that sketched in Figure 2.6, are mechanically not sufficiently stable to enable packing into columns for continuous-flow synthesis. To overcome this problem polyacrylamides can be generated on kieselguhr (macroporous silicon dioxide); this results in more rigid particles, which are suitable for continuous-flow peptide synthesis [151,152]. This material is called Pepsyn K or Macrosorb SPR, and has been used for the solid-phase synthesis of peptides [153,154], oligonucleotides [155], and glycopeptides [156]. Macroporous polystyrene (Polyhipe [157]) has also been used as a mechanical support for polyacrylamides.

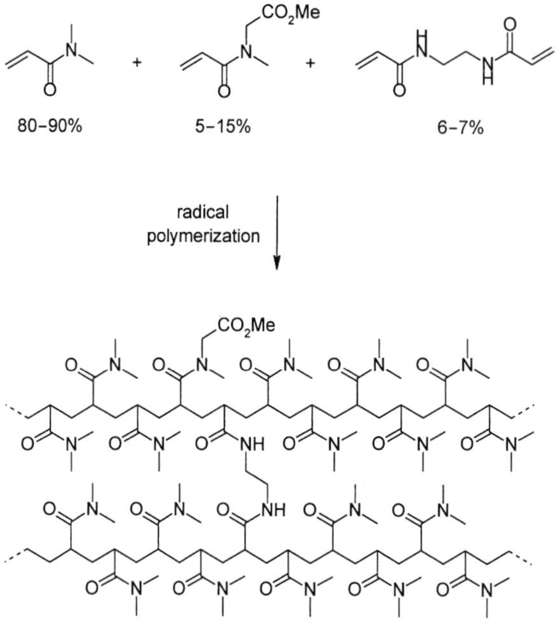

Fig. 2.6. Preparation of a functionalized, cross-linked poly-*N,N*-dimethylacrylamide (Pepsyn).

Polyacrylamides are chemically stable towards acids (TFA, hydrogen fluoride [140]), bases, and weak oxidants or reducing agents. Problems, which might be related to cleavage of the amide bonds in poly(*N,N*-dialkylacrylamides), were encountered upon treatment of such supports with sodium in liquid ammonia [146].

2.5 Polyacrylamide–PEG Copolymers

Meldal and coworkers developed polyacrylamides cross-linked with poly(ethylene glycol) as supports for solid-phase synthesis and on-bead enzymatic assays, called PEGA [158–160]. Functionalization of the polymer was performed in a similar fashion as in other polyacrylamides, i.e. either by copolymerization with *N*-acryloylsarcosine ethyl ester followed by aminolysis with ethylenediamine, or by copolymerization with an amino group containing monomer. The monomers used for a high-capacity (0.4–0.8 mmol/g [159]) and a low-capacity (0.2–0.4 mmol/g [158]) PEGA support are sketched in Figure 2.7.

PEGA supports swell in a broad choice of solvents. Most notably, enzymes can diffuse efficiently into these polymers, making PEGA the support of choice for on-bead-screening [161–163] and enzyme-mediated solid-phase chemistry [164]. PEGA is, furthermore, stable towards hydrogen fluoride and therefore suitable for the solid-phase synthesis of peptides using the Boc methodology [165]. The main disadvantage of PEGA-based supports is their limited mechanical stability, which occasionally leads to problems during the filtration of the polymer.

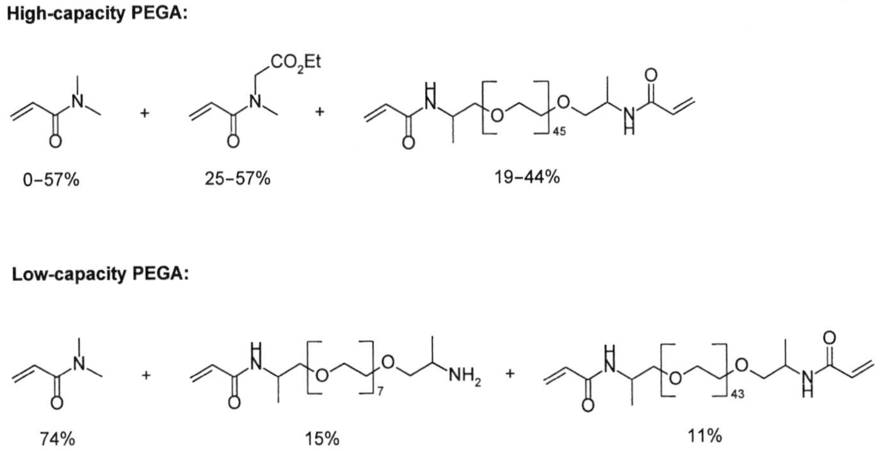

High-capacity PEGA:

0–57%   25–57%   19–44%

Low-capacity PEGA:

74%   15%   11%

Fig. 2.7. Monomers and approximate composition of two types of PEGA support [158,159].

2.6 Silica

Different forms of silicon dioxide have been used as supports for solid-phase organic synthesis. Silica gel is a rigid, insoluble material which does not swell in organic solvents. Commercially available silica gel differs in particle size, pore size (typically 2–10 nm), and surface area (typically 200–800 m^2/g). Like macroporous, highly cross-linked polystyrene, silica gel enables efficient and rapid transfer of solvents and reagents to its entire surface. Because the synthetic intermediates are only located on the surface of the support, enzyme-mediated reactions can be realized on silica [166,167]. Silica gel is particularly well suited for continuous-flow synthesis because its volume stays constant and diffusion rates are high.

In 1970 Bayer and coworkers investigated the synthesis of peptides on silica gel [168]. The first amino acid was attached to the support as the ester of 1,4-bis(hydroxymethyl)benzene (Figure 2.8); loadings of 0.006–0.06 mmol/g were attained. Although higher coupling rates than in polystyrene were reported [168], silica gel never became an established support for solid-phase peptide synthesis.

For the automated solid-phase synthesis of oligonucleotides, however, silica was found to be the support of choice [169–174]. Silica with large pore size (25–300 nm), so-called 'controlled pore glass' (CPG), is generally used for this purpose. The main advantages of CPG, compared with silica gel, are its more regular particle size and shape, and greater mechanical stability.

Attachment of suitable linkers to the surface of silica can be achieved by transesterification with (3-aminopropyl)triethoxysilane, whereby the support **2** (Figure 2.8) results [175–177]. Alternatively, silica can be functionalized by reaction with alkyltrichlorosilanes [178]. For the solid-phase synthesis of oligonucleotides, supports with a longer spacer, such as in **3**, have proven more convenient than **2** [179–183]. Supports **3**, so-called LCAA-CPG (long chain alkylamine CPG [171,172]), are commercially available (typical loading: 0.1 mmol/g) and are currently the most common supports for the synthesis of oligonucleotides. For this purpose protected nucleosides are converted into succinic acid monoesters, and then coupled to LCAA-CPG. CPG functionalized with a 3-mercaptopropyl linker has been used for the solid-phase synthesis of oligosaccharides [184].

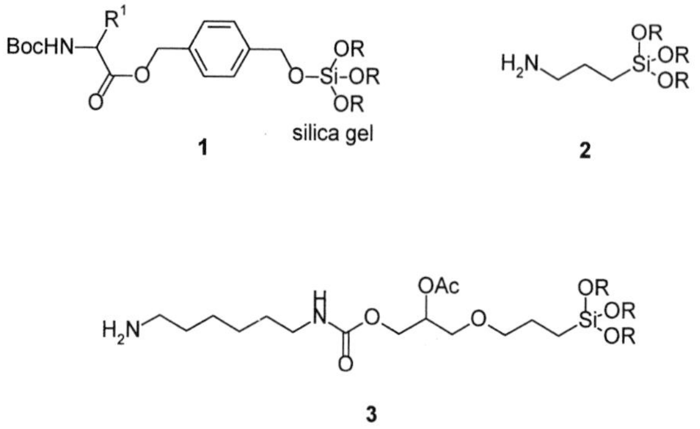

Fig. 2.8. Linkers used to attach synthetic intermediates to silica. **1** [168], **2** [175], **3** [179].

The main disadvantages of CPG are its low loading, its high price, and its high hydrophilicity. Water is difficult to remove quantitatively from this support, which can reduce the yield of acylation reactions [78].

2.7 Polysaccharides

Cellulose is an unbranched polysaccharide consisting of 1,4-linked β-D-glucose and can attain lengths of up to 15000 pyranose units. It neither dissolves nor swells in most solvents, but can be hydrolyzed upon prolonged treatment with acids. Cellulose powder was the first type of support used (with limited success) for solid-phase peptide synthesis [185]. Limitations were mainly the low loading (0.1 mmol/g) attained in these initial experiments, and the high reactivity of cellulose. Despite these problems, cellulose has been occasionally used for solid-phase synthesis, e.g. in the form of cotton [186], paper [187–192], and Perloza (beaded cellulose) [193].

Cellulose contains two secondary and one primary hydroxyl groups per pyranose, which form intra- and interstrand hydrogen bonds and thereby stabilize the compact structure of this polysaccharide. The acylation of all three hydroxyl groups therefore requires harsh reaction conditions [186]. The usual strategy for derivatization of a polysaccharide consists in its partial O-acylation or O-alkylation with a suitable linker or intermediate, followed by acetylation of the remaining hydroxyl groups. Loadings of up to 0.27 mmol/g have been attained by acylation of cotton with Fmoc glycine (DIC, HOBt, N-methylimidazole, DMF, 2 × 4 h [186]). Mixed [188] and symmetric anhydrides [194] can also be used to acylate cellulose. Some strategies for the functionalization of polysaccharides are sketched in Figure 2.9.

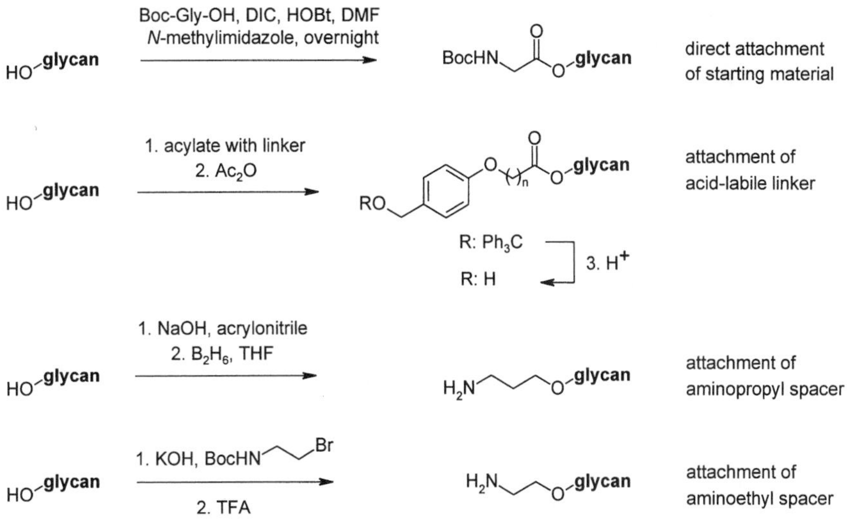

Fig. 2.9. Strategies for the derivatization of polysaccharides [186,188,193–195].

After capping (i.e. acetylation of the remaining hydroxyl groups), the resulting func-tionalized cellulose can be used for the synthesis of peptides using both Boc and Fmoc methodology. Because cellulose is hydrolyzed by hydrogen fluoride (but not by TFA) nucleophilic saponification must be used to cleave the peptides from the support [186]. Alternatively the support can be derivatized with a TFA-labile linker [188].

Another polysaccharide which has proven suitable as a support for the synthesis of peptides [195] and oligonucleotides [196] is sephadex (dextran, a branched glycan con-sisting of 1,6-α-linked glucopyranose). Sepharose (agarose, an unbranched glycan con-sisting of D-galactose and 3,6-anhydro-L-galactose) has been used with success as sup-port for the enzyme-mediated synthesis of oligosaccharides [197].

Polysaccharides are chemically less stable than most other supports presented in this chapter. The chemical modification of polysaccharides can increase the solubility or change the mechanical properties of these materials. This can lead to a deteriora-tion and partial loss of the support during longer synthetic sequences.

2.8 Miscellaneous Supports

Polyethylene has been grafted with acrylic acid, 2-hydroxyethyl methacrylate/N,N-dimethylacrylamide, or methacrylic acid/N,N-dimethylacrylamide to yield supports (e.g. Figure 2.10) suitable for the synthesis of peptides [2,198–200] and other com-pounds [201,202]. A similar support, also suitable for the synthesis of peptides, is poly-propylene grafted with hydroxypropyl acrylate [203]. These supports can be used as membranes or crown-shaped pinheads (Multipin; 2–4 mm diameter) with a loading of 1.2–2.2 μmol per crown.

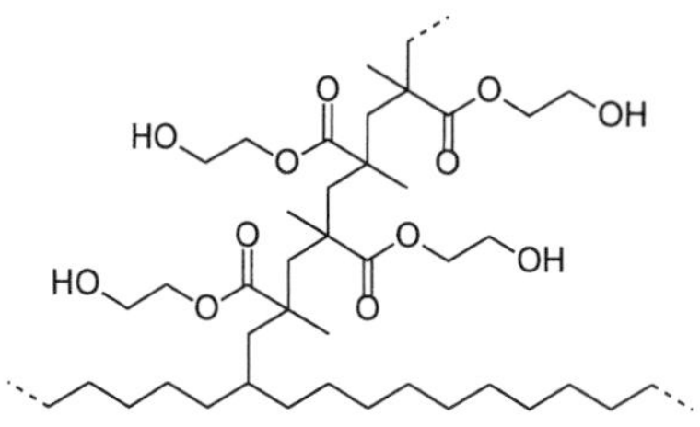

Fig. 2.10. Hydroxyethyl methacrylate grafted on to poly-ethylene [2,200].

Treatment of polyethylene with $Cr_2O_3/H_2SO_4/H_2O$ leads to partial oxidation of its surface and the generation of carboxyl groups [204]. These have been used as attachment points for a functionalized polysaccharide (carboxymethyldextran) to yield a hydrophilic support, suitable for the preparation of peptide libraries [204].

Copolymers of methacrylates with vinyl alcohol (Toyopearl, Fractogel) have, moreover, been successfully used as supports for the synthesis of peptides [205], oligonucleotides [206], and PNA [207]. Cross-linked poly(methyl methacrylates) [10,80,208] have also been investigated and used for solid-phase peptide synthesis.

Further materials which have been evaluated as supports for solid-phase synthesis include phenol–formaldehyde polymers [209,210], proteins (bovine serum albumin) [211], polylysine [212], soluble poly(vinyl alcohol) [213], and different copolymers of vinyl alcohol [4,214,215].

References for Chapter 2

[1] Lebl, M. *Biopolymers* **1998**, *47*, 397–404.
[2] Maeji, N. J.; Valerio, R. M.; Bray, A. M.; Campbell, R. A.; Geysen, H. M. *Reactive Polymers* **1994**, *22*, 203–212.
[3] Hird, N.; Hughes, I.; Hunter, D.; Morrison, M. G. J. T.; Sherrington, D. C.; Stevenson, L. *Tetrahedron* **1999**, *55*, 9575–9584.
[4] Gravert, D. J.; Janda, K. D. *Chem. Rev.* **1997**, *97*, 489–509.
[5] Bayer, E. *Angew. Chem. Int. Ed. Engl.* **1991**, *30*, 113–129.
[6] Mutter, M.; Uhmann, R.; Bayer, E. *Liebigs Ann. Chem.* **1975**, *901*–915.
[7] Bayer, E.; Dengler, M.; Hemmasi, B. *Int. J. Pept. Prot. Res.* **1985**, *25*, 178–186.
[8] Sherrington, D. C. *Chem. Commun.* **1998**, 2275–2286.
[9] Andreatta, R. H.; Rink, H. *Helv. Chim. Acta* **1973**, *56*, 1205–1218.
[10] Labadie, J. W. *Curr. Opinion Chem. Biol.* **1998**, *2*, 346–352.
[11] Blackburn, C. *Biopolymers* **1998**, *47*, 311–351.
[12] Shemyakin, M. M.; Ovchinnikov, Y. A.; Kinyushkin, A. A.; Kozhevnikova, I. V. *Tetrahedron Lett.* **1965**, 2323–2327.
[13] Chiu, S. H. L.; Anderson, L. *Carbohyd. Res.* **1976**, *50*, 227–238.
[14] Chen, S.; Janda, K. D. *J. Am. Chem. Soc.* **1997**, *119*, 8724–8725.
[15] Hayatsu, H.; Khorana, H. G. *J. Am. Chem. Soc.* **1967**, *89*, 3880–3887.
[16] Narita, M. *Bull. Chem. Soc. Jpn.* **1978**, *51*, 1477–1480.
[17] Cramer, F.; Helbig, R.; Hettler, H.; Scheit, K. H.; Seliger, H. *Angew. Chem.* **1966**, *78*, 640–641.
[18] Pepper, K. W.; Paisley, H. M.; Young, M. A. *J. Chem. Soc.* **1953**, 4097–4105.
[19] Pickup, S.; Blum, F. D.; Ford, W. T.; Periyasami, M. *J. Am. Chem. Soc.* **1986**, *108*, 3987–3990.
[20] Balakrishnan, T.; Ford, W. T. *J. Appl. Pol. Sci.* **1982**, *27*, 133–138.
[21] Merrifield, R. B. *J. Am. Chem. Soc.* **1963**, *85*, 2149–2154.
[22] Letsinger, R. L.; Kornet, M. J. *J. Am. Chem. Soc.* **1963**, *85*, 3045–3046.
[23] Sarin, V. K.; Kent, S. B. H.; Merrifield, R. B. *J. Am. Chem. Soc.* **1980**, *102*, 5463–5470.
[24] Shea, K. J.; Stoddard, G. J. *Macromolecules* **1991**, *24*, 1207–1209.
[25] Köster, H.; Geussenhainer, S. *Angew. Chem. Int. Ed. Engl.* **1972**, *11*, 713–714.

[26] Santini, R.; Griffith, M. C.; Qi, M. *Tetrahedron Lett.* **1998**, *39*, 8951–8954.
[27] Merrifield, R. B. *Brit. Pol. J.* **1984**, *16*, 173–178.
[28] Yan, B.; Fell, J. B.; Kumaravel, G. *J. Org. Chem.* **1996**, *61*, 7467–7472.
[29] Naumann, G. In *Ullmanns Encyklopädie der Technischen Chemie*; Verlag Chemie: Weinheim, New York, **1977**; p 303.
[30] Kanda, P.; Kennedy, R. C.; Sparrow, J. T. *Int. J. Pept. Prot. Res.* **1991**, *38*, 385–391.
[31] Akelah, A.; Sherrington, D. C. *Chem. Rev.* **1981**, *81*, 557–587.
[32] Stranix, B. R.; Gao, J. P.; Barghi, R.; Salha, J.; Darling, G. D. *J. Org. Chem.* **1997**, *62*, 8987–8993.
[33] Stranix, B. R.; Darling, G. D. *J. Org. Chem.* **1997**, *62*, 9001–9004.
[34] Darling, G. D.; Fréchet, J. M. J. *J. Org. Chem.* **1986**, *51*, 2270–2276.
[35] Farrall, M. J.; Fréchet, J. M. J. *J. Org. Chem.* **1976**, *41*, 3877–3882.
[36] Seliger, H. *Makromol. Chem.* **1973**, *169*, 83–93.
[37] Dowling, L. M.; Stark, G. R. *Biochemistry* **1969**, *8*, 4728–4734.
[38] Mizoguchi, T.; Shigezane, K.; Takamura, N. *Chem. Pharm. Bull.* **1970**, *18*, 1465–1474.
[39] Matsueda, G. R.; Stewart, J. M. *Peptides* **1981**, *2*, 45–50.
[40] Ajayaghosh, A.; Pillai, V. N. R. *Tetrahedron* **1988**, *44*, 6661–6666.
[41] Kobayashi, S.; Moriwaki, M. *Tetrahedron Lett.* **1997**, *38*, 4251–4254.
[42] Scarr, R. B.; Findeis, M. A. *Pept. Res.* **1990**, *3*, 238–241.
[43] Wang, S.; Merrifield, R. B. *J. Am. Chem. Soc.* **1969**, *91*, 6488–6491.
[44] Park, B. D.; Lee, H. I.; Ryoo, S. J.; Lee, Y. S. *Tetrahedron Lett.* **1997**, *38*, 591–594.
[45] Tomoi, M.; Kori, N.; Kakiuchi, H. *Reactive Polymers* **1985**, *5*, 341–349.
[46] Pinnell, R. P.; Khune, G. D.; Khatri, N. A.; Manatt, S. L. *Tetrahedron Lett.* **1984**, *25*, 3511–3514.
[47] Gerigk, U.; Gerlach, M.; Neumann, W. P.; Vieler, R.; Weintritt, V. *Synthesis* **1990**, 448–452.
[48] Ford, W. T.; Yacoub, S. A. *J. Org. Chem.* **1981**, *46*, 819–821.
[49] Arshady, R.; Kenner, G. W.; Ledwith, A. *Makromol. Chem.* **1976**, *177*, 2911–2918.
[50] Sheng, Q.; Stöver, H. D. H. *Macromolecules* **1997**, *30*, 6712–6714.
[51] Mohanraj, S.; Ford, W. T. *Macromolecules* **1986**, *19*, 2470–2472.
[52] Mitchell, A. R.; Kent, S. B. H.; Erickson, B. W.; Merrifield, R. B. *Tetrahedron Lett.* **1976**, 3795–3798.
[53] Zikos, C. C.; Ferderigos, N. G. *Tetrahedron Lett.* **1995**, *36*, 3741–3744.
[54] Mitchell, A. R.; Kent, S. B. H.; Engelhard, M.; Merrifield, R. B. *J. Org. Chem.* **1978**, *43*, 2845–2852.
[55] Adams, J. H.; Cook, R. M.; Hudson, D.; Jammalamadaka, V.; Lyttle, M. H.; Songster, M. F. *J. Org. Chem.* **1998**, *63*, 3706–3716.
[56] Sparrow, J. T. *J. Org. Chem.* **1976**, *41*, 1350–1353.
[57] Bernard, M.; Ford, W. T. *J. Org. Chem.* **1983**, *48*, 326–332.
[58] O'Brien, R. A.; Chen, T.; Rieke, R. D. *J. Org. Chem.* **1992**, *57*, 2667–2677.
[59] Fyles, T. M.; Leznoff, C. C. *Can. J. Chem.* **1976**, *54*, 935–942.
[60] Lochmann, L.; Fréchet, J. M. J. *Macromolecules* **1996**, *29*, 1767–1771.
[61] Ruel, G.; The, N. K.; Dumartin, G.; Delmond, B.; Pereyre, M. *J. Organomet. Chem.* **1993**, *444*, C18–C20.
[62] Ruhland, T.; Andersen, K.; Pedersen, H. *J. Org. Chem.* **1998**, *63*, 9204–9211.
[63] Nicolaou, K. C.; Pastor, J.; Barluenga, S.; Winssinger, N. *Chem. Commun.* **1998**, 1947–1948.
[64] Nicolaou, K. C.; Winssinger, N.; Pastor, J.; DeRoose, F. *J. Am. Chem. Soc.* **1997**, *119*, 449–450.
[65] Neumann, W. P.; Peterseim, M. *Reactive Polymers* **1993**, *20*, 189–205.
[66] Deleuze, H.; Sherrington, D. C. *J. Chem. Soc. Perkin Trans. 2* **1995**, 2217–2221.
[67] Tam, J. P.; Lu, Y. A. *J. Am. Chem. Soc.* **1995**, *117*, 12058–12063.
[68] Epton, R.; Wellings, D. A.; Williams, A. *Reactive Polymers* **1987**, *6*, 143–157.
[69] Dixit, D. M.; Leznoff, C. C. *J. Chem. Soc. Chem. Commun.* **1977**, 798–799.
[70] Farrall, M. J.; Fréchet, J. M. J. *J. Am. Chem. Soc.* **1978**, *100*, 7998–7999.
[71] Biagini, S. C. G.; Gibson, S. E.; Keen, S. P. *J. Chem. Soc. Perkin Trans. 1* **1998**, 2485–2499.
[72] Guyot, A. *Pure Appl. Chem.* **1988**, *60*, 365–376.
[73] Svec, F.; Fréchet, J. M. J. *Science* **1996**, *273*, 205–211.
[74] Millar, J. R.; Smith, D. G.; Marr, W. E.; Kressman, T. R. E. *J. Chem. Soc.* **1963**, 218–224.
[75] Kunin, R.; Meitzner, E.; Bortnick, N. *J. Am. Chem. Soc.* **1962**, *84*, 305–306.
[76] Hori, M.; Gravert, D. J.; Wentworth, P.; Janda, K. D. *Bioorg. Med. Chem. Lett.* **1998**, *8*, 2363–2368.
[77] Malenfant, P. R. L.; Fréchet, J. M. J. *Chem. Commun.* **1998**, 2657–2658.
[78] McCollum, C.; Andrus, A. *Tetrahedron Lett.* **1991**, *32*, 4069–4072.
[79] Köster, H.; Cramer, F. *Liebigs Ann. Chem.* **1974**, 946–958.
[80] Sophiamma, P. N.; Sreekumar, K. *Indian J. Chem. Sect. B (Org. Chem.)* **1997**, *36*, 995–999.
[81] Varkey, J. T.; Pillai, V. N. R. *J. Pept. Res.* **1998**, *51*, 49–54.

30   2 *Supports for Solid-Phase Organic Synthesis*

[82] Varkey, J. T.; Pillai, V. N. R. *J. Appl. Pol. Sci.* **1999**, *71*, 1933–1939.
[83] Toy, P. H.; Janda, K. D. *Tetrahedron Lett.* **1999**, *40*, 6329–6332.
[84] Renil, M.; Pillai, V. N. R. *J. Appl. Pol. Sci.* **1996**, *61*, 1585–1594.
[85] Renil, M.; Nagaraj, R.; Pillai, V. N. R. *Tetrahedron* **1994**, *50*, 6681–6688.
[86] Renil, M.; Pillai, V. N. R. *Tetrahedron Lett.* **1994**, *35*, 3809–3812.
[87] Kumar, K. S.; Pillai, V. N. R. *Tetrahedron* **1999**, *55*, 10437–10446.
[88] Wilson, M. E.; Paech, K.; Zhou, W.; Kurth, M. J. *J. Org. Chem.* **1998**, *63*, 5094–5099.
[89] Renil, M.; Meldal, M. *Tetrahedron Lett.* **1996**, *37*, 6185–6188.
[90] Buchardt, J.; Meldal, M. *Tetrahedron Lett.* **1998**, *39*, 8695–8698.
[91] Berg, R. H.; Almdal, K.; Pedersen, W. B.; Holm, A.; Tam, J. P.; Merrifield, R. B. *J. Am. Chem. Soc.* **1989**, *111*, 8024–8026.
[92] Kent, S. B. H.; Merrifield, R. B. *Israel J. Chem.* **1978**, *17*, 243–247.
[93] Birch-Hirschfeld, E.; Földes-Papp, Z.; Gührs, K. H.; Seliger, H. *Nucleic Acids Res.* **1994**, *22*, 1760–1761.
[94] Birch-Hirschfeld, E.; Földes-Papp, Z.; Gührs, K. H.; Seliger, H. *Helv. Chim. Acta* **1996**, *79*, 137–150.
[95] Witkowski, W.; Birch-Hirschfeld, E.; Weiss, R.; Zarytova, V. F.; Gorn, V. V. *J. Prakt. Chem.* **1984**, *326*, 320–328.
[96] Weiss, R.; Birch-Hirschfeld, E.; Witkowski, W.; Friese, K. *Z. Chem.* **1986**, *26*, 127–130.
[97] Zhao, C.; Shi, S.; Mir, D.; Hurst, D.; Li, R.; Xiao, X. Y.; Lillig, J.; Czarnik, A. W. *J. Comb. Chem.* **1999**, *1*, 91–95.
[98] Devivar, R. V.; Koontz, S. L.; Peltier, W. J.; Pearson, J. E.; Guillory, T. A.; Fabricant, J. D. *Bioorg. Med. Chem. Lett.* **1999**, *9*, 1239–1242.
[99] Bray, A. M.; Jhingran, A. G.; Valerio, R. M.; Maeji, N. J. *J. Org. Chem.* **1994**, *59*, 2197–2203.
[100] Juby, C. D.; Richardson, C. D.; Brousseau, R. *Tetrahedron Lett.* **1991**, *32*, 879–882.
[101] Takahashi, T.; Tomida, S.; Inoue, H.; Doi, T. *Synlett* **1998**, 1261–1263.
[102] Tommasi, R. A.; Nantermet, P. G.; Shapiro, M. J.; Chin, J.; Brill, W. K. D.; Ang, K. *Tetrahedron Lett.* **1998**, *39*, 5477–5480.
[103] Takahashi, T.; Ebata, S.; Doi, T. *Tetrahedron Lett.* **1998**, *39*, 1369–1372.
[104] Sucholeiki, I.; Perez, J. M. *Tetrahedron Lett.* **1999**, *40*, 3531–3534.
[105] Szymonifka, M. J.; Chapman, K. T. *Tetrahedron Lett.* **1995**, *36*, 1597–1600.
[106] Hellermann, H.; Lucas, H. W.; Maul, J.; Pillai, V. N. R.; Mutter, M. *Makromol. Chem.* **1983**, *184*, 2603–2617.
[107] Kates, S. A.; McGuinness, B. F.; Blackburn, C.; Griffin, G. W.; Solé, N. A.; Barany, G.; Albericio, F. *Biopolymers* **1998**, *47*, 365–380.
[108] Wright, P.; Lloyd, D.; Rapp, W.; Andrus, A. *Tetrahedron Lett.* **1993**, *34*, 3373–3376.
[109] Gooding, O. W.; Baudart, S.; Deegan, T. L.; Heisler, K.; Labadie, J. W.; Newcomb, W. S.; Porco, J. A.; van Eikeren, P. *J. Comb. Chem.* **1999**, 113–122.
[110] Porco, J. A.; Deegan, T.; Devonport, W.; Gooding, O. W.; Heisler, K.; Labadie, J. W.; Newcomb, B.; Nguyen, C.; van Eikeren, P.; Wong, J.; Wright, P. *Mol. Diversity* **1997**, *2*, 197–206.
[111] Keifer, P. A. *J. Org. Chem.* **1996**, *61*, 1558–1559.
[112] Pursch, M.; Schlotterbeck, G.; Tseng, L. H.; Albert, K. *Angew. Chem. Int. Ed. Engl.* **1997**, *35*, 2867–2869.
[113] Bettinger, T.; Remy, J. S.; Erbacher, P.; Behr, J. P. *Bioconjugate Chem.* **1998**, *9*, 842–846.
[114] Keifer, P. A.; Baltusis, L.; Rice, D. M.; Tymiak, A. A.; Shoolery, J. N. *J. Magn. Resonance Series A* **1996**, *119*, 65–75.
[115] Wehler, T.; Westman, J. *Tetrahedron Lett.* **1996**, *37*, 4771–4774.
[116] Anderson, R. C.; Jarema, M. A.; Shapiro, M. J.; Stokes, J. P.; Ziliox, M. *J. Org. Chem.* **1995**, *60*, 2650–2651.
[117] Li, W.; Xiao, X.; Czarnik, A. W. *J. Comb. Chem.* **1999**, *1*, 127–129.
[118] Li, W.; Yan, B. *J. Org. Chem.* **1998**, *63*, 4092–4097.
[119] Swali, V.; Wells, N. J.; Langley, G. J.; Bradley, M. *J. Org. Chem.* **1997**, *62*, 4902–4903.
[120] Hutchins, S. M.; Chapman, K. T. *Tetrahedron Lett.* **1996**, *37*, 4869–4872.
[121] Nouvet, A.; Binard, M.; Lamaty, F.; Martinez, J.; Lazaro, R. *Tetrahedron* **1999**, *55*, 4685–4698.
[122] Harris, J. M. *J. Macromol. Sci. C. Rev. Macromol. Chem. Phys.* **1985**, *C25*, 325–373.
[123] Mutter, M.; Hagenmaier, H.; Bayer, E. *Angew. Chem.* **1971**, *83*, 883–884.
[124] Novabiochem Catalog and Peptide Synthesis Handbook, Läufelfingen, **1999**.
[125] Simmonds, R. G. *Int. J. Pept. Prot. Res.* **1996**, *47*, 36–41.
[126] Bonora, G. M.; Biancotto, G.; Maffini, M.; Scremin, C. L. *Nucleic Acids Res.* **1993**, *21*, 1213–1217.
[127] Wang, Y.; Zhang, H.; Voelter, W. *Chem. Lett.* **1995**, 273–274.
[128] Brandstetter, F.; Schott, H.; Bayer, E. *Tetrahedron Lett.* **1973**, 2997–3000.

[129] Ito, Y.; Kanie, O.; Ogawa, T. *Angew. Chem. Int. Ed. Engl.* **1996**, *35*, 2510–2512.

[130] Douglas, S. P.; Whitfield, D. M.; Krepinsky, J. J. *J. Am. Chem. Soc.* **1995**, *117*, 2116–2117.

[131] Bayer, E.; Mutter, M.; Uhmann, R.; Polster, J.; Mauser, H. *J. Am. Chem. Soc.* **1974**, *96*, 7333–7336.

[132] Molteni, V.; Annunziata, R.; Cinquini, M.; Cozzi, F.; Benaglia, M. *Tetrahedron Lett.* **1998**, *39*, 1257–1260.

[133] Park, W. K. C.; Auer, M.; Jaksche, H.; Wong, C. H. *J. Am. Chem. Soc.* **1996**, *118*, 10150–10155.

[134] Far, A. R.; Tidwell, T. T. *J. Org. Chem.* **1998**, *63*, 8636–8637.

[135] Yoon, J.; Cho, C. W.; Han, H.; Janda, K. D. *Chem. Commun.* **1998**, 2703–2704.

[136] Shey, J. Y.; Sun, C. M. *Synlett* **1998**, 1423–1425.

[137] Benaglia, M.; Annunziata, R.; Cinquini, M.; Cozzi, F.; Ressel, S. *J. Org. Chem.* **1998**, *63*, 8628–8629.

[138] Rademann, J.; Meldal, M.; Bock, K. *Chem. Eur. J.* **1999**, *5*, 1218–1225.

[139] Rademann, J.; Grøtli, M.; Meldal, M.; Bock, K. *J. Am. Chem. Soc.* **1999**, *121*, 5459–5466.

[140] Arshady, R.; Atherton, E.; Clive, D. L. J.; Sheppard, R. C. *J. Chem. Soc. Perkin Trans. 1* **1981**, 529–537.

[141] McMurray, J. S. *Biopolymers* **1998**, *47*, 405–411.

[142] Bergbreiter, D. E.; Case, B. L.; Liu, Y. S.; Caraway, J. W. *Macromolecules* **1998**, *31*, 6053–6062.

[143] Bonora, G. M.; Baldan, A.; Schiavon, O.; Ferruti, P.; Veronese, F. M. *Tetrahedron Lett.* **1996**, *37*, 4761–4764.

[144] Calas, B.; Méry, J.; Parello, J.; Cave, A. *Tetrahedron* **1985**, *41*, 5331–5339.

[145] Smith, C. W.; Stahl, G. L.; Walter, R. *Int. J. Pept. Prot. Res.* **1979**, *13*, 109–112.

[146] Stahl, G. L.; Walter, R.; Smith, C. W. *J. Am. Chem. Soc.* **1979**, *101*, 5383–5394.

[147] Mendre, C.; Sarrade, V.; Calas, B. *Int. J. Pept. Prot. Res.* **1992**, *39*, 278–284.

[148] Epton, R.; Goddard, P.; Marr, G.; McLaren, J. V.; Morgan, G. J. *Polymer* **1979**, *20*, 1444–1446.

[149] Atherton, E.; Clive, D. L. J.; Sheppard, R. C. *J. Am. Chem. Soc.* **1975**, *97*, 6584–6585.

[150] Sparrow, J. T.; Knieb-Cordonier, N. G.; Obeyseskere, N. U.; McMurray, J. S. *Pept. Res.* **1996**, *9*, 297–304.

[151] Dryland, A.; Sheppard, R. C. *J. Chem. Soc. Perkin Trans. 1* **1986**, 125–137.

[152] Atherton, E.; Brown, E.; Sheppard, R. C.; Rosevear, A. *J. Chem. Soc. Chem. Commun.* **1981**, 1151–1152.

[153] Dryland, A.; Sheppard, R. C. *Tetrahedron* **1988**, *44*, 859–876.

[154] Atherton, E.; Sheppard, R. C. *Solid Phase Peptide Synthesis; A Practical Approach*; Oxford University Press: Oxford, **1989**.

[155] Gait, M. J.; Matthes, H. W. D.; Singh, M.; Sproat, B. S.; Titmas, R. C. *Nucleic Acids Res.* **1982**, *10*, 6243–6254.

[156] Chadwick, R. J.; Thompson, J. S.; Tomalin, G. *Biochem. Soc. Trans.* **1991**, *19*, 406S.

[157] Small, P. W.; Sherrington, D. C. *J. Chem. Soc. Chem. Commun.* **1989**, 1589–1591.

[158] Meldal, M. *Tetrahedron Lett.* **1992**, *33*, 3077–3080.

[159] Renil, M.; Meldal, M. *Tetrahedron Lett.* **1995**, *36*, 4647–4650.

[160] Renil, M.; Ferreras, M.; Delaisse, J. M.; Foged, N. T.; Meldal, M. *J. Pept. Sci.* **1998**, *4*, 195–210.

[161] Hilaire, P. M. S.; Lowary, T. L.; Meldal, M.; Bock, K. *J. Am. Chem. Soc.* **1998**, *120*, 13312–13320.

[162] Meldal, M.; Svendsen, I. *J. Chem. Soc. Perkin Trans. 1* **1995**, 1591–1596.

[163] Meldal, M.; Svendsen, I.; Breddam, K.; Auzanneau, F. I. *Proc. Natl. Acad. Sci. USA* **1994**, *91*, 3314–3318.

[164] Meldal, M.; Auzanneau, F. I.; Hindsgaul, O.; Palcic, M. M. *J. Chem. Soc. Chem. Commun.* **1994**, 1849–1850.

[165] Camarero, J. A.; Cotton, G. J.; Adeva, A.; Muir, T. W. *J. Pept. Res.* **1998**, *51*, 303–316.

[166] Singh, J.; Allen, M. P.; Ator, M. A.; Gainor, J. A.; Whipple, D. A. *J. Med. Chem.* **1995**, *38*, 217–219.

[167] Schuster, M.; Wang, P.; Paulson, J. C.; Wong, C. H. *J. Am. Chem. Soc.* **1994**, *116*, 1135–1136.

[168] Bayer, E.; Jung, G.; Halász, I.; Sebastian, I. *Tetrahedron Lett.* **1970**, 4503–4505.

[169] Köster, H. *Tetrahedron Lett.* **1972**, 1527–1530.

[170] Pon, R. T.; Ogilvie, K. K. *Tetrahedron Lett.* **1984**, *25*, 713–716.

[171] Engels, J. W.; Uhlmann, E. *Angew. Chem. Int. Ed. Engl.* **1989**, *28*, 716–734.

[172] Ghosh, P. K.; Kumar, P.; Gupta, K. C. *J. Indian Chem. Soc.* **1998**, *75*, 206–218.

[173] Ogilvie, K. K.; Nemer, M. J. *Tetrahedron Lett.* **1980**, *21*, 4159–4162.

[174] Köster, H.; Biernat, J.; McManus, J.; Wolter, A.; Stumpe, A.; Narang, C. K.; Sinha, N. D. *Tetrahedron* **1984**, *40*, 103–112.

[175] Matteucci, M. D.; Caruthers, M. H. *J. Am. Chem. Soc.* **1981**, *103*, 3185–3191.

[176] Matteucci, M. D.; Caruthers, M. H. *Tetrahedron Lett.* **1980**, *21*, 719–722.

[177] Keana, J. F. W.; Shimizu, M.; Jernsted, K. K. *J. Org. Chem.* **1986**, *51*, 1641–1644.

[178] Parr, W.; Grohmann, K. *Angew. Chem. Int. Ed. Engl.* **1972**, *11*, 314–315.

[179] Adams, S. P.; Kavka, K. S.; Wykes, E. J.; Holder, S. B.; Galluppi, G. R. *J. Am. Chem. Soc.* **1983**, *105*, 661–663.
[180] Gough, G. R.; Brunden, M. J.; Gilham, P. T. *Tetrahedron Lett.* **1981**, *22*, 4177–4180.
[181] Usman, N.; Ogilvie, K. K.; Jiang, M. Y.; Cedergren, R. J. *J. Am. Chem. Soc.* **1987**, *109*, 7845–7854.
[182] Katzhendler, J.; Cohen, S.; Rahamim, E.; Weisz, M.; Ringel, I.; Deutsch, J. *Tetrahedron* **1989**, *45*, 2777–2792.
[183] Van Aerschot, A.; Herdewijn, P.; van der Haeghe, H. *Nucleosides and Nucleotides* **1988**, *7*, 75–90.
[184] Heckel, A.; Mross, E.; Jung, K. H.; Rademann, J.; Schmidt, R. R. *Synlett* **1998**, 171–173.
[185] Merrifield, B. In *Peptides; Synthesis, Structures, and Applications*; Gutte, B. Ed.; Academic Press: London, **1995**.
[186] Eichler, J.; Bienert, M.; Stierandova, A.; Lebl, M. *Pept. Res.* **1991**, *5*, 296–307.
[187] Blankemeyer-Menge, B.; Frank, R. *Tetrahedron Lett.* **1988**, *29*, 5871–5874.
[188] Frank, R.; Döring, R. *Tetrahedron* **1988**, *44*, 6031–6040.
[189] Ott, J.; Eckstein, F. *Nucleic Acids Res.* **1984**, *12*, 9137–9142.
[190] Frank, R.; Heikens, W.; Heisterberg-Moutsis, G.; Blöcker, H. *Nucleic Acids Res.* **1983**, *11*, 4365–4377.
[191] Frank, R.; Meyerhans, A.; Schwellnus, K.; Blöcker, H. *Methods Enzymol.* **1987**, *154*, 221–249.
[192] Kramer, A.; Schuster, A.; Reineke, U.; Malin, R.; Volkmer-Engert, R.; Landgraf, C.; Schneider-Mergener, J. *Methods: A Companion to Methods in Enzymology* **1994**, *6*, 388–395.
[193] Englebretsen, D. R.; Harding, D. R. K. *Int. J. Pept. Prot. Res.* **1992**, *40*, 487–496.
[194] Dittrich, F.; Tegge, W.; Frank, R. *Bioorg. Med. Chem. Lett.* **1998**, *8*, 2351–2356.
[195] Annis, I.; Chen, L.; Barany, G. *J. Am. Chem. Soc.* **1998**, *120*, 7226–7238.
[196] Köster, H.; Heyns, K. *Tetrahedron Lett.* **1972**, 1531–1534.
[197] Blixt, O.; Norberg, T. *J. Org. Chem.* **1998**, *63*, 2705–2710.
[198] Geysen, H. M.; Rodda, S. J.; Mason, T. J.; Tribbick, G.; Schoofs, P. G. *J. Immunol. Meth.* **1987**, *102*, 259–274.
[199] Valerio, R. M.; Bray, A. M.; Maeji, N. J. *Int. J. Pept. Prot. Res.* **1994**, *44*, 158–165.
[200] Valerio, R. M.; Bray, A. M.; Campbell, R. A.; Dipasquale, A.; Margellis, C.; Rodda, S. J.; Geysen, H. M.; Maeji, N. J. *Int. J. Pept. Prot. Res.* **1993**, *42*, 1–9.
[201] Valerio, R. M.; Bray, A. M.; Patsiouras, H. *Tetrahedron Lett.* **1996**, *37*, 3019–3022.
[202] Bray, A. M.; Chiefari, D. S.; Valerio, R. M.; Maeji, N. J. *Tetrahedron Lett.* **1995**, *36*, 5081–5084.
[203] Daniels, S. B.; Bernatowicz, M. S.; Coull, J. M.; Köster, H. *Tetrahedron Lett.* **1989**, *30*, 4345–4348.
[204] Luo, K. X.; Zhou, P.; Lodish, H. F. *Proc. Natl. Acad. Sci. USA* **1995**, *92*, 11761–11765.
[205] Buettner, J. A.; Dadd, C. A.; Baumbach, G. A.; Masecar, B. L.; Hammond, D. J. *Int. J. Pept. Prot. Res.* **1996**, *47*, 70–83.
[206] Reddy, M. P.; Michael, M. A.; Farooqui, F.; Girgis, S. *Tetrahedron Lett.* **1994**, *35*, 5771–5774.
[207] Stetsenko, D. A.; Lubyako, E. N.; Potapov, V. K.; Azhikina, T. L.; Sverdlov, E. D. *Tetrahedron Lett.* **1996**, *37*, 3571–3574.
[208] Kempe, M.; Barany, G. *J. Am. Chem. Soc.* **1996**, *118*, 7083–7093.
[209] Inukai, N.; Nakano, K.; Murakami, M. *Bull. Chem. Soc. Jpn.* **1968**, *41*, 182–186.
[210] Flanigan, E.; Marshall, G. R. *Tetrahedron Lett.* **1970**, 2403–2406.
[211] Hansen, P. R.; Holm, A.; Houen, G. *Int. J. Pept. Prot. Res.* **1993**, *41*, 237–245.
[212] Chapman, T. M.; Kleid, D. G. *J. Chem. Soc. Chem. Commun.* **1973**, 193–194.
[213] Schott, H.; Brandstetter, F.; Bayer, E. *Makromol. Chem.* **1973**, *173*, 247–251.
[214] Eynde, J. J. V.; Rutot, D. *Tetrahedron* **1999**, *55*, 2687–2694.
[215] Seliger, H.; Aumann, G. *Tetrahedron Lett.* **1973**, 2911–2914.

3 Linkers for Solid-Phase Organic Synthesis

Linkers are molecules which keep the intermediates in solid-phase synthesis bound to the support. Linkers should enable the easy attachment of the starting material to the support, be stable under a broad variety of reaction conditions, and yet enable selective cleavage at the end of a synthesis, without damage to the product. Several types of linker have been developed; these meet these conflicting requirements to different extents [1].

These linkers enable the attachment of a variety of functional groups to a solid support and upon cleavage either the originally attached or a new functional group may be generated. In the following chapters the known linkers have been organized according to the functional group *resulting upon cleavage* from the support.

In many of the resins used for solid-phase synthesis, the linkers are attached to the support by means of a spacer (Figure 3.1). It has been proposed [2] that flexible spacers facilitate the diffusion of reagents to the resin-bound substrate, by increasing its distance from the support. Although this might be of importance for the synthesis of certain 'difficult' peptides, there is little evidence that spacers improve the synthesis of small molecules on insoluble supports [3]. For the characterization of compounds attached to a polymeric support by NMR spectroscopy, however, long spacers are an advantage, because they increase the mobility of the substrate and reduce the line-broadening usually observed in the NMR spectra of polymers.

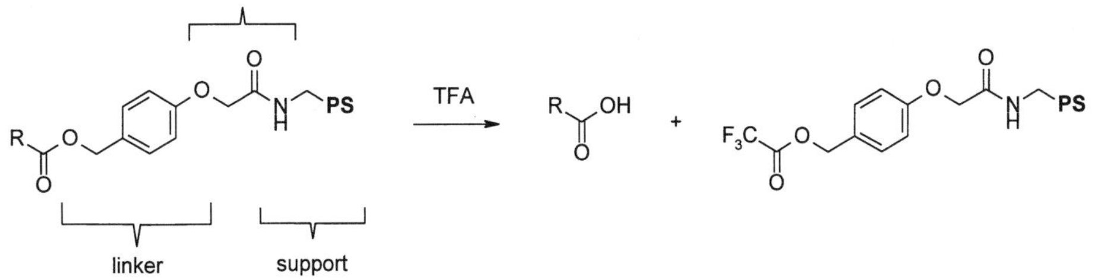

Fig. 3.1. Example of a resin used for solid-phase synthesis.

Linkers must be attached to the support in such a way that no release of the linker occurs either during a synthetic sequence or upon cleavage of the linker–product bond. For this purpose linkers often contain a carboxyl group which enables irreversible attachment of the linker to amino group containing supports as amides. In the

case of polystyrene, aminomethyl polystyrene or amino(4-methylphenyl)methyl polystyrene ('methylbenzhydrylamine', MBHA resin, Figure 3.2) are generally used to
attach other, more elaborate linkers. Aminomethyl polystyrene can be prepared
directly from polystyrene by acid-catalyzed aminomethylation with *N*-(hydroxymethyl)phthalimide or *N*-(chloromethyl)phthalimide [4,5], or by ammonolysis of
chloromethyl polystyrene [6]. MBHA resin is prepared by Friedel–Crafts acylation of
polystyrene with 4-methylbenzoyl chloride, followed by reductive amination of the
resulting benzophenone [7], or by amidoalkylation of polystyrene with *N*-[1-chloro-
1(4-tolyl)methyl]phthalimide [5]. Unlike N-acylated aminomethyl polystyrene, N-acylated MBHA resin is not stable towards strong acids, and can, in fact, be used as support for amides cleavable by hydrogen fluoride (Section 3.3).

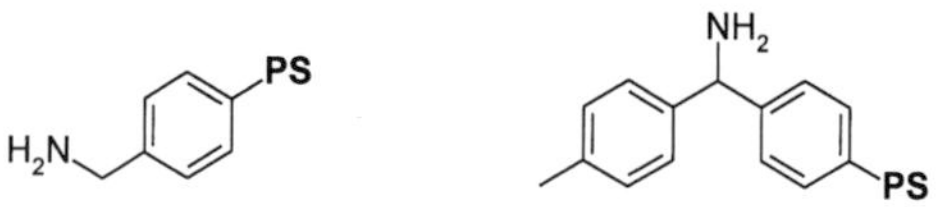

Fig. 3.2. Polystyrene derivatives suitable for the irreversible attachment of linkers.

Amino group bearing derivatives of most other commonly used supports for solid-
phase synthesis (PEG-grafted polystyrene, PEG, polyacrylamides, CPG) are commercially available; these enable the irreversible attachment of different linkers.

For some applications it might be desirable to cleave the product from a support in
two or more portions. This can be realized by derivatizing a functionalized support
with a mixture of different linkers which enable a sequential cleavage [8]. The resulting support can, for instance, be used to prepare and screen combinatorial peptide
libraries by the mix-and-split method ([9–11]; one different peptide on each bead).
The first portion of peptide released would be tested for biological activity, and, once
an active peptide has been identified, the remaining peptide on the support could be
used for structure elucidation.

Before cleaving products from a support, either extensive washing of the resin with
volatile solvents or drying of the resin is required, because certain solvents (e.g. NMP,
DMF) and reagents (e.g. DIPEA, DBU) tend to stick strongly to some supports.
These reagents can inhibit the cleavage reaction or lead to inpure products. Extensive
washing is particularly relevant if the products are to be analyzed or screened without
prior purification.

Most acid-labile linkers and protective groups used in solid-phase synthesis lead to
the formation of carbocations during acidolytic cleavage. Products containing electron-rich structural elements, such as pyrroles, indoles, thiols, phenols, etc. (but not
amines) can be irreversibly alkylated by these carbocations, with low yields or product
mixtures as a result. This can be avoided by adding scavengers to the cleavage reagent.
Typical carbocation scavengers include thiols (EDT, PhSH), thioethers (Me$_2$S,
PhSMe), phenols (cresol, PhOH), arylethers (PhOMe), and silanes (Et$_3$SiH, *i*Pr$_3$SiH).
If non-volatile scavengers are used, or when non-volatile byproducts are formed during cleavage and deprotection (e.g. during detritylation) the crude products can
usually not be used directly but need to be further purified.

3.1 Linkers for Carboxylic Acids

Carboxylic acids are generally attached to polymeric supports as esters or amides. Depending on the type of linker and on the cleavage conditions used, cleavage can lead either to the regeneration of a carboxylic acid, or to the formation of a new product, such as esters, amides, ketones, or alcohols. This section covers only linkers which lead, upon cleavage, to the release of carboxylic acids.

3.1.1 Linkers for Acids Cleavable by Acids or Other Electrophiles

Acid-labile linkers are the oldest and still most commonly used linkers for carboxylic acids. Most are based on the acidolysis of benzylic C–O bonds.

3.1.1.1 Benzyl Alcohol Linkers

Benzyl esters cleavable under acidic conditions were the first type of linker investigated in detail. The reason for this was probably the initial choice of polystyrene as insoluble support for solid-phase synthesis [12]. Polystyrene-derived benzyl esters were initially prepared by treatment of partially chloromethylated polystyrene with salts of carboxylic acids (Figure 3.3).

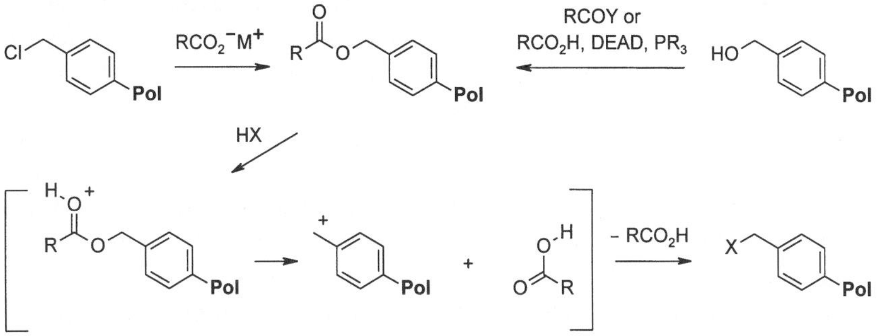

Fig. 3.3. Immobilization of carboxylic acids as benzyl esters and acidolytic cleavage. M^+: Cs^+, NMe_4^+, Na^+, Zn^{2+}; Y: leaving group; HX: HBF_4, HBr, HF, TFA, TfOH.

Acidolysis of benzylic C–O bonds becomes easier with increasing stability of the corresponding benzylic cation. The sensitivity of benzylic linkers towards acids can therefore be fine-tuned by varying the substitution pattern at the arene. Electron-donating substituents will increase the sensitivity towards acids, and electron-acceptors will diminish acid-sensitivity. This order of reactivity towards electrophiles is the opposite of that towards nucleophiles: the more electron-rich the aromatic, the more resistant are the corresponding benzyl esters towards nucleophilic attack.

Table 3.1 lists the most common types of acid-sensitive benzyl alcohol linker. All these can be attached to various supports by use of different types of spacer. Because resins with these linkers are commercially available, their preparation will not be discussed here.

The 4-alkylbenzyl alcohol linker was the first type of linker used in solid-phase pep-tide synthesis [12]. Esters of this alcohol can readily be prepared from chloromethyl polystyrene (Merrifield resin) and carboxylates [12] or by acylation of hydroxymethyl polystyrene (see Section 13.4). Acidolysis of this linker requires acids with high ioniz-ing power, such as hydrogen fluoride [13–15], trifluoromethanesulfonic (triflic) acid [16], trimethylsilyl triflate/TFA [17,18], trimethylsilyl bromide/TFA [19], hydrogen bromide/AcOH [20,21], or aluminium chloride/DCM/MeNO$_2$ [22]. The stability of 4-alkylbenzyl esters towards weak acids enables the deprotection of Boc-protected amino groups with TFA without cleavage of the product from the resin.

Esters of the PAM linker are slightly more resistant towards acids than the corre-sponding 4-alkylbenzyl esters [4,23–25] (Table 3.1). The PAM linker is particularly well suited for solid-phase peptide synthesis using *N*-Boc amino acids, because less than 0.02 % cleavage of the peptide from the support occurs during the acidolytic deprotection steps [25]. Esters of both the 4-alkylbenzyl alcohol linker and the PAM linker can also be cleaved by nucleophiles (see Sections 3.1.2 and 3.3.3).

4-Alkoxybenzyl alcohol was first used for solid-phase synthesis by Wang in 1973 [26], and has become one of the most widely used linkers (Wang linker). Cross-linked polystyrene with Wang linker is often referred to as Wang resin. Esters of the Wang linker are difficult to cleave from the support by treatment with nucleophiles or with weak acids (< 10 % cleavage occurs upon treatment with neat acetic acid for 4 h at 100 °C [27]), but are readily hydrolyzed by treatment with 50 % TFA in DCM or with Lewis acids such as Et$_2$AlCl [28]. The Wang linker is well suited for the preparation of peptides using the Fmoc methodology (Section 16.1.3).

Dialkoxy- and trialkoxybenzyl esters are even more acid-labile than the Wang lin-ker, and can, e.g., be cleaved with dilute TFA, acetic acid, or hexafluoroisopropanol

Table 3.1. Acid-labile benzyl alcohol linkers. See text for additional cleavage reagents.

Name of linker/resin	Structure	Cleavage reagent for benzyl esters	Literature
4-alkylbenzyl alcohol (e.g. hydroxymethyl polystyrene)		HF/Me$_2$S/*p*-cresol 25:65:10, 0 °C, 2 h	[15]
4-(hydroxymethyl)phenylacetamide (PAM linker)		HF/*p*-cresol 9:1, 0 °C, 1 h	[4,15,23,30]
4-alkoxybenzyl alcohol (Wang linker)		TFA/DCM 1:1, 20 °C, 0.5 h	[26,31]
4-alkoxy-2-methoxybenzyl alcohol [e.g. Sasrin (Pol = PS); *super acid-sensitive resin*]		1% TFA in DCM, 5 min	[29,31,32]
4-alkoxy-2,6-dimethoxybenzyl alcohol (HAL linker)		0.1% TFA in DCM, 25 °C, 1 h	[33]

without simultaneous acidolysis of Boc groups. These linkers thus enable the solid-phase synthesis of protected peptide fragments or other acid-sensitive products [29].

Three different strategies are generally used for the attachment of carboxylic acids to resins as benzyl esters: (a) acylation of resin-bound benzyl alcohols [34–36], (b) O-alkylation of carboxylates by resin-bound benzylic halides [37–39], or (c) O-alkylation of carboxylic acids under Mitsunobu conditions [40,41] (Figure 3.3). These reactions are treated in detail in Section 13.4.

3.1.1.2 Diarylmethanol (Benzhydrol) Linkers

The trialkoxy benzhydrol linker, developed by Rink in 1987 [42] ('Rink acid resin', Figure 3.4) is a further acid-labile linker for carboxylic acids. Esters of this linker can, like trityl esters, even be cleaved with acetic acid or HOBt [43], so care must be taken to avoid loss of product during synthetic operations.

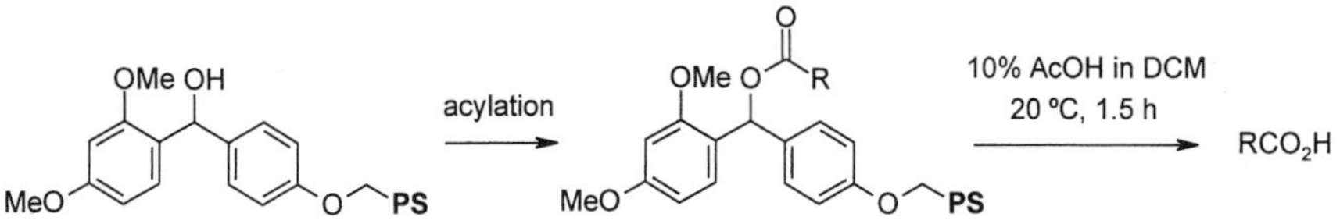

Fig. 3.4. The Rink acid linker [42].

Neither the trialkoxybenzhydryl alcohol linker nor other types of benzhydryl alcohols [40,41] have found widespread use as linkers for carboxylic acids. These linkers do not seem to offer special advantages compared with benzyl alcohol or trityl linkers.

3.1.1.3 Trityl Alcohol Linkers

Esters of unsubstituted, polystyrene-bound trityl alcohol are too acid-sensitive to be useful for solid-phase synthesis [44]. Even treatment with alcohols or HOBt can lead to significant product release from the support. More stable towards solvolysis is the 2-chlorotrityl alcohol linker [43], and numerous examples of the use of this linker for the attachment of carboxylic acids to polymeric supports have been reported [45–52]. Cleavage can be effected by treatment with dilute TFA, acetic acid, or hexafluoroisopropanol [53]. Trityl alcohol linkers substituted with electron-withdrawing groups have also been described (e.g. commercially available, resin-bound 4-carboxytrityl alcohol [54,55] and 9-hydroxy-9-(4-carboxyphenyl)fluorene [56]); these are sufficiently stable to enable the attachment of carboxylic acids as triarylmethyl esters.

The most valuable property of 2-chlorotrityl esters is their high stability towards nucleophiles and bases. 2-Chlorotrityl esters have been used with success at elevated temperatures (e.g. 80 °C, 5 h, toluene [57]), but prolonged heating can lead to cleavage [57].

Attachment of carboxylic acids to supports as trityl esters is achieved by treatment of the corresponding trityl chloride resin with the acid in the presence of an excess of a tertiary amine (Figure 3.5; see also Section 13.4.2). This esterification usually proceeds more quickly than the acylation of benzyl alcohol linkers. Less racemization is generally observed during the esterification of *N*-protected α-amino acids with trityl linkers than with benzyl alcohol linkers [43]. If valuable acids are to be linked to insoluble supports, quantitative esterification can be accomplished by using excess 2-chlorotrityl chloride resin, followed by displacement of the remaining chloride with methanol [58].

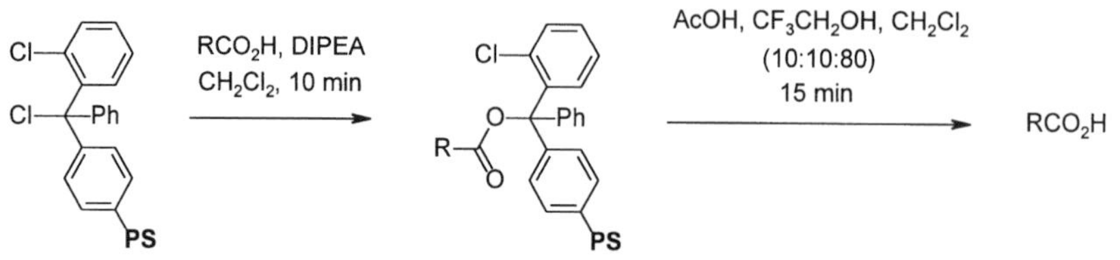

Fig. 3.5. Immobilization of carboxylic acids as 2-chlorotrityl esters [43].

3.1.1.4 Non-Benzylic Alcohol Linkers

Few examples have been reported of the use of resin-bound, tertiary alcohols as linkers for carboxylic acids. Some examples are sketched in Figure 3.6. Esters of the tertiary alcohol linkers **1**, **2**, and **4** can be cleaved by 50 % trifluoroacetic acid, and can therefore be used for peptide synthesis with the Fmoc method but not with the Boc method. In linker **3** only the primary hydroxyl group is utilized for the attachment of carboxylic acids. Cleavage requires previous dehydratization by TFA, leading to a vinyl ester which is readily hydrolyzed [59] (no examples given). The secondary alcohol linker **5** is stable towards trifluoroacetic acid, but can be cleaved by hydrogen fluoride, thereby enabling the synthesis of peptides by Boc methodology [60].

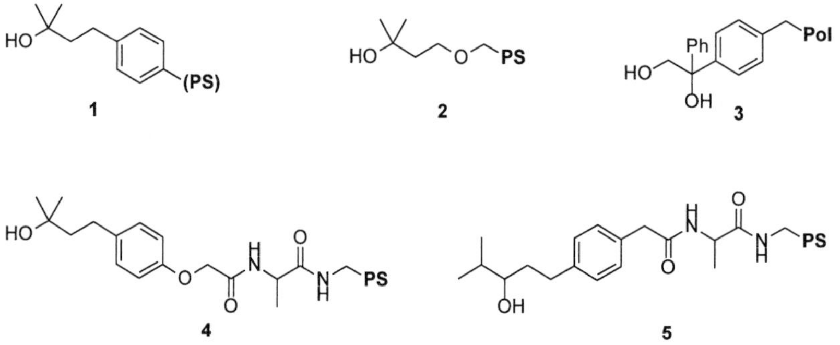

Fig. 3.6. Non-benzylic alcohols as acid-labile linkers for carboxylic acids (**1**: [61–64]; **2**: [65]; **3**: [59]; **4**: [66]; **5**: [60]).

The main advantage of *tert*-alkyl esters as linkers is their stability towards nucleophiles. For instance, no diketopiperazine formation was observed during the preparation of peptides containing carboxy-terminal proline, an otherwise common side reaction when using benzyl alcohol linkers (Section 15.22.1).

Tertiary aliphatic alcohol linkers have only occasionally been used in solid-phase organic synthesis [67]. This might be because of the vigorous conditions required for their acylation. Esterification of resin-bound linker **4** with *N*-Fmoc-proline [66,68] could not be achieved with the symmetric anhydride in the presence of DMAP (20 h), but required the use of *N*-Fmoc-prolyl chloride (10–40 % pyridine in DCM, 25 °C, 10–20 h [66]). A further problem with these linkers is that they can undergo elimination, a side reaction which cannot occur with benzyl or trityl linkers. Hence, for most applications where a nucleophile-resistant linker for carboxylic acids is needed, 2-chlorotrityl- or 4-acyltrityl esters will probably be a better choice than *tert*-alkyl esters.

Resin-bound (4-acyloxy-2-buten-1-yl)silanes, which can be prepared from resin-bound allylsilanes and allyl esters by cross metathesis, react with dilute TFA to yield free carboxylic acids (Figure 3.7). The scope of this strategy remains, however, to be explored.

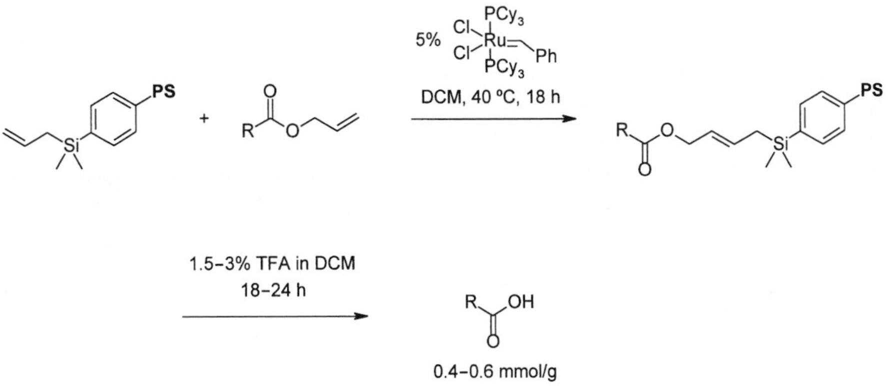

Fig. 3.7. (4-Acyloxy-2-buten-1-yl)silanes as acid-labile linkers for carboxylic acids [69].

3.1.2 Linkers for Acids Cleavable by Bases or Nucleophiles

Linkers for carboxylic acids cleavable by nucleophiles were mainly developed for the preparation of peptide amides or cyclic peptides. These linkers are, however, increasingly being used for the preparation of non-peptides. Several different carboxylic acid derivatives are available by nucleophilic cleavage, the most common being carboxylates, esters, and amides. In addition to these, ketones or primary alcohols can be obtained by use of organometallic reagents or reducing agents to effect nucleophilic cleavage. These applications will be discussed in chapters dealing with linkers for the resulting functional groups.

Most of the reagents required for the saponification of resin-bound esters are, unfortunately, non-volatile, and purification of the products is generally necessary. This is probably the main reason for the scarcity of examples of cleavage by saponification, compared with other methods of cleavage.

The type of support chosen can have an impact on the facility with which nucleophilic cleavage takes place. Polystyrene, a very hydrophobic support, is difficult to perfuse with small ions, such as hydroxide, and for this reason the saponification of poly-styrene-bound esters usually proceeds more slowly than in solution. Tentagel, poly-acrylamides, or other, more hydrophilic, supports are generally a better choice if sapo-nifications or other reactions involving salts are to be performed.

3.1.2.1 Benzyl Alcohol Linkers

Polystyrene-supported benzyl esters can be cleaved by treatment with different types of nucleophile. Benzyl esters become more sensitive towards nucleophiles with decreasing electron density of the benzene ring [70]. On the other hand, acid-sensitive benzyl alcohol linkers, in particular those with electron-donating alkoxy groups (e.g. the Wang linker), are rather resistant towards nucleophilic attack. 4-(Hydroxy-methyl)benzoic acid (HMBA-linker) was especially designed for nucleophilic clea-vage and enables, e.g., the preparation of peptides by use of Boc methodology and cleavage of the peptides from the support by treatment with different nucleophiles. Similarly sensitive towards nucleophilic attack are esters of 4-hydroxymethyl-3-nitro-benzoic acid (a photocleavable linker, Entry **5**, Table 3.2). Fluoride-sensitive benzyl-esters have been developed, which enable nucleophilic cleavage under mild conditions (Entries **6** and **7**, Table 3.2 [71,72]).

3.1.2.2 Non-Benzylic Alcohol Linkers

Support-bound non-benzylic alcohols can also be used to immobilize carboxylic acids as esters (Table 3.3). The advantage of this type of linker is the stability towards electrophiles. Attachment of carboxylic acids is usually realized by acylation of the resin-bound alcohol with a reactive acid derivative.

PEG-grafted polystyrene resins (e.g. Tentagel) enable the direct attachment of car-boxylic acids as 2-alkoxyalkyl esters. PEG esters are usually cleaved with nucleo-philes, to yield carboxylates, esters, amides, or hydrazides.

Entries **4** and **5** (Table 3.3) are interesting examples of linkers sensitive towards reducing agents. Treatment of the azide-containing linker (Entry **5**) with tributyl-phosphine in the presence of water leads to the reduction of the azido group to an amino group, which cleaves the benzamide by intramolecular transamidation. The di-sulfide-based linker (Entry **4**) also undergoes reductive cleavage upon treatment with phosphines to yield a 2-mercaptoethyl ester. This ester is hydrolyzed under mildly basic conditions because of intramolecular nucleophilic catalysis by the mercapto group [85].

Entry **6** in Table 3.3 is an additional linker based on intramolecular nucleophilic cleavage. In this case it is an imidazole which efficiently catalyzes the saponification of the ester linkage.

A special group of base-labile linkers for carboxylic acids is based on cleavage by β-elimination. Here, the resin-bound alcohol must bear an electron-withdrawing group in the β position (Figure 3.8) which facilitates elimination by acidifying this

Table 3.2. Saponification of resin-bound benzyl esters.

Entry	Loaded resin	Cleavage conditions	Product, yield (purity)	Ref.
1		Bu$_4$NOH, THF, MeOH, 70 °C, 24 h	100%	[73] see also [74,75]
2		THF/satd aq K$_2$CO$_3$ 10:1, [Bu$_4$N][HSO$_4$], 25 °C, 2–24 h	70–100% (85–90%)	[76]
3	(PAM linker)	DBU (2 eq), LiBr (5 eq), THF/H$_2$O 20:1, 25 °C, 4 h	81%	[77] see also [78]
4		TBAF·3 H$_2$O (0.05 mol/L), DMF, 0.5 h	48%	[70] see also [79,80]
5		TBAF (8 eq), MeCN, 25 °C, 40 min; or LiOH (15 eq), H$_2$O, 25 °C, 10 min	54%	[81]
6		TBAF (3 eq), DMF, 65 °C, 1 h	78%	[82] see also [83]
7		TBAF·3 H$_2$O (1 eq), PhSH (1.2 eq), DMF, 5 min; or TFA/DCM/Me$_2$S 5:4:1, PhSH (1.2 eq), 0.5 h	100%	[84]

Table 3.3. Saponification of non-benzylic, resin-bound esters.

Entry	Loaded resin	Cleavage conditions	Product, yield (purity)	Ref.
1	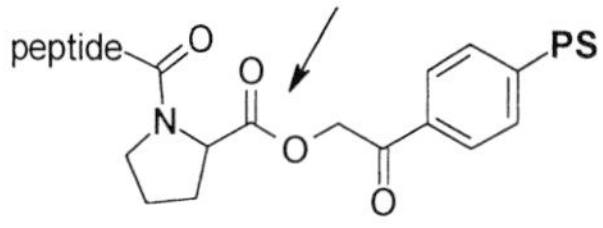 peptide … PS	TBAF (0.1 mol/L), EDT (1 eq), DMF, 25 °C, 0.5 h	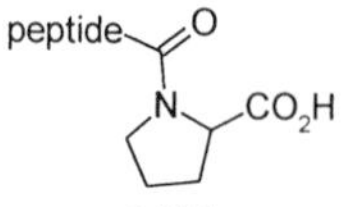 peptide … CO_2H 35%	[79] see also [86]
2	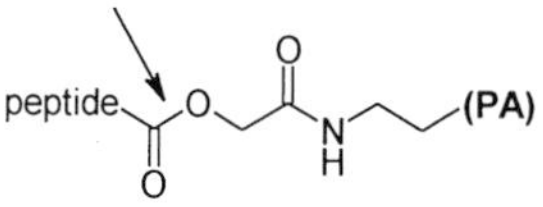 peptide … (PA)	NaOH (0.01 mol/L), H_2O/iPrOH 3:7 or H_2O, 3 h	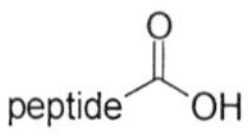 peptide—OH	[87] see also [88]
3	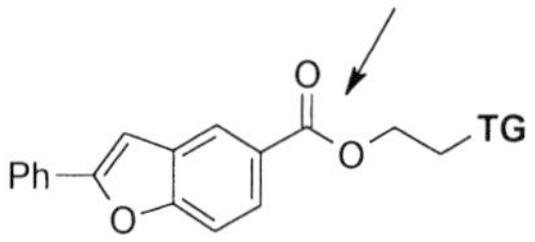 Ph … TG	aq NaOH (1 mol/L)/ iPrOH 5:2, 50 °C, 8 h	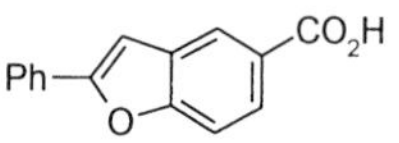 Ph … CO_2H 71% (88%)	[89]
4	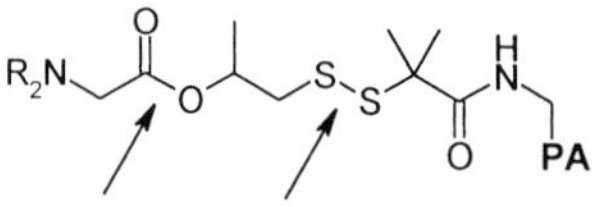 R_2N … S–S … PA	1. PR_3, pH 4.5 2. pH 8–9	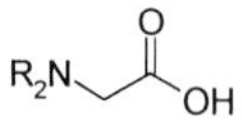 R_2N … OH	[85]
5	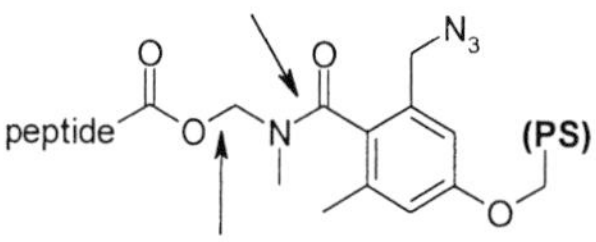 peptide … N_3 (PS)	PBu_3 (3 eq), buffer (pH 7), DMF, 10 h	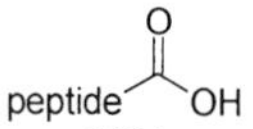 peptide—OH 60%	[90]
6	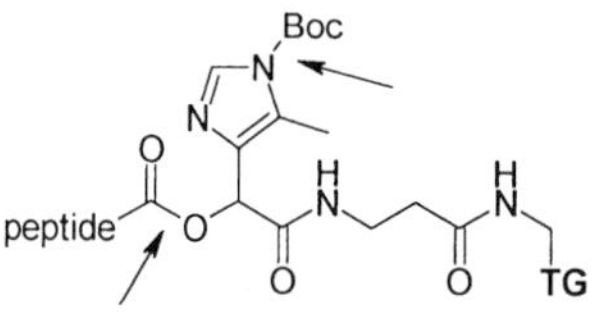 Boc N … peptide … TG	1. TFA/DCM 1:1 2. phosphate buffer (0.01 mol/L, pH 7.5), 50 °C, 25 min	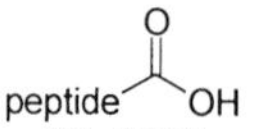 peptide—OH 33–80%	[91] see also [92]

position. Mechanistically these linkers are closely related to the Fmoc protective group, and some of these linkers are indeed based on 9-fluorenylmethanol (e.g. Entries **1** and **2**, Table 3.4). The stability of these linkers towards acids enables their use for the preparation of peptides with Boc methodology.

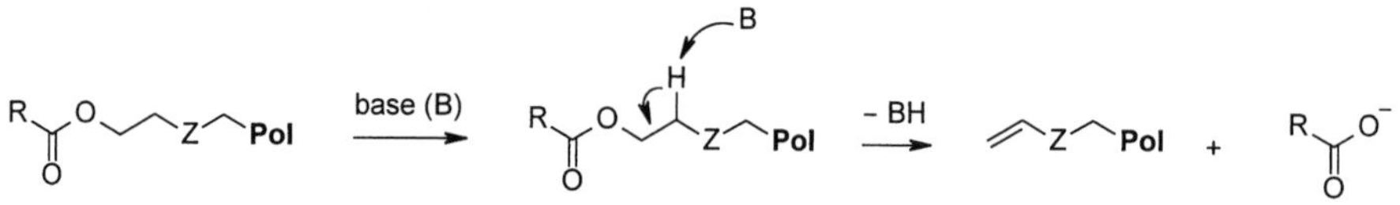

Z = electron-withdrawing group

Fig. 3.8. Cleavage of linkers by β-elimination.

Table 3.4. Linkers for carboxylic acids cleavable by base-induced β-elimination.

Entry	Loaded resin	Cleavage conditions	Product, yield (purity)	Ref.
1		morpholine/DMF 1:4, 25 °C, 0.5 h	peptide—OH 100% (> 95%)	[93] see also [94]
2		piperidine/DMF 15:85, 3 × 5 min	peptide—OH 72% (90%)	[95]
3		satd aq NH₃/dioxane 9:1, 50 °C, overnight	R—OH	[96]
4		NaOH (4 mol/L)/ MeOH/dioxane 1:9:30, 0.5 h	peptide—OH 54–60%	[97] see also [98]

Support-bound phenols, oximes and related compounds yield, upon acylation, esters which are highly susceptible towards nucleophilic cleavage. These esters are often used as insoluble acylating agents for the preparation of amides or esters, but not as linkers for carboxylic acids. These linkers are treated in Sections 3.3.3 (amides) and 3.5.1 (esters).

3.1.2.3 Miscellaneous Linkers Cleavable by Bases or Nucleophiles

Carboxylic acids can also be attached to solid supports as amides, imides, and thiol esters. Illustrative examples of the saponification of such linkers are listed in Table 3.5.

Thiol esters are more sensitive towards nucleophilic attack than the corresponding esters, and can be readily saponified. Resin-bound thiol esters have, however, mainly been used for the preparation of amides by nucleophilic cleavage with amines (see Section 3.3.3).

Resin-bound amides generally need to be activated to become susceptible to saponification under acceptably mild reaction conditions [99] (Table 3.5). Particularly elegant are those linkers which enable this activation to be realized as last synthetic step before cleavage (safety-catch linkers [100,101]). The activation of some amide-based safety-catch linkers is sketched in Figure 3.9.

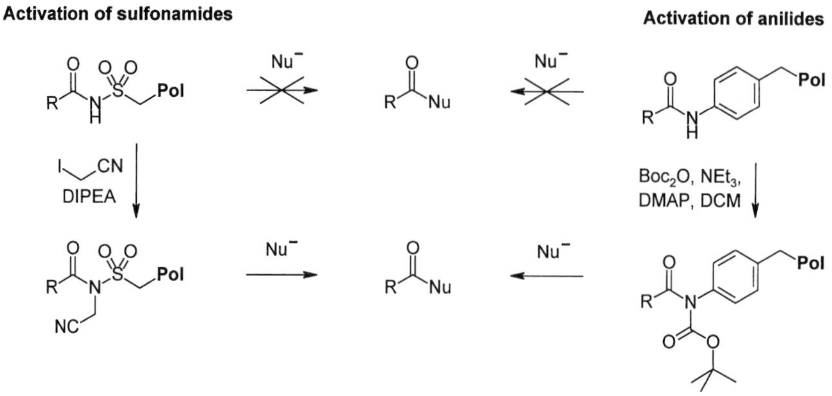

Activation of sulfonamides   **Activation of anilides**

Fig. 3.9. Activation and nucleophilic cleavage of amide-based safety-catch linkers [102,103].

Table 3.5. Saponification of resin-bound thiol esters, imides, and related compounds.

Entry	Loaded resin	Cleavage conditions	Product, yield (purity)	Ref.
1	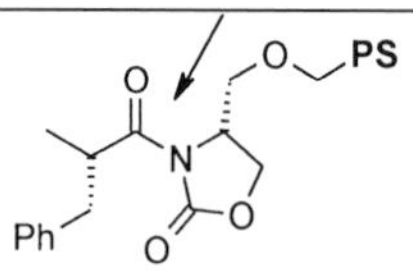	LiOH, THF/H$_2$O 3:1, 12 h	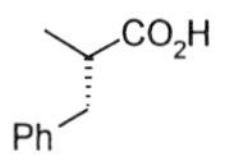	[104] see also [105–107]
2	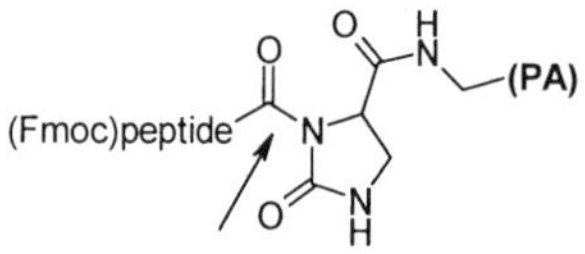	NaOH (1 eq), CaCl$_2$, iPrOH/H$_2$O 7:3, 2 × 1.5 h	 72%	[108] see also [109]
3	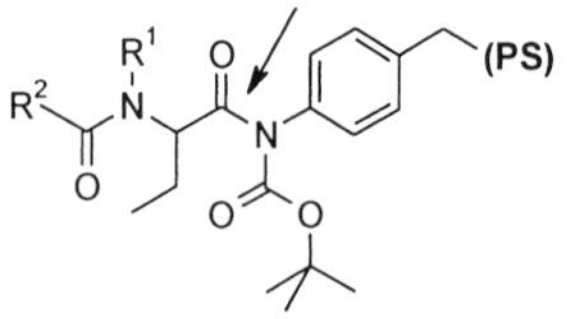	LiOH, 5% H$_2$O$_2$, H$_2$O, THF	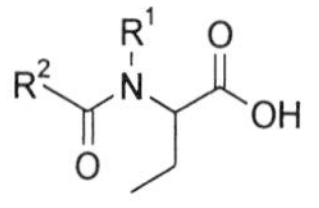	[103]
4	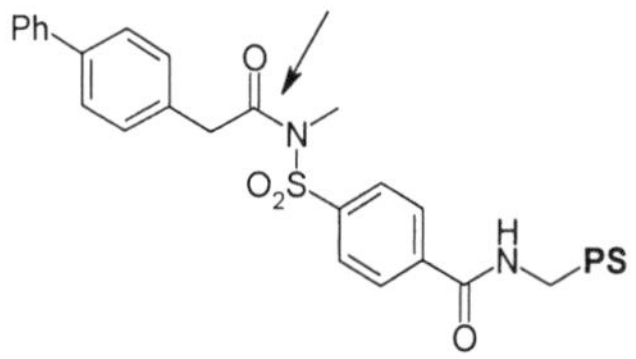	OH$^-$, H$_2$O, 20 °C	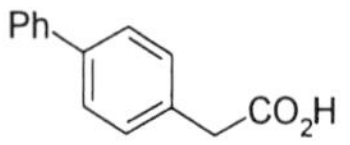 93%	[110] see also [100]
5	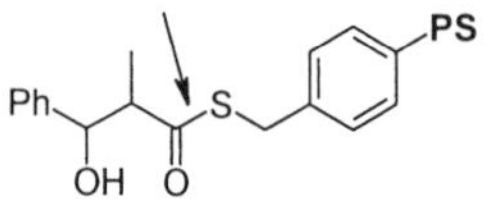	aq NaOH (1 mol/L)/ dioxane 1:4, 100 °C, 6 h	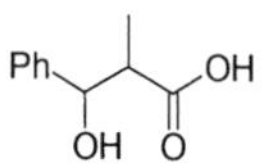 59%	[111]

3.1.3 Photocleavable Linkers for Carboxylic Acids

The development of light-sensitive protective groups began in the early 1960s [112–114] and led to the identification of several functionalities which could be selectively cleaved by UV radiation (for a review, see [115]). Some of these protective groups, such as 2-nitrobenzyl esters, carbonates, or carbamates [114,116–118], benzoin [119–122] and other phenacyl esters [123] were found to be also useful as photocleavable linkers.

Photocleavable linkers can be stable both towards bases and towards acids, and overcome thereby some of the inherent limitations of other types of linker. Cleavage by photolysis is, however, complicated by other problems. UV-Absorbing byproducts, which are, unfortunately, often formed during synthetic transformations, can completely obstruct the passage of light through the polymer and thereby inhibit photolytic cleavage. For this reason the use of photocleavable linkers is restricted to synthetic sequences in which no, or only insignificant amounts of, UV-absorbing products are formed.

Most photocleavable linkers for carboxylic acids used today are based on the photoisomerization of 2-nitrobenzyl esters and on the light-induced cleavage of phenacyl esters (Figure 3.10).

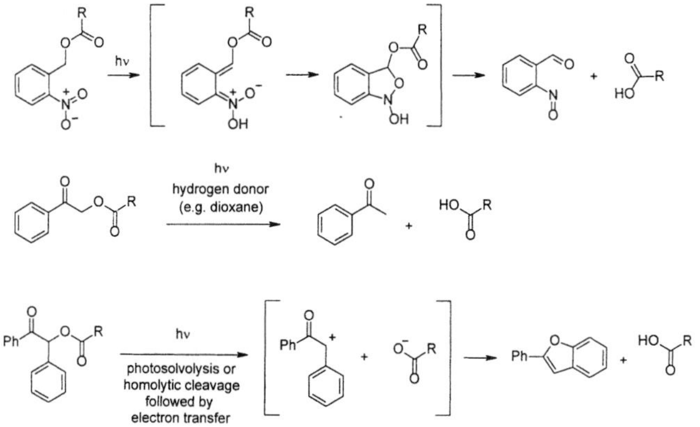

Fig. 3.10. Light-induced photolysis of 2-nitrobenzyl esters and phenacyl esters [119,121,123].

The first type of light-sensitive 2-nitrobenzyl ester used in solid-phase synthesis were derived from 4-hydroxymethyl-3-nitrobenzoic acid (Entry **1**, Table 3.6 [6,124,125]). This linker suffers, however, from several limitations. During photolysis a nitrosobenzaldehyde is formed, which strongly absorbs UV radiation and thereby prevents the quantitative release of product. Generally long cleavage times (> 10 h) are required; this can lead to other undesired photochemical reactions of the products. Photodimerization of thymine has, e.g., been observed when photolabile linkers were used for the solid-phase synthesis of oligonucleotides [126,127]. In addition, thioethers, such as methionine-containing peptides, might be oxidized to the corresponding sulfoxides during prolonged irradiation. These problems were in part solved by the development of 4,5-dialkoxy-2-nitrobenzyl alcohol linkers (Entry **4**, Table 3.6

[128–131]), which can be cleaved photolytically within 2 h to yield crude products of high purity. These linkers are stable towards both TFA and piperidine, thereby enabling the preparation of peptides by use of either Boc or Fmoc methodology. It has been observed that the addition of hydrazine or ethanolamine as scavengers of resin-bound nitrosoarenes formed during photolysis often improves the yield of released product [132].

Phenacyl and benzoin esters can also serve as photolabile linkers (Entries 5 and 6, Table 3.6), but have not received as much attention as nitrobenzyl derivatives. Entry 6 in Table 3.6 is an example of a 'safety-catch' photolinker, in which the ketone, which enables photolytic cleavage, is masked as 1,3-dithiane. Several possible mechanisms

Table 3.6. Photolytic cleavage of resin-bound esters.

Entry	Loaded resin	Cleavage conditions	Product, yield (purity)	Ref.
1		hv, CF₃CH₂OH/ DCM 1:4, 15 h	82%	[133]
2		hv (350 nm), CuSO₄, EtOH/DCM 1:1, 24 h	40–48%	[134] see also [135]
3		hv (350 nm), CuSO₄, EtOH, 24 h	60%	[136]
4		hv (354 nm)	92%	[137]
5		hv (350 nm), DMF, 72 h	77%	[138] see also [123]
6		1. PhI(O₂CCF₃)₂ or Hg(ClO₄)₂ or MeOTf or HIO₄; THF/H₂O, 20 °C, 18 h 2. hv (350 nm), 2 h, THF/MeOH, 28 °C	65–75%	[139] [140]
7		hv (320–340 nm), THF/H₂O, 12 min	85% (93%)	[141]

have been proposed for the photolytic cleavage of benzoin esters, one of the most recent is the dissociation of the excited phenacyl ester into a carboxylate and an phenacyl cation ('photosolvolysis', Figure 3.10 [119]).

In 1998 a new type of light-sensitive linker based on the photolysis of *tert*-butylketones was developed by Giese and coworkers (Entry **7**, Table 3.6). The proposed mechanism of cleavage is outlined in Figure 3.11.

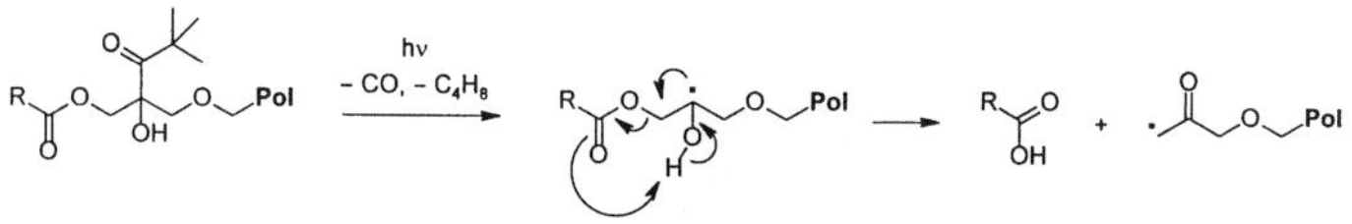

Fig. 3.11. Mechanism of photolytic cleavage of 2-acyloxyethyl *tert*-butyl ketones [141].

This linker proved to be stable under a variety of reaction conditions (e.g. 50 % TFA or 5 % BF$_3$·Et$_2$O in DCM, 20 °C, 2 h; 5 % DBU in toluene, 80 °C, 2 h). An additional advantage of this linker, compared with 2-nitrobenzyl alcohol derivatives, is that no UV-absorbing products are formed during photolysis. This enables fast and complete photolytic detachment of the resin-bound product. Relatively short wavelengths (< 340 nm) are, however, required to effect cleavage.

3.1.4 Linkers for Acids Cleavable by Transition Metal Catalysis

Benzyl alcohol linkers, such as those described in Section 3.1.1.1, can also be cleaved by palladium-catalyzed hydrogenolysis. Carboxylic acids have, e.g., been obtained by hydrogenolysis of insoluble benzyl esters with Pd(OAc)$_2$/DMF/H$_2$ [80,142]. Resin-bound benzylic carbamates [143] and amides [144] can also be released by treatment with Pd(OAc)$_2$ in DMF in the presence of a hydrogen source, such as 1,4-cyclohexadiene or ammonium formate. These reactions are quite surprising, because they require the formation of metallic palladium within the gelated beads.

Allyl esters, carbonates, or carbamates readily undergo C–O bond cleavage upon reaction with palladium(0) to yield allyl palladium(II) complexes. These complexes are electrophilic and can react with nucleophiles, forming products of allylic nucleophilic substitution, and palladium(0) is regenerated. Linkers based on this reaction have been designed which are cleavable by treatment with catalytic amounts of palladium complexes [145,146]. For the immobilization of carboxylic acids support-bound allyl alcohols have proven suitable. (Figure 3.12, Table 3.7).

The two most common types of allyl alcohol linker are 4-hydroxycrotonic acid derivatives (Entry **1**, Table 3.7) and (*Z*)- or (*E*)-2-butene-1,4-diol derivatives (Entries **2** and **3**, Table 3.7). The former are well suited for solid-phase peptide synthesis using Boc methodology, but give poor results when using the Fmoc technique, probably because of Michael addition of piperidine to the α,β-unsaturated carbonyl compound [147]. Butene-1,4-diol derivatives, however, tolerate well acids, bases, and weak nucleophiles, and are therefore suitable linkers for a broad range of solid-phase chemistry.

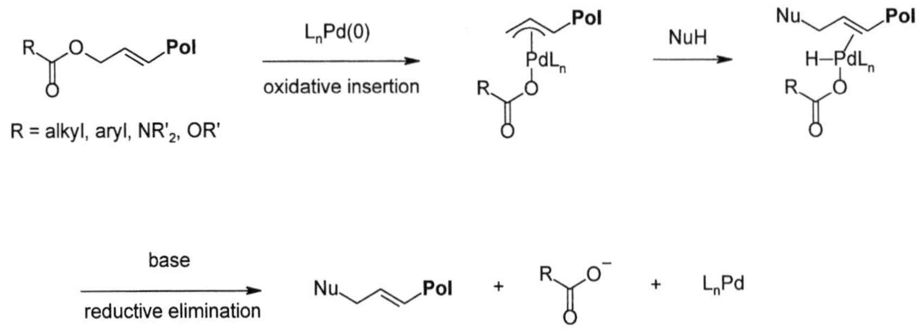

Fig. 3.12. Cleavage of support-bound allyl esters via allyl complex formation.

Table 3.7. Linkers cleavable by palladium(0)-catalyzed allylic nucleophilic substitution.

Entry	Loaded resin	Cleavage conditions	Product, yield (purity)	Ref.
1	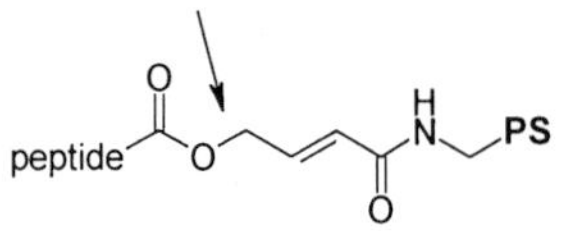	Pd(PPh$_3$)$_4$ (1.4 mmol/L), THF/ morpholine 10:1, 20 °C, 2 h	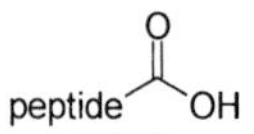 peptide OH 98%	[148]
2	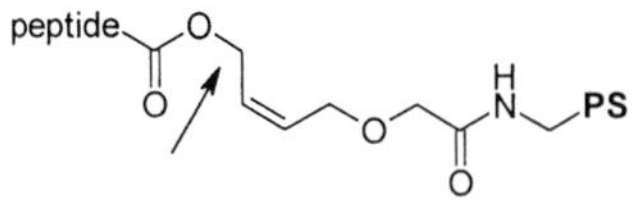	PdCl$_2$(PPh$_3$)$_2$ (0.7 mmol/L), Bu$_3$SnH (0.05 mol/L), DCM, 20 °C, 20 min; then AcOEt/HCl	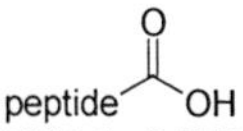 peptide OH 80% (> 95%)	[149] see also [150]
3	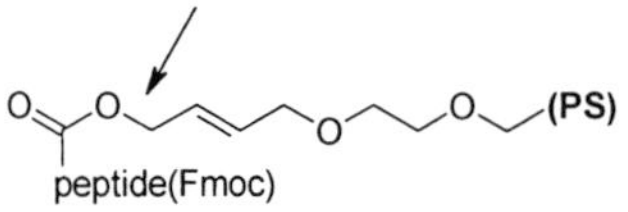	Pd(PPh$_3$)$_4$ (2.9 mmol/L), PhNHMe (0.34 mol/L), DCM/DMSO 1:1, 15 h	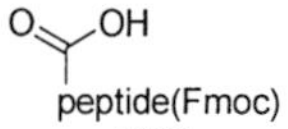 peptide(Fmoc) 62%	[147] see also [151]

Hydrazides have been used as oxidant-labile linkers for carboxylic acids (Figure 3.13 [152–154]). Cleavage is effected by copper(II)-catalyzed or NBS-mediated oxidation of the hydrazide to an azo compound, which decomposes to yield the free carboxylic acid or amides, depending on the nucleophile present during cleavage. The same type of cleavage methodology has been used for the generation of arenes ('traceless linker', see Section 3.16).

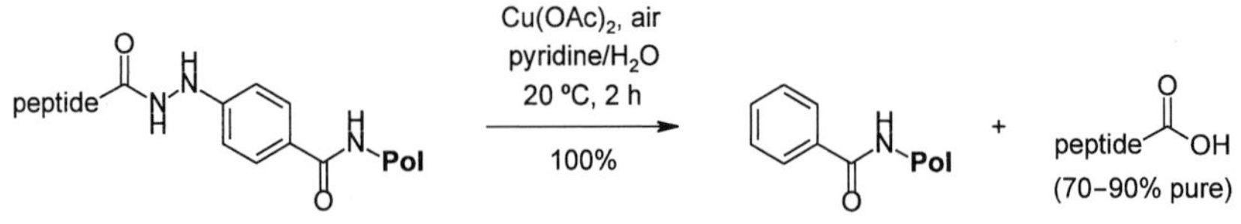

Fig. 3.13. Oxidative cleavage of hydrazides [155].

3.1.5 Linkers for Acids Cleavable by Enzymes

The efficiency of enzyme catalysis on solid supports depends strongly on the type of support and the size of the enzyme chosen. Enzymes can generally only be functional in water or in solvent mixtures with high water content, and can therefore be used with water-compatible supports only. Microporous polystyrene, being a highly hydrophobic support, is not well suited for enzyme-catalyzed reactions [156,157]. The suitablility of PEG-grafted polystyrene supports, such as Tentagel or Argogel, for reactions involving enzymes seems to depend on the size of the enzyme used [158], and both positive results (lipase RB 001-05 [159], penicillin amidase [160], porcine pancreas lipase VI-S [157]) and negative results (papain [161], chymotrypsin, elastase, pepsin [162]) have been reported.

Authors who examined the suitablility of different types of support for the realization of enzyme-catalyzed reactions concluded that supports such as Tentagel or Argo-

Table 3.8. Linkers for carboxylic acids cleavable by enzymes.

Entry	Loaded resin	Cleavage conditions	Product, yield (purity)	Ref.
1	(racemic)	lipase VI-S (porcine pancreas), *t*BuOH/THF 5:1, phosphate buffer (pH 7.8), 30 °C, 13 h	16% (88% ee)	[157]
2	(racemic)	lipase VI-S (porcine pancreas), *t*BuOH/THF 5:1, phosphate buffer (pH 7.8), 30 °C, 6 h	30% (90% ee)	[157]
3		α-chymotrypsin, H₂O (pH 7.0)	> 95%	[167]
4		penicillin amidase, phosphate buffer (pH 7.5), 25 °C, 16 h (also cleavable by TFA, HCl, or NaOH)	50% (25% on PEGA)	[160]
5		lipase RB 001-05, MES buffer (0.05 mol/L, pH 5.8), 30 °C	70–80%	[159]

gel are not generally suitable for these reactions. Selective and quantitative hydrolysis of peptides with papain could, e.g., only be achieved on PEGA but not on PEG-grafted polystyrene [161]. Similar results were obtained by Meldal and coworkers, who used PEGA as support for combinatorial peptide and glycopeptide libraries. On-bead screening of these libraries with enzymes enabled the rapid identification of new enzyme substrates [163–166].

Further supports, which have been claimed to be compatible with enzymatic catalysis are CPG [156,167], polyacrylamides [165,168,169], and sepharose [170]. Macroporous polystyrene or other supports with the attachment sites located exclusively on the surface should also be suitable for enzyme-mediated reactions.

There have been a few reports of linkers for carboxylic acids, which could be cleaved with enzymes from PEG-grafted polystyrene supports, such as Tentagel (Table 3.8). Some of these linkers are also sensitive towards acids or bases, and will, therefore, only remain uncleaved under a narrow selection of reaction conditions.

3.2 Linkers for Thiocarboxylic Acids

Support-bound thiols can be suitable linkers for thiocarboxylic acids. Peptide thiocarboxylic acids are interesting synthetic intermediates which smoothly react with various alkylating agents, including bromoacetyl peptides, to yield the corresponding thiol esters, which have been used as peptide mimetics [171,172]. Furthermore, thiocarboxylic acids react with diaryl disulfides to yield acyl disulfides, which can be used as acylating agents [173]. Amines are directly acylated by thiocarboxylic acids in the presence of silver nitrate to yield amides [174].

As linkers for thiocarboxylic acids mostly 4-alkoxybenzhydrylthiols (Figure 3.14) have been used. Cleavage of the benzylic C–S bond of these linkers requires strongly ionizing acids, such as HF.

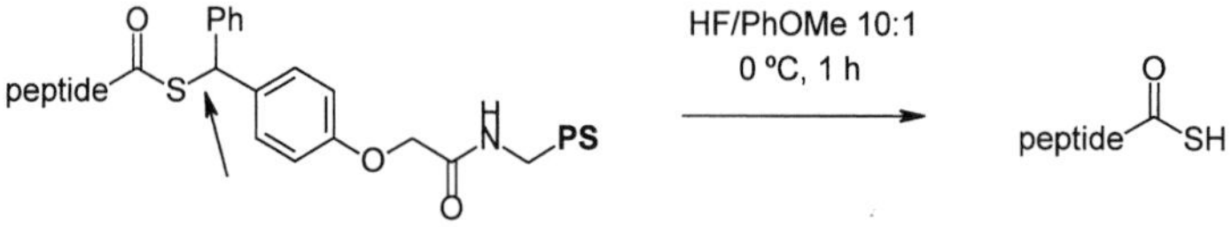

Fig. 3.14. Benzhydrylthiol-based linker for thiocarboxylic acids [173,175].

3.3 Linkers for Amides, Sulfonamides, Carbamates, and Ureas

Several different types of linker have been developed which yield amides upon cleavage. These linkers can often also be used to prepare sulfonamides, carbamates, or ureas.

Essentially three different strategies enable the release of amides from insoluble supports: (a) cleavage of the benzylic C–N bond of resin-bound *N*-alkyl-*N*-benzyl amides (backbone amide linkers, BAL-linkers), (b) nucleophilic cleavage of resin-

bound acylating agents with amines, and (c) acylation/debenzylation of resin-bound
N-benzyl-N,N-dialkylamines.

3.3.1 Benzylamine Linkers

A series of linkers has been developed for the solid-phase preparation of peptide
amides (peptide–CONH$_2$); most of these are based on the scission of benzylic C–N
bonds (Figure 3.15). Because of the lower polarization of C–N bonds compared with
C–O bonds benzylic amines or amides are generally more difficult to cleave with electrophiles than benzylic ethers or esters. The reactivity of the N-benzylamides towards
electrophiles, however, depends to a great extent on the electronic properties of all
the substituents. Electron-withdrawing substituents at nitrogen will increase the polarization and thereby promote solvolysis of benzylic C–N bonds. Anilides (R: Ar, Figure
3.15) are, therefore, usually more easy to cleave from a support by acidolysis than the
corresponding N-alkylamides. Strongly acidic amides, such as sulfonamides, are also
more readily cleaved than amides from an acid-labile linker [176].

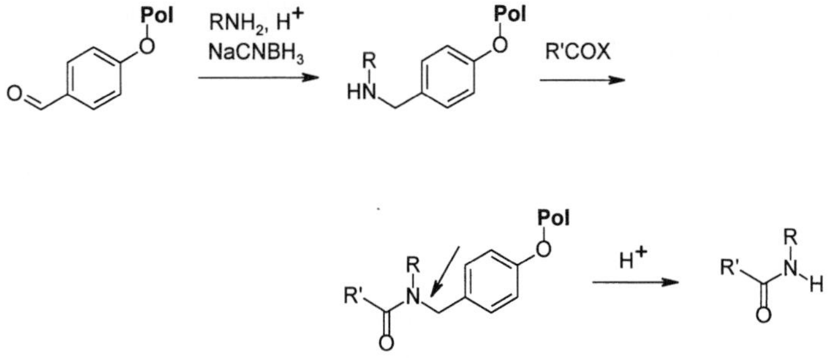

Fig. 3.15. Backbone amide linking.

As illustrated by the examples in Table 3.9, resin-bound 4-alkoxybenzylamides
often require higher concentrations of TFA and longer reaction times than Wang resin
bound carboxylic acids. For this reason, the more acid-sensitive di- or (trialkoxybenzyl)amines are generally preferred as backbone amide linkers. The required resin-
bound, secondary benzylamines can readily be prepared by reductive amination of
resin-bound benzaldehydes (Figure 3.15 [177]) or by N-alkylation of primary amines
with resin-bound benzyl halides or sulfonates. Sufficiently acidic amides can also be
N-alkylated by resin-bound benzyl alcohols under Mitsunobu conditions (see, e.g.,
[178]; attachment to Sasrin of Fmoc cycloserine, an O-alkyl hydroxamic acid).

One problem occasionally encountered with BAL linkers attached to polystyrene as
aryl benzyl ethers (Table 3.9, Figure 3.16) is that cleavage of the whole linker can compete
with C–N bond cleavage. Surprisingly, cleavage of polystyrene-bound aryl ethers can
even occur under mildly acidic conditions, under which such cleavage would not have
been expected. The cleavage sketched in Figure 3.16, e.g., should only have occurred
upon treatment with hydrogen fluoride (acidolysis of 4-alkylbenzyl ester; see Table 3.1).
Similar effects have been observed with the Rink amide linker (Section 3.3.2). Because of

this problem, amide backbone linkers are often attached to polystyrene not directly as benzyl ethers but with a spacer, such as an ω-aryloxyalkanoic acid, to aminomethyl polystyrene or similar supports (see, e.g., Entries **7** and **12**, Table 3.9).

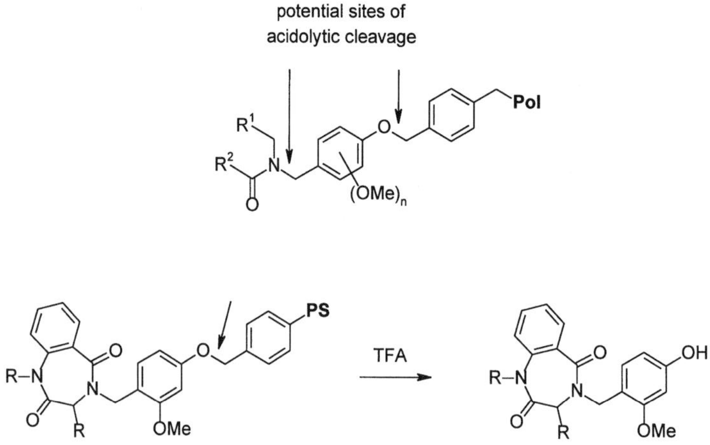

Fig. 3.16. Acidolytic cleavage of aryl benzyl ethers as competing reaction during the acidolysis of *N*-benzylamides [179].

A further problematic substrate are amides containing a basic functional group near the attachment site of the backbone linker (Figure 3.17). These are, for instance, non-acylated α-amino acid amides or other amides containing a tertiary amine or a pyridine ring close to the carboxamido group (Entries **5**, **7**, **8**, and **12**, Table 3.9). Protonation of the basic functionality reduces the basicity of the amide which needs to be protonated for cleavage to occur. Because the concentration of protonated amide is reduced, cleavage of amides containing additional basic groups proceeds more slowly than the cleavage of similar, non-basic amides (Figure 3.17). As illustrated by Entries **7** and **8** (Table 3.9), this inhibition of the cleavage reaction can be rather strong. In such cases the use of neat TFA as cleavage reagent and long cleavage times may be required [180].

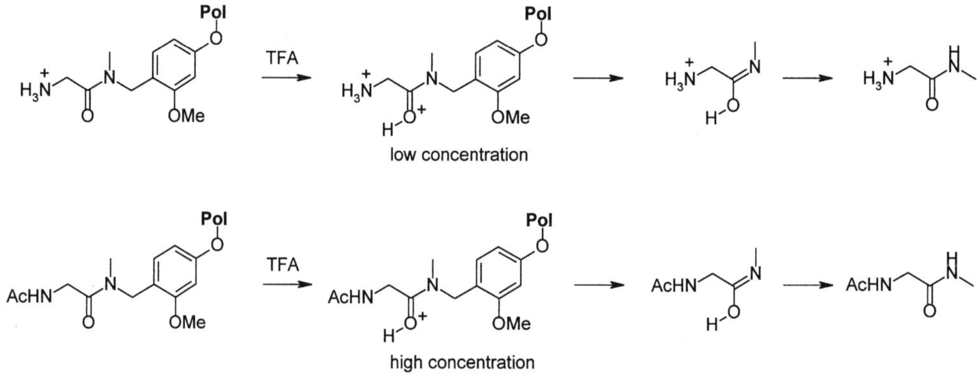

Fig. 3.17. Inhibition of the cleavage of benzylic C–N bonds by basic functional groups in the support-bound amide.

Table 3.9. Acidolytic cleavage of resin-bound *N*-benzylamine derivatives.

Entry	Loaded resin	Cleavage conditions	Product, yield (purity)	Ref.
1		TFA/DCM 1:1, 0.5 h	no cleavage	[181]
2		HF/*p*-cresol 10:1, −5 °C, 1 h (no cleavage with neat TFA, 25 °C, 2 h)		[182]
3		TFA/H$_2$O 95:5, 20 °C, 24 h	53% (95%)	[183] see also [180]
4		TFA/H$_2$O 95:5, 20 °C, 24 h	56% (96%)	[183]
5		TFA/Et$_3$SiH 95:5	100% (> 90%)	[184]
6		TFA/DCM > 2:8		[185]
7		TFA/DCM 7:3, 25 °C, 2 h	24% (incomplete cleavage)	[186]
8		2.5% Et$_3$SiH in TFA, overnight (incomplete cleavage after 4 h)	99% (> 90%)	[180]
9		TFA/DCM 5:95, 20 °C, 10 min Ar: 4-(MeO)C$_6$H$_4$	79% (90%)	[176] see also [178] [179] [187] [188]

Table 3.9. continued.

Entry	Loaded resin	Cleavage conditions	Product, yield (purity)	Ref.
10		TFA/DCM 5:95, 20 °C, 10 min	67% (95%)	[176]
11		TFA/DCM 5:95, 20 °C, 10 min	63% (90%)	[176] see also [189]
12	(PAL linker)	TFA/DCM 7:3, 25 °C, 2 h	91%	[186] see also [190]
13		TFA/Me₂S/H₂O 90:5:5, 36 h	75%	[179]
14		TFA/H₂O/Me₂S 100:5:5, 85 °C, 12 h; or TFA/H₂O/iPr₃SiH 100:6:6, 85 °C, 12 h		[191]
15		TFA/EDT/PhOH/PhSMe 90:5:3:2, 20 °C, 1 h (no cleavage with TBAF (0.2 mol/L), DMF, 2 h)	43% (> 90%)	[192]
16		TFA/DCM 1:1, 4 h	88%	[193]

Treatment of tertiary benzylamines with acylating agents can lead to debenzylation. If the benzyl group is linked to an insoluble polymer, acylation and debenzylation will lead to the release of an acylated amine into solution (Entry **1**, Table 3.10). Similarly, polystyrene-bound triazenes can be cleaved with acyl halides in the absence of bases to yield amides (Entry **2**, Table 3.10). These cleavage reactions generally yield products contaminated with acylating agent which require further purification.

Photolabile linkers for amides are most often based on 2-nitrobenzyl derivatives (Entries **3–6**, Table 3.10; for preparation of linkers see also [130,194]). The mechanism of photolysis is the same as for the related 2-nitrobenzyl alcohol linkers (Section 3.1.3).

Table 3.10. Acidolytic and photolytic release of amides from supports.

Entry	Loaded resin	Cleavage conditions	Product, yield (purity)	Ref.
1	Ph–N piperazine–N–CH$_2$–C$_6$H$_4$–O–PS	PhCOCl (0.4 mol/L, 7 eq), DCM, 26 h	Ph–N piperazine N–C(=O)–Ph, 69% (> 95%)	[195]
2	(formyl)–N piperazine–N–N=N–C$_6$H$_4$–O–PS	AcCl, THF, 20 °C, 12 h	(formyl)–N piperazine N–C(=O)CH$_3$, (> 90%)	[196]
3	R–CH=CH–S(O$_2$)–N(CH)– (OMe, NO$_2$ benzyl) –O–(CH$_2$)$_3$–C(=O)–NH–TG	hv (354 nm), MeOH, 40 h	R–CH=CH–S(O$_2$)–NH–, 50%	[197]
4	R–CH$_2$–C(=O)–NH–(CH)– (OMe, NO$_2$ benzyl) –O–(CH$_2$)$_3$–C(=O)–NH–Pol	hv (365 nm), 5% DMSO in phosphate-buffered saline (pH 7.4), 3 h	R–CH$_2$–C(=O)–NH$_2$, > 90% (95%)	[128] [132]
5	O$_2$N–C$_6$H$_4$–CH(NH–C(=O)–pyridyl)–C(CH$_3$)–C(=O)–NH–TG	hv (365 nm), MeOH/H$_2$O 1:4, 4 h	pyridyl–C(=O)NH$_2$, 86%	[198]
6	BocHN–CH(CH$_2$CH(CH$_3$)$_2$)–C(=O)–N–CH$_2$–C$_6$H$_3$(NO$_2$)–C(=O)–NH–PS	hv (350 nm), MeOH, 18 h	BocHN–CH(CH$_2$CH(CH$_3$)$_2$)–C(=O)–NH–, 70%	[199] see also [200] [201]

3.3.2 (Diarylmethyl)amine and Tritylamine Linkers

One of the oldest linkers for amides is the (4-methylbenzhydryl)amine linker (MBHA; Entry **1**, Table 3.11). In contrast with the corresponding benzhydrol linker (cleavable by 5 % TFA in DCM, 5 min [41]), acidolysis of the benzylic C–N bond of the MBHA linker requires treatment with hydrogen fluoride or a similar acid. As for

Table 3.11. Acidolytic cleavage of resin-bound (diarylmethyl)amine and tritylamine derivatives.

Entry	Loaded resin	Cleavage conditions	Product, yield (purity)	Ref.
1	 (MBHA linker)	HF/PhOMe 93:7		[204] see also [205] [206]
2		TFA/DCM/H$_2$O 9:90:1, 20 °C, 1 h	 49% (90%)	[207]
3		TFA/H$_2$O/PhSMe/ EtSMe/EDT/PhSH 82:5:5:3:3:2, 20 °C, 6 h	 54%	[208] see also [209]
4		TFA/PhSMe/EDT 90:5:5	 43%	[210] [211]
5		TFA/PhSMe 4:1, 3 h	 100%	[181]
6		PhSMe (1 mol/L), TFA, 28 °C, 1 h	 41%	[212]
7	 (XAL or Sieber linker)	TFA/DCM 1:99, 0.5 h	 84% (> 90%)	[213]
8		TFA/DCM/H$_2$O 57:40:3, 20 °C, 1 h	Fmoc-NH$_2$ 100%	[214]

Table 3.11. continued.

Entry	Loaded resin	Cleavage conditions	Product, yield (purity)	Ref.
9	(Rink amide linker)	TFA/DCM 1:1, 20 °C, 15 min	80%	[42]
10		TFA/H₂O 9:1	70–90%	[215] ureas: [216]
11		TFA/iPr₃SiH//H₂O 90:2:8, 20 °C, 0.5 h	55% (98%)	[202]
12		TFA/DCM 20:80	71%	[217] see also [218]
13		TFA/DCM/H₂O 95:5:1, 23 °C, 40 min	99% (> 90%)	[219]
14		TFA/DCM 5:95, 0.5 h	75%	[220]
15		TFA/DCM 2:8, 15 min	(73%)	[221]
16		TFA/DCM/PhOMe/ PhSMe/EDT/SiCl₄ 150:15:3:8:6:15, 25 °C, 3 h	62%	[222] see also [223]

Table 3.11. continued.

Entry	Loaded resin	Cleavage conditions	Product, yield (purity)	Ref.
17	FmocHN— (structure) Ph PS	1% TFA in DCM, 2 h	FmocHN— (structure) NH, 94% (91%)	[178]
18	RO— (structure) Ph PS Cl	10% TFA in DCM, 0.5 h	RO— (structure) NH, 80–90%	[224]

N-benzylamides, the acid-lability of *N*-(diarylmethyl)amides increases with the number of electron-donating substituents on the aryl groups.

Several different alkoxy-substituted (diarylmethyl)amine linkers have been described (Table 3.11), but it is the 'Rink' amide linker (Entry **9**, Table 3.11) which is most frequently used for the synthesis of amides R-CONH$_2$. Because this linker is attached to polystyrene as a phenyl ether, strong acids can also lead to a cleavage of the linker from the support. With some types of product this (undesirable) cleavage of the support–linker bond can even occur upon treatment with trifluoroacetic acid. This happens particularly readily with amides containing free amino groups or other basic functional groups close to the attachment site (see Figure 3.17). The Rink linker has also been attached to amine-functionalized polystyrene (e.g. MBHA or aminomethyl polystyrene) as an aryloxyacetamide, resulting in more robust linkage to the support.

Entry **16** in Table 3.11 is an example of a safety-catch linker: cleavage of the benzylic C–N bond requires a reducing agent (e.g. EDT + silyl halides) to convert the electron-withdrawing sulfoxide into an electron-donating thioether.

The attachment of amides to supports as *N*-(diarylmethyl)amides can be achieved either by acylation of resin-bound (diarylmethyl)amines, or by acid-catalyzed N-alkylation of amides with resin-bound benzhydryl alcohols [42]. The former strategy is by far the more general.

For the preparation of secondary amides RNHCOR by backbone amide linking, benzylamine linkers (Table 3.9) are more appropriate than (diarylmethyl)amine linkers, because *N*-alkyl-*N*-(diarylmethyl)amines are often difficult to acylate because of steric hindrance [202]. A few examples of the preparation of secondary amides by backbone amide linking to (diarylmethyl)amine linkers have been reported (Table 3.11).

Tritylamine linkers have not been extensively used for the attachment of amides, probably because *N*-tritylamides are difficult to prepare. This is not the case for strongly acidic amides and cyclic imides, which are readily N-tritylated with trityl chloride resins [203] to yield sufficiently stable, resin-bound amides (Entries **17** and **18**, Table 3.11).

3.3.3 Linkers for Amides Cleavable by Nucleophiles

Amides, carbamates, and ureas can be generated by nucleophilic cleavage of resin-bound esters, carbonates, and carbamates with amines (Figure 3.18). These reactions only proceed well if sufficiently reactive resin-bound derivatives are used.

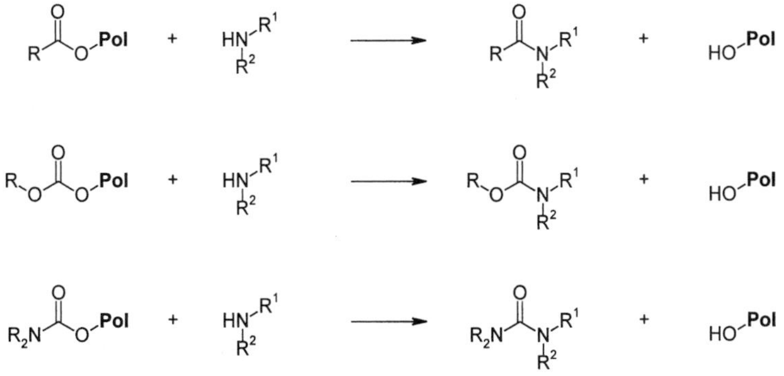

Fig. 3.18. Formation of amides, carbamates, and ureas by nucleophilic cleavage with amines.

3.3.3.1 Nucleophilic Cleavage of Alkyl Esters

The examples of nucleophilic cleavage of resin-bound benzyl-, PEG-, or other alkyl esters by amines (Table 3.12) show that this reaction generally requires forcing conditions to proceed. For this reason it is often possible to perform reactions of amines or other nucleophiles with compounds linked to benzyl alcohol resins as esters, without significant loss of product (e.g. removal of Fmoc protective groups with piperidine from Wang resin bound peptides). If peptides are cleaved from benzyl alcohol linkers by aminolysis, partial racemization of the C-terminal amino acid is often observed [225].

More susceptible towards nucleophilic cleavage are thiol esters (Entry **8**, Table 3.12 [226]). Thiol ester attachment can, e.g., not be used for standard solid-phase peptide synthesis using Fmoc methodology [227].

A particularly interesting variant of nucleophilic cleavage is the intramolecular attack of the linking ester by a substrate-bound nucleophile. Thereby simultaneous cleavage from the support and formation of a cyclic acid derivative is realized (Figure 3.19).

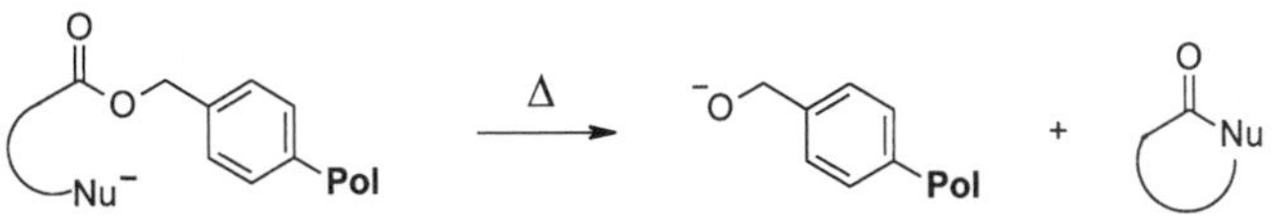

Fig. 3.19. Intramolecular nucleophilic cleavage of esters.

Numerous examples of different variants of this cyclization/cleavage protocol have been reported. Diketopiperazine formation (Section 15.22.1), an unwanted side reaction in solid-phase peptide synthesis, also belongs to this class of compound release. Because intramolecular processes generally take place more readily than the corresponding intermolecular reactions, cyclization/cleavage can occur with alkyl esters under conditions which would not lead to intermolecular nucleophilic cleavage. In particular five- or six-membered rings are readily formed. This cleavage strategy is discussed in the chapters dedicated to the type of heterocycle formed upon cleavage.

Table 3.12. Aminolysis of resin-bound alkyl esters.

Entry	Loaded resin	Cleavage conditions	Product, yield (purity)	Ref.
1		neat BnNH$_2$, 20 °C, 48 h; or	benzylamide, 43% (80%)	[225]
		neat PrNH$_2$, 20 °C, 22 h	propylamide, 96% (96%)	[225]
2		1-methylpiperazine/ AlCl$_3$ 4:1, DCM, 20 °C, 18 h	45% (84%)	[228]
3		70% EtNH$_2$ in H$_2$O/THF 1:1, 20 °C, overnight	98% (94%)	[229]
4		(neat), 55 °C, 18 h	61%	[230] see also [231] [232]
5		HexNH$_2$ (0.5 mol/L), DMF, 25 °C, 16 h; or excess Me$_2$NH in DCM, 25 °C, 15 h	hexylamide, 25% dimethylamide, 2%	[81]
6		NH$_3$ (satd in CF$_3$CH$_2$OH), 20 °C, 17 h	peptide—CONH$_2$ > 98% (> 95%)	[88] see also [233]
7		BnNH$_2$, AlMe$_3$, DCM, PhMe, 22 °C, 20 h		[234] see also [235]
8		AgNO$_3$, piperidine (each 0.07 mol/L), DMF, 1 h	97%	[236] see also [237] [238]

3.3.3.2 Nucleophilic Cleavage of Aryl Esters

Aryl esters are generally more readily cleaved by nucleophiles than alkyl esters. This sensitivity towards nucleophiles becomes more pronounced with increasing acidity of the corresponding phenol. Resin-bound phenyl esters are convenient reagents for the acylation of amines (Table 3.13). A particularly interesting type of phenol linker is the the 4-(alkylthio)phenol linker, which enables further activation of the corresponding phenyl esters by oxidation of the thioether to a sulfoxide or sulfone. This type of linker is also suitable for the preparation of cyclic peptides by intramolecular, nucleophilic cleavage [239]. However, as shown by the examples in Table 3.13, the additional activation by conversion into a sulfone is not always required, because the unoxidized 4-(alkylthio)phenol is a good leaving group.

Immobilized, highly reactive phenyl esters can be prepared by acylating resin-bound 4-acyl-2-nitrophenol (Entry **4**, Table 3.13 [240–243]). These esters are similar to

Table 3.13. Nucleophilic cleavage of resin-bound aryl esters.

Entry	Loaded resin	Cleavage conditions	Product, yield (purity)	Ref.
1		BuNH$_2$, pyridine, DIPEA, 20 °C, 24–36 h		80%	[244]
2		NEt$_3$/DCM 1:1, 27 °C, 4 h		66%	[245]
3		ethyl glycinate, DMF, 24 h		38%	[246]
4		3-aminobenzophenone, NEt$_3$, MeCN, 70 °C, 24 h		[247] see also [248] [249]
5		DCM, 20 °C, 40 h		95%	[241]

oxime esters (see Section 3.3.3.3), and even react with weak nucleophiles, such as anilines or alcohols. This type of linker is not, therefore, well suited for long synthetic sequences on insoluble supports, but only for the preparation of simple acid derivatives. The preparation of cyclic peptides by intramolecular nucleophilic cleavage of polystyrene-bound 2-nitrophenyl esters has been reported [242].

Table 3.14. Nucleophilic cleavage of resin-bound oxime, HOBt, and related esters.

Entry	Loaded resin	Cleavage conditions	Product, yield (purity)	Ref.
1		satd NH₃ in MeOH/ DCM 1:1, 2 h	96%	[267]
2		BnNH₂ (0.5 mol/L), CHCl₃, 20 °C, 4 h	87%	[254]
3		PhNH₂ (0.03 mol/L), CHCl₃, 20 °C, 3 h	50%	[268]
4		PhNH₂ (0.1 mol/L, 1 eq), DMF, 20 °C, 20 h	64%	[266]
5		ethyl glycinate (0.05 mol/L, 0.5 eq), CHCl₃, 30 °C, 14 h	53%	[269]
6		*tert*-butyl glycinate (7 mmol/L, 0.16 eq), DMF/CHCl₃ 1:1, overnight	98%	[270] see also [271]
7		*N*-hydroxy-succinimide (0.8 eq), DCM, 20 °C, 7 h	86%	[272]

3.3.3.3 Nucleophilic Cleavage of Oxime and Related Esters

A further type of linker, especially designed for facile nucleophilic cleavage, is the oxime linker. Oximes of resin-bound ketones or aldehydes can be acylated to yield *O*-acyl oximes, which are readily cleaved by a variety of nucleophiles (Table 3.14). A widely used support of this type is *p*-nitrobenzophenone oxime resin (Entries **1** and **2**, Tabel 3.14; for preparation, see [250]), which has been used with success for the synthesis of protected peptide fragments [251–254], diketopiperazines [255], macrocyclic peptides [256–258], and macrocyclic lactams [259].

Resin-bound 1-hydroxybenzotriazole (HOBt) has also been prepared [260–262], and yields, upon O-acylation, highly reactive esters [260,263,264]. These are cleaved even by weak nucleophiles, such as anilines or even *N*-hydroxysuccinimide (Entry **7**, Table 3.14). The reactivity of resin-bound HOBt esters is greater than that of resin-bound 4-acyl-2-nitrophenyl esters [248]. Similarly, resin-bound *N*-hydroxysuccinimide has also been used for the preparation of polymeric acylating agents (Entry **6**, Table 3.14 [265]).

O-Acylated, resin-bound hydroxybenzotriazoles and related compounds are, because of their sensitivity towards nucleophiles, not suitable for multistep synthetic sequences on solid phase. These intermediates can, however, be used as insoluble acylating agents, which are a practical alternative to soluble reagents for the derivatization of amines or alcohols [266].

3.3.3.4 Nucleophilic Cleavage of Amides and Carbamates

Amides are generally very resistant towards nucleophilic cleavage. 'Safety-catch' linkers as those described in Section 3.1.2.3 can, however, be cleaved by amines to yield amides (Entries **1** and **2**, Table 3.15). Carbamates are generally more resistant towards nucleophilic cleavage than amides or carboxylic esters. Some phenol- or oxime-derived carbamates, however, react readily with amines, and resin-bound carbamates of this type have been successfully used for the conversion of amines into ureas (Table 3.15).

Support-bound *N*-sulfonyl carbamates, which can be prepared by N-sulfonylation of resin-bound carbamates, are susceptible to nucleophilic cleavage. These intermediates enable the solid-phase preparation of *N*-aryl- or *N*-alkylsulfonamides using inexpensive hydroxymethyl polystyrene (Entry **8** and **9**, Table 3.15).

3.4 Linkers for Hydroxamic Acids and Hydrazides

Several strategies enable the generation of hydroxamic acids or hydrazides upon cleavage of a carboxylic acid derivative from a support (Figure 3.20).

Table 3.15. Generation of amides by nucleophilic cleavage of resin-bound amides and carbamates.

Entry	Loaded resin	Cleavage conditions	Product, yield (purity)	Ref.
1		RNH$_2$, DMF, 60 °C R: alkyl	 100% (97%)	[273]
2		PhNH$_2$ (1 mol/L, 14 eq), dioxane, 100 °C, 15 h (no reaction with 4-nitroaniline)	 84%	[274] see also [102]
3		NH$_3$ (satd in *i*PrOH), 24 h	 60–80%	[109]
4		4-(MeO)C$_6$H$_4$NH$_2$ (0.13 mol/L, 4 eq), DCM/PhMe, 80 °C, overnight	 88% (87%)	[275] see also [276]
5		morpholine (0.13 mol/L, 4 eq), DCM/PhMe, 80 °C, overnight	 77% (92%)	[275] [277]
6		BnNH$_2$ (1.3 eq), NEt$_3$ (4 eq), MeCN, 60 °C, 24 h	 89% (> 98%)	[278]
7		BnNH$_2$ (0.74 eq), THF, 66 °C, overnight	 80% (47%)	[279]
8		LiOH, H$_2$O, THF, 20 °C, 54 h	 63% (96%)	[280]
9		NaOMe (1.5 mol/L), MeOH/THF 1:1, 20 °C, 16 h	 88% (98%)	[280]

Hydroxylamine can be linked to supports via the oxygen atom as the benzyl ether [281–283], as the benzhydryl ether [284], or as the trityl ether [285–287]. These intermediates can then be acylated at nitrogen and, upon cleavage from the support, hydroxamic acids result (Table 3.16). Resin-bound hydroxylamine can be prepared by treatment of an appropriate resin-bound benzyl halide or sulfonate [282] with *N*-hydroxyphthalimide in the presence of a base, or by Mitsunobu reaction of *N*-hydroxyphthalimide with resin-bound benzyl alcohols [281]. The phthaloyl protective group of the resulting intermediate is removed by treatment with hydrazine. Alternatively, resin-bound hydroxylamine can be prepared from 2-chlorotrityl chloride resin and Fmoc- or Dde-protected hydroxylamine [285]. Hydroxylamine can also be attached to a backbone amide linker via the nitrogen atom (Entry 6, Table 3.16).

Hydrazine can readily be linked to acid-labile benzyl alcohol resins as carbamate (Entries 7 and 8, Table 3.16). The unacylated amino group can then be acylated, resulting in resin-bound hydrazides. These are stable towards nucleophilic attack, but can be cleaved by acids (depending on the type of benzyl alcohol resin chosen) to yield hydrazides. Resin-bound hydrazides can also be cleaved by oxidants (see Section 3.1.4) to yield carboxylic acids, amides, or esters [155].

Hydroxamic acids and hydrazides can also be prepared by nucleophilic cleavage of resin-bound esters or activated amides with hydroxylamine and hydrazine, respectively.

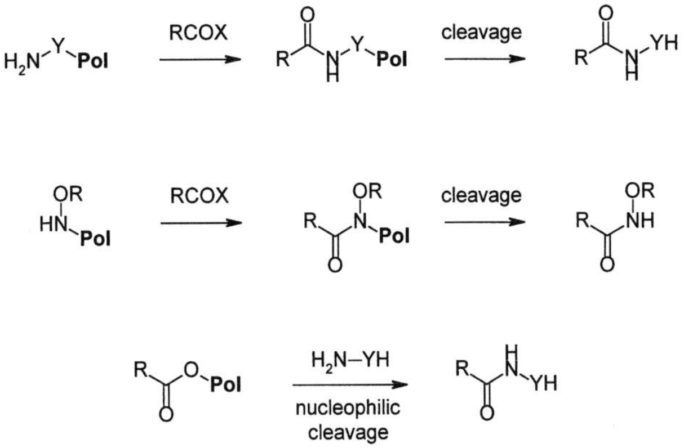

Fig. 3.20. Strategies for the solid-phase preparation of hydroxamic acids and hydrazides. X: leaving group, Y: NH, O.

Table 3.16. Linkers for hydroxamic acids and hydrazides.

Entry	Loaded resin	Cleavage conditions	Product, yield (purity)	Ref.
1		TFA/*i*Pr$_3$SiH/DCM 50:5:45		[282] see also [281]
2		TFA/DCM 5:95, ultrasound, 15 min		[288]
3		TFA/*i*Pr$_3$SiH/EDT/ H$_2$O 90:1:4:5, 30 °C, 4 h	80–90%	[284]
4		HCO$_2$H/THF 1:3, 1 h	50–78% (> 84%)	[286]
5		TFA/DCM 5:95, 25 °C, 1 h	89% (> 95%)	[287] see also [285]
6		1. TFA/DCM/H$_2$O 3:96:1, 1 h 2. TFA/DCM/H$_2$O 50:49:1, 1 h	88%	[289]
7		TFA/DCM 1:1, 20 °C, 0.5 h	76%	[61] see also [62]
8		TFA/DCM 1:1, 20 °C, 0.5 h	42%	[26]
9		aq NH$_2$OH (50%, 25 eq), H$_2$O, THF, 2 d	81% (87%)	[290]

Table 3.16. continued.

Entry	Loaded resin	Cleavage conditions	Product, yield (purity)	Ref.
10	peptide— (PS-linked benzyl ester)	N$_2$H$_4$/DMF 1:9, 24 h; or N$_2$H$_4$·H$_2$O/EtOH 25:100, 6 h	peptide—(hydrazide) 30–78%	[291] [292] see also [293]
11	(Boc-protected Phe on PS-linked resin)	2.5% N$_2$H$_4$·H$_2$O in MeOH, 20 °C, 3 h	(Boc-protected Phe hydrazide) 88%	[86]

3.5 Linkers for Carboxylic Esters

Esters, which have no possible site of attachment, cannot be directly linked to supports, but may be generated upon cleavage from a support. This cleavage can be mediated by electrophiles, nucleophiles, or oxidants. A few examples have been reported of the preparation of esters by O-alkylation of carboxylates by resin-bound alkylating agents, such as sulfonic esters [294,295].

3.5.1 Linkers for Esters Cleavable by Nucleophiles

Resin-bound carboxylic acid derivatives can be susceptible towards nucleophilic attack by alcohols. The preparation of esters by nucleophilic cleavage with alcohols is generally only practical when using low-boiling alcohols, which can be readily removed by evaporation after cleavage.

The nucleophilic cleavage of acid derivatives by alcohols can be catalyzed by tertiary amines [296,297], alcoholates, alkali metal cyanides, or by acids (Table 3.17). As in the saponification or aminolysis of resin-bound carboxylic acid derivatives, nucleophilic cleavage with alcohols under basic conditions proceeds more readily with increasing electrophilicity of the resin-bound acid derivative (hydroxybenzotriazole esters > aryl esters > alkyl esters > alkoxy-substituted benzyl esters > trityl esters). Attempts to prepare macrocyclic lactones by intramolecular alcoholysis of resin-bound ω-hydroxyalkanoic acids, bound to polystyrene as 4-sulfonylphenyl esters [298] or thiol esters [226] led to mixtures of cyclic oligoesters.

Table 3.17. Generation of esters by alcoholysis of resin-bound carboxylic acid derivatives.

Entry	Loaded resin	Cleavage conditions	Product, yield (purity)	Ref.
1		NaOMe (0.25 mol/L), MeOH/THF 1:4, 20 °C, 10 h	79%	[299] see also [77]
2		MeOH/NEt₃/ DMF 9:1:1, 60 °C	95%	[300]
3		NaOMe (0.02 mol/L), MeOH/THF 1:4, reflux overnight	> 95% (> 90%)	[301] see also [302]
4		NEt₃/MeOH 1:4, 20 °C, 20 h	58%	[303] see also [304]
5		Ti(OEt)₄ (5 eq), neat allyl alcohol, 120 °C, 2 h	42%	[305]
6		*i*PrOH/MeOH/ Me₃SiCl 7:3:1	(> 95%)	[306]
7	(HMBA linker)	NEt₃/MeOH 1:9, 50 °C, 20 h	98% (> 96%)	[307] see also [308]
8		KCN (0.3 mol/L), MeOH, 25 °C, 6 h	49%	[81]

Table 3.17. continued.

Entry	Loaded resin	Cleavage conditions	Product, yield (purity)	Ref.
9		NEt$_3$/MeOH 2:8, 20 °C, 23 h	 63%	[86]
10		NEt$_3$/MeOH 2:8, 20 °C, 24 h	 > 98% (> 95%)	[88]
11		NaCN, MeOH, 20 °C, 16 h	 12–29% (83–95%)	[309] see also [310]
12		NaOMe, MeOH/THF 1:1, 20 h		[103]
13		LiOCH$_2$Ph	 26%	[28]
14		EtOH/THF/H$_2$SO$_4$/ H$_2$O 90:100:4:5, reflux, 7 d	 68%	[311]
15		TBAF, AcOH, THF, 40 °C, 6 h	 61%	[312]
16		TFA/DCM 3:97 (Wang resin)	 70%	[313]
17		TBAF, AcOH, THF, 40 °C, 14 h	 41% 14%	[314]

3.5.2 Linkers for Lactones Cleavable by Electrophiles or Oxidants

An interesting variant of electrophilic cleavage is intramolecular attack at the linking carboxylic acid derivative by a substrate-bound electrophile (Figure 3.21). Lactones are the products generally obtained from this cleavage strategy.

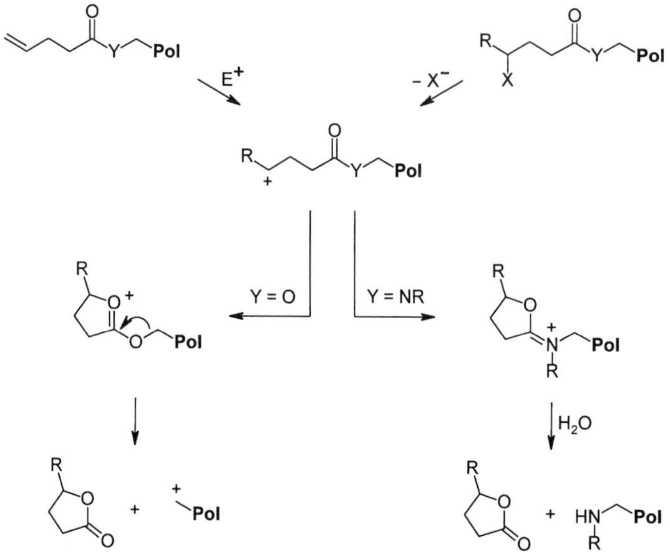

Fig. 3.21. Intramolecular electrophilic cleavage of carboxylic acid derivatives from polymeric supports. X: leaving group; Y: NR, O.

Because of the special structural requirements for the resin-bound substrate, this type of cleavage reaction lacks general applicability. Some of the few examples reported are listed in Table 3.18. Lactones have furthermore been obtained by acid-catalyzed lactonization of resin-bound 4-hydroxy or 3-oxiranyl carboxylic acids [315]. Treatment of polystyrene-bound cyclic acetals with Jones reagent also leads to the release of lactones into solution (Entry **5**, Table 3.18).

Table 3.18. Formation of lactones during cleavage of carboxylic acid derivatives from insoluble supports.

Entry	Loaded resin	Cleavage conditions	Product, yield (purity)	Ref.
1	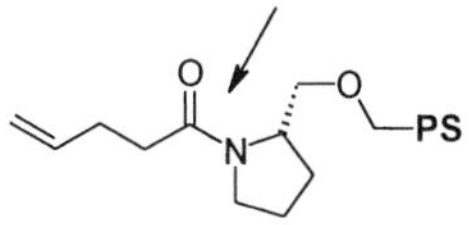	I_2, THF/H_2O, 20 °C, 3 d	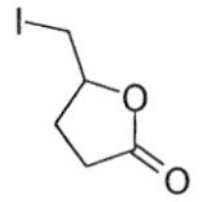 40%; 30% ee	[316] see also [317]
2	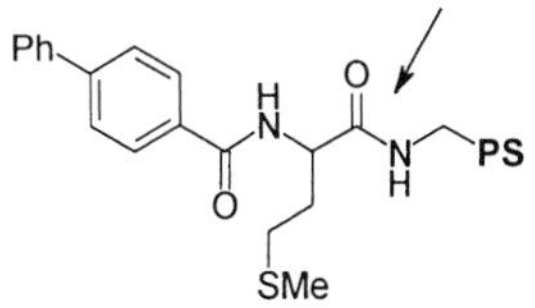	BrCN, CHCl$_3$/H_2O, TFA, 20 °C, 24 h	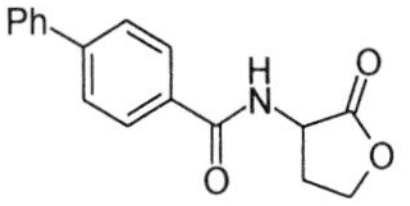	[318]
3	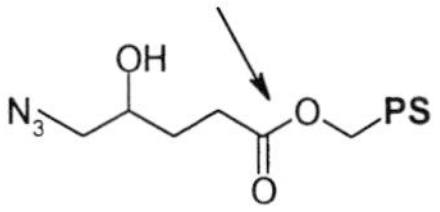	TFA/DCM 1:1, 20 °C, 2 h	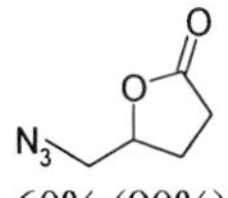 60% (90%)	[315]
4	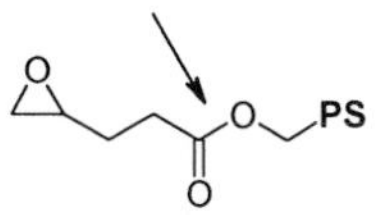	TFA/DCM 1:1, 20 °C, 2 h	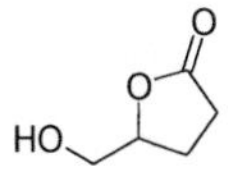 57% (90%)	[315]
5	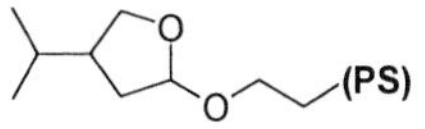	Jones reagent (CrO$_3$/H_2SO_4; 2 eq), acetone, 20 °C, 3 h	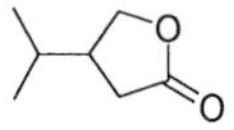	[319]

3.6 Linkers for Primary and Secondary Aliphatic Amines

Primary or secondary aliphatic amines can be linked to polymeric supports by acid-labile linkers and by linkers sensitive to nucleophiles. Linkers cleavable by light or by transition metal catalysis have also been described. The main types of linker for amines are sketched in Figure 3.22.

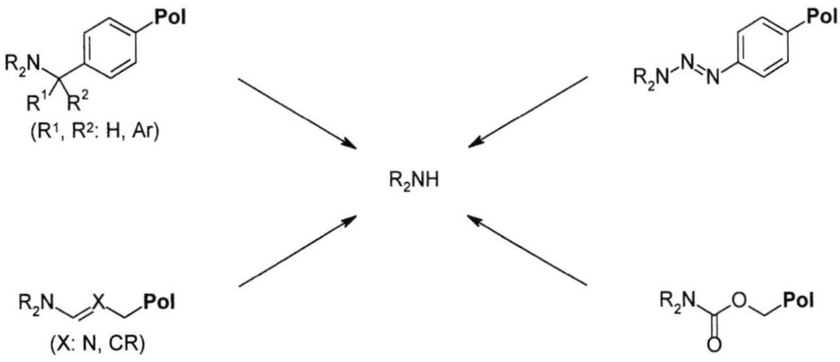

Fig. 3.22. Strategies for the attachment of amines to insoluble supports.

3.6.1 Benzylamine, (Diarylmethyl)amine, and Tritylamine Linkers

Aliphatic primary or secondary amines can be linked to insoluble supports as benzylamines by reductive alkylation with support-bound benzaldehydes or by N-alkylation with support-bound benzyl halides or sulfonates (Figure 3.23; see also Chapter 10.1). Benzhydrylamines and tritylamines are usually prepared by N-alkylation with the corresponding halides.

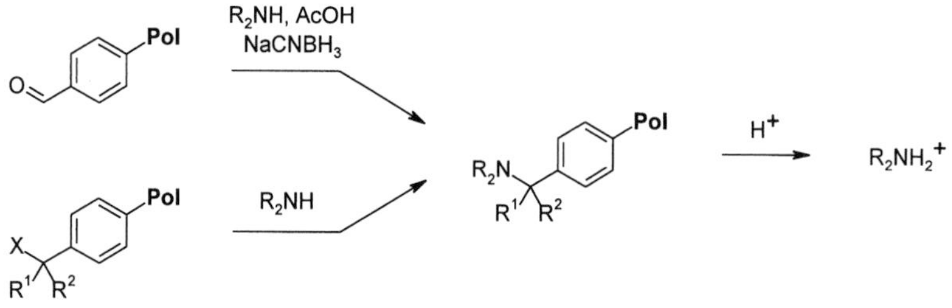

(R¹, R²: H, Ar; X: Cl, Br, RSO₃)

Fig. 3.23. Immobilization of amines as benzylamines.

Benzylic C–N bonds in amines are generally more difficult to cleave with electrophiles than those of *N*-benzylamides. Therefore, as illustrated by the examples in Table 3.19, most of the linkers suitable for backbone amide linking (Chapter 3.3) cannot be used for amines, unless more vigorous cleavage conditions are applied. As an

Table 3.19. Benzylamine linkers for aliphatic amines.

Entry	Loaded resin	Cleavage conditions	Product, yield (purity)	Ref.
1	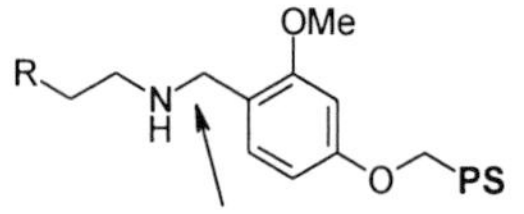	TFA/H₂O 95:5	no cleavage	[176]
2	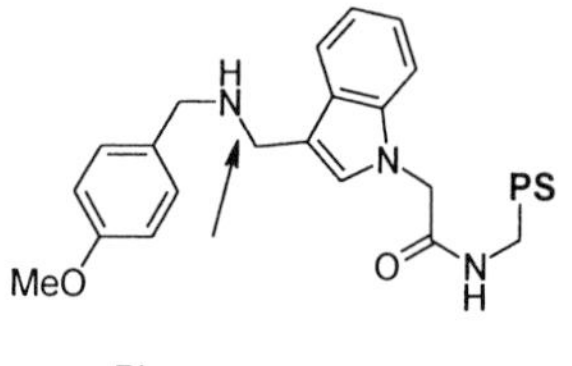	TFA/DCM 95:5, 4 h	no cleavage	[193]
3	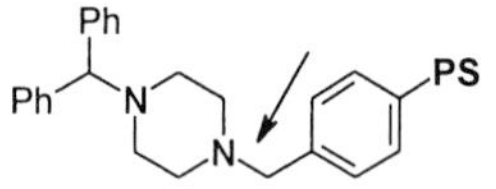	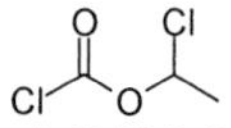 (10 eq), DCM, 20 °C, 3 h; then MeOH, 65 °C, 3 h	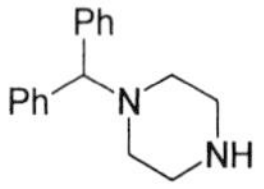 95% (95%)	[320]
4	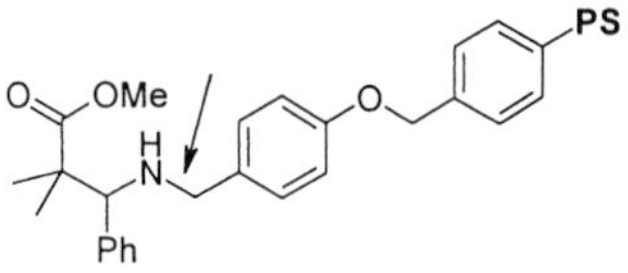	DDQ (0.1 mol/L), C₆H₆, 20 °C, 3 h	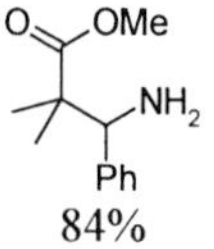 84%	[321]

Table 3.20. Benzhydrylamine and tritylamine linkers for aliphatic amines.

Entry	Loaded resin	Cleavage conditions	Product, yield (purity)	Ref.
1	(MBHA linker)	HF/PhOMe, 0 °C, 9 h	> 70% (> 95%)	[324] see also [325]
2		TFA/DCM 5:95, 0.5 h	95% (96%)	[220] see also [323]
3	Ar: 4-FC$_6$H$_4$	TFA/DCM 1:1, 3 h	45% (96%)	[322] see also [326]
4		TFA/H$_2$O/DCM 5:5:90, 20 °C, 5 h	82% (78%)	[327]
5		TFA/iPr$_3$SiH/ DCM 50:2:50, 0.5 h	81%	[328]
6		TFA/DCM 1:3, 1 min	(> 90%)	[329] see also [330-332]
7		TFA/DCM 2:98, 0 °C, 5 min	82%	[333]
8		95% TFA, 16 h	66%	[334]
9		TFA/DCM 1:99, 20 °C, 5 min (repeat three times)		[335]

alternative to cleavage with acids, tertiary benzylamines can be debenzylated by treatment with alkyl chloroformates. If 1-chloroethyl chloroformate is used as debenzylating agent, the 1-chloroethylcarbamates formed during cleavage from the support undergo facile hydrolysis in methanol (Entry **3**, Table 3.19). Electron-rich benzylamines can also be cleaved by treatment with DDQ (Entry **4**, Table 3.19).

Support-bound 2-nitrobenzylamines can be cleaved photolytically in the presence of small amounts of TFA (J.C. Tomesch, personal communication). As in the related 2-nitrobenzylalcohol linkers for acids, photolytic cleavage of 2-nitrobenzylamines proceeds via intramolecular disproportionation (Figure 3.10).

Benzhydrylamines are better suited than benzylamines as acid-labile linkers for amines. The MBHA linker ('methylbenzhydrylamine'), which is usually used to prepare peptide amides (see Chapter 3.3), can also be used as linker for amines (Entry **1**, Table 3.20). Hydrogen fluoride is, however, required as cleavage reagent. Easier to cleave are alkoxy-substituted benzhydrylamines (Entries **2–5**, Table 3.20), which can be prepared from the corresponding benzhydryl chlorides [220] or by reductive alkylation [322] or solvolysis [323] of the Rink amide linker. In the case of benzhydrylamines linked to polystyrene as benzylethers treatment with TFA can lead to the release of the linker into solution (acidolysis of the benzylic C–O bond, see Figure 3.16).

Tritylamines can serve as both linker and protective group for aliphatic amines, because tritylamines (unlike benzhydrylamines) do not usually undergo acylation when treated with activated acid derivatives. Tritylation of aliphatic amines is readily accomplished by adding excess amine to a support-bound trityl chloride. Illustrative cleavage reactions are listed in Table 3.20.

3.6.2 Carbamate Attachment

Amines can be linked to benzyl alcohol or tertiary alcohol linkers as carbamates. Carbamate attachment of amines can be achieved by reaction of isocyanates with alcohol linkers, or by treatment of alcohol linkers with phosgene [280,336,337] or a synthetic equivalent thereof, and then with an amine (Figure 3.24). The reagents most commonly used for the 'activation' of alcohol linkers are 4-nitrophenyl chloroformate [63,338–345] and carbonyl diimidazole [336,346–349]. The preparation of support-bound carbamates is discussed in Chapter 14.6.

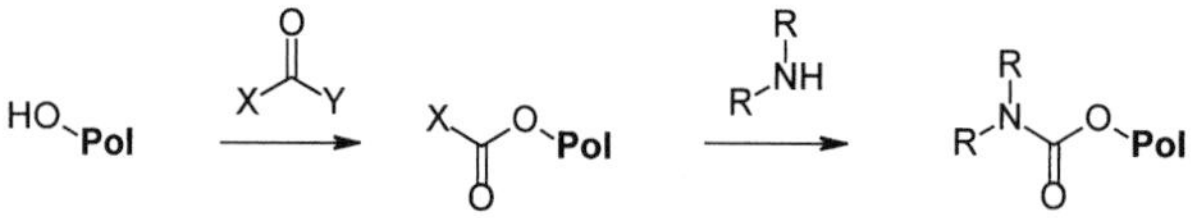

Fig. 3.24. Carbamate attachment of amines to polymeric alcohols. X, Y: leaving groups.

Carbamate attachment enables the broad variety of cleavage strategies developed for carboxylic acids (for the synthesis of peptides on solid phase; acidolysis, nucleophilic cleavage, β-elimination, photolysis, and transition metal catalysis) to be extended to the solid-phase synthesis of amines (Table 3.21). Benzyl carbamates are usually more stable than amides towards nucleophilic attack, and are therefore well suited as linkers if reactions with nucleophiles or bases are to be performed. The cleavage conditions for carbamates prepared from resin-bound benzyl alcohols are usually the same as for the corresponding carboxylic esters (Table 3.21). For instance, carba-

Table 3.21. Carbamate attachment of amines.

Entry	Loaded resin	Cleavage conditions	Product, yield (purity)	Ref.
1		TFA/DCM 1:1, 20 °C, 1 h	$R\!-\!NH_2$ 80% (85%)	[350]
2		LiAlH$_4$ (10 eq), THF, 60 °C, 14 h	84% (> 95%)	[343] see also [351]
3		Et$_3$SiH (1 eq), TFA/DCM 2:1, 48 h	21% (88%)	[352]
4		3% TFA in DCM, 25 °C, 69 h	83%	[353]
5		TFA/DCM 1:9, 20 °C, 4.5 h	$R\!-\!NH_2$	[67] see also [63]
6		hv (350 nm), 3 h, MeCN/H$_2$O 9:1	$R\!-\!NH_2$ up to 98%	[354] see also [355]
7		(Ph$_3$P)$_2$PdCl$_2$, Bu$_3$SnH, AcOH, DMSO/DCM 1:1, 1 h	$R\!-\!NH_2$ 92%	[356]
8		NaOMe (0.05 mol/L), THF, 1 h	65% (98%)	[357]
9		NaOH, phosphate buffer (pH 12), 3 min		[358] see also [351]

mates prepared from Wang resin (Entries **1** and **2**, Table 3.21) can be cleaved by treatment with TFA/DCM 1:1 for 20 min at room temperature, to yield the amine as trifluoroacetate salt. Carbamate attachment of amines to Wang resin is a widely used and convenient way of immobilizing amines.

An alternative method of cleaving carbamates is exhaustive reduction with LiAlH$_4$ to yield methylamines (Entry **2**, Table 3.21). Entry **8** in Table 3.21 is an example of the nucleophilic cleavage of a carbamate with sodium methoxide. The mild reaction conditions required are attributable to the structure of the amine (a vinylogous amide), and will probably not lead to the cleavage of *N*-alkyl- or *N*-dialkylcarbamates. *N*-Arylcarbamates, however, are also susceptible to nucleophilic cleavage (Entry **6**, Table 3.25).

3.6.3 Miscellaneous Linkers for Aliphatic Amines

A series of special linkers and cleavage strategies has been developed for the release of amines from insoluble supports (Table 3.22). These include the attachment of amines as triazenes, enamines, amidines, sulfonamides, and amides.

Triazenes have been prepared by treatment of resin-bound aromatic diazonium salts with secondary amines (Figure 3.25). Regeneration of the amine can be effected by mild acidolysis (Entry **1**, Table 3.22). Triazenes have been shown to be stable towards bases such as TBAF or potassium hydroxide, and under the conditions of the Heck reaction [359].

Support-bound triacylmethanes (e.g. 2-acetyldimedone) readily react with primary aliphatic amines to yield enamines. These are stable towards weak acids and bases, and can be used as linker for solid-phase peptide synthesis with either Boc or Fmoc methodology. Cleavage of these enamines can be achieved by treatment with primary amines or hydrazine (Entries **2** and **3**, Table 3.22; see also section 10.1.10.4).

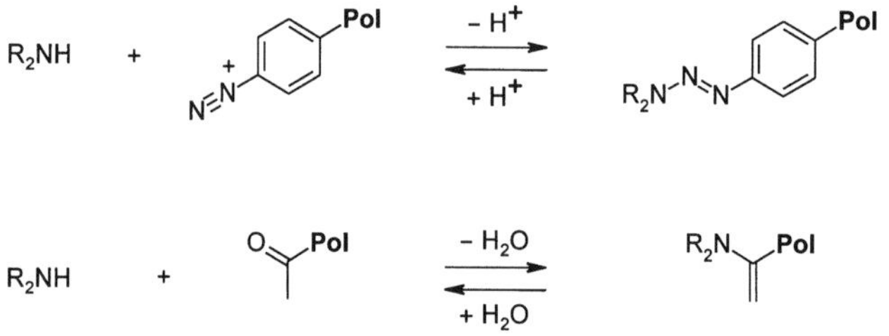

Fig. 3.25. Immobilization of amines as triazenes and as enamines.

Amidines and sulfonamides have also been used as linkers for primary or secondary aliphatic amines (Entry **4**, Table 3.22). These derivatives are stable under basic and acidic reaction conditions, and can only be cleaved by strong nucleophiles. Phenylalanine amides can be hydrolyzed by treatment with certain enzymes (Entry **6**, Table 3.22), and can therefore be used for linking amines to supports compatible with enzyme-mediated reactions (CPG, some polyacrylamides, macroporous polystyrene, etc.).

Table 3.22. Special linkers for aliphatic amines.

Entry	Loaded resin	Cleavage conditions	Product, yield (purity)	Ref.
1	Ph, N-N=N-O-PS, CO_2Me	TFA/DCM 1:9, 20 °C, 5 min	Ph, NH, CO_2Me (> 90%)	[196]
2	O...O, R-N(H)-Pol	5% $N_2H_4 \cdot H_2O$ or 10% $PrNH_2$ in THF/H_2O 1:1 (no cleavage with piperidine/DMF 2:8 or TFA/DCM 1:1, 24 h)	R-NH_2 (> 90%)	[360]
3	MeO_2C, Ph, N...N-PS, O	2% $N_2H_4 \cdot H_2O$, DMF, 5 min	Ph, MeO_2C-NH_2 100%	[361]
4	N=N-O-PS, Ar-O-Ph	N_2H_4/AcOH/EtOH/THF 1:0.7:40:40, 60 °C, overnight Ar: 3,5-dimethoxy-phenyl	NH, Ar-O-Ph 40%	[362] see also [363]
5	Ph, O O NO_2, N-S, Tol, N-(PS), O	PhSK (0.08 mol/L), MeCN, 20 h	Ph, NH, Tol 65%	[364]
6	Ph, R-N(H)-O-N(H)-(PA), O	α-chymotrypsin, TRIS-HCl buffer (pH 7.8), 40 °C, 24 h	R-NH_2 72%	[168]

3.7 Linkers for Tertiary Amines

The preparation of tertiary amines with the aid of insoluble supports has mostly been performed by β-elimination of support-bound quaternary ammonium salts and by N-alkylation of secondary amines with support-bound alkylating agents (Figure 3.26).

Base-induced β-elimination of quaternized amines can be mediated by amines or by basic ion-exchange resins [365]. Because non-quaternized amines, which are potential byproducts in this synthetic sequence (Figure 3.26), do not undergo β-elimination as readily as the quaternized ammonium salts, pure tertiary amines are generally obtained by this technique.

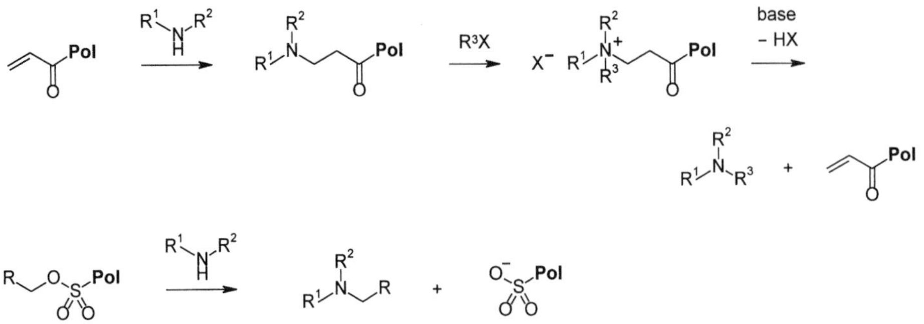

Fig. 3.26. Solid-phase preparation of tertiary amines.

Tertiary amines have also been prepared by N-alkylating primary or secondary amines with resin-bound alkylating agents, such as sulfonates, allyl esters (Entries **3** and **4**, Table 3.23) or Michael acceptors (Table 15.21 [366]). Furthermore, mixed aminals of support-bound benzotriazole and secondary amines can be cleaved with carbon nucleophiles to yield tertiary amines (Entry **5**, Table 3.23).

Table 3.23. Linkers for tertiary amines.

Entry	Loaded resin	Cleavage conditions	Product, yield (purity)	Ref.
1	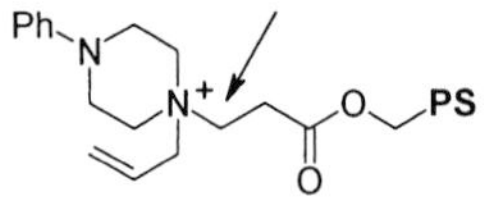	DIPEA/DCM 4:96, 20 °C, 2.5 h	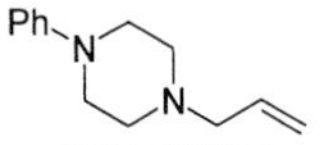 75% (99%)	[367] see also [365] [368]
2	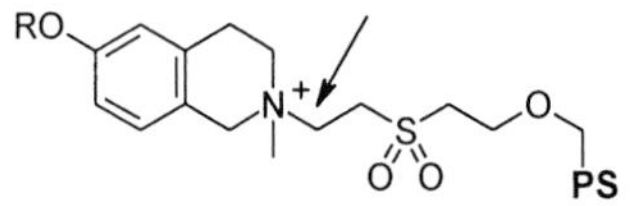	DIPEA/DCM 3:10, 20 °C, 0.5 h	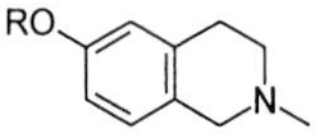 (0.25 mmol/g)	[203] see also [369] [370]
3	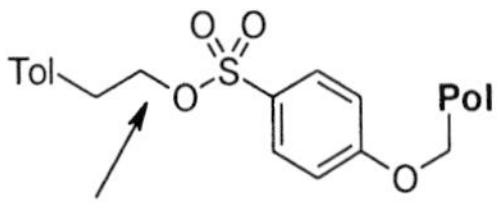	piperidine (0.5 mol/L), MeCN, 60 °C, 18 h (anilines were not alkylated by this reagent)	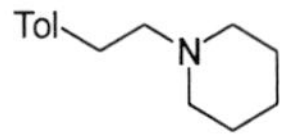 68% (84%)	[371] see also [372]
4	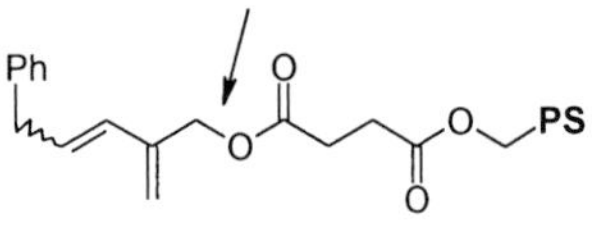	morpholine (2–3 eq), 7% Pd(PPh$_3$)$_4$, THF, 50 °C, 8 h	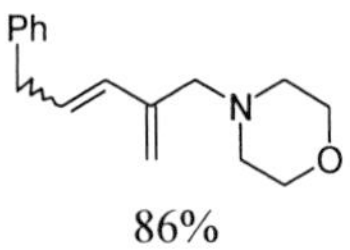 86%	[373]
5	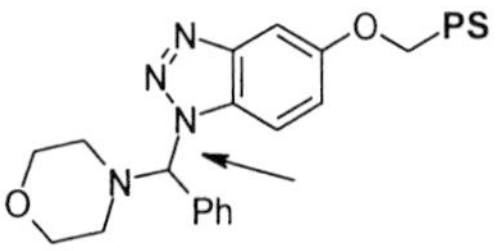	BuMgBr, THF, 67 °C, 4 h	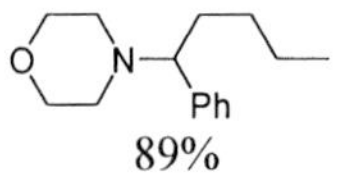 89%	[374] see also [262]

3.8 Linkers for Aryl- and Heteroarylamines

Aromatic and heteroaromatic amines can be linked to insoluble supports by use of strategies similar to those used for aliphatic amines. Because of the lower basicity of aromatic amines, however, their *N*-benzyl derivatives will usually be more susceptible to acidolytic cleavage than aliphatic *N*-benzylamines. For the same reason, *N*-acyl derivatives of aromatic amines will generally be more sensitive towards nucleophiles than the corresponding derivatives of aliphatic amines.

Illustrative examples of cleavage reactions of *N*-arylbenzylamine derivatives are listed in Table 3.24. Aromatic amines can be immobilized as *N*-benzylanilines by

Table 3.24. Benzylamine and benzhydrylamine linkers for aromatic amines.

Entry	Loaded resin	Cleavage conditions	Product, yield (purity)	Ref.
1		TFA/H$_2$O/Me$_2$S 95:5:5, 1 h	51–85%	[376]
2		TFA/DCM 7:3, 1 h	90%	[323] see also [220]
3		3% TFA in DCM, 45 min Ar: 4-(NO$_2$)C$_6$H$_4$	75%	[377] see also [378]
4		TFA/DCM 1:1, 4 h	> 99%	[193]
5		TFA/H$_2$O 95:5, 50 °C, 4 h Ar: 4-ClC$_6$H$_4$	76% (94%)	[379]
6		TFA/H$_2$O 95:5, 2 h	87% (99%)	[379]

reductive amination of resin-bound aldehydes or by nucleophilic substitution of resin-bound benzyl halides (Chapter 10). Furthermore, anilines have been linked to resin-bound dihydropyran as aminals [375].

The attachment of anilines to benzyl alcohol linkers as carbamates can be achieved either by reaction of aryl isocyanates with a resin-bound alcohol [380–382] or by treatment of alcohol-functionalized supports with phosgene [337,383] or a synthetic equivalent thereof, and then with the aromatic amine (Chapter 14.6). The latter strategy generally requires the use of more reactive activating agents than fo aliphatic amines. For instance, resin-bound benzyloxycarbonyl imidazoles, obtained by the reaction of resin-bound benzyl alcohols with carbonyldiimidazole, must be activated by N-alkylation (e.g. with MeOTf [67]), to yield an imidazolium salt which is sufficiently electrophilic to undergo reaction with anilines.

Table 3.25 lists illustrative examples of cleavage reactions of support-bound N-arylcarbamates. N-Arylcarbamates are more susceptible to attack by nucleophiles than N-alkylcarbamates, and, if strong bases or nucleophiles are to be used in a reaction sequence it might be a better choice to link the aniline as N-benzyl derivative to the support.

Table 3.25. Carbamate attachment of aromatic and heteroaromatic amines.

Entry	Loaded resin	Cleavage conditions	Product, yield (purity)	Ref.
1		HF/PhOMe 9:1, 0 °C, 1 h		[383]
2		TFA/DCM 1:1, 5 min	> 95% (90%)	[382]
3		TFA/DCM 1:1, 5 min	> 95% (90%)	[382]
4		TFA/DCM 1:9, 20 °C, 4.5 h		[67]
5		10% NH$_4$OH in CF$_3$CH$_2$OH, 40 °C, 4 h	43–74%	[381]
6		NaOH (0.5 mol/L), 90 °C, 0.5 h	95–97%	[384]

Some heteroarylamines have been prepared by aromatic nucleophilic substitution of suitable support-bound arylating agents with amines (Table 3.26). This technique has been successfully employed in the synthesis of 2-(alkylamino)pyrimidines [385,386], 2-(arylamino)pyrimidines [387], and 1,3,5-triazines [388]. When the heteroarene is bound to the support as a thioether, nucleophilic cleavage is facilitated by oxidation to a sulfone (e.g. with MCPBA or *N*-benzenesulfonyl-3-phenyloxaziridine). If these cleavage reactions are to be performed with non-volatile amines, then less than one equivalent of amine should be used. This will generally enable complete conversion of the amine, and crude products of high purity can be obtained.

Table 3.26. Generation of heteroaryl amines by simultaneous aromatic nucleophilic substitution and cleavage.

Entry	Loaded resin	Cleavage conditions	Product, yield (purity)	Ref.
1		1. MCPBA (1.2 eq), DCM, 20 °C, 18 h 2. pyrrolidine (0.3 mol/L), dioxane, 20 °C, 3 h	32%	[385]
2		pyrrolidine (1.5 eq), dioxane, 20 °C, 6 h	90% (98%)	[386]
3		pyrrolidine (0.2 mol/L), dioxane, 60 °C, 6 h	85%	[388]
4		4-methoxyaniline (0.3 mol/L), dioxane, 80 °C, 15 h		[388]

3.9 Linkers for Guanidines and Amidines

Several strategies enable the attachment of guanidines and amidines to insoluble supports (Figure 3.27, Table 3.27). These include attachment as *N*-benzyl, *N*-acyl, and *N*-alkoxycarbonyl derivatives. Furthermore, support-bound isothioureas can be used to convert amines into guanidines. The synthesis of support-bound guandines is treated in Chapter 14.3.

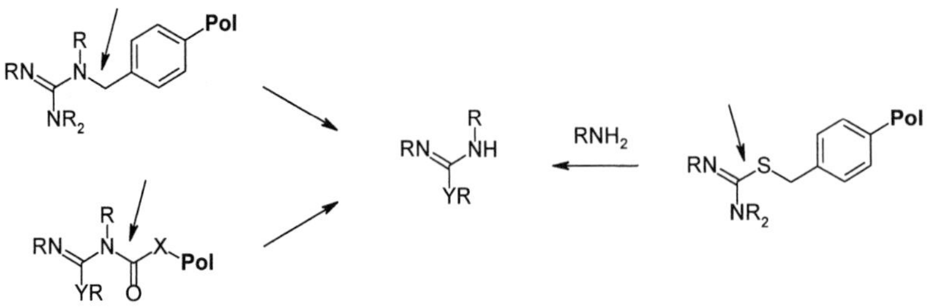

Fig. 3.27. Cleavage of guanidines and amidines from insoluble supports. Y: CR$_2$, NR; X: CR$_2$, O.

Few examples of benzylamine-type linkers for guanidines have been described. It has been reported that N-benzylguanidines undergo acidolysis more easily than the corresponding N-benzylamines, but more slowly than comparable N-benzylamides [324].

The acidic desulfonylation of N-(arylsulfonyl)guanidines generally requires long treatment with strong acids [389], unless the arene is substituted with electron-donating groups. The N–S bond of (4-alkoxybenzene)sulfonylguanidines can be cleaved by treatment with hydrogen fluoride, and such derivatives have proven useful as linkers for guanidines (Entry **4**, Table 3.27). Resin-bound (alkoxyarene)sulfonylguanidines have been prepared by reaction of support-bound sulfonyl chlorides with guanidines in the presence of a base. This reaction is slow, however, and requires several days to attain completion (e.g. attachment of N-Boc arginine as sulfonamide: KOH (0.8 mol/L) in dioxane, 75 °C, 2 d [390]).

Carbamate-bound guanidines have been prepared by condensation of amines with resin-bound thioureas [391]. The direct reaction of guanidines with resin-bound carbonates or other alkoxycarbonylating agents has not yet been used to link guanidines to insoluble supports, but will probably require the use of chloroformates or other reactive carbonic acid derivatives.

Carbamate-bound amidines (Entries **6** and **7**, Table 3.27) have been prepared by reaction of amidines with resin-bound 4-nitrophenyl carbonates (0.9 mol/L amidine in DMF/DIPEA, > 4 h [392–394]).

An additional strategy for generating guanidines from insoluble supports is the nucleophilic cleavage of resin-bound isothioureas with amines (Entries **8** and **9**, Table 3.27). This reaction is closely related to the preparation of 2-aminopyrimidines by nucleophilic cleavage (Table 3.26), and is generally limited to the use of volatile, low-molecular-weight amines, if the aim is to obtain crude products of high purity.

Table 3.27. Linkers for guanidines and amidines.

Entry	Loaded resin	Cleavage conditions	Product, yield (purity)	Ref.
1		TFA/DCM 1:1, 4 h Ar: 4-(MeO)C$_6$H$_4$	96%	[193]
2		HF, PhOMe, 0 °C, 9 h; (100% TFA, 20 °C, 0.5 h yielded 30–50% guanidine)	> 70%	[324]
3		TFA/CHCl$_3$/MeOH 1:1:1, 60 °C, 24–72 h	48% (90%)	[395]
4		HF/PhOMe 8:2, 0 °C, 4 h		[390] see also [396] [397]
5		TFA/DCM/*i*Pr$_3$SiH 49:49:2	> 85% (> 90%)	[391]
6		TFA/H$_2$O 95:5, 40 min	> 90% (> 70%)	[392] see also [393] [394]
7		hν (350 nm), dioxane	20% (> 65%)	[393]
8		NH$_3$/MeOH/DMF, 15 h	95% (100%)	[398]
9		BnNH$_2$ (0.1 mol/L), DMF, 50 °C, 16 h (no reaction with secondary or aromatic amines)	90%	[398]

3.10 Linkers for Pyrroles, Imidazoles, Triazoles, and Tetrazoles

Heterocycles containing an N–H group, such as, e.g., pyrroles, indoles, imidazoles, triazoles, etc., can be linked to insoluble supports as *N*-alkyl-, *N*-aryl, or *N*-acyl derivatives (Table 3.28). The optimal choice depends mainly on the N–H acidity of the heterocycle in question. Increasing acidity will facilitate the acidolytic cleavage of *N*-benzyl groups and the nucleophilic cleavage of *N*-acyl groups from these heterocycles.

Histidine and histamine derivatives have been successfully immobilized by N-tritylation of the imidazole ring with trityl chloride resin [399] (Entry 2, Table 3.28; see

Table 3.28. Linkers for pyrroles, imidazoles, triazoles, and tetrazoles.

Entry	Loaded resin	Cleavage conditions	Product, yield (purity)	Ref.
1		AcOH, 100 °C, 2 h (no cleavage with TFA/H$_2$O 9:1, 1 h)	72% (94%)	[403]
2		TFA/H$_2$O 95:5		[404] see also [399]
3		2-mercaptoethanol/ NEt$_3$/DMF 9:0.6:90, 12 h	82%	[80,400]
4		TFA/DCM 1:9, 2 × 15 min	63%	[375]
5		TFA/DCM 1:9, 2 × 15 min		[375]
6		TFA/DCM 1:8, 20 °C, 10 min	95% (97%)	[401]
7		3% HCl in MeOH, 24 h	58%	[402]

also Chapter 15.8). Histidine can also be linked to insoluble supports as the *N*-dinitro-phenyl derivative (Entry **3**, Table 3.28), which is prepared by sequential nucleophilic substitutions at 1,5-difluoro-2,4-dinitrobenzene [80,400].

Some heterocycles can be linked to supports as tetrahydropyranyl derivatives. Attachment of indoles, purines, or tetrazoles (Table 3.28) has been achieved by treatment of a support-bound dihydropyran with the heterocycle in the presence of catalytic amounts of pyridinium tosylate [375], camphorsulfonic acid [401], or TFA [402] in DCE at 60–80 °C for 16–24 h. Imidazole derivatives have further been linked to supports by N-alkylation of imidazole with support-bound alkylating agents [371].

3.11 Linkers for Alcohols and Phenols

The attachment of alcohols to insoluble supports has been intensively investigated, in particular with regard to the solid-phase synthesis of oligosaccharides and oligonucleotides. The linking strategies and cleavage methods most commonly used are sketched in Figure 3.28.

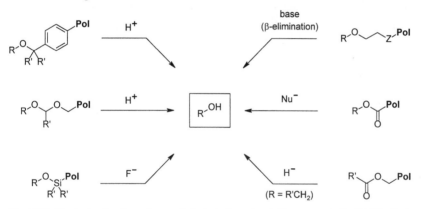

Fig. 3.28. Strategies for the release of alcohols from insoluble supports. R, R': H, alkyl, aryl; Z: electron-withdrawing group.

3.11.1 Attachment as Ethers

Most acid-labile benzyl alcohol linkers suitable for the attachment of carboxylic acids to insoluble supports can also be used to attach aliphatic or aromatic alcohols as ethers. The attachment of alcohols as ethers is less easily accomplished than esterification, and might require the use of strong bases (Williamson ether synthesis; [311,405]) or acids. These harsh reaction conditions limit the choice of additional functional groups present in the alcohol. Some suitable etherification strategies are sketched in Figure 3.29. Etherifications are treated in detail in Chapter 7.2.

Phenols can be etherified with resin-bound benzyl alcohols by the Mitsunobu reaction [406,407], or, alternatively, by nucleophilic substitution of resin-bound benzyl halides or sulfonates [408,409]. Both reactions proceed smoothly under mild reaction conditions.

Aliphatic alcohols have been etherified with Wang resin by conversion of the latter into a trichloroacetimidate (Cl$_3$CCN/DCM/DBU 15:100:1, 0 °C, 40 min), followed by nucleophilic substitution with the alcohol under slightly acidic conditions (0.07 mol/L ROH in DCM/C$_6$H$_{12}$ 1:1, 0.17 % (volume) BF$_3$ · OEt$_2$, 10 min) [410–413]. A similar protocol can also be performed on hydroxymethyl polystyrene [414]. Support-bound aryl diazomethanes, which can be prepared by oxidation of hydrazones or by thermo-lysis of the sodium salt of sulfonyl hydrazones (Chapter 10.5) react with alcohols in the presence of Lewis acids to yield ethers. Resin-bound benzylic thiocarbonates react with aliphatic alcohols in the presence of silver(I) salts to yield resin-bound benzyl ethers [313,415]. The nucleophilic substitution of resin-bound benzyl halides with ali-phatic alcoholates requires strong bases, such as sodium hydride [405], and might therefore be difficult to automate.

Few examples have been reported of the etherification of alcohols with resin-bound diarylmethyl alcohols (Entry **5**, Table 3.29; Entry **5**, Table 3.30 [416]). Diarylmethyl ethers do not seem to offer advantages compared with the more readily accessible tri-tyl ethers, which are widely used linkers for both phenols and aliphatic alcohols. Attachment of alcohols to trityl linkers is usually effected by treating trityl chloride resin or 2-chlorotrityl chloride resin with the alcohol in the presence of a base (phe-nols: pyridine/THF, 50 °C [417] or DIPEA/DCM [418]; aliphatic alcohols: pyridine, 20–70 °C, 3 h–5 d [419–424] or collidine, Bu$_4$NI, DCM, 20 °C, 65 h [73]). Aliphatic or aromatic alcohols can be attached as ethers to the same type of light-sensitive linker used for carboxylic acids (Chapter 3.1.3).

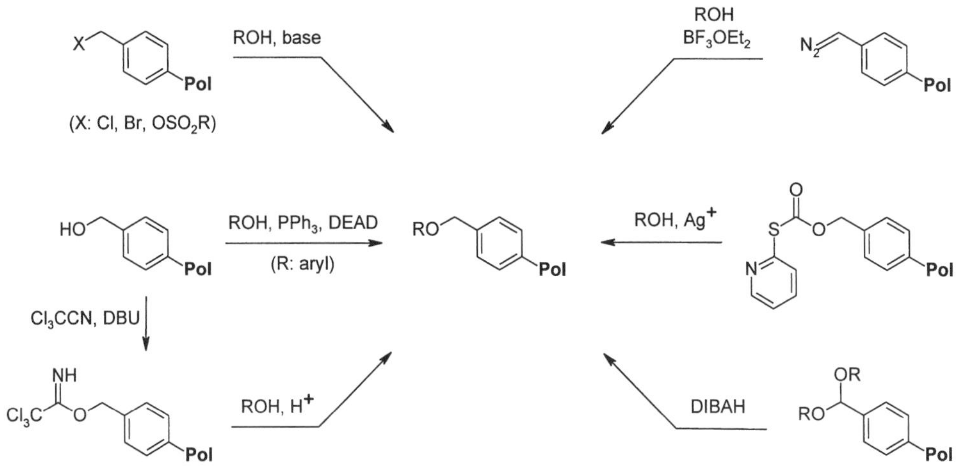

Fig. 3.29. Strategies for the etherification of alcohols with insoluble supports.

Ethers are generally inert towards nucleophilic attack and therefore suitable linkers for solid-phase chemistry involving strong nucleophiles. Wang linker derived ethers can, however, be oxidized to acetals under mild conditions (0.06 mol/L DDQ in DCM, 20 °C, 3 h [425]), which are easy to hydrolyze.

Table 3.29. Cleavage of support-bound ethers to yield aliphatic alcohols.

Entry	Loaded resin	Cleavage conditions	Product, yield (purity)	Ref.
1		HCl, dioxane		[311]
2		1% TFA in DCM, 2–4 h; or 10% TFA in DCM, 0.5 h	98%	[411] see also [410]
3		TFA/DCM 1:1, 0.5 h	45–64%	[405]
4		BF$_3$OEt$_2$ (0.075 mol/L), DCM, 20 °C, 3 h	26–30%, 67% de	[412]
5		TFA/DCM 7:3, 1 h	59%	[323] see also [220] [426]
6		TFA vapor, 20 °C, overnight	73–84% (76–96%)	[427]
7		HCl (0.35 mol/L), dioxane, 20 °C, 48 h; or TsOH (0.2 mol/L), THF/MeOH 1:1, 22 h; or DIBAH, C$_6$H$_6$, 80 °C, 16 h	38–60%	[73] [428] see also [420] [421]
8		0.1% Cl$_2$CHCO$_2$H in CHCl$_3$, 25 °C, 1 h; or AcOH/CHCl$_3$ 2:8, 25 °C, 72 h	100%	[422]

Table 3.29. continued.

Entry	Loaded resin	Cleavage conditions	Product, yield (purity)	Ref.
9		NEt$_3$/DCM 2:8	(90%)	[429]
10		hv, THF, 25 °C	95%	[430] see also [431]
11		CAN (5 eq), MeCN/H$_2$O 10:1	70%	[432]
12		DDQ (0.03 mol/L, 1.2 eq), DCM/H$_2$O 20:1, 20 °C, 4 h (repeat once)	91%	[433]
13		CAN, MeCN/H$_2$O 1:1, hexane, ultrasound		[434]

Cleavage conditions for alkyl benzyl ethers prepared from acid-labile benzyl alcohols are similar to those for the corresponding benzyl esters (Table 3.29). Aryl benzyl ethers, however, are generally cleaved more easily by acidolysis than esters or alkyl ethers. Phenols etherified with hydroxymethyl polystyrene, for instance, can even be released by treatment with trifluoroacetic acid (Entry **1**, Table 3.30). It has also been shown that Wang resin derived phenyl ethers are less stable than Wang resin derived esters towards acetic acid under reflux [27].

Illustrative examples of the cleavage of support-bound ethers are listed in Tables 3.29 and 3.30. Acidolytic cleavage is the most common strategy, but base-mediated, photolytic, and oxidative cleavage have also been reported.

TFA-mediated cleavage of alcohols from supports occasionally leads to the formation of trifluoroacetic acid esters of the released alcohol. This esterification can sometimes be avoided by using wet TFA (e.g. with 5 % water) instead of anhydrous TFA.

Table 3.30. Cleavage of support-bound ethers to yield phenols.

Entry	Loaded resin	Cleavage conditions	Product, yield (purity)	Ref.
1		TFA/DCM 65:35, 20 °C, 3 h R: alkyl, acyl	51–70%	[435] see also [436]
2		TMSOTf (0.3 mol/L), DCM, 3 h	98%	[321]
3		1% TFA in DCM, 0.5 h	98%	[411] see also [407] [437-440]
4		TFA/DCM/Me₂S 45:50:5, 1.5 h	72%	[441]
5		TFA/DCM 5:95, 0.5 h	96% (91%)	[220]
6		TFA/DCM 1:99; or TFA/DCM/MeOH 2:7:1	60–95%	[417] [418]
7		hv (350 nm)		[442] see also [443]

3.11.2 Attachment as Silyl Ethers

Alcohols can be linked to insoluble supports as silyl ethers (Figure 3.30). This form of attachment can be realized by treatment of support-bound silyl chlorides [444–448] or silyl trifluoroacetates [449] with alcohols in the presence of a base (imidazole, DIPEA, or DMAP in DCM [73,445,450,451]). Alternatively, silyl ethers can be prepared by heating resin-bound silanes R₃SiH [445,446] with aromatic or aliphatic alcohols, ketones, or aldehydes in the presence of catalytic amounts of a rhodium complex [(PPh₃)₃RhCl [452], rhodium(II) perfluorobutyrate [445]]. Dialkoxysilanes (e.g. Entry

3, Table 3.31) have been prepared by first treating an alcohol in solution with a di-alkyldihalosilane, and then coupling the resulting monohalosilane with a suitable polymeric alcohol [453].

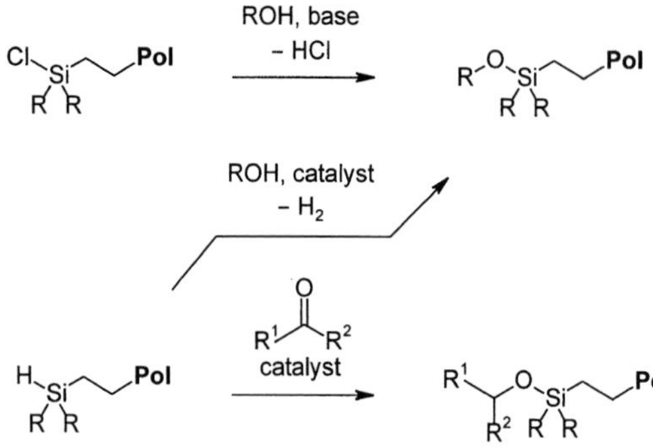

Fig. 3.30. Preparation of support-bound silyl ethers.

Silyl ethers are inert towards strong bases, oxidants (ozone [73], Dess–Martin peri-odinane [450], iodonium salts [454], sulfur trioxide–pyridine complex [314]), and weak acids (e.g. 1 mol/L HCO_2H in DCM [450]), but can be selectively cleaved by treat-ment with HF in pyridine or with TBAF (Table 3.31). Silyl ether attachment has been successfully used for the solid-phase synthesis of oligosaccharides [447,448,455,456].

Table 3.31. Silyl ethers as linkers for alcohols.

Entry	Loaded resin	Cleavage conditions	Product, yield (purity)	Ref.
1	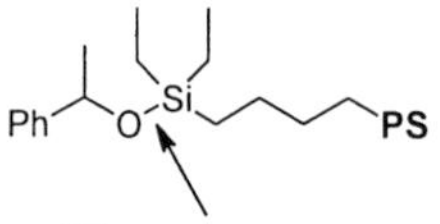	HF/pyridine (0.4 mol/L), THF, 2 h	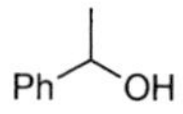 50%	[452] see also [446] [450]
2	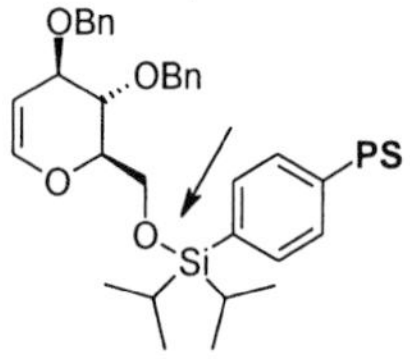	TBAF (0.25 mol/L), AcOH (0.13 mol/L), THF, 40 °C, 18 h	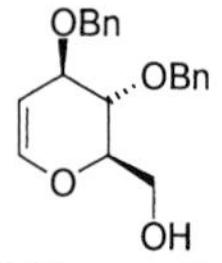 0.61 mmol/g	[447] see also [456] [457]
3	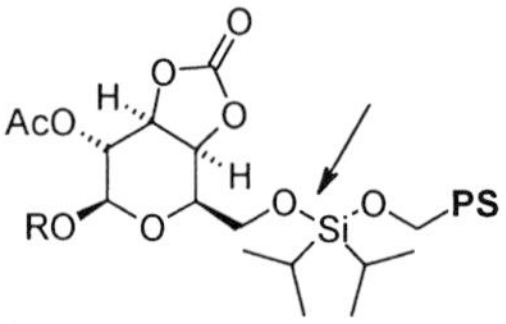	HF/pyridine (0.57 mol/L), PhOMe (0.15 mol/L), DCM, −10 °C, 4 h, 20 °C, 24 h	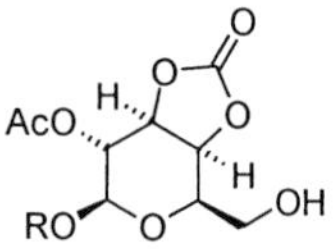	[453] see also [458]
4	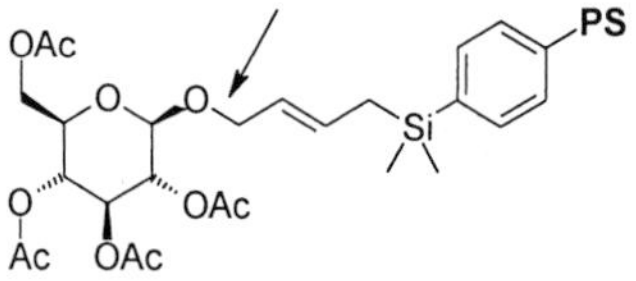	3% TFA in DCM, 18 h	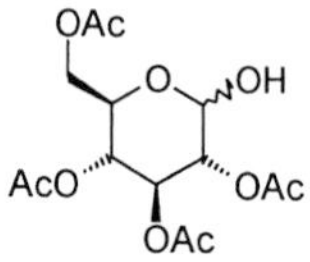 0.41 mmol/g	[69]

3.11.3 Attachment as Acetals

Acetals are a further functional group suitable for linking alcohols to insoluble supports (Table 3.32). A frequently used linker of this class is resin-bound dihydropyran [459–463], which forms mixed acetals (tetrahydropyranyl ethers) with aliphatic [464] or aromatic alcohols [465] upon acid catalysis (e.g. PPTS, DCE, 80 °C, 16 h). The resulting acetals are stable towards strongly basic or nucleophilic reagents, such as organolithium compounds [465], organocuprates [459], or Grignard reagents [460,462].

Mixed acetals of a support-bound and a non-support-bound alcohol with acetaldehyde have also been used as linker (Entry **3**, Table 3.32). Such non-cyclic, mixed acetals are, however, not easy to prepare on solid supports and are more conveniently synthesized by conventional solution-phase chemistry and then loaded on the support [466]. One strategy for preparing mixed, non-cyclic acetals on insoluble supports is the oxidative haloalkoxylation of support-bound enol ethers (Entry **5**, Table 6.1).

Resin-bound aldehydes and ketones have been used as linkers for 1,2- or 1,3-diols (Entries **4–6**, Table 3.32). Cleavage of acetal-based linkers is usually effected by acid-catalyzed transacetalization or by hydrolysis.

3.11.4 Attachment as Esters

Both aliphatic alcohols and phenols have been immobilized as esters of support-bound carboxylic acids. The esterification can be achieved by treatment of resin-bound acids with alcohols and a carbodiimide, under Mitsunobu conditions, or by acylation of alcohols with support-bound acyl halides (see Chapter 13.4).

Ester attachment of alcohols is particularly useful when acidic reaction conditions are to be employed in a synthetic sequence. The solid-phase synthesis of oligosaccharides [471] and oligonucleotides (Chapter 16) is often performed with ester linkage to the support, because glycosylations, for instance, also generally require acid catalysis. Valuable alcohols can be esterified quantitatively with excess support-bound acylating agent [472]. The excess of acylating agent can be capped by treatment of the support with methanol.

Cleavage of support-bound esters can be effected by a variety of reagents (Table 3.33). Saponification with alkali metal hydroxides or transesterification with alcoholates [473,474] are efficient, but yield products contaminated with the (non-volatile) cleavage reagent. If crude products of high purity are to be obtained and used without further purification (as, e.g., for arrays of compounds prepared by parallel solid-phase synthesis) volatile nucleophiles should be used for ester cleavage. These include ammonia, which is the common cleavage agent for resin-bound oligonucleotide hemisuccinates (Entry **5**, Table 3.33), other low-molecular-weight amines, or hydrazine [475]. Aryl esters are generally more easily cleaved than esters of aliphatic alcohols.

In 1999 a linker for alcohols was described, which can be cleaved by reducing agents (Entry **8**, Table 3.33). This linker is based on a quinone which, after reduction to the corresponding hydroquinone, undergoes intramolecular nucleophilic cleavage of an ester. The alcohol is thereby released and a resin-bound lactone is formed [476].

Table 3.32. Acetals as linkers for alcohols and phenols.

Entry	Loaded resin	Cleavage conditions	Product, yield (purity)	Ref.
1		BuOH/DCE 1:1, PPTS (2 eq), 60 °C, 16 h	95%	[464]
2		TFA/DCM/MeOH 1:5:1	32–50%	[465]
3		TFA/DCM 3:7, 3 h	(30–70%)	[466]
4		TFA/DCM 5:95, 20 °C, 1.5 h	73–98% (70–95%)	[467] see also [466]
5		TFA/DCM/MeOH 10:90:1, 20 °C, 1 h	88%	[468] see also [469]
6		TFA/H₂O 90:2.5, scavengers	(octreotide) 74%	[470]
7		MeOTf, MeSSMe, DTBMP, DCE, 40 °C, 21 h (ROH: 2,3,6-tri-*O*-benzyl-β-D-glucose)	50%	[425]
8		HCl (2 mol/L); or NaOH (2 mol/L); or penicillin amidase, 25 °C, 16 h; or TFA/DCM/H₂O 9:10:1, 3 h		[160]

Table 3.33. Ester attachment of alcohols and phenols cleavable by nucleophiles.

Entry	Loaded resin	Cleavage conditions	Product, yield (purity)	Ref.
1		NaOMe (satd in MeOH)/THF 1:4		[477] see also [478]
2		NaOH (0.5 mol/L), dioxane/H₂O 1:1, 20 °C, 20 h or 60 °C, 3 h	74%	[479] see also [73]
3		NEt₃/MeOH/Me₂S 15:75:10, 20 °C, 3 h (six repetitions)	92%	[441] see also [480]
4		NaOMe (0.3 mol/L), THF/MeOH 60:3, 2–4 h		[481] [482] see also [483]
5		30% NH₃ in H₂O, 20 °C, 1 h		[484] see also [485]
6		NaOH/MeCN/H₂O, 3 h		[486]
7		NaBH₄ (0.01 mol/L, 2 eq), hot EtOH, 3 h	76% (mayor triazole regioisomer shown)	[487]
8		NaHSO₃, H₂O/THF 5:8, 20 °C, 2.5 h	70–89%	[476]

Alcohols can also be generated by hydride reduction of esters of support-bound alcohols or thiols [312,488], or by reduction of resin-bound imides (Table 3.34). This cleavage strategy generally requires aqueous workup and is not well-suited for parallel synthesis of large numbers of compounds.

Table 3.34. Linkers for the preparation of alcohols by reductive cleavage.

Entry	Loaded resin	Cleavage conditions	Product, yield (purity)	Ref.
1		LiBH₄, 25 °C		[489]
2		DIBAH (10 eq), PhMe, 0 °C, 12 h	51%	[490]
3		NaBH₄/LiBr (3 eq) in THF/EtOH 6:1, 20 °C, 24 h (N-Fmoc groups are partially cleaved)	93% (91%)	[491]
4		DIBAH, PhMe, 0 °C	26%	[492]
5		DIBAH, PhMe, −78 °C	85%	[493]

3.11.5 Miscellaneous Linkers for Alcohols and Phenols

Alcohols can be attached to support-bound alcohol linkers as carbonates [494], but few examples have been reported. For the preparation of carbonates the support-bound alcohol needs to be converted into a reactive carbonic acid derivative by reaction with phosgene or a synthetic equivalent thereof (e.g. disuccimidyl carbonate [494] or carbonyl diimidazole [139], see Chapter 14.7). Best results are usually obtained with support-bound chloroformates. The resulting intermediate is then treated with an alcohol and a base (DIPEA, DMAP, or DBU), whereby the unsymmetric carbonate results.

Carbonates are generally more resistant towards nucleophilic cleavage than esters, but less stable than carbamates. Aryl carbonates are easily cleaved by nucleophiles and are therefore not suitable as linkers for phenols.

Examples of the cleavage of support-bound carbonates are given in Table 3.35. Depending on the structure of the carbonate, acidolytic, base-induced, nucleophilic, or photolytic cleavage can be used to release the alcohol. Acidolysis of the benzylic

Table 3.35. Carbonate, carbamate, and phosphate attachment of alcohols and phenols.

Entry	Loaded resin	Cleavage conditions	Product, yield (purity)	Ref.
1		HF/PhOMe 9:1, 0 °C, 1 h	> 95%	[496]
2		TBAF (1 mol/L), THF, 1 h	78% (92%)	[497]
3		DBU (0.5 mol/L), dioxane; or NH₃, 55 °C, 5 h; or piperidine/DMF 2:8, 20 °C, 3 h		[498] see also [499]
4	3-cholesteryl (SG)	PBu₃/NEt₃/DMF 2:1:4, 80 °C, 5 h	cholesterol (0.14 mmol/g)	[353]
5		hv (365 nm), MeCN/H₂O 9:1, 2 h R: oligonucleotide	83%	[500] see also [496]
6		hv (350 nm), THF/MeOH 3:1, 28 °C, 3 h (ROH: cholesterol)	cholesterol 72% (> 95%)	[139]
7		1. bovine spleen phosphodiesterase, pH 5.7 2. alkaline phospho-monoesterase	> 83%	[169]

C–O bond of resin-bound benzyl carbonates leads to the release of an unstable carbonic acid ester, which decarboxylates to yield the alcohol.

Few examples have been described of nucleophilic cleavage of carbonate- or carbamate-linked alcohols from insoluble supports. A serine-based linker for phenols releases the phenol upon fluoride-induced, intramolecular nucleophilic cleavage of an aryl carbamate (Entry **2**, Table 3.35). Furthermore, a linker for oligonucleotides has been reported, in which the carbohydrate is bound as carbonate to resin-bound 2-(2-nitrophenyl)ethanol, and which is cleaved by base-induced β-elimination (Entry **3**, Table 3.35). Trichloroethyl carbonates, susceptible to cleavage by reducing agents such as zinc or phosphines, have been used with success to link aliphatic alcohols to silica gel (Entry **4**, Table 3.35). These carbonates can also be cleaved by acidolysis (Table 3.21).

Entry **7** in Table 3.35 is a rare example for the use of a phosphodiester as linker for alcohols. This linker, when used in combination with an enzyme-compatible support, can be selectively cleaved with a phosphodiesterase. To obtain the free alcohol, the released phosphate must be subjected to an additional enzymatic dephosphorylation.

Polystyrene-derived phenylboronic acids have been used for immobilizing diols (carbohydrates) as boronic esters [495]. Cleavage was effected by treatment with acetone/water or THF/water. This high lability towards water and alcohols severly limits the choice of reactions which can be performed without premature cleavage of this linker.

3.12 Linkers for Thiols

Thiols have been linked to insoluble supports as acid-labile benzyl thioethers, as aryl thioethers, as *S*-carbamoyl derivatives, and as unsymmetrical disulfides (Table 3.36). Because thiols often undergo oxidative dimerization at the air to yield symmetric disulfides, the latter might be the only product isolated if cleavage is not conducted under inert gas (e.g. Entry **3**, Table 3.36).

Benzylic thioethers can be significantly more stable towards acidolytic solvolysis than the corresponding benzylic ethers. As illustrated by Entry **1** in Table 3.36, Wang linker derived thioethers are not cleaved by TFA but only by acids with high ionizing power, such as HF.

The mercapto group of cysteine has been attached to insoluble supports as the thiocarbamate and as the aryl thioether (Entries **4** and **5**, Table 3.36). Both types of linker are susceptible to nucleophilic cleavage.

Thiols can be linked to insoluble supports as disulfides by disulfide interchange. Mixed disulfides can be prepared on insoluble supports by treating support-bound thiols with excess 'activated' disulfide (e.g. 2-benzothiazolyl, 2-nitrophenyl, or 3-nitro-2-pyridyl disulfides [55,501]; Figure 3.31), or by treating a support-bound disulfide (e.g. a 2-pyridyl disulfide [170]) with a thiol. Resin-bound disulfides are stable under the conditions of standard Fmoc peptide synthesis, but can be cleaved by reducing agents (Entries **6** and **7**, Table 3.36 [170,502,503]).

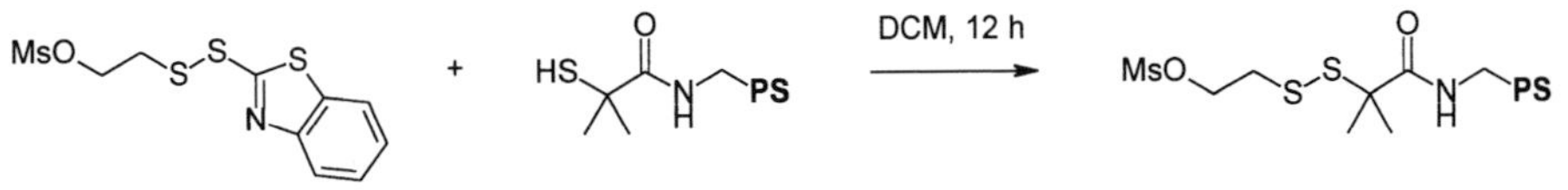

Fig. 3.31. Preparation of support-bound unsymmetric disulfides [501].

Table 3.36. Linkers for thiols.

Entry	Loaded resin	Cleavage conditions	Product, yield (purity)	Ref.
1	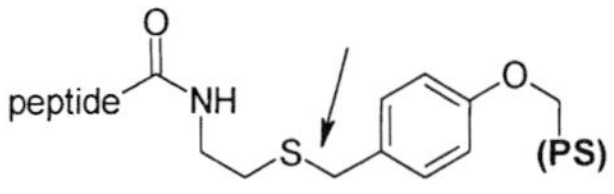	HF/*p*-cresol 9:1, −5 °C, 1 h (no cleavage with neat TFA, 60 h)	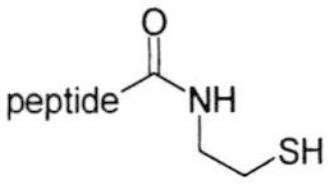 > 30%	[504]
2	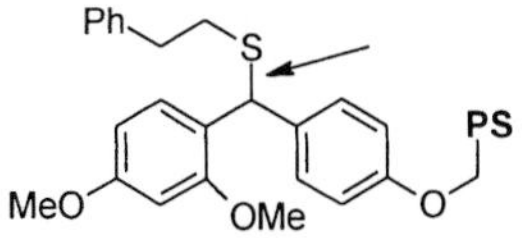	TFA/DCM 5:95, 0.5 h	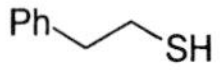 92% (95%)	[220]
3	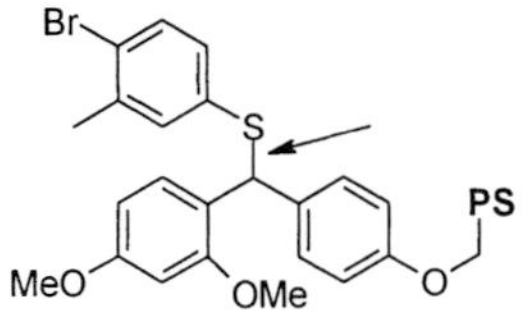	TFA/H$_2$O 95:5, 3 h	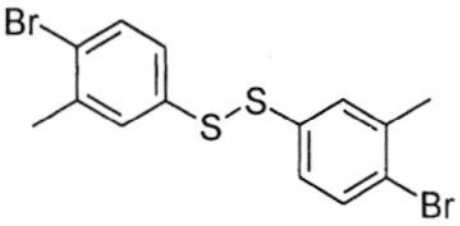 100%	[426] see also [220]
4	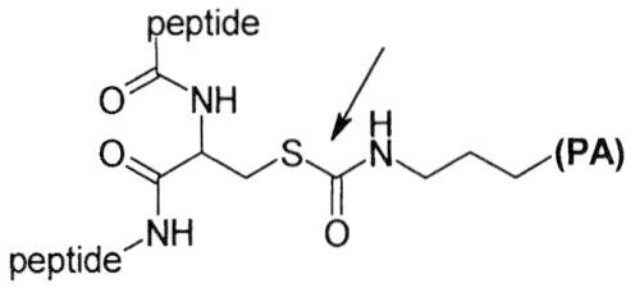	NaOH (0.1 mol/L), MeOH/H$_2$O 9:1, 15 min; or liquid ammonia	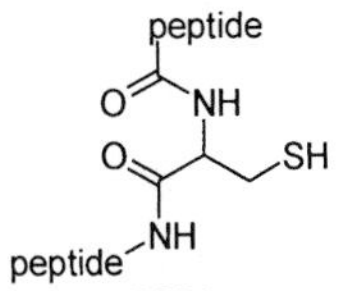 59%	[80]
5	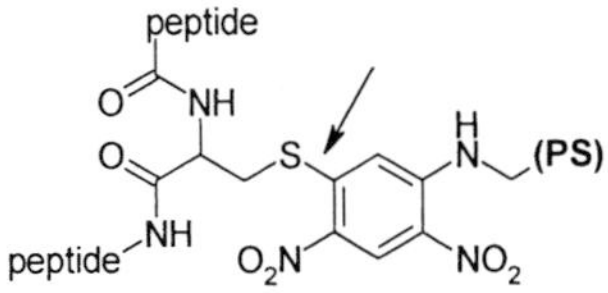	2-mercaptoethanol, NMM, AcOH, DMF, 2 × 24 h	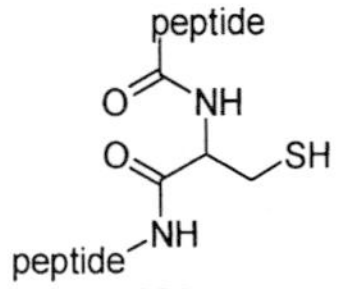 58%	[505]
6	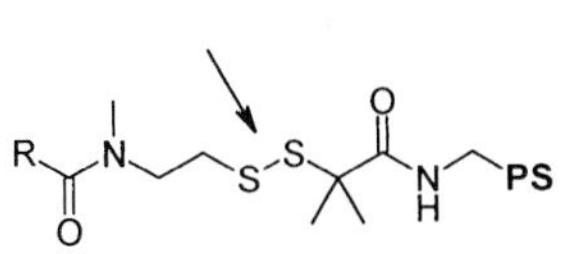	P(CH$_2$CH$_2$CO$_2$H)$_3$ (4 mmol/L), dioxane/H$_2$O 9:1, 8 h	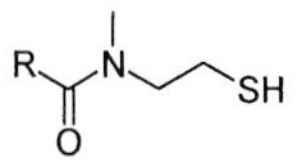 (not isolated)	[506] see also [501] [507]
7	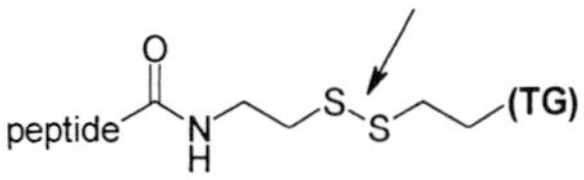	dithiothreitol (0.01 mol/L), TRIS buffer (pH 7.5), 20 °C, 3 × 1 h	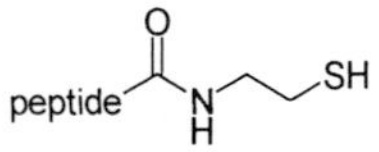	[508]

Table 3.37. Generation of alkyl, vinyl, and aryl halides upon cleavage from insoluble supports.

Entry	Loaded resin	Cleavage conditions	Product, yield (purity)	Ref.
1		ICl (3 eq), DCM, 10 min	> 90%	[510]
2		Br$_2$ (6 eq), pyridine (3 eq), DCM, 0 °C, 2 h	97%	[510]
3		Br$_2$ (4 eq), DCM, 2 × 5 min Ar: 3-(MeO)C$_6$H$_4$	59%	[511] see also [512]
4		I$_2$ (1.1 eq), THF, 23 °C, 2 h	100%	[513]
5		30% HBr in AcOH, CCl$_4$, 10 min	30%	[419]
6		NaI, 2-butanone, 65 °C (cleavage also succeeded with N$_3^-$ or AcO$^-$)	85–91%	[295] see also [294]
7		MeI, NaI, DMF, 75 °C, 20 h	82%	[514]
8		MeI, 110 °C, 12 h	94% (97%)	[359]
9		ICl, DCM, −78 °C, 1.5 h	0.26 mmol/g	[515] [516]

3.13 Linkers for Alkyl and Aryl Halides

Alkyl or aryl halides can be generated upon cleavage from a polymeric support by treatment of silanes, organogermanium compounds, or stannanes with halogens (Entries **1–4**, Table 3.37), or by solvolysis of resin-bound alcohols with hydrogen halides (Entry **5**, Table 3.37). Nucleophilic cleavage with iodide of resin-bound sulfonic esters and dialkyl aryl sulfonium salts (prepared in situ from alkyl aryl sulfides and methyl iodide) has, furthermore, been used to prepare iodinated carbohydrates and alkyl iodides, respectively (Entries **6** and **7**, Table 3.37). Aryl iodides have been prepared by thermolysis of resin-bound triazenes with methyl iodide (Entry **8**, Table 3.37 [509]). This reaction probably proceeds via methylation of the trisubstituted nitrogen atom of the triazene, followed by release of an aryldiazonium iodide into solution. Thermal decomposition of this diazonium salt yields the observed aryl iodide.

In addition to the procedures listed in Table 3.37, further reactions have been used to generate halides upon cleavage. In Chapter 3.5.2 the iodolactonization is presented as a method for the preparation of iodomethyl lactones from resin-bound pentenoic- or hexenoic acid derivatives. Closely related to the iodolactonization is the iodine-mediated formation of 2-(iodomethyl)tetrahydrofurans from resin-bound isoxazolidines (Entry **9**, Table 3.37; for mechanism see Figure 15.5).

3.14 Linkers for Aldehydes and Ketones

Interest in linkers for carbonyl compounds has only slowly emerged in recent years. The main driving force for the development of such linkers was the need for methods to prepare peptide aldehydes and related compounds (e.g. peptide trifluoromethyl ketones), which can be highly specific and valuable enzyme inhibitors [517,518], and are potentially useful for the treatment of various diseases.

The main strategies for the release of aldehydes or ketones from insoluble supports are sketched in Figure 3.32. These include the hydrolysis of acetals and related derivatives, the treatment of support-bound carboxylic acid derivatives with carbon nucleophiles, and the ozonolysis of resin-bound alkenes.

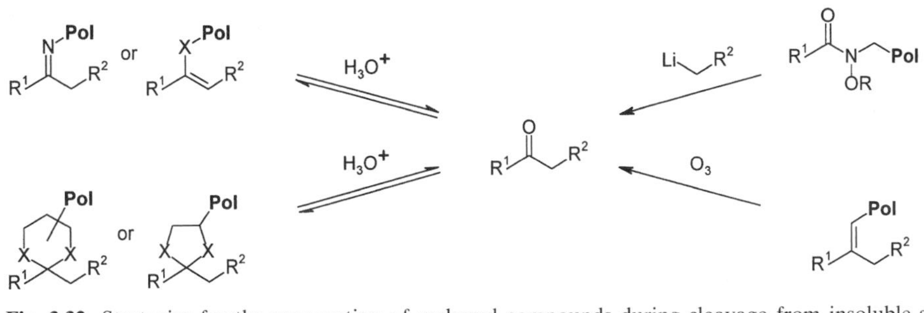

Fig. 3.32. Strategies for the preparation of carbonyl compounds during cleavage from insoluble supports. X: NR, O, S.

3.14.1 Attachment as Enol Ethers, Enamines, and Imines

Insoluble supports bearing hydroxyl groups can be used to immobilize aldehydes and ketones, either as acetals or as enol ethers (Table 3.38). 1,3-Dicarbonyl compounds react smoothly with resin-bound alcohols to yield enol ethers upon azeotropic removal of water [519,520]. Enol ethers can, furthermore, be prepared by carbonyl methylenation of resin-bound esters with the Tebbe reagent [521]. The preparation of support-bound silyl enol ethers from resin-bound silyl triflates and silyl esters [522], and ketones or aldehydes [523] has also been reported. Most of these enol ethers can be cleaved by mild acidic hydrolysis, whereby the corresponding carbonyl compounds are released into solution (Table 3.38).

Resin-bound amines can be converted into imines [524,525] or enamines by reaction with carbonyl compounds (Entries **6** and **7**, Table 3.38). Resin-bound enamines have also been prepared by Michael addition of resin-bound secondary amines to acceptor-substituted alkynes [526], or by chemical modification of other resin-bound enamines [526,527]. Acceptor-substituted enamines ('push–pull' olefins) are not always susceptible to hydrolytic cleavage by TFA alone, but might require aqueous acids to undergo hydrolysis [528].

3.14.2 Attachment of Carbonyl Compounds as Acetals

Support-bound alcohols and thiols can be used to immobilize aldehydes and ketones as acetals. Mixed acetals of carbonyl compounds with support-bound alcohols can be prepared by transacetalization of a symmetric acetal under acidic conditions [530]. The formation of mixed acetals on solid phase is, however, not always easy to perform and control, and a preferred scheme is often the preparation of a mixed acetal in solution, followed by its loading on to a support [466,472]. Carbohydrates can be linked to resin-bound alcohols or thiols as glycosides (Table 3.39).

Resin-bound diols, aminoalcohols, and dithiols, which reversibly form cyclic acetals with aldehydes and ketones, have been successfully used as linkers for carbonyl compounds (Entries **5–11**, Table 3.39). Acetal formation on insoluble supports can be achieved by azeotropic removal of water (C_6H_6, TsOH, reflux [531]), whereas dithioacetals can be prepared by acid-catalysis alone (BF_3OEt_2 or TMSCl; $CHCl_3$, 0 °C, 2 h [532]). Acetals are usually easy to cleave by acid-catalyzed transacetalization or hydrolysis (Table 3.39). Dithioacetals, on the other hand, tend to be more resistant towards hydrolysis, but cleavage can be achieved by treatment with mercury(II) salts or by oxidation with [bis(trifluoroacetoxy)iodo]benzene [533] or periodic acid. The latter reagent can, however, also lead to the conversion of methyl ketones into iodomethyl ketones [532].

Table 3.38. Attachment of aldehydes and ketones as enol ethers, enamines, and semicarbazides.

Entry	Loaded resin	Cleavage conditions	Product, yield (purity)	Ref.
1		TFA/DCM 3:97, 20 min	80% (> 95%)	[519]
2		TFA/acetone 5:95, 0.5 h	42% (> 95%)	[519]
3		TFA/CDCl$_3$ 1:99, 25 °C	30% (> 90%; endo/exo 97:3)	[521] see also [529]
4		TFA/DCM 1:99, 25 °C	74% (> 90%)	[521]
5		TFA/DCM 1:9	77% (> 90%)	[523]
6		TFA/DCM 3:97, 10 min	78% (99%)	[526]
7		TFA/DCM 3:97, 10 min Ar: 4-BrC$_6$H$_4$	63% (90%)	[527]
8		aq HCl (1 mol/L)/ THF/AcOH/H$_2$O 1:75:2:6, 65 °C, 4 h	40%	[518]
9		dilute aqueous acid, HCHO		[517]

Table 3.39. Attachment of aldehydes and ketones as acetals.

Entry	Loaded resin	Cleavage conditions	Product, yield (purity)	Ref.
1		NBS (4 eq), DTBP, THF/MeOH 10:1, 1.5 h		[534] see also [535]
2		Hg(O₂CCF₃)₂, DCM, H₂O, 20 °C, 5 h	64%	[536]
3		CSA (3 eq), DCM/H₂O 2:1, 25 °C, 40 h	up to 75%	[472]
4		TFA/DCM/H₂O 6:3:1	17–83% (95–98%)	[537]
5		aq HCl (3 mol/L)/dioxane 1:1, 80 °C, 48 h	> 95%	[538] see also [539–542]
6		TFA/H₂O 95:5, 15 min	96%	[543]
7		PPTS, dioxane/H₂O 8:2, 95 °C, 9 h	0.68 mmol/g (92%)	[531] see also [544]
8		AcOH/H₂O 5:95, 60 °C, 0.5 h (no cleavage with 95% TFA)	(94%)	[545]

Table 3.39. continued.

Entry	Loaded resin	Cleavage conditions	Product, yield (purity)	Ref.
9	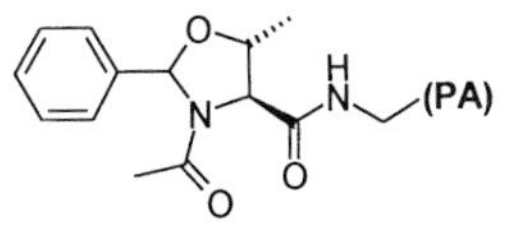	AcOH/H$_2$O 5:95, 60 °C, 0.5 h (no cleavage with 95% TFA)	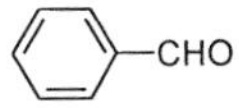 —CHO 10%	[545]
10	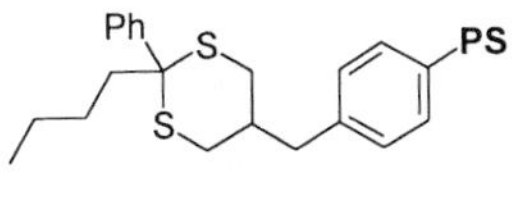	H$_5$IO$_6$ (2.8 eq), THF, 5 h	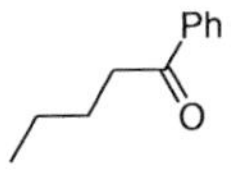 92%	[532]
11	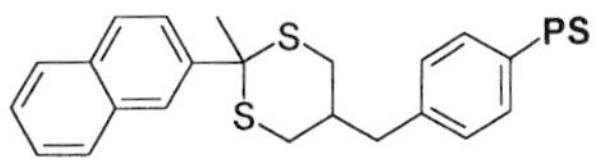	Hg(ClO$_4$)$_2$·3 H$_2$O (3 eq)	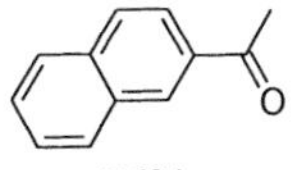 76%	[532]

3.14.3 Miscellaneous Linkers for Aldehydes and Ketones

Aldehydes and ketones have also been prepared by nucleophilic cleavage of resin-bound *O*-alkyl hydroxamic acids (Weinreb amides [546]) with lithium aluminum hydride [547] or Grignard reagents (Entries **1** and **2**, Table 3.40). Similarly, support-bound thiol esters can be cleaved with Grignard reagents to yield ketones [237], or with reducing agents to yield aldehydes (Entry **3**, Table 3.40). Intramolecular Dieckmann cyclization of polystyrene-bound pimelates has been used to prepare β-keto esters (Entry **4**, Table 3.40). Oxidative cleavage reactions leading to the formation of aldehydes include the ozonolysis of resin-bound alkenes and the periodate-mediated cleavage of 1,2-diols (Entries **5** and **6**, Table 3.40).

3.15 Linkers for Olefins

The main strategies used for the preparation of olefins during cleavage from insoluble supports are β-elimination and olefin metathesis (Figure 3.33). Because some of these linkers enable the preparation of pure alkenes, devoid of additional functional groups, these linkers are sometimes also called 'traceless' linkers, although the C–C double bond reveals the original point of attachment to the support.

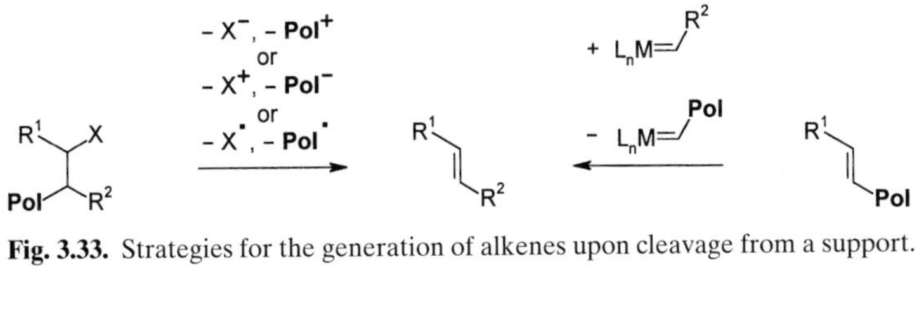

Fig. 3.33. Strategies for the generation of alkenes upon cleavage from a support.

Table 3.40. Formation of aldehydes and ketones by cleavage of carboxylic acid derivatives and alkenes.

Entry	Loaded resin	Cleavage conditions	Product, yield (purity)	Ref.
1		LiAlH$_4$, THF, 0 °C, 0.5 h	21% (80%)	[283]
2		PhMgCl (15 eq), THF, 60 °C, 15 h	33%	[548]
3		DIBAH, DCM, –78 °C, 19 h	73%	[111]
4		KOCEt$_3$, PhMe, 110 °C, 2 min	46%	[549]
5		O$_3$, DCM, –78 °C, 5 min, then Me$_2$S, DCM, 3 h	45%	[550] see also [551]
6		1. TFA/H$_2$O/*i*Pr$_3$SiH 95:2.5:2.5, 20 °C, 3 h 2. NaIO$_4$ (6 eq), H$_2$O/AcOH 5:1, 2 min R: peptide	38%	[552]

3.15.1 Linkers for Olefins Cleavable by β-Elimination

During the release of alkenes by β-elimination the polymeric support might act as a leaving group for nucleophilic displacement (e.g. **Pol**–SO$_2^-$, **Pol**–P(O)R$_2$, **Pol**–O$^-$), as a group capable of yielding a stable cation (e.g. **Pol**–SiR$_2^+$), or as a group with a weak covalent bond to carbon, prone to undergo homolytic cleavage and yield a radical (e.g. **Pol**–SnR$_2$). Most examples of these cleavage strategies reported to date are Wittig reactions, in which phosphorus is irreversibly bound to the support (Table 3.41 [553,554]). Treatment of the immobilized ylide precursor with a base and a carbonyl compound leads to carbonyl olefination and simultaneous release of the alkene into solution.

Table 3.41. Generation of olefins from support-bound phosphonium ylides.

Entry	Loaded resin	Cleavage conditions	Product, yield (purity)	Ref.
1	NHBoc	Me$_2$CHCHO (10 eq), LiBr, NEt$_3$, MeCN, 24 h	NHBoc 72%	[555]
2	MeO Ph Ph PS	OHC–C$_6$H$_4$–CO$_2$Me NaOMe, MeOH, 65 °C, 2 h Ar: 4-(MeO$_2$C)C$_6$H$_4$	MeO 82%	[556]
3	MeO Ph Ph PS	KOtBu, DMF, PhMe, 110 °C, 45 min	MeO 78%	[556]
4	Ph Ph Pol	1. NaN(SiMe$_3$)$_2$, THF, 20 °C, 0.5 h 2. ArCHO (0.5 eq), THF, 20 °C, 20 min	70–95%	[557]
5	R OMe (PS)	PhCHO (10 eq), K$_2$CO$_3$, 18-crown-6, PhMe, 65 °C, 3 h R: alkyl	R Ph 87%	[558]
6	OMe (PS)	K$_2$CO$_3$, 18-crown-6, PhH, 65 °C, 12 h	E 35–65%	[558]

If excess carbonyl compound is used in the product-releasing Wittig reaction, the product will be contaminated by the carbonyl compound. Removal of the excess carbonyl compound can be accomplished by extraction with aqueous bisulfite [555], by imine formation with an aminomethyl resin [556], or by formation of a water soluble hydrazone with Girard's Reagent T [(carboxymethyl)trimethylammonium chloride hydrazide] [556]. A more elegant strategy is, however, the use of excess resin-bound Wittig reagent [557], whereby pure olefins can be obtained directly.

Few examples have been reported of cleavage by β-elimination, in which the support acts as anionic leaving group (Table 3.42; see also Table 15.21). Resin-bound benzocyclobutane has been used as precursor to o-quinodimethanes, which readily undergo Diels–Alder reaction with a variety of dienophiles (Figure 3.34; Entries **2** and **3**, Table 3.42).

Olefines can also be generated by reductive cleavage of resin-bound allyl sulfones or allyl esters with hydride or with carbon nucleophiles (see, e.g., Table 3.46).

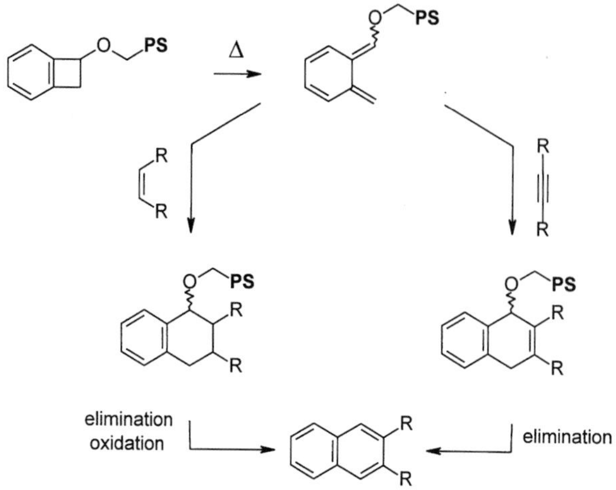

Fig. 3.34. Preparation of substituted naphthalenes from resin-bound *o*-quinodimethanes [414].

The palladium-catalyzed coupling of aryl iodides with vinylstannanes (Stille coupling) leads to the formation of styrenes. With resin-bound vinylstannanes this reaction can be conducted in such a way that simultaneous detachment from the support of the newly formed styrenes occurs. This has been realized intramolecularly in the preparation of macrocyclic lactones (Entry **4**, Table 3.42). The required resin-bound vinylstannanes were prepared either by hydrostannylation of alkynes with a resin-bound stannane HR_2Sn–**Pol** or by treatment of a resin-bound trialkyltin chloride with vinyl lithium compounds (Chapter 4.3). The Suzuki reaction has been used in a similar approach to prepare macrocyclic biphenyl derivatives ([559], see Chapter 3.16.2).

Resin-bound selenium has been used as a linker for olefins in two ways: (a) as an oxidant-sensitive linker (selenoxides readily undergo β-elimination; Entries **5** and **6**, Table 3.42), or (b) as a linker cleavable by tin radicals (Figure 3.35; Entry **7**, Table 3.42; see also Chapter 9). The main advantage of selenides as linkers is their stability under a broad variety of (non-oxidizing) reaction conditions, including high temperatures and treatment with acids or bases, and the mild conditions required for their cleavage (Table 3.42).

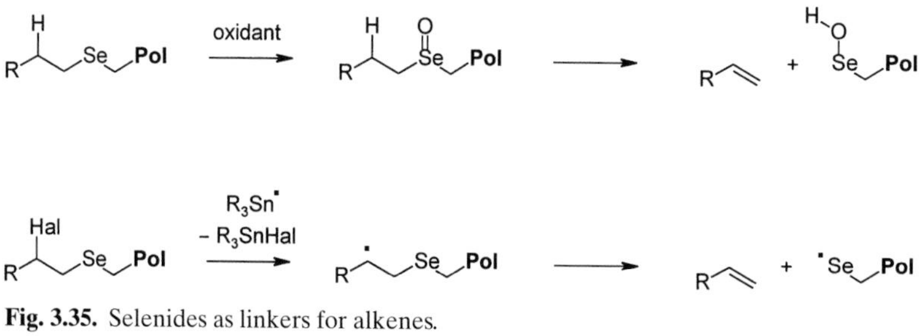

Fig. 3.35. Selenides as linkers for alkenes.

Table 3.42. Generation of olefins upon cleavage from supports by β-elimination and by vinylic substitution.

Entry	Loaded resin	Cleavage conditions	Product, yield (purity)	Ref.
1		DBU (1 eq), DCM, 25 °C, 5 min	 86% (96%)	[560]
2		DMAD (1 eq), PhMe, 105 °C, 14 h	 41%	[414]
3		benzoquinone (1 eq), PhMe, 105 °C, 14 h	 39%	[414]
4		Pd(PPh₃)₄ (0.1 eq), PhMe, 100 °C, 48 h	 51%	[513]
5		30% H₂O₂/THF 1:10, 30 °C, 2 h	 91% (95%)	[561]
6		30% H₂O₂ (1 eq), THF, 23 °C, 12 h	 78%	[562]
7		Bu₃SnH (4 eq), AIBN (0.01 eq), PhMe, 100 °C, 6 h	 92%	[562]
8		HO⌒OH Na, 198 °C, 2 h	 77%	[563]

Support-bound sulfonylhydrazones can also be used as linkers for olefins. Cleavage is effected by heating in the presence of an alcoholate (Entry **8**, Table 3.42; Bamford–Stevens reaction).

3.15.2 Linkers for Olefins Cleavable by Olefin Metathesis

Since the discovery of ruthenium and molybdenum carbene complexes which efficiently catalyze olefin metathesis under mild reaction conditions and which are compatible with a broad range of functional groups, olefin metathesis is increasingly being used for the preparation of alkenes on solid supports. In particular the ruthenium complexes $Cl_2(PCy_3)_2Ru=CHR$, developed by Grubbs, have sufficient catalytic activity even in the presence of air and water [564] and are well suited for solid-phase synthesis.

For the cleavage of olefins from a support by metathesis several strategies can be envisaged. In most of the examples reported, ring-closing metathesis of resin-bound dienes was used to release either a cycloalkene or an acyclic alkene into solution (Figure 3.36, Table 3.43). Further metathesis of the products in solution only occurs to a small extent when internal olefins are the products initially released, because these normally react more slowly with the catalytically active carbene complex than with terminal olefins. If, however, terminal alkenes are to be prepared, self metathesis of the product (to yield ethylene and a symmetrically substituted ethylene) is likely to become a serious side reaction.

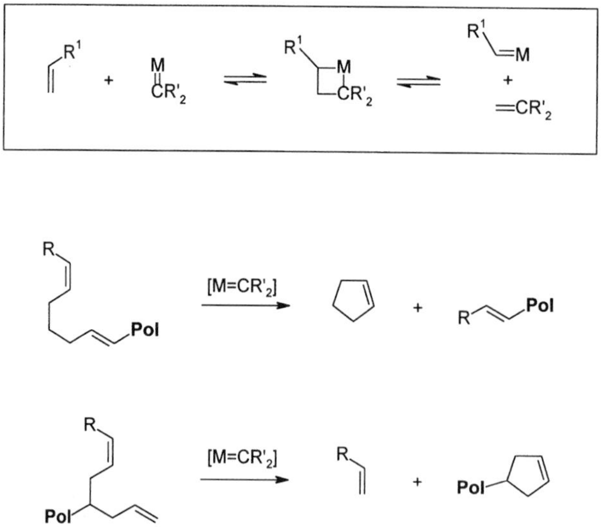

Fig. 3.36. Mechanism of olefin metathesis and strategies for the cleavage of alkenes from polymeric supports by olefin metathesis.

Although five- and six-membered carbo- or heterocycles are most easily formed by ring-closing metathesis, macrocyclizations with simultaneous cleavage from the support have also successfully been performed [565]. Illustrative examples are listed in Table 3.43.

Table 3.43. Generation of olefins upon cleavage from insoluble supports by olefin metathesis.

Entry	Loaded resin	Cleavage conditions	Product, yield (purity)	Ref.
1	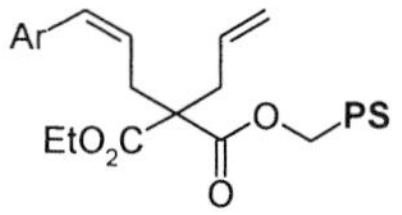	$Cl_2(PCy_3)_2Ru=CHPh$ (0.03–0.23 eq), DCM, 20 °C, 12 h	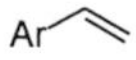 24–55%	[566]
2	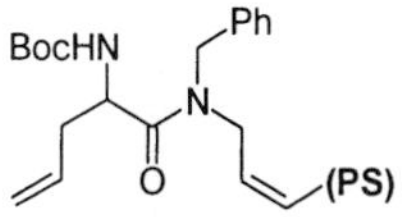	$Cl_2(PCy_3)_2Ru=CHPh$ (1 eq), PhMe, 50 °C, 16 h	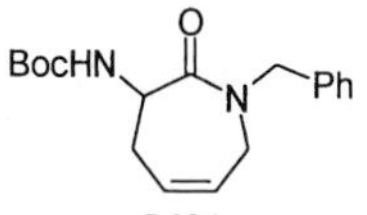 54%	[567] see also [568] [569]
3	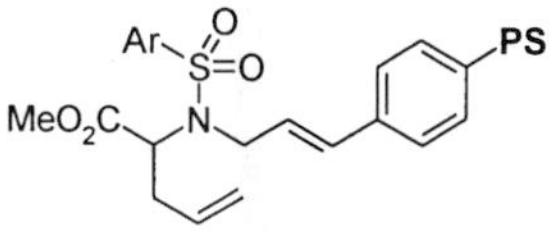	$Cl_2(PCy_3)_2Ru=CHPh$ (0.05 eq), DCM, 20 °C, 16 h Ar: 2,4-dinitrophenyl	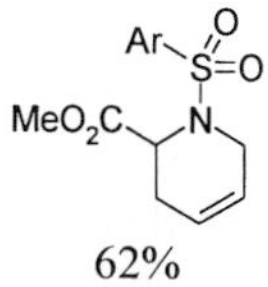 62%	[570] see also [571]

For the preparation of cycloalkenes (and heterocycles; see the corresponding chapters) by ring-closing metathesis with simultaneous cleavage from the support, the addition of terminal olefins (e.g. styrene) to the reaction mixture can lead to increased yields of cycloalkene [571]. This effect is probably related to the regeneration of the catalyst by reaction of support-bound carbene-complexes with the alkene (Figure 3.37). As the rate of catalyst regeneration depends directly on the concentration of alkene $RCH=CH_2$ (Figure 3.37), high concentrations of this alkene will lead to fast

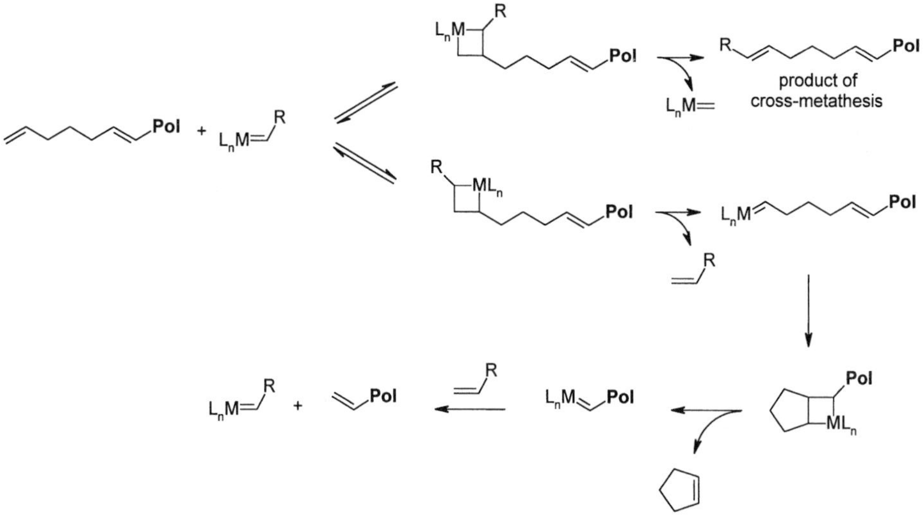

Fig. 3.37. Ring-closing metathesis with simultaneous cleavage from the support, and the mechanism of catalyst regeneration.

catalyst regeneration. Excessively high concentrations of an additional alkene can, however, be detrimental, because cross metathesis of the support-bound diene with the added alkene might compete with ring-closing metathesis. Cross metathesis will become a serious side reaction when 'difficult' ring-formations are to be performed (e.g. synthesis of seven- or eight-membered or larger rings) and when large amounts of catalyst are used.

3.16 Linkers for Alkanes and Arenes

To expand the scope of products available by solid-phase synthesis, a series of strategies have been developed which enable the generation of C–H and C–C bonds upon cleavage from a support, and in this way enable the preparation of unfunctionalized hydrocarbons. These linkers are sometimes also called 'traceless' linkers, because in some types of product the attachment point to the support can no longer be located.

The strategies described to date for the generation of C–H and C–C bonds during cleavage include decarboxylative cleavage, acidolysis of silanes, reductive cleavage of acetals, thioethers, selenides, sulfones, sulfonates, triazenes, sulfonylhydrazones, or organometallic compounds, the nucleophilic cleavage of resin-bound alkylating agents by carbon nucleophiles, and the oxidative cleavage of hydrazides (Figure 3.38).

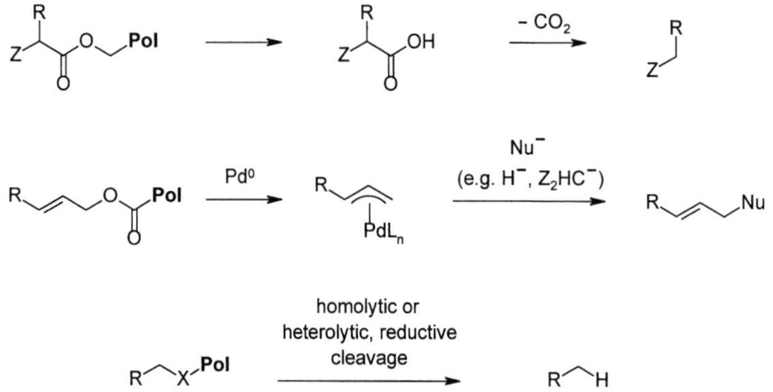

Fig. 3.38. Generation of C–C and C–H bonds upon cleavage from supports. Z: electron-withdrawing group; X: metal, N_2NR, PR_2^+, O, SO_n, Se, etc.

3.16.1 Cleavage followed by Decarboxylation

Some types of acceptor-substituted carboxylic acid readily undergo thermal decarboxylation. If such acids are released from a support under acidic conditions, decarboxylation can ensue spontaneously or upon heating to yield a compound lacking an obvious 'attachment point'. This cleavage strategy has mainly been used for the preparation of ketones and nitriles (Table 3.44).

Table 3.44. Generation of C–H bonds upon decarboxylative cleavage from supports.

Entry	Loaded resin	Cleavage conditions	Product, yield (purity)	Ref.
1		AcOH (neat), 8 h	(95%)	[47]
2		TFA/DCM/Et$_3$SiH 7:2:1, 1 h	80%	[572]
3		TFA/DCM/Et$_3$SiH 7:2:1, 1 h	42%	[572] see also [573]
4		TFA/DCM 1:1, 20 °C, 35 min	71% (71%)	[574]
5		(Me$_3$Si)$_2$NH, TFA/CDCl$_3$ 1:1	17%	[575]
6		NaI (0.17 mol/L), Me$_3$SiCl (0.4 mol/L), dioxane/MeCN 1:1, 75 °C, 72 h	69% (95%)	[576]

3.16.2 Cleavage of Silanes and related Compounds

The C–Si bond of aryl-, vinyl-, and allylsilanes can be cleaved under mild conditions by treatment with acids or fluoride to yield a hydrocarbon and a silyl ester or silyl fluoride. Several linkers of this type have been tested and have proven useful for the preparation of unfunctionalized arenes and alkenes upon cleavage from insoluble supports. Typical loading procedures for these linkers are sketched in Figure 3.39.

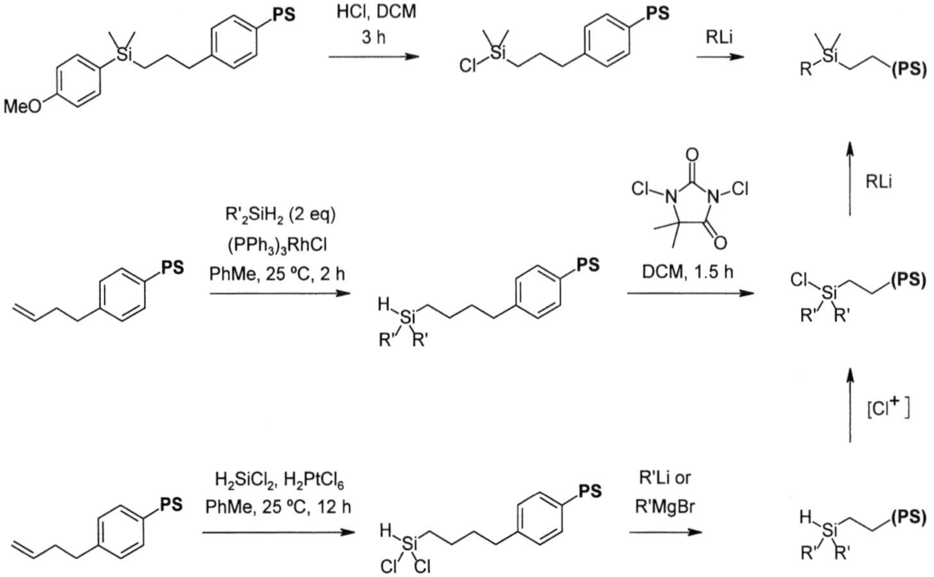

Fig. 3.39. Preparation and loading of silane-based linkers for hydrocarbons [445,577,578].

The optimal conditions for cleavage of resin-bound arylsilanes depend on the substitution pattern of the arene. Some donor-substituted arenes can already be cleaved from silyl linkers by treatment with trifluoroacetic acid. Particularly acid-sensitive are resin-bound 3-(dialkylarylsilyl)propionamides (Entry 8, Table 3.45). Arylsilanes bearing electron-withdrawing groups on the arene are more difficult to desilylate with weak acids (compare, e.g., Entries 1 and 5, Table 3.45). When trifluoroacetic acid fails to promote protodesilylation, hydrogen fluoride, cesium fluoride, or TBAF might bring about the cleavage (Table 3.45).

Support-bound allylsilanes can also be cleaved by treatment with carbon electrophiles. Entry 11 in Table 3.45 is an example for such a cleavage, in which the electrophile is an α-alkoxy carbocation generated from an acetal and TiCl₄.

Arylboronic acids esterified with support-bound 1,2-diols undergo Suzuki reaction with aryl iodides, whereby biaryls are release into solution (Entry 12, Table 3.45). This technique has also been used to prepare β-turn mimetics by simultaneous macrocyclization and cleavage from the support [559].

Table 3.45. Generation of C–H and C–C bonds upon cleavage of support-bound silanes, organogermanium compounds, and boronic esters.

Entry	Loaded resin	Cleavage conditions	Product, yield (purity)	Ref.
1		TFA/DCM 1:1, 25 °C, 3 h	80%	[445]
2		TBAF (1 mol/L), THF, 12 h	58%	[445] see also [577]
3		HF, 12 h (no cleavage by TFA/Me$_2$S/H$_2$O 85:10:5) Ar: 4-(MeO)C$_6$H$_4$	68%	[511]
4		TFA, 60 °C, 24 h Ar: 4-(MeO)C$_6$H$_4$	58%	[511]
5		CsF, DMF/H$_2$O 4:1, 110 °C (no cleavage by neat TFA, 25 °C) Ar: 4-formylphenyl	66%	[579]
6		TFA/DCM 1:1, 5% Me$_2$S, 24 h Ar: 4-(MeO)C$_6$H$_4$	60%	[510]
7		TBAF, DMF, 65 °C, 1 h Ar: 4-(MeO)C$_6$H$_4$		[580]
8		TFA/DCM 1:1, 20 °C, 2 h Ar: 1-naphthyl	100%	[581]
9		1.5% TFA in DCM	0.35–0.52 mmol/g	[582]

Table 3.45. continued.

Entry	Loaded resin	Cleavage conditions	Product, yield (purity)	Ref.
10	Ph, Si, PS	3% TFA in DCM, 18 h	Ph, 0.5 mmol/g	[69]
11	Ph, Si, PS	MeCH(OEt)$_2$, TiCl$_4$ (both 0.07 mol/L), DCM, −78 °C, 22 h	Ph, OEt, 0.5 mmol/g	[69]
12	Ph–B, O, O, (PS)	4-iodoanisol (5 eq), aq K$_3$PO$_4$ (2 mol/L, 3 eq), PdCl$_2$BINAP (0.05 eq), DMF, 60 °C, 24 h	MeO, Ph, 85% (> 95%)	[559]

3.16.3 Reductive Cleavage of Carbon–Oxygen and Carbon–Nitrogen Bonds

The direct homolytic or heterolytic, reductive cleavage of carbon–heteroatom bonds can be used to release products from polymeric supports. C–O Bonds are too strong to undergo homolytic cleavage under acceptably mild reaction conditions, but acetals and allyl esters can smoothly be cleaved heterolytically. When resin-bound acetals are treated with Lewis acids in the presence of a reducing agent or a carbon nucleophile, reductive cleavage from the support can occur. Ethers and sulfonamides have been prepared using this cleavage strategy (Entries **1** and **2**, Table 3.46).

Esters of allylic alcohols with resin-bound carboxylic acids can be converted into palladium allyl complexes, which react with carbon nucleophiles and with hydride sources to yield the formally reduced allyl derivatives (Entries **3** and **4**, Table 3.46). Also arylsulfonates can be reduced by treatment with catalytic amounts of Pd(OAc)$_2$ and formic acid as hydride source (Entry **5**, Table 3.46).

Support-bound triazenes, which can be prepared from resin-bound secondary, aliphatic amines and aromatic diazonium salts [359], undergo cleavage upon treatment with acids. The aromatic diazonium salts are thereby regenerated; in cross-linked polystyrene these decompose to yield, preferentially, radical-derived products. If the acidolysis of polystyrene-bound triazenes is conducted in the presence of hydrogen-atom donors (e.g. THF), the reduced arenes can be obtained (Entries **6** and **7**, Table 3.46). In the presence of alkenes or alkynes and Pd(OAc)$_2$ the initially formed diazonium salts undergo Heck reaction to yield vinylated or alkynylated arenes (Entry **8**, Table 3.46).

Similarly, unsubstituted arenes can be obtained by oxidative cleavage of support-bound *N*-aryl-*N'*-acylhydrazines (Entry **9**, Table 3.46). Oxidation leads to the formation of *N*-aryl-*N'*-acyldiazenes, which in the presence of nucleophiles undergo deacy-

Table 3.46. Formation of C–H and C–C bonds upon reductive cleavage of C–O and C–N bonds.

Entry	Loaded resin	Cleavage conditions	Product, yield (purity)	Ref.
1		TFA (5 eq), Et$_3$SiH (10 eq), DCM, 20 °C, 16–24 h	41%	[414]
2		SnCl$_4$ (1.1 eq), allyltrimethylsilane (2.5 eq), DCM, 20 °C, 16–24 h	47%	[414]
3		THF, triethyl ammonium formate (5 eq), 7% Pd(PPh$_3$)$_4$, 70 °C	69%	[373]
4		dimethyl malonate sodium salt (3 eq), 7% Pd(PPh$_3$)$_4$, THF, 50 °C, 8 h	78%	[373]
5		HCO$_2$H (7.5 eq), NEt$_3$ (8 eq), Pd(OAc)$_2$ (0.2 eq), dppp, 110 °C, 12 h	36–74%	[583]
6		THF/conc HCl 10:1, 50 °C, ultrasound, 5 min	53%	[584]
7		HCl/THF or H$_3$PO$_2$/Cl$_2$HCCO$_2$H	81%	[584]
8		Pd(OAc)$_2$ (5%), TFA, MeOH, 40 °C, 2–12 h	53% (85%)	[585]

Table 3.46. continued.

Entry	Loaded resin	Cleavage conditions	Product, yield (purity)	Ref.
9		Cu(OAc)$_2$ (0.5 eq), pyridine (10 eq), air, MeOH, 20 °C, 2 h	93% (> 90%)	[154]
10		NaBH$_4$ (1.1 mol/L, 8 eq), THF, 67 °C, 8 h	27%	[563]

lation to yield acid derivatives and aryldiazenes. The latter are unstable and decompose into arenes and nitrogen. Air in the presence of catalytic amounts of Cu(OAc)$_2$, or NBS [154,155] can be used as oxidants for hydrazides.

Support-bound sulfonylhydrazones can be reduced to alkanes by sodium borohydride (Entry **10**, Table 3.46). This reaction, which has not yet been fully explored for solid-phase synthesis, enables the conversion of ketones into alkanes under mild reaction conditions.

3.16.4 Reductive Cleavage of Carbon–Phosphorus, Carbon–Sulfur, and Carbon–Selenium Bonds

Phosphonium salts can be dealkylated by treatment with alkoxides to yield alkanes. Although the hydrolytic cleavage of phosphonium salts in solution is well known, the solid-phase variant of this reaction has not yet found broad application. One example, in which traceless linking was based on the alkoxide-induced dealkylation of a resin-bound phosphonium salt, is given in Table 3.47 (Entry **1**).

Hydrocarbons can be generated by nucleophilic cleavage of resin-bound allyl sulfones with carbon nucleophiles (e.g. Entry **3**, Table 3.47), whereby the resin-bound sulfinate acts as the leaving group. Thioethers, sulfoxides, and sulfones can also undergo C–S bond cleavage upon photolysis or upon treatment with reducing agents such as tin hydrides, sodium amalgam, or Raney nickel (Entries **4–6**, Table 3.47). These reducing agents are, unfortunately, non-volatile, and further purification of the crude products will be necessary in most instances, making this cleavage strategy unsuitable for parallel synthesis.

Selenides are more readily cleaved by tin radicals than thioethers. Two examples of the tin radical mediated cleavage of selenides are listed in Table 3.47 (Entries **7** and **8**). Radical-mediated cleavage proceeds under mild, essentially neutral reaction conditions and is well suited for the release of sensitive organic compounds. Purification of the resulting products will generally be required, however.

Table 3.47. Formation of C–H and C–C bonds by reductive cleavage of C–P, C–S, and C–Se bonds.

Entry	Loaded resin	Cleavage conditions	Product, yield (purity)	Ref.
1		NaOMe (0.11 mol/L), MeOH, 65 °C, 4.5 h	 81%	[556]
2		 (0.61 mol/L, 0.9 eq), NaH (0.9 eq), DMSO, 100 °C, 3.5 h	 62%	[586]
3		PhLi, THF, CuI, 0 °C, 4 h	 20%	[587]
4		Bu₃SnH, AIBN, PhH, 80 °C, 18 h; or H₂, Raney Ni, MeOH/EtOH 1.5:1, 20 °C, 3 h	$R\diagdown$ 40% (Bu₃SnH) 94% (H₂)	[588] see also [589]
5		hν (350 nm), MeCN, 5 h Ar: 4-PhC₆H₄	 58%	[503] see also [590]
6		5% Na/Hg, Na₂HPO₄, MeOH, −40 °C to 0 °C, 2 h	 98%	[591]
7		Bu₃SnH (2 eq), AIBN (0.005 eq), PhMe, 110 °C, 6 h	 89%	[562] see also [592]
8		Bu₃SnH (2 eq), AIBN (0.005 eq), PhMe, 110 °C, 8 h		[562]

3.17 Non-Covalent Linkers

3.17.1 Ion-Exchange Resins

Charged organic compounds can be immobilized on ion-exchange resins. Release is achieved by displacement with salts, acids, or bases. The facility with which displacement of ionic products from ion-exchange resins occurs severely limits the choice of reactions to be performed on these supports without premature product release. For this reason, ion-exchange resins are generally only used either as convenient means of purifying charged [593] or neutral [594,595] organic products, or for the immobilization of reagents (e.g. alcoholates [596], thiolates [597], carboxylates [598], phosphorus ylides [599], diazonium arenes [600], or thiocyanate [601]).

3.17.2 Transition Metal Complexes

Kinetically stable complexes can be used as linkers for solid-phase synthesis. The cobalt(III) complex shown in Table 3.48 (Entry **1**) was prepared in solution and then loaded on to polystyrene. Less than 5 % cleavage occurred upon treatment of this support-bound complex with TFA/DCM 1:1 for 12 h or with 20 % piperidine in DMF for 6 h, and it appears, therefore, to be suitable for the solid-phase preparation of small peptides.

Similarly, chromium(0) arene complexes can be used as linkers for arenes (Entry **2**, Table 3.48). Attachment is achieved by photolyzing a chromium(0) arene tricarbonyl complex in the presence of polystyrene-bound triphenylphosphine, whereby one carbonyl ligand is replaced by the phosphine. The linker is stable towards reducing agents (LiAlH$_4$) and acylating agents (acetyl chloride), but the arene can be cleaved selectively from the support either by ligand exchange with pyridine or by photolysis in the presence of air [602].

Table 3.48. Transition metal complexes as linkers.

Entry	Loaded resin	Cleavage conditions	Product, yield (purity)	Ref.
1		DMF, dithiothreitol (0.5 mol/L), DIPEA (0.5 mol/L), 0.5 h	74–97%	[603] [604]
2		pyridine, heat, 2 h	90%	[602] see also [605]

3.17.3 Miscellaneous Non-Covalent Linkers

Tetrabenzo[*a,c,g,i*]fluorene has been used to link synthetic intermediates selectively to charcoal, for the purpose of their purification. In polar solvents the tetrabenzofluorene is strongly adsorbed by charcoal; this enables efficient separation of the intermediate from reagents. After centrifugation and washing, the intermediate is displaced from charcoal into solution by addition of a non-polar solvent, and a new synthetic operation in solution can be conducted (Figure 3.40). Tetrabenzofluorene has also been used for the purification of peptides [606] and oligonucleotides [607].

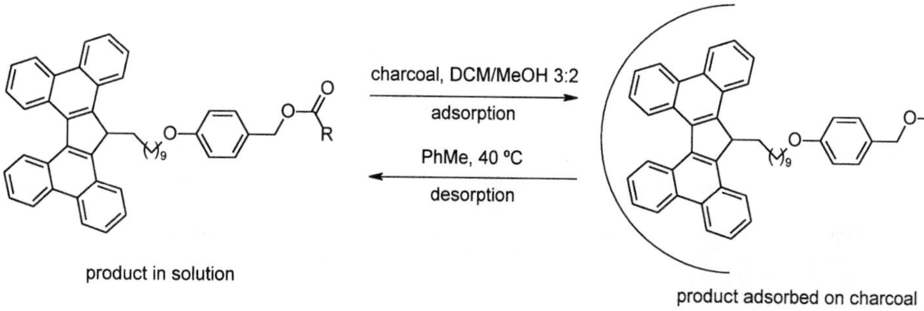

Fig. 3.40. Use of tetrabenzofluorene derivatives for the reversible adsorption of compounds on charcoal [608].

References for Chapter 3

[1] Blackburn, C. *Biopolymers* **1998**, *47*, 311–351.
[2] Sparrow, J. T. *J. Org. Chem.* **1976**, *41*, 1350–1353.
[3] Sarin, V. K.; Kent, S. B. H.; Mitchell, A. R.; Merrifield, R. B. *J. Am. Chem. Soc.* **1984**, *106*, 7845–7850.
[4] Mitchell, A. R.; Kent, S. B. H.; Engelhard, M.; Merrifield, R. B. *J. Org. Chem.* **1978**, *43*, 2845–2852.
[5] Adams, J. H.; Cook, R. M.; Hudson, D.; Jammalamadaka, V.; Lyttle, M. H.; Songster, M. F. *J. Org. Chem.* **1998**, *63*, 3706–3716.
[6] Rich, D. H.; Gurwara, S. K. *J. Am. Chem. Soc.* **1975**, *97*, 1575–1579.
[7] Matsueda, G. R.; Stewart, J. M. *Peptides* **1981**, *2*, 45–50.
[8] Cardno, M.; Bradley, M. *Tetrahedron Lett.* **1996**, *37*, 135–138.
[9] Furka, A.; Sebestyén, F.; Asgedom, M.; Dibó, G. *Int. J. Pept. Prot. Res.* **1991**, *37*, 487–493.
[10] Zhao, P. L.; Zambias, R.; Bolognese, J. A.; Boulton, D.; Chapman, K. *Proc. Natl. Acad. Sci. USA* **1995**, *92*, 10212–10216.
[11] Burgess, K.; Liaw, A. I.; Wang, N. *J. Med. Chem.* **1994**, *37*, 2985–2987.
[12] Merrifield, R. B. *J. Am. Chem. Soc.* **1963**, *85*, 2149–2154.
[13] Meutermans, W. D. F.; Alewood, P. F. *Tetrahedron Lett.* **1995**, *36*, 7709–7712.
[14] Merrifield, R. B.; Vizioli, L. D.; Boman, H. G. *Biochemistry* **1982**, *21*, 5020–5031.
[15] Tam, J. P.; Heath, W. F.; Merrifield, R. B. *J. Am. Chem. Soc.* **1983**, *105*, 6442–6455.
[16] Yajima, H.; Fujii, N.; Ogawa, H.; Kawatani, H. *J. Chem. Soc. Chem. Commun.* **1974**, 107–108.
[17] Yajima, H.; Fujii, N.; Funakoshi, S.; Watanabe, T.; Murayama, E.; Otaka, A. *Tetrahedron* **1988**, *44*, 805–819.
[18] Fujii, N.; Otaka, A.; Ikemura, O.; Hatano, M.; Okamachi, A.; Funakoshi, S.; Sakurai, M.; Shioiri, T.; Yajima, H. *Chem. Pharm. Bull.* **1987**, *35*, 3447–3452.
[19] Nomizu, M.; Inagaki, Y.; Yamashita, T.; Ohkubo, A.; Otaka, A.; Fujii, N.; Roller, P. P.; Yajima, H. *Int. J. Pept. Prot. Res.* **1991**, *37*, 145–152.
[20] Yan, B.; Gstach, H. *Tetrahedron Lett.* **1996**, *37*, 8325–8328.

[21] Blake, J.; Li, C. H. *J. Am. Chem. Soc.* **1968**, *90*, 5882–5884.
[22] Mata, E. G. *Tetrahedron Lett.* **1997**, *38*, 6335–6338.
[23] Mitchell, A. R.; Erickson, B. W.; Ryabtsev, M. N.; Hodges, R. S.; Merrifield, R. B. *J. Am. Chem. Soc.* **1976**, *98*, 7357–7362.
[24] Mitchell, A. R.; Erickson, B. W.; Ryabtsev, M. N.; Hodges, R. S.; Merrifield, R. B. *J. Am. Chem. Soc.* **1976**, *98*, 7357–7362.
[25] Kent, S. B. H.; Merrifield, R. B. *Israel J. Chem.* **1978**, *17*, 243–247.
[26] Wang, S. *J. Am. Chem. Soc.* **1973**, *95*, 1328–1333.
[27] Sarshar, S.; Siev, D.; Mjalli, A. M. M. *Tetrahedron Lett.* **1996**, *37*, 835–838.
[28] Winkler, J. D.; McCoull, W. *Tetrahedron Lett.* **1998**, *39*, 4935–4936.
[29] Mergler, M.; Gosteli, J.; Grogg, P.; Nyfeler, R.; Tanner, R. *Chimia* **1999**, *53*, 29–34.
[30] Tam, J. P.; Kent, S. B. H.; Wong, T. W.; Merrifield, R. B. *Synthesis* **1979**, 955–957.
[31] Sheppard, R. C.; Williams, B. J. *Int. J. Pept. Prot. Res.* **1982**, *20*, 451–454.
[32] Mergler, M.; Tanner, R.; Gosteli, J.; Grogg, P. *Tetrahedron Lett.* **1988**, *29*, 4005–4008.
[33] Albericio, F.; Barany, G. *Tetrahedron Lett.* **1991**, *32*, 1015–1018.
[34] Blankemeyer-Menge, B.; Nimtz, M.; Frank, R. *Tetrahedron Lett.* **1990**, *31*, 1701–1704.
[35] Sieber, P. *Tetrahedron Lett.* **1987**, *28*, 6147–6150.
[36] Albericio, F.; Barany, G. *Int. J. Pept. Prot. Res.* **1998**, *26*, 92–97.
[37] Gisin, B. F. *Helv. Chim. Acta* **1973**, *56*, 1476–1482.
[38] Collini, M. D.; Ellingboe, J. W. *Tetrahedron Lett.* **1997**, *38*, 7963–7966.
[39] Nugiel, D. A.; Wacker, D. A.; Nemeth, G. A. *Tetrahedron Lett.* **1997**, *38*, 5789–5790.
[40] Barlos, K.; Gatos, D.; Kallitsis, J.; Papaioannou, D.; Sotiriu, P.; Schäfer, W. *Liebigs Ann. Chem.* **1987**, 1031–1035.
[41] Barlos, K.; Gatos, D.; Hondrelis, J.; Matsoukas, J.; Moore, G. J.; Schäfer, W.; Sotiriu, P. *Liebigs Ann. Chem.* **1989**, 951–955.
[42] Rink, H. *Tetrahedron Lett.* **1987**, *28*, 3787–3790.
[43] Barlos, K.; Chatzi, O.; Gatos, D.; Stavropoulos, G. *Int. J. Pept. Prot. Res.* **1991**, *37*, 513–520.
[44] Barlos, K.; Gatos, D.; Kallitsis, J.; Papaphotiou, G.; Sotiriu, P.; Wenqing, Y.; Schäfer, W. *Tetrahedron Lett.* **1989**, *30*, 3943–3946.
[45] Yang, L.; Chiu, K. *Tetrahedron Lett.* **1997**, *38*, 7307–7310.
[46] Richter, H.; Jung, G. *Tetrahedron Lett.* **1998**, *39*, 2729–2730.
[47] Garibay, P.; Nielsen, J.; Høeg-Jensen, T. *Tetrahedron Lett.* **1998**, *39*, 2207–2210.
[48] Ede, N. J.; Ang, K. H.; James, I. W.; Bray, A. M. *Tetrahedron Lett.* **1996**, *37*, 9097–9100.
[49] Barlos, K.; Gatos, D.; Kapolos, S.; Poulos, C.; Schäfer, W.; Wenqing, Y. *Int. J. Pept. Prot. Res.* **1991**, *38*, 555–561.
[50] Barlos, K.; Gatos, D.; Kutsogianni, S.; Papaphotiou, G.; Poulos, C.; Tsegenidis, T. *Int. J. Pept. Prot. Res.* **1991**, *38*, 562–568.
[51] Barlos, K.; Gatos, D.; Kapolos, S.; Papaphotiou, G.; Schäfer, W.; Wenqing, Y. *Tetrahedron Lett.* **1989**, *30*, 3947–3950.
[52] Barlos, K.; Gatos, D.; Papaphotiou, G.; Schäfer, W. *Liebigs Ann. Chem.* **1993**, 215–220.
[53] Bollhagen, R.; Schmiedberger, M.; Barlos, K.; Grell, E. *J. Chem. Soc. Chem. Commun.* **1994**, 2559–2560.
[54] Zikos, C. C.; Ferderigos, N. G. *Tetrahedron Lett.* **1994**, *35*, 1767–1768.
[55] Novabiochem Catalog and Peptide Synthesis Handbook, Läufelfingen, **1999**.
[56] Henkel, B.; Bayer, E. *Tetrahedron Lett.* **1998**, *39*, 9401–9402.
[57] Gordeev, M. F. *Biotechnology and Bioengineering* **1998**, *61*, 13–16.
[58] Xiao, X. Y.; Parandoosh, Z.; Nova, M. P. *J. Org. Chem.* **1997**, *62*, 6029–6033.
[59] Wieland, T.; Birr, C.; Fleckenstein, P. *Liebigs Ann. Chem.* **1972**, *756*, 14–19.
[60] Rosenthal, K.; Erlandsson, M.; Undén, A. *Tetrahedron Lett.* **1999**, *40*, 377–380.
[61] Wang, S.; Merrifield, R. B. *J. Am. Chem. Soc.* **1969**, *91*, 6488–6491.
[62] Wang, S. *J. Org. Chem.* **1975**, *40*, 1235–1239.
[63] Léger, R.; Yen, R.; She, M. W.; Lee, V. J.; Hecker, S. J. *Tetrahedron Lett.* **1998**, *39*, 4171–4174.
[64] Wolters, E. T. M.; Tesser, G. I.; Nivard, R. J. F. *J. Org. Chem.* **1974**, *39*, 3388–3392.
[65] Hernández, A. S.; Hodges, J. C. *J. Org. Chem.* **1997**, *62*, 3153–3157.
[66] Akaji, K.; Kiso, Y.; Carpino, L. A. *J. Chem. Soc. Chem. Commun.* **1990**, 584–586.
[67] Wilson, M. W.; Hernández, A. S.; Calvet, A. P.; Hodges, J. C. *Mol. Diversity* **1998**, *3*, 95–112.
[68] Blackburn, C.; Pingali, A.; Kehoe, T.; Herman, L. W.; Wang, H. Q.; Kates, S. A. *Bioorg. Med. Chem. Lett.* **1997**, *7*, 823–826.
[69] Schuster, M.; Lucas, N.; Blechert, S. *Chem. Commun.* **1997**, 823–824.
[70] Ueki, M.; Kai, K.; Amemiya, M.; Horino, H.; Oyamada, H. *J. Chem. Soc. Chem. Commun.* **1988**, 414–415.

[71] Chao, H. G.; Bernatowicz, M. S.; Reiss, P. D.; Klimas, C. E.; Matsueda, G. R. *J. Am. Chem. Soc.* **1994**, *116*, 1746–1752.

[72] Ramage, R.; Barron, C. A.; Bielecki, S.; Holden, R.; Thomas, D. W. *Tetrahedron* **1992**, *48*, 499–514.

[73] Gennari, C.; Ceccarelli, S.; Piarulli, U.; Aboutayab, K.; Donghi, M.; Paterson, I. *Tetrahedron* **1998**, *54*, 14999–15016.

[74] Yedidia, V.; Leznoff, C. C. *Can. J. Chem.* **1980**, *58*, 1144–1150.

[75] ApSimon, J. W.; Dixit, D. M. *Can. J. Chem.* **1982**, *60*, 368–370.

[76] Anwer, M. K.; Spatola, A. F. *Tetrahedron Lett.* **1992**, *33*, 3121–3124.

[77] Seebach, D.; Thaler, A.; Blaser, D.; Ko, S. Y. *Helv. Chim. Acta* **1991**, *74*, 1102–1118.

[78] Whitney, D. B.; Tam, J. P.; Merrifield, R. B. *Tetrahedron* **1984**, *40*, 4237–4244.

[79] Kiso, Y.; Kimura, T.; Fujiwara, Y.; Shimokura, M.; Nishitani, A. *Chem. Pharm. Bull.* **1988**, *36*, 5024–5027.

[80] Stahl, G. L.; Walter, R.; Smith, C. W. *J. Am. Chem. Soc.* **1979**, *101*, 5383–5394.

[81] Nicolás, E.; Clemente, J.; Ferrer, T.; Albericio, F.; Giralt, E. *Tetrahedron* **1997**, *53*, 3179–3194.

[82] Routledge, A.; Stock, H. T.; Flitsch, S. L.; Turner, N. J. *Tetrahedron Lett.* **1997**, *38*, 8287–8290.

[83] Ramage, R.; Barron, C. A.; Bielecki, S.; Thomas, D. W. *Tetrahedron Lett.* **1987**, *28*, 4105–4108.

[84] Mullen, D. G.; Barany, G. *J. Org. Chem.* **1988**, *53*, 5240–5248.

[85] Aldrian-Herrada, G.; Rabié, A.; Wintersteiger, R.; Brugidou, J. *J. Pept. Sci.* **1998**, *4*, 266–281.

[86] Mizoguchi, T.; Shigezane, K.; Takamura, N. *Chem. Pharm. Bull.* **1970**, *18*, 1465–1474.

[87] Baleux, F.; Calas, B.; Méry, J. *Int. J. Pept. Prot. Res.* **1986**, *28*, 22–28.

[88] Baleux, F.; Daunis, J.; Jacquier, R.; Calas, B. *Tetrahedron Lett.* **1984**, *25*, 5893–5896.

[89] Fancelli, D.; Fagnola, M. C.; Severino, D.; Bedeschi, A. *Tetrahedron Lett.* **1997**, *38*, 2311–2314.

[90] Osborn, N. J.; Robinson, J. A. *Tetrahedron* **1993**, *49*, 2873–2884.

[91] Hoffmann, S.; Frank, R. *Tetrahedron Lett.* **1994**, *35*, 7763–7766.

[92] Panke, G.; Frank, R. *Tetrahedron Lett.* **1998**, *39*, 17–18.

[93] Rabanal, F.; Giralt, E.; Albericio, F. *Tetrahedron* **1995**, *51*, 1449–1458.

[94] Rabanal, F.; Giralt, E.; Albericio, F. *Tetrahedron Lett.* **1992**, *33*, 1775–1778.

[95] Mutter, M.; Bellof, D. *Helv. Chim. Acta* **1984**, *67*, 2009–2016.

[96] de la Torre, B. G.; Aviñó, A.; Tarrason, G.; Piulats, J.; Albericio, F.; Eritja, R. *Tetrahedron Lett.* **1994**, *35*, 2733–2736.

[97] Katti, S. B.; Misra, P. K.; Haq, W.; Mathur, K. B. *J. Chem. Soc. Chem. Commun.* **1992**, 843–844.

[98] Tesser, G. I.; Buis, J. T. W. A. R. M.; Wolters, E. T. M.; Bothé-Helmes, E. G. A. M. *Tetrahedron* **1976**, *32*, 1069–1072.

[99] Flynn, D. L.; Zelle, R. E.; Grieco, P. A. *J. Org. Chem.* **1983**, *48*, 2424–2426.

[100] Kenner, G. W.; McDermott, J. R.; Sheppard, R. C. *J. Chem. Soc. Chem. Commun.* **1971**, 636–637.

[101] Pátek, M.; Lebl, M. *Biopolymers* **1998**, *47*, 353–363.

[102] Backes, B. J.; Virgilio, A. A.; Ellman, J. A. *J. Am. Chem. Soc.* **1996**, *118*, 3055–3056.

[103] Hulme, C.; Peng, J.; Morton, G.; Salvino, J. M.; Herpin, T.; Labaudiniere, R. *Tetrahedron Lett.* **1998**, *39*, 7227–7230.

[104] Allin, S. M.; Shuttleworth, S. J. *Tetrahedron Lett.* **1996**, *37*, 8023–8026.

[105] Purandare, A. V.; Natarajan, S. *Tetrahedron Lett.* **1997**, *38*, 8777–8780.

[106] Sola, R.; Méry, J.; Pascal, R. *Tetrahedron Lett.* **1996**, *37*, 9195–9198.

[107] Phoon, C. W.; Abell, C. *Tetrahedron Lett.* **1998**, *39*, 2655–2658.

[108] Pascal, R.; Sola, R. *Tetrahedron Lett.* **1998**, *39*, 5031–5034.

[109] Sola, R.; Saguer, P.; David, M. L.; Pascal, R. *J. Chem. Soc. Chem. Commun.* **1993**, 1786–1788.

[110] Backes, B. J.; Ellman, J. A. *J. Am. Chem. Soc.* **1994**, *116*, 11171–11172.

[111] Kobayashi, S.; Hachiya, I.; Yasuda, M. *Tetrahedron Lett.* **1996**, *37*, 5569–5572.

[112] Birr, C.; Lochinger, W.; Stahnke, G.; Lang, P. *Liebigs Ann. Chem.* **1972**, *763*, 162–172.

[113] Chamberlin, J. W. *J. Org. Chem.* **1966**, *31*, 1658–1660.

[114] Barltrop, J. A.; Plant, P. J.; Schofield, P. *J. Chem. Soc. Chem. Commun.* **1966**, 822–823.

[115] Pillai, V. N. R. *Synthesis* **1980**, 1–26.

[116] Ramesh, D.; Wieboldt, R.; Billington, A. P.; Carpenter, B. K.; Hess, G. P. *J. Org. Chem.* **1993**, *58*, 4599–4605.

[117] Amit, B.; Zehavi, U.; Patchornik, A. *J. Org. Chem.* **1974**, *39*, 192–196.

[118] Patchornik, A.; Amit, B.; Woodward, R. B. *J. Am. Chem. Soc.* **1970**, *92*, 6333–6335.

[119] Pirrung, M. C.; Shuey, S. W. *J. Org. Chem.* **1994**, *59*, 3890–3897.

[120] Corrie, J. E. T.; Trentham, D. R. *J. Chem. Soc. Perkin Trans. 1* **1992**, 2409–2417.

[121] Givens, R. S.; Athey, P. S.; Kueper, L. W.; Matuszewski, B.; Xue, J. Y. *J. Am. Chem. Soc.* **1992**, *114*, 8708–8710.

[122] Sheehan, J. C.; Wilson, R. M.; Oxford, A. W. *J. Am. Chem. Soc.* **1971**, *93*, 7222–7228.

[123] Sheehan, J. C.; Umezawa, K. *J. Org. Chem.* **1973**, *38*, 3771–3774.
[124] Baldwin, J. J.; Burbaum, J. J.; Henderson, I.; Ohlmeyer, M. H. J. *J. Am. Chem. Soc.* **1995**, *117*, 5588–5589.
[125] Lloyd-Williams, P.; Gairí, M.; Albericio, F.; Giralt, E. *Tetrahedron* **1991**, *47*, 9867–9880.
[126] Venkatesan, H.; Greenberg, M. M. *J. Org. Chem.* **1996**, *61*, 525–529.
[127] Greenberg, M. M.; Gilmore, J. L. *J. Org. Chem.* **1994**, *59*, 746–753.
[128] Holmes, C. P.; Jones, D. G. *J. Org. Chem.* **1995**, *60*, 2318–2319.
[129] Teague, S. J. *Tetrahedron Lett.* **1996**, *37*, 5751–5754.
[130] Åkerblom, E. B.; Nygren, A. S.; Agback, K. H. *Mol. Diversity* **1998**, *3*, 137–148.
[131] Yoo, D. J.; Greenberg, M. M. *J. Org. Chem.* **1995**, *60*, 3358–3364.
[132] Holmes, C. P.; Jones, D. G.; Frederick, B. T.; Dong, L. C. *Peptides:Chemistry, Structure and Biology; Proceedings of the 14th American Peptide Symposium,* **1995**; Mayflower Scientific Ltd. **1996**, 44–45.
[133] Lloyd-Williams, P.; Gairí, M.; Albericio, F.; Giralt, E. *Tetrahedron* **1993**, *49*, 10069–10078.
[134] Ajayaghosh, A.; Pillai, V. N. R. *Tetrahedron* **1988**, *44*, 6661–6666.
[135] Rich, D. H.; Gurwara, S. K. *J. Chem. Soc. Chem. Commun.* **1973**, 610–611.
[136] Ajayaghosh, A.; Pillai, V. N. R. *J. Org. Chem.* **1987**, *52*, 5714–5717.
[137] Whitehouse, D. L.; Savinov, S. N.; Austin, D. J. *Tetrahedron Lett.* **1997**, *38*, 7851–7852.
[138] Wang, S. *J. Org. Chem.* **1976**, *41*, 3258–3261.
[139] Routledge, A.; Abell, C.; Balasubramanian, S. *Tetrahedron Lett.* **1997**, *38*, 1227–1230.
[140] Lee, H. B.; Balasubramanian, S. *J. Org. Chem.* **1999**, *64*, 3454–3460.
[141] Peukert, S.; Giese, B. *J. Org. Chem.* **1998**, *63*, 9045–9051.
[142] Schlatter, J. M.; Mazur, R. H. *Tetrahedron Lett.* **1977**, 2851–2852.
[143] Pande, C. S.; Gupta, N.; Bhardwaj, A. *J. Appl. Pol. Sci.* **1995**, *56*, 1127–1131.
[144] Colombo, R. *J. Chem. Soc. Chem. Commun.* **1981**, 1012–1013.
[145] Seitz, O.; Kunz, H. *Angew. Chem. Int. Ed. Engl.* **1995**, *34*, 803–805.
[146] Zhang, X. H.; Jones, R. A. *Tetrahedron Lett.* **1996**, *37*, 3789–3790.
[147] Seitz, O.; Kunz, H. *J. Org. Chem.* **1997**, *62*, 813–826.
[148] Kunz, H.; Dombo, B. *Angew. Chem. Int. Ed. Engl.* **1988**, *27*, 711–713.
[149] Guibé, F.; Dangles, O.; Balavoine, G.; Loffet, A. *Tetrahedron Lett.* **1989**, *30*, 2641–2644.
[150] Lloyd-Williams, P.; Jou, G.; Albericio, F.; Giralt, E. *Tetrahedron Lett.* **1991**, *32*, 4207–4210.
[151] Seitz, O. *Tetrahedron Lett.* **1999**, *40*, 4161–4164.
[152] Semenov, A. N.; Gordeev, K. Y. *Int. J. Pept. Prot. Res.* **1995**, *45*, 303–304.
[153] Wieland, T.; Lewalter, J.; Birr, C. *Liebigs Ann. Chem.* **1970**, *740*, 31–47.
[154] Stieber, F.; Grether, U.; Waldmann, H. *Angew. Chem. Int. Ed. Engl.* **1999**, *38*, 1073–1077.
[155] Millington, C. R.; Quarrell, R.; Lowe, G. *Tetrahedron Lett.* **1998**, *39*, 7201–7204.
[156] Singh, J.; Allen, M. P.; Ator, M. A.; Gainor, J. A.; Whipple, D. A. *J. Med. Chem.* **1995**, *38*, 217–219.
[157] Nanda, S.; Rao, A. B.; Yadav, J. S. *Tetrahedron Lett.* **1999**, *40*, 5905–5908.
[158] Quarrell, R.; Claridge, T. D. W.; Weaver, G. W.; Lowe, G. *Mol. Diversity* **1996**, *1*, 223–232.
[159] Sauerbrei, B.; Jungmann, V.; Waldmann, H. *Angew. Chem. Int. Ed. Engl.* **1998**, *37*, 1143–1146.
[160] Böhm, G.; Dowden, J.; Rice, D. C.; Burgess, I.; Pilard, J. F.; Guilbert, B.; Haxton, A.; Hunter, R. C.; Turner, N. J.; Flitsch, S. L. *Tetrahedron Lett.* **1998**, *39*, 3819–3822.
[161] Leon, S.; Quarrell, R.; Lowe, G. *Bioorg. Med. Chem. Lett.* **1998**, *8*, 2997–3002.
[162] Vagner, J.; Barany, G.; Lam, K. S.; Krchnák, V.; Sepetov, N. F.; Ostrem, J. A.; Strop, P.; Lebl, M. *Proc. Natl. Acad. Sci. USA* **1996**, *93*, 8194–8199.
[163] Hilaire, P. M. S.; Lowary, T. L.; Meldal, M.; Bock, K. *J. Am. Chem. Soc.* **1998**, *120*, 13312–13320.
[164] Renil, M.; Ferreras, M.; Delaisse, J. M.; Foged, N. T.; Meldal, M. *J. Pept. Sci.* **1998**, *4*, 195–210.
[165] Meldal, M.; Svendsen, I.; Breddam, K.; Auzanneau, F. I. *Proc. Natl. Acad. Sci. USA* **1994**, *91*, 3314–3318.
[166] Meldal, M.; Svendsen, I. *J. Chem. Soc. Perkin Trans. 1* **1995**, 1591–1596.
[167] Schuster, M.; Wang, P.; Paulson, J. C.; Wong, C. H. *J. Am. Chem. Soc.* **1994**, *116*, 1135–1136.
[168] Yamada, K.; Nishimura, S. I. *Tetrahedron Lett.* **1995**, *36*, 9493–9496.
[169] Elmore, D. T.; Guthrie, D. J. S.; Wallace, A. D.; Bates, S. R. E. *J. Chem. Soc. Chem. Commun.* **1992**, 1033–1034.
[170] Blixt, O.; Norberg, T. *J. Org. Chem.* **1998**, *63*, 2705–2710.
[171] Schnölzer, M.; Kent, S. B. H. *Science* **1992**, *256*, 221–225.
[172] Dawson, P. E.; Kent, S. B. H. *J. Am. Chem. Soc.* **1993**, *115*, 7263–7266.
[173] Yamashiro, D.; Li, C. H. *Int. J. Pept. Prot. Res.* **1988**, *31*, 322–334.
[174] Blake, J. *Int. J. Pept. Prot. Res.* **1981**, *17*, 273–274.
[175] Canne, L. E.; Walker, S. M.; Kent, S. B. H. *Tetrahedron Lett.* **1995**, *36*, 1217–1220.
[176] Fivush, A. M.; Willson, T. M. *Tetrahedron Lett.* **1997**, *38*, 7151–7154.

[177] Bui, C. T.; Rasoul, F. A.; Ercole, F.; Pham, Y.; Maeji, N. J. *Tetrahedron Lett.* **1998**, *39*, 9279–9282.
[178] Gordeev, M. F.; Luehr, G. W.; Hui, H. C.; Gordon, E. M.; Patel, D. V. *Tetrahedron* **1998**, *54*, 15879–15890.
[179] Boojamra, C. G.; Burow, K. M.; Thompson, L. A.; Ellman, J. A. *J. Org. Chem.* **1997**, *62*, 1240–1256.
[180] Swayze, E. E. *Tetrahedron Lett.* **1997**, *38*, 8465–8468.
[181] Penke, B.; Rivier, J. *J. Org. Chem.* **1987**, *52*, 1197–1200.
[182] Bourne, G. T.; Meutermans, W. D. F.; Alewood, P. F.; McGeary, R. P.; Scanlon, M.; Watson, A. A.; Smythe, M. L. *J. Org. Chem.* **1999**, *64*, 3095–3101.
[183] Raju, B.; Kogan, T. P. *Tetrahedron Lett.* **1997**, *38*, 4965–4968.
[184] Swayze, E. E. *Tetrahedron Lett.* **1997**, *38*, 8643–8646.
[185] Bui, C. T.; Bray, A. M.; Ercole, F.; Pham, Y.; Rasoul, F. A.; Maeji, N. J. *Tetrahedron Lett.* **1999**, *40*, 3471–3474.
[186] Albericio, F.; Barany, G. *Int. J. Pept. Prot. Res.* **1987**, *30*, 206–216.
[187] Sarantakis, D.; Bicksler, J. J. *Tetrahedron Lett.* **1997**, *38*, 7325–7328.
[188] Yu, K. L.; Civiello, R.; Roberts, D. G. M.; Seiler, S. M.; Meanwell, N. A. *Bioorg. Med. Chem. Lett.* **1999**, *9*, 663–666.
[189] Ngu, K.; Patel, D. V. *Tetrahedron Lett.* **1997**, *38*, 973–976.
[190] Albericio, F.; Kneib-Cordonier, N.; Biancalana, S.; Gera, L.; Masada, R. I.; Hudson, D.; Barany, G. *J. Org. Chem.* **1990**, *55*, 3730–3743.
[191] Brummond, K. M.; Lu, J. *J. Org. Chem.* **1999**, *64*, 1723–1726.
[192] Chao, H. G.; Bernatowicz, M. S.; Matsueda, G. R. *J. Org. Chem.* **1993**, *58*, 2640–2644.
[193] Estep, K. G.; Neipp, C. E.; Stramiello, L. M. S.; Adam, M. D.; Allen, M. P.; Robinson, S.; Roskamp, E. J. *J. Org. Chem.* **1998**, *63*, 5300–5301.
[194] Hammer, R. P.; Albericio, F.; Gera, L.; Barany, G. *Int. J. Pept. Prot. Res.* **1990**, *36*, 31–45.
[195] Miller, M. W.; Vice, S. F.; McCombie, S. W. *Tetrahedron Lett.* **1998**, *39*, 3429–3432.
[196] Bräse, S.; Köbberling, J.; Enders, D.; Lazny, R.; Wang, M.; Brandtner, S. *Tetrahedron Lett.* **1999**, *40*, 2105–2108.
[197] Gennari, C.; Longari, C.; Ressel, S.; Salom, B.; Piarulli, U.; Ceccarelli, S.; Mielgo, A. *Eur. J. Org. Chem.* **1998**, 2437–2449.
[198] Sternson, S. M.; Schreiber, S. L. *Tetrahedron Lett.* **1998**, *39*, 7451–7454.
[199] Ajayaghosh, A.; Pillai, V. N. R. *J. Org. Chem.* **1990**, *55*, 2826–2829.
[200] Renil, M.; Pillai, V. N. R. *Tetrahedron Lett.* **1994**, *35*, 3809–3812.
[201] Kumar, K. S.; Pillai, V. N. R. *Tetrahedron* **1999**, *55*, 10437–10446.
[202] Chan, W. C.; Mellor, S. L. *J. Chem. Soc. Chem. Commun.* **1995**, 1475–1477.
[203] Heinonen, P.; Lönnberg, H. *Tetrahedron Lett.* **1997**, *38*, 8569–8572.
[204] Dörner, B.; Husar, G. M.; Ostresh, J. M.; Houghten, R. A. *Bioorg. Med. Chem.* **1996**, *4*, 709–715.
[205] Nefzi, A.; Ostresh, J. M.; Meyer, J. P.; Houghten, R. A. *Tetrahedron Lett.* **1997**, *38*, 931–934.
[206] Wang, S. S.; Wang, B. S. H.; Hughes, J. L.; Leopold, E. J.; Wu, C. R.; Tam, J. P. *Int. J. Pept. Prot. Res.* **1992**, *40*, 344–349.
[207] Brown, D. S.; Revill, J. M.; Shute, R. E. *Tetrahedron Lett.* **1998**, *39*, 8533–8536.
[208] Noda, M.; Yamaguchi, M.; Ando, E.; Takeda, K.; Nokihara, K. *J. Org. Chem.* **1994**, *59*, 7968–7975.
[209] Patterson, J. A.; Ramage, R. *Tetrahedron Lett.* **1999**, *40*, 6121–6124.
[210] Breipohl, G.; Knolle, J.; Stüber, W. *Tetrahedron Lett.* **1987**, *28*, 5651–5654.
[211] Stüber, W.; Knolle, J.; Breipohl, G. *Int. J. Pept. Prot. Res.* **1989**, *34*, 215–221.
[212] Funakoshi, S.; Murayama, E.; Guo, L.; Fujii, N.; Yajima, H. *J. Chem. Soc. Chem. Commun.* **1988**, 382–384.
[213] Han, Y. X.; Bontems, S. L.; Hegyes, P.; Munson, M. C.; Minor, C. A.; Kates, S. A.; Albericio, F.; Barany, G. *J. Org. Chem.* **1996**, *61*, 6326–6339.
[214] Breipohl, G.; Knolle, J.; Stüber, W. *Int. J. Pept. Prot. Res.* **1989**, *34*, 262–267.
[215] Paikoff, S. J.; Wilson, T. E.; Cho, C. Y.; Schultz, P. G. *Tetrahedron Lett.* **1996**, *37*, 5653–5656.
[216] Kim, J. M.; Bi, Y. Z.; Paikoff, S. J.; Schultz, P. G. *Tetrahedron Lett.* **1996**, *37*, 5305–5308.
[217] Kim, S. W.; Bauer, S. M.; Armstrong, R. W. *Tetrahedron Lett.* **1998**, *39*, 6993–6996.
[218] Tempest, P. A.; Brown, S. D.; Armstrong, R. W. *Angew. Chem. Int. Ed. Engl.* **1996**, *35*, 640–642.
[219] Brown, E. G.; Nuss, J. M. *Tetrahedron Lett.* **1997**, *38*, 8457–8460.
[220] Garigipati, R. S. *Tetrahedron Lett.* **1997**, *38*, 6807–6810.
[221] Beaver, K. A.; Siegmund, A. C.; Spear, K. L. *Tetrahedron Lett.* **1996**, *37*, 1145–1148.
[222] Kimura, T.; Fukui, T.; Tanaka, S.; Akaji, K.; Kiso, Y. *Chem. Pharm. Bull.* **1997**, *45*, 18–26.
[223] Pátek, M.; Lebl, M. *Tetrahedron Lett.* **1991**, *32*, 3891–3894.
[224] Tomkinson, N. C. O.; Sefler, A. M.; Plunket, K. D.; Blanchard, S. G.; Parks, D. J.; Willson, T. M. *Bioorg. Med. Chem. Lett.* **1997**, *7*, 2491–2496.

[225] Mergler, M.; Nyfeler, R. In *Solid Phase Synthesis*; Epton, R. Ed.; Intercept: Andover, **1992**; pp 429–432.
[226] Mohanraj, S.; Ford, W. T. *J. Org. Chem.* **1985**, *50*, 1616–1620.
[227] Li, X.; Kawakami, T.; Aimoto, S. *Tetrahedron Lett.* **1998**, *39*, 8669–8672.
[228] Barn, D. R.; Morphy, J. R.; Rees, D. C. *Tetrahedron Lett.* **1996**, *37*, 3213–3216.
[229] Yang, L. H.; Guo, L. Q. *Tetrahedron Lett.* **1996**, *37*, 5041–5044.
[230] Baird, E. E.; Dervan, P. B. *J. Am. Chem. Soc.* **1996**, *118*, 6141–6146.
[231] Lelièvre, D.; Chabane, H.; Delmas, A. *Tetrahedron Lett.* **1998**, *39*, 9675–9678.
[232] Fathi, R.; Patel, R.; Cook, A. F. *Mol. Diversity* **1997**, *2*, 125–134.
[233] Bray, A. M.; Jhingran, A. G.; Valerio, R. M.; Maeji, N. J. *J. Org. Chem.* **1994**, *59*, 2197–2203.
[234] Rölfing, K.; Thiel, M.; Künzer, H. *Synlett* **1996**, 1036–1038.
[235] Prien, O.; Rölfing, K.; Thiel, M.; Künzer, H. *Synlett* **1997**, 325–326.
[236] Kaljuste, K.; Tam, J. P. *Tetrahedron Lett.* **1998**, *39*, 9327–9330.
[237] Vlattas, I.; Dellureficio, J.; Dunn, R.; Sytwu, I. I.; Stanton, J. *Tetrahedron Lett.* **1997**, *38*, 7321–7324.
[238] Camarero, J. A.; Cotton, G. J.; Adeva, A.; Muir, T. W. *J. Pept. Res.* **1998**, *51*, 303–316.
[239] Flanigan, E.; Marshall, G. R. *Tetrahedron Lett.* **1970**, 2403–2406.
[240] Fridkin, M.; Patchornik, A.; Katchalski, E. *J. Am. Chem. Soc.* **1966**, *88*, 3164–3165.
[241] Kalir, R.; Fridkin, M.; Patchornik, A. *Eur. J. Biochem.* **1974**, *42*, 151–156.
[242] Fridkin, M.; Patchornik, A.; Katchalski, E. *J. Am. Chem. Soc.* **1965**, *87*, 4646–4648.
[243] Fridkin, M.; Hazum, E.; Kalir, R.; Rotman, M.; Koch, Y. *J. Solid-Phase Biochem.* **1977**, *2*, 175–182.
[244] Breitenbucher, J. G.; Johnson, C. R.; Haight, M.; Phelan, J. C. *Tetrahedron Lett.* **1998**, *39*, 1295–1298.
[245] Fantauzzi, P. P.; Yager, K. M. *Tetrahedron Lett.* **1998**, *39*, 1291–1294.
[246] Marshall, D. L.; Liener, I. E. *J. Org. Chem.* **1970**, *35*, 867–868.
[247] Parlow, J. J.; Normansell, J. E. *Mol. Diversity* **1996**, *1*, 266–269.
[248] Cohen, B. J.; Karoly-Hafeli, H.; Patchornik, A. *J. Org. Chem.* **1984**, *49*, 922–924.
[249] Hahn, H. G.; Chang, K. H.; Nam, K. D.; Bae, S. Y.; Mah, H. *Heterocycles* **1998**, *48*, 2253–2261.
[250] Scarr, R. B.; Findeis, M. A. *Pept. Res.* **1990**, *3*, 238–241.
[251] Sasaki, T.; Findeis, M. A.; Kaiser, E. T. *J. Org. Chem.* **1991**, *56*, 3159–3168.
[252] DeGrado, W. F.; Kaiser, E. T. *J. Org. Chem.* **1980**, *45*, 1295–1300.
[253] DeGrado, W. F.; Kaiser, E. T. *J. Org. Chem.* **1982**, *47*, 3258–3261.
[254] Voyer, N.; Lavoie, A.; Pinette, M.; Bernier, J. *Tetrahedron Lett.* **1994**, *35*, 355–358.
[255] Smith, R. A.; Bobko, M. A.; Lee, W. *Bioorg. Med. Chem. Lett.* **1998**, *8*, 2369–2374.
[256] Ösapay, G.; Taylor, J. W. *J. Am. Chem. Soc.* **1990**, *112*, 6046–6051.
[257] Ösapay, G.; Profit, A.; Taylor, J. W. *Tetrahedron Lett.* **1990**, *31*, 6121–6124.
[258] Nishino, N.; Xu, M.; Mihara, H.; Fujimoto, T.; Ohba, M.; Ueno, Y.; Kumagai, H. *J. Chem. Soc. Chem. Commun.* **1992**, 180–181.
[259] Hodge, P.; Peng, P. P. *Polymer* **1999**, *40*, 1871–1879.
[260] Kalir, R.; Warshawsky, A.; Fridkin, M.; Patchornik, A. *Eur. J. Biochem.* **1975**, *59*, 55–61.
[261] Cohen, B. J.; Kraus, M. A.; Patchornik, A. *J. Am. Chem. Soc.* **1981**, *103*, 7620–7629.
[262] Schiemann, K.; Showalter, H. D. H. *J. Org. Chem.* **1999**, *64*, 4972–4975.
[263] Mokotoff, M.; Zhao, M.; Roth, S. M.; Shelley, J. A.; Slavosky, J. N.; Kouttab, N. M. *J. Med. Chem.* **1990**, *33*, 354–360.
[264] Mokotoff, M.; Patchornik, A. *Int. J. Pept. Prot. Res.* **1983**, *21*, 145–154.
[265] Adamczyk, M.; Fishpaugh, J. R.; Mattingly, P. G. *Bioorg. Med. Chem. Lett.* **1999**, *9*, 217–220.
[266] Pop, I. E.; Déprez, B. P.; Tartar, A. L. *J. Org. Chem.* **1997**, *62*, 2594–2603.
[267] Mohan, R.; Chou, Y. L.; Morrissey, M. M. *Tetrahedron Lett.* **1996**, *37*, 3963–3966.
[268] Kumari, K. A.; Sreekumar, K. *Polymer* **1996**, *37*, 171–176.
[269] Sophiamma, P. N.; Sreekumar, K. *Indian J. Chem. Sect. B (Org. Chem.)* **1997**, *36*, 995–999.
[270] Adamczyk, M.; Fishpaugh, J. R.; Mattingly, P. G. *Tetrahedron Lett.* **1999**, *40*, 463–466.
[271] Katoh, M.; Sodeoka, M. *Bioorg. Med. Chem. Lett.* **1999**, *9*, 881–884.
[272] Dendrinos, K. G.; Kalivretenos, A. G. *Tetrahedron Lett.* **1998**, *39*, 1321–1324.
[273] Link, A.; van Calenbergh, S.; Herdewijn, P. *Tetrahedron Lett.* **1998**, *39*, 5175–5176.
[274] Backes, B. J.; Ellman, J. A. *J. Org. Chem.* **1999**, *64*, 2322–2330.
[275] Scialdone, M. A.; Shuey, S. W.; Soper, P.; Hamuro, Y.; Burns, D. M. *J. Org. Chem.* **1998**, *63*, 4802–4807.
[276] Hamuro, Y.; Marshall, W. J.; Scialdone, M. A. *J. Comb. Chem.* **1999**, *1*, 163–172.
[277] Scialdone, M. A. *Tetrahedron Lett.* **1996**, *37*, 8141–8144.
[278] Dressman, B. A.; Singh, U.; Kaldor, S. W. *Tetrahedron Lett.* **1998**, *39*, 3631–3634.

[279] Fitzpatrick, L. J.; Rivero, R. A. *Tetrahedron Lett.* **1997**, *38*, 7479–7482.
[280] Raju, B.; Kogan, T. P. *Tetrahedron Lett.* **1997**, *38*, 3373–3376.
[281] Floyd, C. D.; Lewis, C. N.; Patel, S. R.; Whittaker, M. *Tetrahedron Lett.* **1996**, *37*, 8045–8048.
[282] Richter, L. S.; Desai, M. C. *Tetrahedron Lett.* **1997**, *38*, 321–322.
[283] Salvino, J. M.; Mervic, M.; Mason, H. J.; Kiesow, T.; Teager, D.; Airey, J.; Labaudiniere, R. *J. Org. Chem.* **1999**, *64*, 1823–1830.
[284] Mellor, S. L.; Chan, W. C. *Chem. Commun.* **1997**, 2005–2006.
[285] Mellor, S. L.; McGuire, C.; Chan, W. C. *Tetrahedron Lett.* **1997**, *38*, 3311–3314.
[286] Bauer, U.; Ho, W. B.; Koskinen, A. M. P. *Tetrahedron Lett.* **1997**, *38*, 7233–7236.
[287] Khan, S. I.; Grinstaff, M. W. *Tetrahedron Lett.* **1998**, *39*, 8031–8034.
[288] Barlaam, B.; Koza, P.; Berriot, J. *Tetrahedron* **1999**, *55*, 7221–7232.
[289] Ngu, K.; Patel, D. V. *J. Org. Chem.* **1997**, *62*, 7088–7089.
[290] Dankwardt, S. M. *Synlett* **1998**, 761
[291] Chang, J. K.; Shimizu, M.; Wang, S. S. *J. Org. Chem.* **1976**, *41*, 3255–3258.
[292] Kessler, W.; Iselin, B. *Helv. Chim. Acta* **1966**, *49*, 1330–1344.
[293] Ohno, M.; Anfinsen, C. B. *J. Am. Chem. Soc.* **1967**, *89*, 5994–5995.
[294] Takahashi, T.; Tomida, S.; Inoue, H.; Doi, T. *Synlett* **1998**, 1261–1263.
[295] Hunt, J. A.; Roush, W. R. *J. Am. Chem. Soc.* **1996**, *118*, 9998–9999.
[296] Beyerman, H. C.; Hindriks, H.; De Leer, E. W. B. *J. Chem. Soc. Chem. Commun.* **1968**, 1668
[297] de Bont, D. B. A.; Moree, W. J.; Liskamp, R. M. J. *Bioorg. Med. Chem.* **1996**, *4*, 667–672.
[298] Rothe, M.; Zieger, M. *Tetrahedron Lett.* **1994**, *35*, 9011–9012.
[299] Bhalay, G.; Blaney, P.; Palmer, V. H.; Baxter, A. D. *Tetrahedron Lett.* **1997**, *38*, 8375–8378.
[300] Annis, D. A.; Helluin, O.; Jacobsen, E. N. *Angew. Chem. Int. Ed. Engl.* **1998**, *37*, 1907–1909.
[301] Frenette, R.; Friesen, R. W. *Tetrahedron Lett.* **1994**, *35*, 9177–9180.
[302] Shimizu, H.; Ito, Y.; Kanie, O.; Ogawa, T. *Bioorg. Med. Chem. Lett.* **1996**, *6*, 2841–2846.
[303] Beyerman, H. C.; Kranenburg, P.; Syrier, J. L. M. *Recl. Trav. Chim. Pays-Bas* **1971**, *90*, 791–800.
[304] Barton, M. A.; Lemieux, R. U.; Savoie, J. Y. *J. Am. Chem. Soc.* **1973**, *95*, 4501–4506.
[305] O'Donnell, M. J.; Zhou, C. Y.; Scott, W. L. *J. Am. Chem. Soc.* **1996**, *118*, 6070–6071.
[306] Sylvain, C.; Wagner, A.; Mioskowski, C. *Tetrahedron Lett.* **1997**, *38*, 1043–1044.
[307] Hutchins, S. M.; Chapman, K. T. *Tetrahedron Lett.* **1996**, *37*, 4869–4872.
[308] Cheng, Y.; Chapman, K. T. *Tetrahedron Lett.* **1997**, *38*, 1497–1500.
[309] Reichwein, J. F.; Liskamp, R. M. J. *Tetrahedron Lett.* **1998**, *39*, 1243–1246.
[310] Far, A. R.; Tidwell, T. T. *J. Org. Chem.* **1998**, *63*, 8636–8637.
[311] Colwell, A. R.; Duckwall, L. R.; Brooks, R.; McManus, S. P. *J. Org. Chem.* **1981**, *46*, 3097–3102.
[312] Kobayashi, S.; Wakabayashi, T.; Yasuda, M. *J. Org. Chem.* **1998**, *63*, 4868–4869.
[313] Hanessian, S.; Ma, J.; Wang, W. *Tetrahedron Lett.* **1999**, *40*, 4631–4634.
[314] Reggelin, M.; Brenig, V.; Welcker, R. *Tetrahedron Lett.* **1998**, *39*, 4801–4804.
[315] Le Hetet, C.; David, M.; Carreaux, F.; Carboni, B.; Sauleau, A. *Tetrahedron Lett.* **1997**, *38*, 5153–5156.
[316] Moon, H.; Schore, N. E.; Kurth, M. J. *J. Org. Chem.* **1992**, *57*, 6088–6089.
[317] Moon, H.; Schore, N. E.; Kurth, M. J. *Tetrahedron Lett.* **1994**, *35*, 8915–8918.
[318] Ko, D. H.; Kim, D. J.; Lyu, C. S.; Min, I. K.; Moon, H. S. *Tetrahedron Lett.* **1998**, *39*, 297–300.
[319] Watanabe, Y.; Ishikawa, S.; Takao, G.; Toru, T. *Tetrahedron Lett.* **1999**, *40*, 3411–3414.
[320] Conti, P.; Demont, D.; Cals, J.; Ottenheijm, H. C. J.; Leysen, D. *Tetrahedron Lett.* **1997**, *38*, 2915–2918.
[321] Kobayashi, S.; Aoki, Y. *Tetrahedron Lett.* **1998**, *39*, 7345–7348.
[322] Purandare, A. V.; Poss, M. A. *Tetrahedron Lett.* **1998**, *39*, 935–938.
[323] Tommasi, R. A.; Nantermet, P. G.; Shapiro, M. J.; Chin, J.; Brill, W. K. D.; Ang, K. *Tetrahedron Lett.* **1998**, *39*, 5477–5480.
[324] Ostresh, J. M.; Schoner, C. C.; Hamashin, V. T.; Nefzi, A.; Meyer, J. P.; Houghten, R. A. *J. Org. Chem.* **1998**, *63*, 8622–8623.
[325] Nefzi, A.; Ostresh, J. M.; Houghten, R. A. *Tetrahedron* **1999**, *55*, 335–344.
[326] Gauzy, L.; Le Merrer, Y.; Depezay, J. C.; Clerc, F.; Mignani, S. *Tetrahedron Lett.* **1999**, *40*, 6005–6008.
[327] Katritzky, A. R.; Xie, L.; Zhang, G.; Griffith, M.; Watson, K.; Kiely, J. S. *Tetrahedron Lett.* **1997**, *38*, 7011–7014.
[328] Boyd, E. A.; Chan, W. C.; Loh, V. M. *Tetrahedron Lett.* **1996**, *37*, 1647–1650.
[329] Youngman, M. A.; Dax, S. L. *Tetrahedron Lett.* **1997**, *38*, 6347–6350.
[330] McNally, J. J.; Youngman, M. A.; Dax, S. L. *Tetrahedron Lett.* **1998**, *39*, 967–970.
[331] Hoekstra, W. J.; Maryanoff, B. E.; Andrade-Gordon, P.; Cohen, J. H.; Costanzo, M. J.; Damiano, B. P.; Haertlein, B. J.; Harris, B. D.; Kauffman, J. A.; Keane, P. M.; McComsey, D. F.; Villani, F. J.; Yabut, S. C. *Bioorg. Med. Chem. Lett.* **1996**, *6*, 2371–2376.

[332] Hoekstra, W. J.; Greco, M. N.; Yabut, S. C.; Hulshizer, B. L.; Maryanoff, B. E. *Tetrahedron Lett.* **1997**, *38*, 2629–2632.
[333] Barlos, K.; Gatos, D.; Kallitsis, I.; Papaioannou, D.; Sotiriu, P. *Liebigs Ann. Chem.* **1988**, 1079–1081.
[334] Bleicher, K. H.; Wareing, J. R. *Tetrahedron Lett.* **1998**, *39*, 4591–4594.
[335] Hidai, Y.; Kan, T.; Fukuyama, T. *Tetrahedron Lett.* **1999**, *40*, 4711–4714.
[336] Hauske, J. R.; Dorff, P. *Tetrahedron Lett.* **1995**, *36*, 1589–1592.
[337] Smith, A. L.; Thomson, C. G.; Leeson, P. D. *Bioorg. Med. Chem. Lett.* **1996**, *6*, 1483–1486.
[338] Marsh, I. R.; Smith, H. K.; Leblanc, C.; Bradley, M. *Mol. Diversity* **1997**, *2*, 165–170.
[339] Dixit, D. M.; Leznoff, C. C. *Israel J. Chem.* **1978**, *17*, 248–252.
[340] Dressman, B. A.; Spangle, L. A.; Kaldor, S. W. *Tetrahedron Lett.* **1996**, *37*, 937–940.
[341] Kaljuste, K.; Undén, A. *Tetrahedron Lett.* **1995**, *36*, 9211–9214.
[342] Dixit, D. M.; Leznoff, C. C. *J. Chem. Soc. Chem. Commun.* **1977**, 798–799.
[343] Ho, C. Y.; Kukla, M. *J. Tetrahedron Lett.* **1997**, *38*, 2799–2802.
[344] Zaragoza, F.; Petersen, S. V. *Tetrahedron* **1996**, *52*, 5999–6002.
[345] Meester, W. J. N.; Rutjes, F. P. J. T.; Hermkens, P. H. H.; Hiemstra, H. *Tetrahedron Lett.* **1999**, *40*, 1601–1604.
[346] Wang, F.; Hauske, J. R. *Tetrahedron Lett.* **1997**, *38*, 6529–6532.
[347] Cuny, G. D.; Cao, J.; Hauske, J. R. *Tetrahedron Lett.* **1997**, *38*, 5237–5240.
[348] Rotella, D. P. *J. Am. Chem. Soc.* **1996**, *118*, 12246–12247.
[349] Hiroshige, M.; Hauske, J. R.; Zhou, P. *J. Am. Chem. Soc.* **1995**, *117*, 11590–11591.
[350] Stephensen, H.; Zaragoza, F. *J. Org. Chem.* **1997**, *62*, 6096–6097.
[351] Veerman, J. J. N.; Rutjes, F. P. J. T.; van Maarseveen, J. H.; Hiemstra, H. *Tetrahedron Lett.* **1999**, *40*, 6079–6082.
[352] Chen, C.; Munoz, B. *Tetrahedron Lett.* **1998**, *39*, 3401–3404.
[353] Keana, J. F. W.; Shimizu, M.; Jernsted, K. K. *J. Org. Chem.* **1986**, *51*, 1641–1644.
[354] McMinn, D. L.; Greenberg, M. M. *Tetrahedron* **1996**, *52*, 3827–3840.
[355] McKeown, S. C.; Watson, S. P.; Carr, R. A. E.; Marshall, P. *Tetrahedron Lett.* **1999**, *40*, 2407–2410.
[356] Kaljuste, K.; Undén, A. *Tetrahedron Lett.* **1996**, *37*, 3031–3034.
[357] Chen, C.; McDonald, I. A.; Munoz, B. *Tetrahedron Lett.* **1998**, *39*, 217–220.
[358] Canne, L. E.; Winston, R. L.; Kent, S. B. H. *Tetrahedron Lett.* **1997**, *38*, 3361–3364.
[359] Nelson, J. C.; Young, J. K.; Moore, J. S. *J. Org. Chem.* **1996**, *61*, 8160–8168.
[360] Chhabra, S. R.; Khan, A. N.; Bycroft, B. W. *Tetrahedron Lett.* **1998**, *39*, 3585–3588.
[361] Bannwarth, W.; Huebscher, J.; Barner, R. *Bioorg. Med. Chem. Lett.* **1996**, *6*, 1525–1528.
[362] Furth, P. S.; Reitman, M. S.; Gentles, R.; Cook, A. F. *Tetrahedron Lett.* **1997**, *38*, 6643–6646.
[363] Furth, P. S.; Reitman, M. S.; Cook, A. F. *Tetrahedron Lett.* **1997**, *38*, 5403–5406.
[364] Kay, C.; Murray, P. J.; Sandow, L.; Holmes, A. B. *Tetrahedron Lett.* **1997**, *38*, 6941–6944.
[365] Yamamoto, Y.; Tanabe, K.; Okonogi, T. *Chem. Lett.* **1999**, 103–104.
[366] Barco, A.; Benetti, S.; De Risi, C.; Marchetti, P.; Pollini, G. P.; Zanirato, V. *Tetrahedron Lett.* **1998**, *39*, 7591–7594.
[367] Brown, A. R.; Rees, D. C.; Rankovic, Z.; Morphy, J. R. *J. Am. Chem. Soc.* **1997**, *119*, 3288–3295.
[368] Ouyang, X. H.; Armstrong, R. W.; Murphy, M. M. *J. Org. Chem.* **1998**, *63*, 1027–1032.
[369] Kroll, F. E. K.; Morphy, R.; Rees, D.; Gani, D. *Tetrahedron Lett.* **1997**, *38*, 8573–8576.
[370] Heinonen, P.; Virta, P.; Lönnberg, H. *Tetrahedron* **1999**, *55*, 7613–7624.
[371] Rueter, J. K.; Nortey, S. O.; Baxter, E. W.; Leo, G. C.; Reitz, A. B. *Tetrahedron Lett.* **1998**, *39*, 975–978.
[372] Bicak, N.; Senkal, B. F. *Reactive and Functional Polymers* **1996**, *29*, 123–128.
[373] Schürer, S. C.; Blechert, S. *Synlett* **1998**, 166–168.
[374] Katritzky, A. R.; Belyakov, S. A.; Tymoshenko, D. O. *J. Comb. Chem.* **1999**, *1*, 173–176.
[375] Smith, A. L.; Stevenson, G. I.; Swain, C. J.; Castro, J. L. *Tetrahedron Lett.* **1998**, *39*, 8317–8320.
[376] Gray, N. S.; Kwon, S.; Schultz, P. G. *Tetrahedron Lett.* **1997**, *38*, 1161–1164.
[377] Gordeev, M. F.; Patel, D. V.; Gordon, E. M. *J. Org. Chem.* **1996**, *61*, 924–928.
[378] Gordeev, M. F.; Patel, D. V.; England, B. P.; Jonnalagadda, S.; Combs, J. D.; Gordon, E. M. *Bioorg. Med. Chem.* **1998**, *6*, 883–889.
[379] Kearney, P. C.; Fernandez, M.; Flygare, J. A. *J. Org. Chem.* **1998**, *63*, 196–200.
[380] Buchstaller, H. P. *Tetrahedron* **1998**, *54*, 3465–3470.
[381] García Echeverría, C. *Tetrahedron Lett.* **1997**, *38*, 8933–8934.
[382] Sunami, S.; Sagara, T.; Ohkubo, M.; Morishima, H. *Tetrahedron Lett.* **1999**, *40*, 1721–1724.
[383] Burdick, D. J.; Struble, M. E.; Burnier, J. P. *Tetrahedron Lett.* **1993**, *34*, 2589–2592.
[384] Han, H. S.; Wolfe, M. M.; Brenner, S.; Janda, K. D. *Proc. Natl. Acad. Sci. USA* **1995**, *92*, 6419–6423.

[385] Masquelin, T.; Sprenger, D.; Baer, R.; Gerber, F.; Mercadal, Y. *Helv. Chim. Acta* **1998**, *81*, 646–660.
[386] Obrecht, D.; Abrecht, C.; Grieder, A.; Villalgordo, J. M. *Helv. Chim. Acta* **1997**, *80*, 65–72.
[387] Gayo, L. M.; Suto, M. J. *Tetrahedron Lett.* **1997**, *38*, 211–214.
[388] Masquelin, T.; Meunier, N.; Gerber, F.; Rossé, G. *Heterocycles* **1998**, *48*, 2489–2505.
[389] Fields, G. B.; Noble, R. L. *Int. J. Pept. Prot. Res.* **1990**, *35*, 161–214.
[390] Zhong, H. M.; Greco, M. N.; Maryanoff, B. E. *J. Org. Chem.* **1997**, *62*, 9326–9330.
[391] Josey, J. A.; Tarlton, C. A.; Payne, C. E. *Tetrahedron Lett.* **1998**, *39*, 5899–5902.
[392] Mohan, R.; Yun, W. Y.; Buckman, B. O.; Liang, A.; Trinh, L.; Morrissey, M. M. *Bioorg. Med. Chem. Lett.* **1998**, *8*, 1877–1882.
[393] Roussel, P.; Bradley, M.; Matthews, I.; Kane, P. *Tetrahedron Lett.* **1997**, *38*, 4861–4864.
[394] Kim, S. W.; Hong, C. Y.; Koh, J. S.; Lee, E. J.; Lee, K. *Mol. Diversity* **1998**, *3*, 133–136.
[395] Wilson, L. J.; Klopfenstein, S. R.; Li, M. *Tetrahedron Lett.* **1999**, *40*, 3999–4002.
[396] Bonnat, M.; Bradley, M.; Kilburn, J. D. *Tetrahedron Lett.* **1996**, *37*, 5409–5412.
[397] Davies, M.; Bonnat, M.; Guillier, F.; Kilburn, J. D.; Bradley, M. *J. Org. Chem.* **1998**, *63*, 8696–8703.
[398] Dodd, D. S.; Wallace, O. B. *Tetrahedron Lett.* **1998**, *39*, 5701–5704.
[399] Eleftheriou, S.; Gatos, D.; Panagopoulos, A.; Stathopoulos, S.; Barlos, K. *Tetrahedron Lett.* **1999**, *40*, 2825–2828.
[400] Glass, J. D.; Schwartz, I. L.; Walter, R. *J. Am. Chem. Soc.* **1972**, *94*, 6209–6211.
[401] Nugiel, D. A.; Cornelius, L. A. M.; Corbett, J. W. *J. Org. Chem.* **1997**, *62*, 201–203.
[402] Yoo, S. E.; Seo, J. S.; Yi, K. Y.; Gong, Y. D. *Tetrahedron Lett.* **1997**, *38*, 1203–1206.
[403] Bilodeau, M. T.; Cunningham, A. M. *J. Org. Chem.* **1998**, *63*, 2800–2801.
[404] Sabatino, G.; Chelli, M.; Mazzucco, S.; Ginanneschi, M.; Papini, A. M. *Tetrahedron Lett.* **1999**, *40*, 809–812.
[405] Lee, C. E.; Kick, E. K.; Ellman, J. A. *J. Am. Chem. Soc.* **1998**, *120*, 9735–9747.
[406] Richter, L. S.; Gadek, T. R. *Tetrahedron Lett.* **1994**, *35*, 4705–4706.
[407] Hamper, B. C.; Dukesherer, D. R.; South, M. S. *Tetrahedron Lett.* **1996**, *37*, 3671–3674.
[408] Hollinshead, S. P. *Tetrahedron Lett.* **1996**, *37*, 9157–9160.
[409] Garigipati, R. S.; Adams, B.; Adams, J. L.; Sarkar, S. K. *J. Org. Chem.* **1996**, *61*, 2911–2914.
[410] Hanessian, S.; Xie, F. *Tetrahedron Lett.* **1998**, *39*, 737–740.
[411] Hanessian, S.; Xie, F. *Tetrahedron Lett.* **1998**, *39*, 733–736.
[412] Furman, B.; Thürmer, R.; Kaluza, Z.; Lysek, R.; Voelter, W.; Chmielewski, M. *Angew. Chem. Int. Ed. Engl.* **1999**, *38*, 1121–1123.
[413] Rottländer, M.; Knochel, P. *J. Comb. Chem.* **1999**, *1*, 181–183.
[414] Craig, D.; Robson, M. J.; Shaw, S. J. *Synlett* **1998**, 1381–1383.
[415] Hanessian, S.; Huynh, H. K. *Tetrahedron Lett.* **1999**, *40*, 671–674.
[416] Mergler, M.; Dick, F.; Gosteli, J.; Nyfeler, R. *Tetrahedron Lett.* **1999**, *40*, 4663–4664.
[417] Shankar, B. B.; Yang, D. Y.; Girton, S.; Ganguly, A. K. *Tetrahedron Lett.* **1998**, *39*, 2447–2448.
[418] Zhu, Z. M.; McKittrick, B. *Tetrahedron Lett.* **1998**, *39*, 7479–7482.
[419] Fréchet, J. M. J.; Nuyens, L. J. *Can. J. Chem.* **1976**, *54*, 926–934.
[420] Fréchet, J. M. J.; Haque, K. E. *Tetrahedron Lett.* **1975**, 3055–3056.
[421] Hayatsu, H.; Khorana, H. G. *J. Am. Chem. Soc.* **1966**, *88*, 3182–3183.
[422] Hayatsu, H.; Khorana, H. G. *J. Am. Chem. Soc.* **1967**, *89*, 3880–3887.
[423] Chen, C.; Randall, L. A. A.; Miller, R. B.; Jones, A. D.; Kurth, M. J. *J. Am. Chem. Soc.* **1994**, *116*, 2661–2662.
[424] Pernerstorfer, J.; Schuster, M.; Blechert, S. *Synthesis* **1999**, 138–144.
[425] Ito, Y.; Ogawa, T. *J. Am. Chem. Soc.* **1997**, *119*, 5562–5566.
[426] Brill, W. K. D.; Schmidt, E.; Tommasi, R. A. *Synlett* **1998**, 906–908.
[427] Krchnák, V.; Weichsel, A. S. *Tetrahedron Lett.* **1997**, *38*, 7299–7302.
[428] Fyles, T. M.; Leznoff, C. C.; Weatherston, J. *Can. J. Chem.* **1977**, *55*, 4135–4143.
[429] Wang, Y.; Zhang, H.; Voelter, W. *Chem. Lett.* **1995**, 273–274.
[430] Nicolaou, K. C.; Winssinger, N.; Pastor, J.; DeRoose, F. *J. Am. Chem. Soc.* **1997**, *119*, 449–450.
[431] Zehavi, U.; Patchornik, A. *J. Am. Chem. Soc.* **1973**, *95*, 5673–5677.
[432] Fukase, K.; Egusa, K.; Nakai, Y.; Kusumoto, S. *Mol. Diversity* **1997**, *2*, 182–188.
[433] Fukase, K.; Nakai, Y.; Egusa, K.; Porco, J. A.; Kusumoto, S. *Synlett* **1999**, 1074–1078.
[434] Nestler, H. P.; Bartlett, P. A.; Still, W. C. *J. Org. Chem.* **1994**, *59*, 4723–4724.
[435] Yamamoto, Y.; Ajito, K.; Ohtsuka, Y. *Chem. Lett.* **1998**, 379–380.
[436] Cabrele, C.; Langer, M.; Beck-Sickinger, A. G. *J. Org. Chem.* **1999**, *64*, 4353–4361.
[437] Tempest, P. A.; Armstrong, R. W. *J. Am. Chem. Soc.* **1997**, *119*, 7607–7608.
[438] Yun, W. Y.; Mohan, R. *Tetrahedron Lett.* **1996**, *37*, 7189–7192.
[439] Phillips, G. B.; Wei, G. P. *Tetrahedron Lett.* **1996**, *37*, 4887–4890.

[440] Plunkett, M. J.; Ellman, J. A. *J. Am. Chem. Soc.* **1995**, *117*, 3306–3307.
[441] Devraj, R.; Cushman, M. *J. Org. Chem.* **1996**, *61*, 9368–9373.
[442] Burbaum, J. J.; Ohlmeyer, M. H. J.; Reader, J. C.; Henderson, I.; Dillard, L. W.; Li, G.; Randle, T. L.; Sigal, N. H.; Chelsky, D.; Baldwin, J. J. *Proc. Natl. Acad. Sci. USA* **1995**, *92*, 6027–6031.
[443] Tremblay, M. R.; Poirier, D. *Tetrahedron Lett.* **1999**, *40*, 1277–1280.
[444] Farrall, M. J.; Fréchet, J. M. J. *J. Org. Chem.* **1976**, *41*, 3877–3882.
[445] Hu, Y.; Porco, J. A.; Labadie, J. W.; Gooding, O. W.; Trost, B. M. *J. Org. Chem.* **1998**, *63*, 4518–4521.
[446] Stranix, B. R.; Liu, H. Q.; Darling, G. D. *J. Org. Chem.* **1997**, *62*, 6183–6186.
[447] Zheng, C.; Seeberger, P. H.; Danishefsky, S. J. *J. Org. Chem.* **1998**, *63*, 1126–1130.
[448] Doi, T.; Sugiki, M.; Yamada, H.; Takahashi, T.; Porco, J. A. *Tetrahedron Lett.* **1999**, *40*, 2141–2144.
[449] Fréchet, J. M. J.; Darling, G. D.; Itsuno, S.; Lu, P. Z.; de Meftahi, M. V.; Rolls, W. A. *Pure Appl. Chem.* **1988**, *60*, 353–364.
[450] Thompson, L. A.; Moore, F. L.; Moon, Y. C.; Ellman, J. A. *J. Org. Chem.* **1998**, *63*, 2066–2067.
[451] Chan, T.; Huang, W. *J. Chem. Soc. Chem. Commun.* **1985**, 909–911.
[452] Hu, Y. H.; Porco, J. A. *Tetrahedron Lett.* **1998**, *39*, 2711–2714.
[453] Savin, K. A.; Woo, J. C. G.; Danishefsky, S. J. *J. Org. Chem.* **1999**, *64*, 4183–4186.
[454] Zheng, C. S.; Seeberger, P. H.; Danishefsky, S. J. *Angew. Chem. Int. Ed. Engl.* **1998**, *37*, 786–789.
[455] Danishefsky, S. J.; McClure, K. F.; Randolph, J. T.; Ruggeri, R. B. *Science* **1993**, *260*, 1307–1309.
[456] Nakamura, K.; Hanai, N.; Kanno, M.; Kobayashi, A.; Ohnishi, Y.; Ito, Y.; Nakahara, Y. *Tetrahedron Lett.* **1999**, *40*, 515–518.
[457] Nakamura, K.; Ishii, A.; Ito, Y.; Nakahara, Y. *Tetrahedron* **1999**, *55*, 11253–11266.
[458] Routledge, A.; Wallis, M. P.; Ross, K. C.; Fraser, W. *Bioorg. Med. Chem. Lett.* **1995**, *5*, 2059–2064.
[459] Chen, S.; Janda, K. D. *Tetrahedron Lett.* **1998**, *39*, 3943–3946.
[460] Wallace, O. B. *Tetrahedron Lett.* **1997**, *38*, 4939–4942.
[461] Koh, J. S.; Ellman, J. A. *J. Org. Chem.* **1996**, *61*, 4494–4495.
[462] Liu, G. C.; Ellman, J. A. *J. Org. Chem.* **1995**, *60*, 7712–7713.
[463] Kick, E. K.; Ellman, J. A. *J. Med. Chem.* **1995**, *38*, 1427–1430.
[464] Ellman, J. A.; Thompson, L. A. *Tetrahedron Lett.* **1994**, *35*, 9333–9336.
[465] Pearson, W. H.; Clark, R. B. *Tetrahedron Lett.* **1997**, *38*, 7669–7672.
[466] Wang, G. T.; Li, S.; Wideburg, N.; Krafft, G. A.; Kempf, D. J. *J. Med. Chem.* **1995**, *38*, 2995–3002.
[467] Wendeborn, S.; De Mesmaeker, A.; Brill, W. K. D. *Synlett* **1998**, 865–868.
[468] Hanessian, S.; Huynh, H. K. *Synlett* **1999**, 102–104.
[469] Fréchet, J. M. J.; Pellé, G. *J. Chem. Soc. Chem. Commun.* **1975**, 225–226.
[470] Wu, Y. T.; Hsieh, H. P.; Wu, C. Y.; Yu, H. M.; Chen, S. T.; Wang, K. T. *Tetrahedron Lett.* **1998**, *39*, 1783–1784.
[471] Zhu, T.; Boons, G. J. *Angew. Chem. Int. Ed. Engl.* **1998**, *37*, 1898–1900.
[472] Nicolaou, K. C.; Winssinger, N.; Vourloumis, D.; Ohshima, T.; Kim, S.; Pfefferkorn, J.; Xu, J. Y.; Li, T. *J. Am. Chem. Soc.* **1998**, *120*, 10814–10826.
[473] Berteina, S.; Wendeborn, S.; De Mesmaeker, A. *Synlett* **1998**, 1231–1233.
[474] Nizi, E.; Botta, M.; Corelli, F.; Manetti, F.; Messina, F.; Maga, G. *Tetrahedron Lett.* **1998**, *39*, 3307–3310.
[475] Swistok, J.; Tilley, J. W.; Danho, W.; Wagner, R.; Mulkerins, K. *Tetrahedron Lett.* **1989**, *30*, 5045–5048.
[476] Zheng, A.; Shan, D.; Wang, B. *J. Org. Chem.* **1999**, *64*, 156–161.
[477] Meyers, H. V.; Dilley, G. J.; Durgin, T. L.; Powers, T. S.; Winssinger, N. A.; Zhu, H.; Pavia, M. R. *Mol. Diversity* **1995**, *1*, 13–20.
[478] Barber, A. M.; Hardcastle, I. R.; Rowlands, M. G.; Nutley, B. P.; Marriott, J. H.; Jarman, M. *Bioorg. Med. Chem. Lett.* **1999**, *9*, 623–626.
[479] Leznoff, C. C.; Dixit, D. M. *Can. J. Chem.* **1977**, *55*, 3351–3355.
[480] Molteni, V.; Annunziata, R.; Cinquini, M.; Cozzi, F.; Benaglia, M. *Tetrahedron Lett.* **1998**, *39*, 1257–1260.
[481] Kurth, M. J.; Randall, L. A. A.; Takenouchi, K. *J. Org. Chem.* **1996**, *61*, 8755–8761.
[482] Kantorowski, E. J.; Kurth, M. J. *J. Org. Chem.* **1997**, *62*, 6797–6803.
[483] Pavia, M. R.; Cohen, M. P.; Dilley, G. J.; Dubuc, G. R.; Durgin, T. L.; Forman, F. W.; Hediger, M. E.; Milot, G.; Powers, T. S.; Sucholeiki, I.; Zhou, S.; Hangauer, D. G. *Bioorg. Med. Chem.* **1996**, *4*, 659–666.
[484] Adams, S. P.; Kavka, K. S.; Wykes, E. J.; Holder, S. B.; Galluppi, G. R. *J. Am. Chem. Soc.* **1983**, *105*, 661–663.
[485] Eckstein, F. *Oligonucleotides and Analogues; A Practical Approach*; Oxford University Press: Oxford, **1991**.

[486] Krchnák, V.; Weichsel, A. S.; Lebl, M.; Felder, S. *Bioorg. Med. Chem. Lett.* **1997**, *7*, 1013–1016.
[487] Moore, M.; Norris, P. *Tetrahedron Lett.* **1998**, *39*, 7027–7030.
[488] Kobayashi, S.; Hachiya, I.; Suzuki, S.; Moriwaki, M. *Tetrahedron Lett.* **1996**, *37*, 2809–2812.
[489] Burgess, K.; Lim, D. *Chem. Commun.* **1997**, 785–786.
[490] Tietze, L. F.; Hippe, T.; Steinmetz, A. *Chem. Commun.* **1998**, 793–794.
[491] Mergler, M.; Nyfeler, R. Rapid Synthesis of fully Protected Peptide Alcohols, Communication from Bachem, Bubendorf **1990**.
[492] Kurth, M. L.; Randall, L. A. A.; Chen, C.; Melander, C.; Miller, R. B.; McAlister, K.; Reitz, G.; Kang, R.; Nakatsu, T.; Green, C. *J. Org. Chem.* **1994**, *59*, 5862–5864.
[493] Ley, S. V.; Mynett, D. M.; Koot, W. J. *Synlett* **1995**, 1017–1020.
[494] Alsina, J.; Rabanal, F.; Chiva, C.; Giralt, E.; Albericio, F. *Tetrahedron* **1998**, *54*, 10125–10152.
[495] Fréchet, J. M. J.; Nuyens, L. J.; Seymour, E. *J. Am. Chem. Soc.* **1979**, *101*, 432–436.
[496] Alsina, J.; Chiva, C.; Ortiz, M.; Rabanal, F.; Giralt, E.; Albericio, F. *Tetrahedron Lett.* **1997**, *38*, 883–886.
[497] Chou, Y. L.; Morrissey, M. M.; Mohan, R. *Tetrahedron Lett.* **1998**, *39*, 757–760.
[498] Eritja, R.; Robles, J.; Fernández Forner, D.; Albericio, F.; Giralt, E.; Pedroso, E. *Tetrahedron Lett.* **1991**, *32*, 1511–1514.
[499] Eritja, R.; Robles, J.; Aviñó, A.; Albericio, F.; Pedroso, E. *Tetrahedron* **1992**, *48*, 4171–4182.
[500] Venkatesan, H.; Greenberg, M. M. *J. Org. Chem.* **1996**, *61*, 525–529.
[501] Souers, A. J.; Virgilio, A. A.; Rosenquist, A.; Fenuik, W.; Ellman, J. A. *J. Am. Chem. Soc.* **1999**, *121*, 1817–1825.
[502] Annis, I.; Chen, L.; Barany, G. *J. Am. Chem. Soc.* **1998**, *120*, 7226–7238.
[503] Sucholeiki, I. *Tetrahedron Lett.* **1994**, *35*, 7307–7310.
[504] Englebretsen, D. R.; Garnham, B. G.; Bergman, D. A.; Alewood, P. F. *Tetrahedron Lett.* **1995**, *36*, 8871–8874.
[505] Glass, J. D.; Talansky, A.; Grzonka, Z.; Schwartz, I. L.; Walter, R. *J. Am. Chem. Soc.* **1974**, *96*, 6476–6480.
[506] Virgilio, A. A.; Schürer, S. C.; Ellman, J. A. *Tetrahedron Lett.* **1996**, *37*, 6961–6964.
[507] Souers, A. J.; Virgilio, A. A.; Schürer, S. S.; Ellman, J. A.; Kogan, T. P.; West, H. E.; Ankener, W.; Vanderslice, P. *Bioorg. Med. Chem. Lett.* **1998**, *8*, 2297–2302.
[508] Lee, T. R.; Lawrence, D. S. *J. Med. Chem.* **1999**, *42*, 784–787.
[509] Moore, J. S.; Weinstein, E. J.; Wu, Z. *Tetrahedron Lett.* **1991**, *32*, 2465–2466.
[510] Han, Y. X.; Walker, S. D.; Young, R. N. *Tetrahedron Lett.* **1996**, *37*, 2703–2706.
[511] Plunkett, M. J.; Ellman, J. A. *J. Org. Chem.* **1997**, *62*, 2885–2893.
[512] Spivey, A. C.; Diaper, C. M.; Rudge, A. J. *Chem. Commun.* **1999**, 835–836.
[513] Nicolaou, K. C.; Winssinger, N.; Pastor, J.; Murphy, F. *Angew. Chem. Int. Ed. Engl.* **1998**, *37*, 2534–2537.
[514] Crosby, G. A.; Kato, M. *J. Am. Chem. Soc.* **1977**, *99*, 278–280.
[515] Beebe, X.; Chiappari, C. L.; Kurth, M. J.; Schore, N. E. *J. Org. Chem.* **1993**, *58*, 7320–7321.
[516] Beebe, X.; Schore, N. E.; Kurth, M. J. *J. Org. Chem.* **1995**, *60*, 4196–4203.
[517] Murphy, A. M.; Dagnino, R.; Vallar, P. L.; Trippe, A. J.; Sherman, S. L.; Lumpkin, R. H.; Tamura, S. Y.; Webb, T. R. *J. Am. Chem. Soc.* **1992**, *114*, 3156–3157.
[518] Poupart, M. A.; Fazal, G.; Goulet, S.; Mar, L. T. *J. Org. Chem.* **1999**, *64*, 1356–1361.
[519] Fraley, M. E.; Rubino, R. S. *Tetrahedron Lett.* **1997**, *38*, 3365–3368.
[520] Gutke, H.-J.; Spitzner, D. *Tetrahedron* **1999**, *55*, 3931–3936.
[521] Ball, C. P.; Barrett, A. G. M.; Commercon, A.; Compère, D.; Kuhn, C.; Roberts, R. S.; Smith, M. L.; Venier, O. *Chem. Commun.* **1998**, 2019–2020.
[522] Hu, Y.; Porco, J. A. *Tetrahedron Lett.* **1999**, *40*, 3289–3292.
[523] Smith, E. M. *Tetrahedron Lett.* **1999**, *40*, 3285–3288.
[524] Worster, P. M.; McArthur, C. R.; Leznoff, C. C. *Angew. Chem. Int. Ed. Engl.* **1979**, *18*, 221–222.
[525] McArthur, C. R.; Worster, P. M.; Jiang, J.; Leznoff, C. C. *Can. J. Chem.* **1982**, *60*, 1836–1841.
[526] Hird, N. W.; Irie, K.; Nagai, K. *Tetrahedron Lett.* **1997**, *38*, 7111–7114.
[527] Crawshaw, M.; Hird, N. W.; Irie, K.; Nagai, K. *Tetrahedron Lett.* **1997**, *38*, 7115–7118.
[528] Wilson, R. D.; Watson, S. P.; Richards, S. A. *Tetrahedron Lett.* **1998**, *39*, 2827–2830.
[529] Chen, C.; Munoz, B. *Tetrahedron Lett.* **1999**, *40*, 3491–3494.
[530] Vojkovsky, T.; Weichsel, A.; Pátek, M. *J. Org. Chem.* **1998**, *63*, 3162–3163.
[531] Chandrasekhar, S.; Padmaja, M. B. *Synthetic Commun.* **1998**, *28*, 3715–3720.
[532] Bertini, V.; Lucchesini, F.; Pocci, M.; De Munno, A. *Tetrahedron Lett.* **1998**, *39*, 9263–9266.
[533] Huwe, C. M.; Künzer, H. *Tetrahedron Lett.* **1999**, *40*, 683–686.
[534] Rademann, J.; Schmidt, R. R. *J. Org. Chem.* **1997**, *62*, 3650–3653.
[535] Rademann, J.; Schmidt, R. R. *Tetrahedron Lett.* **1996**, *37*, 3989–3990.

[536] Yan, L.; Taylor, C. M.; Goodnow, R.; Kahne, D. *J. Am. Chem. Soc.* **1994**, *116*, 6953–6954.
[537] Siev, D. V.; Gaudette, J. A.; Semple, J. E. *Tetrahedron Lett.* **1999**, *40*, 5123–5127.
[538] Chamoin, S.; Houldsworth, S.; Kruse, C. G.; Bakker, W. I.; Snieckus, V. *Tetrahedron Lett.* **1998**, *39*, 4179–4182.
[539] Leznoff, C. C.; Sywanyk, W. *J. Org. Chem.* **1977**, *42*, 3203–3205.
[540] Wong, J. Y.; Manning, C.; Leznoff, C. C. *Angew. Chem. Int. Ed. Engl.* **1974**, *13*, 666–667.
[541] Xu, Z. H.; McArthur, C. R.; Leznoff, C. C. *Can. J. Chem.* **1983**, *61*, 1405–1409.
[542] Leznoff, C. C.; Greenberg, S. *Can. J. Chem.* **1976**, *54*, 3824–3829.
[543] Metz, W. A.; Jones, W. D.; Ciske, F. L.; Peet, N. P. *Bioorg. Med. Chem. Lett.* **1998**, *8*, 2399–2402.
[544] Ren, Q.; Huang, W.; Ho, P. *Reactive Polymers* **1989**, *11*, 237–244.
[545] Ede, N. J.; Bray, A. M. *Tetrahedron Lett.* **1997**, *38*, 7119–7122.
[546] Nahm, S.; Weinreb, S. M. *Tetrahedron Lett.* **1981**, *22*, 3815–3818.
[547] Fehrentz, J. A.; Paris, M.; Heitz, A.; Velek, J.; Liu, C. F.; Winternitz, F.; Martinez, J. *Tetrahedron Lett.* **1995**, *36*, 7871–7874.
[548] Dinh, T. Q.; Armstrong, R. W. *Tetrahedron Lett.* **1996**, *37*, 1161–1164.
[549] Crowley, J. I.; Rapoport, H. *J. Am. Chem. Soc.* **1970**, *92*, 6363–6365.
[550] Hall, B. J.; Sutherland, J. D. *Tetrahedron Lett.* **1998**, *39*, 6593–6596.
[551] Fréchet, J. M.; Schuerch, C. *J. Am. Chem. Soc.* **1971**, *93*, 492–496.
[552] Fruchart, J.-S.; Gras-Masse, H.; Melnyk, O. *Tetrahedron Lett.* **1999**, *40*, 6225–6228.
[553] Bernard, M.; Ford, W. T. *J. Org. Chem.* **1983**, *48*, 326–332.
[554] Claffey, D. J.; Ruth, J. A. *Tetrahedron: Asymmetry* **1997**, *8*, 3715–3716.
[555] Johnson, C. R.; Zhang, B. R. *Tetrahedron Lett.* **1995**, *36*, 9253–9256.
[556] Hughes, I. *Tetrahedron Lett.* **1996**, *37*, 7595–7598.
[557] Bolli, M. H.; Ley, S. V. *J. Chem. Soc. Perkin Trans. 1* **1998**, 2243–2246.
[558] Nicolaou, K. C.; Pastor, J.; Winssinger, N.; Murphy, F. *J. Am. Chem. Soc.* **1998**, *120*, 5132–5133.
[559] Li, W.; Burgess, K. *Tetrahedron Lett.* **1999**, *40*, 6527–6530.
[560] Yamada, M.; Miyajima, T.; Horikawa, H. *Tetrahedron Lett.* **1998**, *39*, 289–292.
[561] Michels, R.; Kato, M.; Heitz, W. *Makromol. Chem.* **1976**, *177*, 2311–2320.
[562] Nicolaou, K. C.; Pastor, J.; Barluenga, S.; Winssinger, N. *Chem. Commun.* **1998**, 1947–1948.
[563] Kamogawa, H.; Kanzawa, A.; Kadoya, M.; Naito, T.; Nanasawa, M. *Bull. Chem. Soc. Jpn.* **1983**, *56*, 762–765.
[564] Zaragoza, F. *Metal Carbenes in Organic Synthesis*; Wiley-VCH: Weinheim, New York, **1999**.
[565] Nicolaou, K. C.; Winssinger, N.; Pastor, J.; Ninkovic, S.; Sarabia, F.; He, Y.; Vourloumis, D.; Yang, Z.; Li, T.; Giannakakou, P.; Hamel, E. *Nature* **1997**, *387*, 268–272.
[566] Peters, J. U.; Blechert, S. *Synlett* **1997**, 348–350.
[567] van Maarseveen, J. H.; den Hartog, J. A. J.; Engelen, V.; Finner, E.; Visser, G.; Kruse, C. G. *Tetrahedron Lett.* **1996**, *37*, 8249–8252.
[568] Piscopio, A. D.; Miller, J. F.; Koch, K. *Tetrahedron Lett.* **1998**, *39*, 2667–2670.
[569] Piscopio, A. D.; Miller, J. F.; Koch, K. *Tetrahedron* **1999**, *55*, 8189–8198.
[570] Piscopio, A. D.; Miller, J. F.; Koch, K. *Tetrahedron Lett.* **1997**, *38*, 7143–7146.
[571] Veerman, J. J. N.; van Maarseveen, J. H.; Visser, G. M.; Kruse, C. G.; Schoemaker, H. E.; Hiemstra, H.; Rutjes, F. P. J. T. *Eur. J. Org. Chem.* **1998**, 2583–2589.
[572] Sim, M. M.; Lee, C. L.; Ganesan, A. *Tetrahedron Lett.* **1998**, *39*, 2195–2198.
[573] Sim, M. M.; Lee, C. L.; Ganesan, A. *Tetrahedron Lett.* **1998**, *39*, 6399–6402.
[574] Zaragoza, F. *Tetrahedron Lett.* **1997**, *38*, 7291–7294.
[575] Hamper, B. C.; Gan, K. Z.; Owen, T. J. *Tetrahedron Lett.* **1999**, *40*, 4973–4976.
[576] Cobb, J. M.; Fiorini, M. T.; Goddard, C. R.; Theoclitou, M. E.; Abell, C. *Tetrahedron Lett.* **1999**, *40*, 1045–1048.
[577] Woolard, F. X.; Paetsch, J.; Ellman, J. A. *J. Org. Chem.* **1997**, *62*, 6102–6103.
[578] Whitfield, D. M.; Ogawa, T. *Glycoconjugate J.* **1998**, *15*, 75–78.
[579] Chenera, B.; Finkelstein, J. A.; Veber, D. F. *J. Am. Chem. Soc.* **1995**, *117*, 11999–12000.
[580] Boehm, T. L.; Showalter, H. D. H. *J. Org. Chem.* **1996**, *61*, 6498–6499.
[581] Hone, N. D.; Davies, S. G.; Devereux, N. J.; Taylor, S. L.; Baxter, A. D. *Tetrahedron Lett.* **1998**, *39*, 897–900.
[582] Schuster, M.; Blechert, S. *Tetrahedron Lett.* **1998**, *39*, 2295–2298.
[583] Jin, S.; Holub, D. P.; Wustrow, D. J. *Tetrahedron Lett.* **1998**, *39*, 3651–3654.
[584] Bräse, S.; Enders, D.; Köbberling, J.; Avemaria, F. *Angew. Chem. Int. Ed. Engl.* **1998**, *37*, 3413–3415.
[585] Bräse, S.; Schroen, M. *Angew. Chem. Int. Ed. Engl.* **1999**, *38*, 1071–1073.
[586] Hennequin, L. F.; Piva-Le Blanc, S. *Tetrahedron Lett.* **1999**, *40*, 3881–3884.
[587] Halm, C.; Evarts, J.; Kurth, M. J. *Tetrahedron Lett.* **1997**, *38*, 7709–7712.

[588] Jung, K. W.; Zhao, X. Y.; Janda, K. D. *Tetrahedron Lett.* **1996**, *37*, 6491–6494.
[589] Jung, K. W.; Zhao, X. Y.; Janda, K. D. *Tetrahedron* **1997**, *53*, 6645–6652.
[590] Forman, F. W.; Sucholeiki, I. *J. Org. Chem.* **1995**, *60*, 523–528.
[591] Zhao, X. Y.; Jung, K. W.; Janda, K. D. *Tetrahedron Lett.* **1997**, *38*, 977–980.
[592] Ruhland, T.; Andersen, K.; Pedersen, H. *J. Org. Chem.* **1998**, *63*, 9204–9211.
[593] Kulkarni, B. A.; Ganesan, A. *Angew. Chem. Int. Ed. Engl.* **1997**, *36*, 2454–2455.
[594] Gayo, L. M.; Suto, M. J. *Tetrahedron Lett.* **1997**, *38*, 513–516.
[595] Siegel, M. G.; Hahn, P. J.; Dressman, B. A.; Fritz, J. E.; Grunwell, J. R.; Kaldor, S. W. *Tetrahedron Lett.* **1997**, *38*, 3357–3360.
[596] Parlow, J. J. *Tetrahedron Lett.* **1996**, *37*, 5257–5260.
[597] Bandgar, B. P.; Ghorpade, P. K.; Shrotri, N. S.; Patil, S. V. *Indian J. Chem. Sect. B (Org. Chem.)* **1995**, *34*, 153–155.
[598] Regen, S. L.; Kimura, Y. *J. Am. Chem. Soc.* **1982**, *104*, 2064–2065.
[599] Cainelli, G.; Contento, M.; Manescalchi, F.; Regnoli, R. *J. Chem. Soc. Perkin Trans. 1* **1980**, 2516–2519.
[600] Khound, S.; Das, P. J. *Tetrahedron* **1997**, *53*, 9749–9754.
[601] Tamami, B.; Kiasat, A. R. *Synthetic Commun.* **1996**, *26*, 3953–3958.
[602] Gibson, S. E.; Hales, N. J.; Peplow, M. A. *Tetrahedron Lett.* **1999**, *40*, 1417–1418.
[603] Arbo, B. E.; Isied, S. S. *Int. J. Pept. Prot. Res.* **1993**, *42*, 138–154.
[604] Mensi, N.; Isied, S. S. *J. Am. Chem. Soc.* **1987**, *109*, 7882–7884.
[605] Semmelhack, M. F.; Hilt, E.; Colley, J. H. *Tetrahedron Lett.* **1998**, *39*, 7683–7686.
[606] Ramage, R.; Raphy, G. *Tetrahedron Lett.* **1992**, *33*, 385–388.
[607] Ramage, R.; Wahl, F. O. *Tetrahedron Lett.* **1993**, *34*, 7133–7136.
[608] Hay, A. M.; Hobbs-DeWitt, S.; MacDonald, A. A.; Ramage, R. *Tetrahedron Lett.* **1998**, *39*, 8721–8724.

4 Preparation of Organometallic Compounds

Organometallic compounds are usually generated on insoluble supports as synthetic intermediates only, and not as target molecules. Organometallic compounds cover a broad range of reactivities, and are highly versatile reagents. In the following chapters the stoichiometric generation of main-group and transition metal derived organometallic compounds on supports is discussed.

4.1 Group I and II Organometallic Compounds

A flexible means of access to functionalized supports for solid-phase synthesis is based on metallated, cross-linked polystyrene, which reacts smoothly with a broad range of electrophiles. Cross-linked polystyrene can be lithiated directly by treatment with n-butyllithium and TMEDA in cyclohexane at 60–70 °C [1–3], to yield a product containing mainly *meta*- and *para*-lithiated phenyl groups [4]. Metallation of non-cross-linked polystyrene with potassium *tert*-amylate/3-(lithiomethyl)heptane has also been reported [5]. The latter type of base can, unlike butyllithium/TMEDA [6], also lead to benzylic metallation [7]. The C-lithiation of more acidic arenes or heteroarenes, such as imidazoles [8], thiophenes [9], and furans [9], has also been performed on insoluble supports (Figure 4.1). These reactions proceed, as in solution, with high regioselectivity.

Alternatively, lithiated polystyrene or lithiated resin-bound arylethers can be prepared from the corresponding bromo- or iodoarenes by halogen–lithium exchange [1,10–14] (Figure 4.2, Experimental Procedure 4.1) under conditions similar to those used in solution. Activated calcium, which can be generated by reduction of calcium(II) iodide with lithium biphenylide, reacts with fluorinated, chlorinated, or brominated polystyrene to yield polymeric arylcalcium compounds, which react with a number of different electrophiles [15]. Resin-bound aryl-, heteroaryl-, or vinylmagnesium reagents have been prepared by treatment of the corresponding iodides or bromides with isopropylmagnesium bromide [16,17] (Figure 4.2), and by transmetallation of lithiated polystyrene with magnesium bromide [13]. 4-Iodobenzoic acid esterified with cross-linked hydroxymethyl polystyrene can be converted into the corresponding *tert*-butyl zincate by halogen–metal exchange with lithium tri-*tert*-butylzincate (THF, 0 °C, 4 h [18]). This zincate can be successfully transmetallated with lithium cyano(2-thienyl)cuprate [18] (Entry **7**, Table 12.4). Metallated polystyrene-bound esters can usually not be warmed to room temperature because nucleophilic attack at the ester will lead to rapid decomposition of these reagents. These metallations must therefore be carefully optimized to identify the most suitable temperature and time for the reaction.

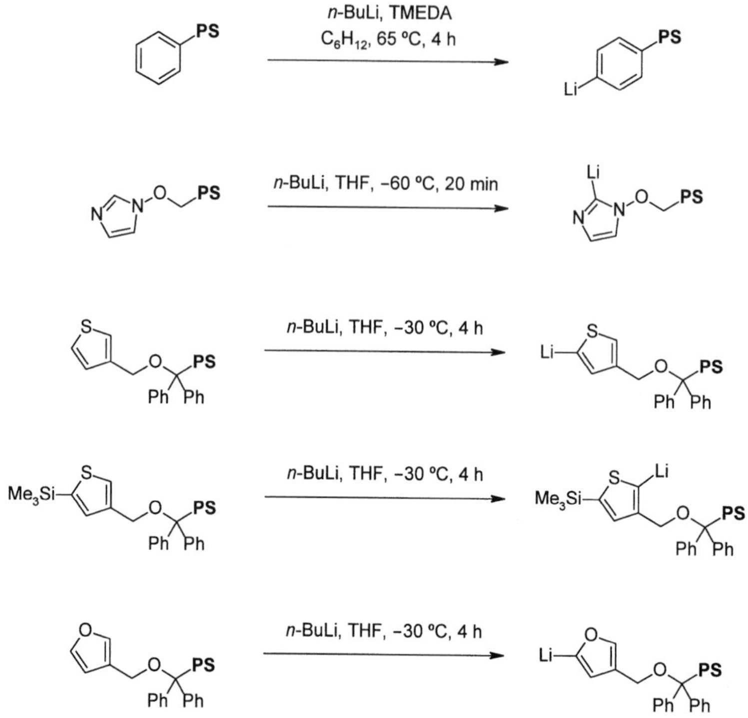

Fig. 4.1. Lithiation of polystyrene-bound arenes and heteroarenes [1,8,9].

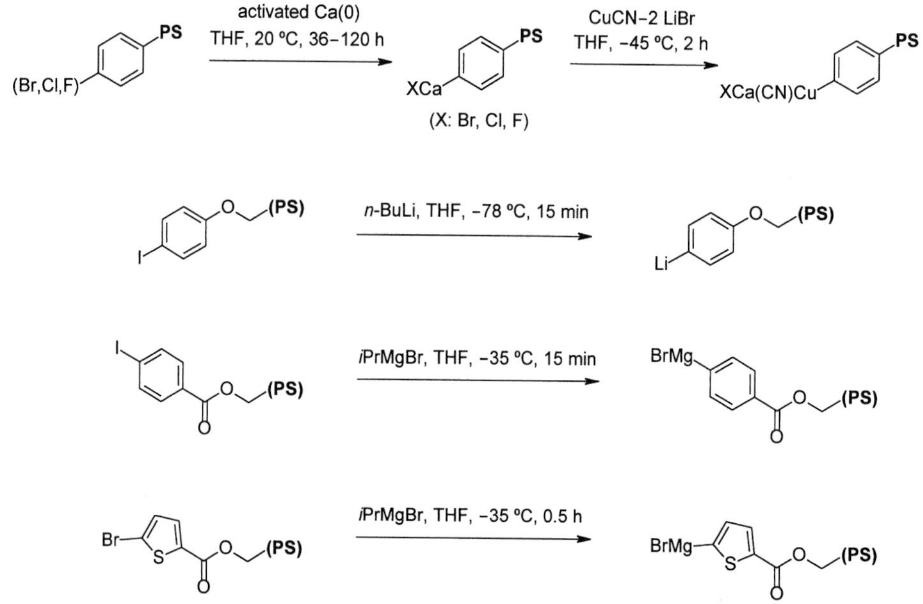

Fig. 4.2. Halogen–metal exchange on cross-linked polystyrene [1,12,15,16]. (PS): Wang resin.

Experimental Procedure 4.1: Lithiation of brominated, cross-linked polystyrene [10]

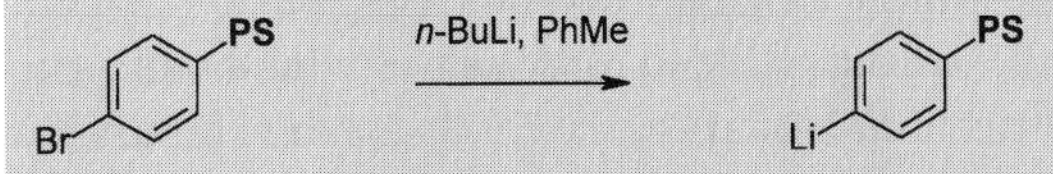

n-Butyllithium (100 mL, 1.6 mol/L in hexane, 160 mmol) was added to a suspension of brominated, cross-linked polystyrene (24.0 g, 70.6 mmol bromide; for preparation see Experimental Procedure 6.2) in toluene (200 mL). The mixture was stirred at room temperature for 2 h, and the resin was allowed to settle. After decantation more butyllithium (200 mL, 1.6 mol/L in hexane, 320 mmol) and toluene (200 mL) were added and the mixture was heated to 60 °C for 3 h. After cooling to room temperature and decantation the lithiated polystyrene was directly used without further washing.

Cross-linked chloromethyl polystyrene cannot be lithiated directly by treatment with lithium or methyllithium, because lithium is insoluble in polystyrene-compatible solvents and methyllithium leads to Wurtz coupling [19]. Lithiomethyl and potassiomethyl polystyrene are, however, available by transmetallation of tributylstannylmethyl polystyrene, which is prepared from Merrifield resin and (tributylstannyl)-lithium [19]. Lithiomethyl polystyrene has also been prepared by ether cleavage of ethoxymethyl polystyrenes with lithium biphenylide in THF [20]. The direct halogen–metal exchange in partially chloromethylated polystyrene can also be achieved with activated calcium [15] (Figure 4.3) and with magnesium/anthracene [21,22].

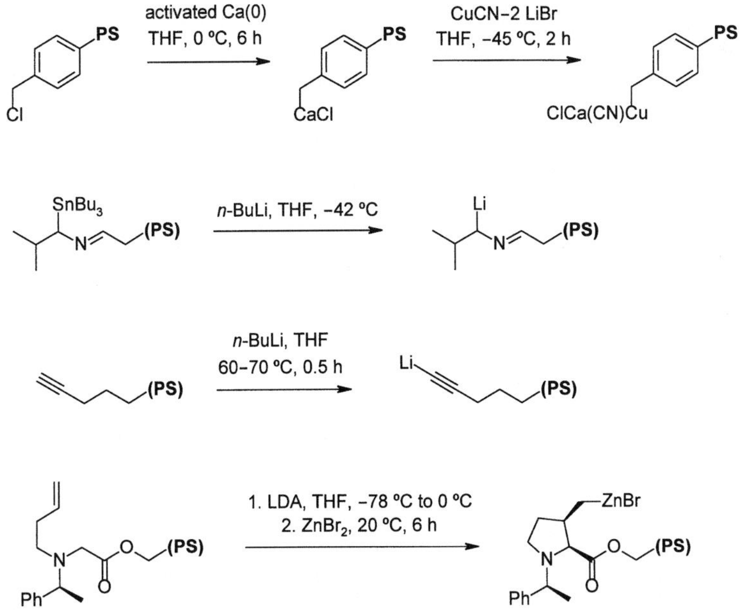

Fig. 4.3. Preparation of polystyrene-bound organometallic compounds by transmetallation and direct lithiation [15,23,24,26].

Non-aromatic organolithium compounds can be prepared by transmetallation of resin-bound stannanes [23] or by deprotonation of alkynes [24], triphenylmethane [25], or other resin-bound C–H acidic compounds with lithium amides or similar bases (Figure 4.3).

Organomercury compounds have been immobilized with polystyrene-bound carboxylates [27]. The resulting product was used as starting material for the preparation of radioactive 6-iodo DOPA (Figure 4.4).

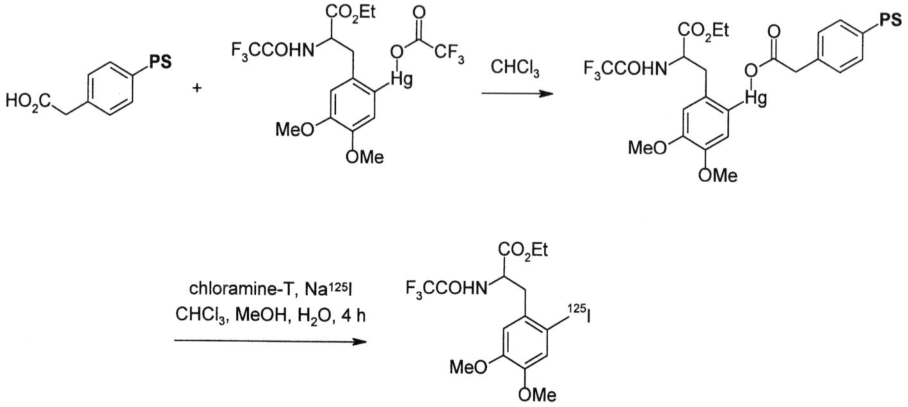

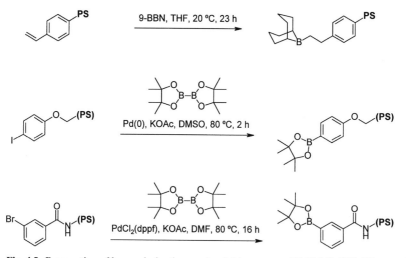

Fig. 4.4. Preparation of radioiodinated DOPA from a polystyrene-bound arylmercury precursor [27].

4.2 Group III Organometallic Compounds

Support-bound boranes have been prepared by hydroboration of vinyl polystyrene with 9-BBN, and used as intermediates for the preparation of hydroxyethyl polystyrene (Figure 4.5 [28]) and alkylated polystyrenes [29]. Hydroboration of vinyl polystyrene with diborane, on the other hand, proceeds with low regioselectivity [30].

Fig. 4.5. Preparation of boron derivatives on insoluble supports [12,30,31]. (PS): Wang resin.

Boronic esters and acids, which are useful intermediates for C–C bond forming reactions, can be obtained from the corresponding aryllithium compounds by treatment with trimethyl borate [1] followed by hydrolysis. Alternatively, support-bound aryl bromides or iodides can be converted into boronates by treatment with diboronates in the presence of a palladium catalyst (Figure 4.5).

4.3 Group IV Organometallic Compounds

Support-bound silanes are useful linkers and intermediates, and can be prepared by several routes. The most versatile approaches include the reaction of resin-bound organolithium compounds with chlorosilanes [32,33] and the hydrosilylation of resin-bound olefins (Figure 4.6). Additional transformations of resin-bound silanes are discussed in Chapter 3.11.2.

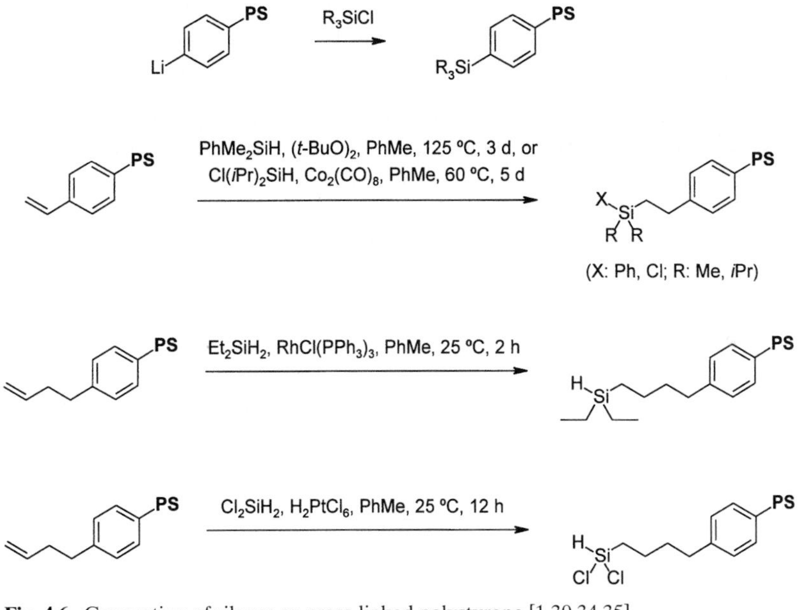

Fig. 4.6. Generation of silanes on cross-linked polystyrene [1,30,34,35].

For the preparation of support-bound stannanes, similar strategies as for silanes can be used (Figure 4.7). The preparation of stannanes on solid phase has not yet received much attention, but with the continuing development of solid-phase protocols for the preparation of elaborate carbon frameworks increased use of immobilized stannanes is to be expected.

Support-bound stannanes have been prepared from phenyllithium bound to macroporous polystyrene and chlorostannanes [13,36], by treatment of support-bound alkyl chlorides with lithiated stannanes [19,36], and by radical addition of stannanes to alkenes and alkynes (Figure 4.7 [37,38]).

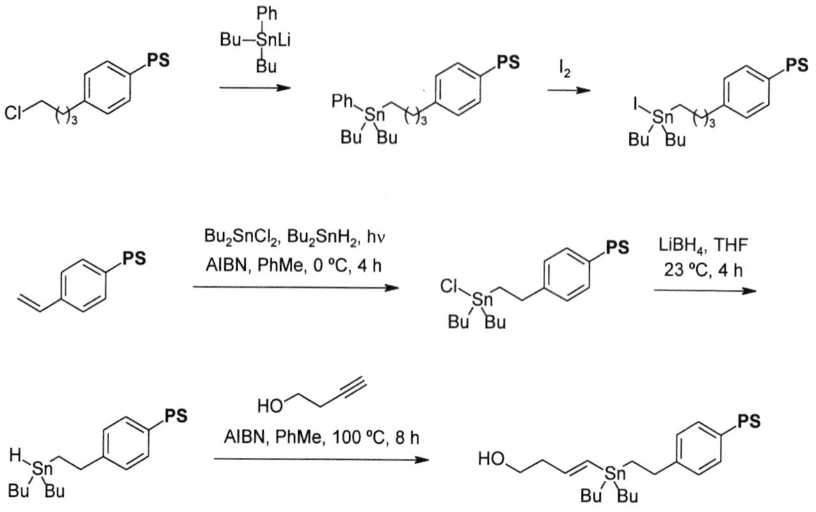

Fig. 4.7. Transformations of polystyrene-bound stannanes [36,39,40].

4.4 Transition Metal Complexes

Support-bound transition metal complexes have mainly been prepared as insoluble catalysts. Table 4.1 lists representative examples of such polymer-bound complexes. Several reviews have covered the preparation and application of support-bound reagents, including transition metal complexes [41–47]. Examples of the preparation and uses of organomercury and organozinc compounds are discussed in Chapter 4.1.

Table 4.1. Support-bound transition metal complexes as heterogenous catalysts.

Metal	Support	Support-bound Ligand	Application	Ref.
Sc	PE	sulfonamide	Lewis acid	[48,49]
Ti	PS-micro[a]	cyclopentadiene	ethylene polymerization	[50]
Ti	PS-macro[a]	cyclopentadiene	hydrogenation	[51]
Cr	PS, PE	pyridine, ammonium	oxidation of alcohols	[41]
Cr	PS, TG	phosphine	preparation of resin-bound chromium(II) carbene complexes	[52]
Mn	PS, PA	salen	asymmetric epoxidation	[53,54]
Ru, Ir	PS-micro	aminosulfonamide	asymmetric reduction of ketones	[55]
RuO_4^-	PS-macro	ammonium	oxidation of alcohols	[56]
OsO_4	PS, PE	pyridine, amine	dihydroxylation of alkenes	[41,57]
Co	PS, silica	salen	kinetic resolution of epoxides by enantioselective hydrolysis	[58]
Rh	PS-macro	cyclopentadiene	hydroformylation, hydrogenation	[59]
Rh	PS-micro	carboxylic acid	hydroformylation, hydrogenation	[60]
Rh	PS-micro	2-aminophosphine	asymmetric hydrogenation	[61]
Rh	PE	phosphine	hydrogenation	[62]
Rh	PS	bis(diphenyl)phosphine	asymmetric hydroformylation, carbonylation of RLi	[63]
Pd	TG	1,1'-binaphthyl-2-phosphine	asymmetric allylic substitution	[64]
Pd	TG	phosphine	Suzuki coupling	[65]
Pd	PS-macro	phosphine	allylic substitution	[66]
Pd	macroporous polysiloxane	thiourea	Suzuki coupling	[67]
Pd	poly(benz-imidazole)	nitrile	oxidation of alkenes to ketones (Wacker reaction)	[68]
Cu	PS	iminodiacetic acid	resolution of amino acids	[63]
Zn	PS-micro	sulfonamide	asymmetric cyclopropanation	[69]
various	PS-micro	imines	asymmetric epoxidation	[70]

[a] PS-micro: microporous, cross-linked polystyrene; PS-macro: macroporous, cross-linked polystyrene.

References for Chapter 4

[1] Farrall, M. J.; Fréchet, J. M. J. *J. Org. Chem.* **1976**, *41*, 3877–3882.
[2] Kobayashi, S.; Furuta, T.; Sugita, K.; Okitsu, O.; Oyamada, H. *Tetrahedron Lett.* **1999**, *40*, 1341–1344.
[3] Halm, C.; Evarts, J.; Kurth, M. J. *Tetrahedron Lett.* **1997**, *38*, 7709–7712.
[4] Broaddus, C. D. *J. Org. Chem.* **1970**, *35*, 10–15.
[5] Lochmann, L.; Fréchet, J. M. J. *Macromolecules* **1996**, *29*, 1767–1771.
[6] Chalk, A. J. *Polymer Lett.* **1968**, *6*, 649–651.
[7] Schlosser, M.; Strunk, S. *Tetrahedron Lett.* **1984**, *25*, 741–744.
[8] Havez, S.; Begtrup, M.; Vedsø, P.; Andersen, K.; Ruhland, T. *J. Org. Chem.* **1998**, *63*, 7418–7420.
[9] Li, Z.; Ganesan, A. *Synlett* **1998**, 405–406.
[10] Ruhland, T.; Andersen, K.; Pedersen, H. *J. Org. Chem.* **1998**, *63*, 9204–9211.
[11] Bernard, M.; Ford, W. T. *J. Org. Chem.* **1983**, *48*, 326–332.
[12] Tempest, P. A.; Armstrong, R. W. *J. Am. Chem. Soc.* **1997**, *119*, 7607–7608.
[13] Weinshenker, N. M.; Crosby, G. A.; Wong, J. Y. *J. Org. Chem.* **1975**, *40*, 1966–1971.
[14] Crosby, G. A.; Weinshenker, N. M.; Uh, H. S. *J. Am. Chem. Soc.* **1975**, *97*, 2232–2235.
[15] O'Brien, R. A.; Chen, T.; Rieke, R. D. *J. Org. Chem.* **1992**, *57*, 2667–2677.

[16] Boymond, L.; Rottländer, M.; Cahiez, G.; Knochel, P. *Angew. Chem. Int. Ed. Engl.* **1998**, *37*, 1701–1703.
[17] Rottländer, M.; Knochel, P. *J. Comb. Chem.* **1999**, *1*, 181–183.
[18] Kondo, Y.; Komine, T.; Fujinami, M.; Uchiyama, M.; Sakamoto, T. *J. Comb. Chem.* **1999**, *1*, 123–126.
[19] Brix, B.; Clark, T. *J. Org. Chem.* **1988**, *53*, 3365–3366.
[20] Mix, H. *Z. Chem.* **1979**, *19*, 148–149.
[21] Itsuno, S.; Darling, G. D.; Stöver, H. D. H.; Fréchet, J. M. J. *J. Org. Chem.* **1987**, *52*, 4644–4645.
[22] Harvey, S.; Raston, C. L. *J. Chem. Soc. Chem. Commun.* **1988**, 652–653.
[23] Pearson, W. H.; Clark, R. B. *Tetrahedron Lett.* **1997**, *38*, 7669–7672.
[24] Fyles, T. M.; Leznoff, C. C.; Weatherston, J. *Can. J. Chem.* **1977**, *55*, 4135–4143.
[25] Cohen, B. J.; Kraus, M. A.; Patchornik, A. *J. Am. Chem. Soc.* **1981**, *103*, 7620–7629.
[26] Karoyan, P.; Triolo, A.; Nannicini, R.; Giannotti, D.; Altamura, M.; Chassaing, G.; Perrotta, E. *Tetrahedron Lett.* **1999**, *40*, 71–74.
[27] Kawai, K.; Ohta, H.; Channing, M. A.; Kubodera, A.; Eckelman, W. C. *Applied Radiation and Isotopes* **1996**, *47*, 37–44.
[28] Sylvain, C.; Wagner, A.; Mioskowski, C. *Tetrahedron Lett.* **1998**, *39*, 9679–9680.
[29] Vanier, C.; Wagner, A.; Mioskowski, C. *Tetrahedron Lett.* **1999**, *40*, 4335–4338.
[30] Stranix, B. R.; Gao, J. P.; Barghi, R.; Salha, J.; Darling, G. D. *J. Org. Chem.* **1997**, *62*, 8987–8993.
[31] Piettre, S. R.; Baltzer, S. *Tetrahedron Lett.* **1997**, *38*, 1197–1200.
[32] Chan, T.; Huang, W. *J. Chem. Soc. Chem. Commun.* **1985**, 909–911.
[33] Schuster, M.; Lucas, N.; Blechert, S. *Chem. Commun.* **1997**, 823–824.
[34] Stranix, B. R.; Liu, H. Q.; Darling, G. D. *J. Org. Chem.* **1997**, *62*, 6183–6186.
[35] Hu, Y.; Porco, J. A.; Labadie, J. W.; Gooding, O. W.; Trost, B. M. *J. Org. Chem.* **1998**, *63*, 4518–4521.
[36] Ruel, G.; The, N. K.; Dumartin, G.; Delmond, B.; Pereyre, M. *J. Organomet. Chem.* **1993**, *444*, C18–C20.
[37] Neumann, W. P.; Peterseim, M. *Reactive Polymers* **1993**, *20*, 189–205.
[38] Whitfield, D. M.; Ogawa, T. *Glycoconjugate J.* **1998**, *15*, 75–78.
[39] Nicolaou, K. C.; Winssinger, N.; Pastor, J.; Murphy, F. *Angew. Chem. Int. Ed. Engl.* **1998**, *37*, 2534–2537.
[40] Gerigk, U.; Gerlach, M.; Neumann, W. P.; Vieler, R.; Weintritt, V. *Synthesis* **1990**, 448–452.
[41] Maud, J. M. In *Traditional Organic Reactions aided by Solid Supports or Catalysts*; **1995**; pp 171–192.
[42] Fréchet, J. M. J.; Darling, G. D.; Itsuno, S.; Lu, P. Z.; de Meftahi, M. V.; Rolls, W. A. *Pure Appl. Chem.* **1988**, *60*, 353–364.
[43] Bailey, D. C.; Langer, S. H. *Chem. Rev.* **1981**, *81*, 109–148.
[44] Shuttleworth, S. J.; Allin, S. M.; Sharma, P. K. *Synthesis* **1997**, 1217–1239.
[45] Burgess, K.; Porte, A. M. *Advances in Catalytic Processes* **1997**, *2*, 69–82.
[46] Bergbreiter, D. E. *Macromol. Symposia* **1996**, *105*, P9–16, 9–16.
[47] Drewry, D. H.; Coe, D. M.; Poon, S. *Med. Res. Rev.* **1999**, *19*, 97–148.
[48] Kobayashi, S.; Nagayama, S. *J. Am. Chem. Soc.* **1996**, *118*, 8977–8978.
[49] Kobayashi, S.; Nagayama, S.; Busujima, T. *Tetrahedron Lett.* **1996**, *37*, 9221–9224.
[50] Barrett, A. G. M.; de Miguel, Y. R. *Chem. Commun.* **1998**, 2079–2080.
[51] Bonds, W. D.; Brubaker, C. H.; Chandrasekaran, E. S.; Gibbons, C.; Grubbs, R. H.; Kroll, L. C. *J. Am. Chem. Soc.* **1975**, *97*, 2128–2132.
[52] Maiorana, S.; Seneci, P.; Rossi, T.; Baldoli, C.; Ciraco, M.; de Magistris, E.; Licandro, E.; Papagni, A.; Provera, S. *Tetrahedron Lett.* **1999**, *40*, 3635–3638.
[53] Canali, L.; Cowan, E.; Deleuze, H.; Gibson, C. L.; Sherrington, D. C. *Chem. Commun.* **1998**, 2561–2562.
[54] Angelino, M. D.; Laibinis, P. E. *Macromolecules* **1998**, *31*, 7581–7587.
[55] ter Halle, R.; Schulz, E.; Lemaire, M. *Synlett* **1997**, 1257–1258.
[56] Ley, S. V.; Bolli, M. H.; Hinzen, B.; Gervois, A. G.; Hall, B. J. *J. Chem. Soc. Perkin Trans. 1* **1998**, 2239–2241.
[57] Nagayama, S.; Endo, M.; Kobayashi, S. *J. Org. Chem.* **1998**, *63*, 6094–6095.
[58] Annis, D. A.; Jacobsen, E. N. *J. Am. Chem. Soc.* **1999**, *121*, 4147–4154.
[59] Dygutsch, D. P.; Eilbracht, P. *Tetrahedron* **1996**, *52*, 5461–5468.
[60] Andersen, J. A. M.; Karodia, N.; Miller, D. J.; Stones, D.; Gani, D. *Tetrahedron Lett.* **1998**, *39*, 7815–7818.
[61] Gilbertson, S. R.; Wang, X. F. *Tetrahedron Lett.* **1996**, *37*, 6475–6478.
[62] Bergbreiter, D. E.; Chandran, R. *J. Am. Chem. Soc.* **1987**, *109*, 174–179.
[63] Akelah, A.; Sherrington, D. C. *Chem. Rev.* **1981**, *81*, 557–587.

[64] Uozumi, Y.; Danjo, H.; Hayashi, T. *Tetrahedron Lett.* **1998**, *39*, 8303–8306.
[65] Uozumi, Y.; Danjo, H.; Hayashi, T. *J. Org. Chem.* **1999**, *64*, 3384–3388.
[66] Trost, B. M.; Warner, R. W. *J. Am. Chem. Soc.* **1983**, *105*, 5940–5942.
[67] Zhang, T. Y.; Allen, M. J. *Tetrahedron Lett.* **1999**, *40*, 5813–5816.
[68] Tang, H. G.; Sherrington, D. C. *J. Mol. Catal.* **1994**, *94*, 7–17.
[69] Halm, C.; Kurth, M. J. *Angew. Chem. Int. Ed. Engl.* **1998**, *37*, 510–512.
[70] Francis, M. B.; Jacobsen, E. N. *Angew. Chem. Int. Ed. Engl.* **1999**, *38*, 937–941.

5 Preparation of Hydrocarbons

5.1 Preparation of Alkanes

The preparation of unfunctionalized alkanes on insoluble supports has only recently received attention. With the aim of preparing ever more elaborate molecules on solid phase, chemists are currently searching for robust methods to assemble complex carbon frameworks on insoluble supports. The synthesis of alkanes has been investigated for this purpose.

Most of the reported preparations of alkanes on insoluble supports can be categorized into hydrolysis of organometallic compounds, reductions, hydrogenations, and coupling reactions.

5.1.1 Preparation of Alkanes by Hydrolysis of Organometallic Compounds

This reaction is only of limited synthetic utility, and has mainly been used to verify if metallation had taken place [1–3], or to introduce regioselectively deuterium or tritium into a molecule. The solvolysis of silanes, organogermanium compounds, and phosphonium salts to yield alkanes with simultaneous cleavage from the support is discussed in Section 3.16.

5.1.2 Preparation of Alkanes by Hydrogenation and Reduction

Heterogeneous catalysts are generally not compatible with insoluble supports. Although the hydrogenolytic cleavage of benzylic C–N [4] and C–O bonds [5] on cross-linked polystyrene with palladium(II) acetate and hydrogen or cyclohexadiene have been reported, catalysis with elemental palladium has not been widely used to hydrogenate alkenes on solid supports. The hydrogenation of C–C double bonds on solid phase is generally accomplished with soluble reagents or catalysts. One reagent which has been successfully used to hydrogenate alkenes and alkynes on cross-linked polystyrene is diimide (Figure 5.1). This reagent can be generated by copper-catalyzed oxidation of hydrazine or by thermolysis of sulfonyl hydrazides. As illustrated by the examples in Figure 5.1, hydrogenations with diimide leave esters and nitro groups intact.

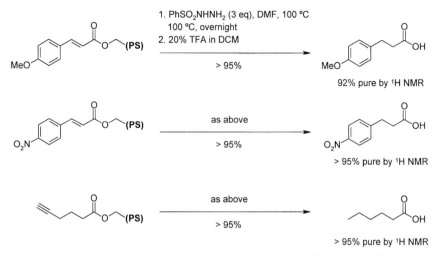

Fig. 5.1. Reduction of Wang resin bound alkenes and alkynes with diimide (HN=NH) [6].

Further reagents suitable for the hydrogenation of alkenes on insoluble supports are silanes in the presence of trifluoroacetic acid [7] and the copper(I) hydride complex [CuH(PPh$_3$)]$_6$ [8] (Figure 5.2). Catalytic hydrogenations can be performed with soluble catalysts such as RhCl(PPh$_3$)$_3$ (Wilkinson catalyst) or related complexes. An example [9] of the use of a chiral rhodium(I) complex as catalyst for the asymmetric hydrogenation of a dehydroamino acid derivative on polystyrene is sketched in Figure 5.2. The yield and stereoselectivity of the hydrogenation were highly dependent on the solvent, but attained, under optimized conditions (Figure 5.2), values similar to those achievable in solution. Unsuitable solvents for the stereoselective hydrogenation were THF, dioxane, and DCM [9].

Polystyrene-bound acetophenones have been directly reduced to the corresponding alkanes with lithium aluminum hydride in the presence of aluminum trichloride (Figure 5.2 [10]).

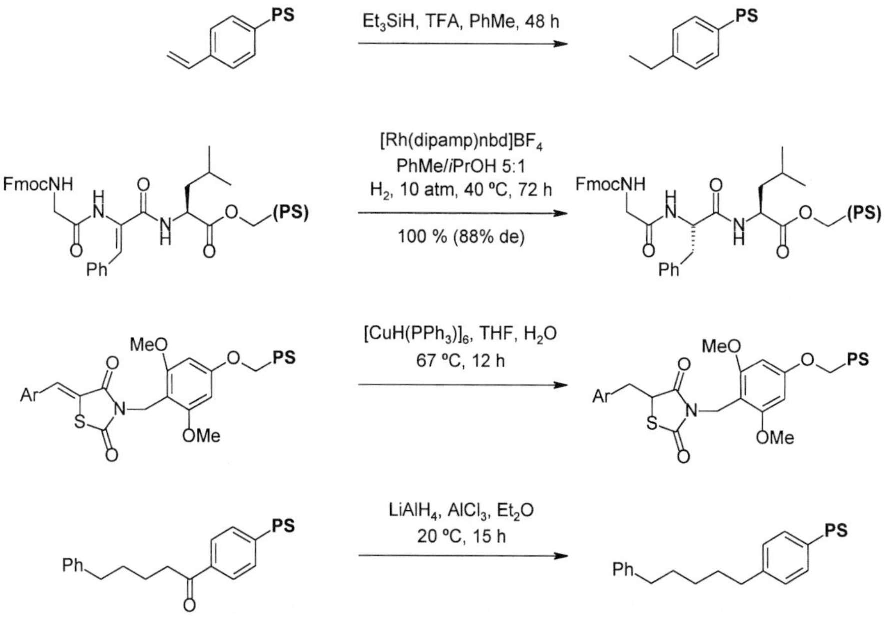

Fig. 5.2. Reduction of polystyrene-bound alkenes and ketones [7–10].

5.1.3 Preparation of Alkanes by Carbon–Carbon Bond Formation

C-Alkylations have been performed both with support-bound carbon nucleophiles and with support-bound carbon electrophiles. Benzyl, allyl, and aryl halides or triflates have mostly been used as carbon electrophiles. Suitable carbon nucleophiles are boranes and organozinc- and organomagnesium compounds. C-Alkylations have also been accomplished by addition of radicals to alkenes. Polystyrene can also be alkylated under harsh conditions, e.g., by Friedel–Crafts alkylation [11–16] in the presence of strong acids. This type of reaction is incompatible with most linkers and only suitable for the preparation of functionalized supports.

Few examples have been reported of the preparation of alkanes by C–C bond formation on solid phase, and general methodology for such preparations is still scarce.

5.1.3.1 Coupling Reactions with Group I Organometallic Compounds

Lithium-, sodium-, and potassium-derived organometallic reagents are highly reactive and can be handled only under strict exclusion of water and oxygen. As discussed in Section 4.1, organolithium and related reagents can be prepared on cross-linked polystyrene by direct lithiation or halogen–metal exchange. The treatment of these immobilized organometallic compounds with alkyl halides has been used to prepare hydrocarbons on insoluble supports. Some of the few examples reported are listed in

Table 5.1. Support-bound cuprates have been prepared by transmetallation of organocalcium compounds [1], and react with alkyl halides to yield alkanes (Entry **4**, Table
5.1). Alternatively, support-bound carbon electrophiles can be alkylated by treatment
with organometallic reagents. This strategy has, e.g., been used to prepare polystyrene-bound cyclopentadienes (Entry **5**, Table 5.1), which can be used as ligands for
transition metals.

Table 5.1. Alkylations with organolithium, organosodium, and organocopper compounds.

Entry	Starting resin	Conditions	Product	Ref.
1		R⌒⌒Br PhMe, 65 °C, 24 h		[17]
2		Cl⌒⌒⌒Br (10 eq), THF, 20 °C, 4 h (macroporous PS)		[18]
3		MeI, THF, −30 °C to 20 °C, 2 h		[19]
4	ClCa(CN)Cu	EtO₂C⌒⌒Br THF, −45 °C, 2 h, 20 °C, 18 h		[1]
5		cyclopentadiene, NaNH₂, THF, 67 °C, 150 h (macroporous PS)	(mixture of isomers)	[20] see also [17,21]

5.1.3.2 Coupling Reactions with Group II Organometallic Compounds

Support-bound arylmagnesium and arylzinc compounds can be prepared from the
corresponding haloarenes (Section 4.1) and alkylated with alkyl halides (Entries **1** and
2, Table 5.2). Alternatively, support-bound carbon electrophiles, such as benzyl
halides or aryl triflates, can be coupled with organomagnesium and organozinc compounds (Entries **3–6**, Table 5.2). Organozinc reagents are less reactive than organomagnesium or organolithium compounds, and are therefore compatible with several
functional groups. However, alkylations with organozinc reagents often require catalysis by palladium complexes. Unfortunately, organozinc compounds are not always
easy to prepare, but, in view of the potential of these reagents, further developments
in this field are to be expected.

Table 5.2. Alkylations with organomagnesium and organozinc compounds.

Entry	Starting resin	Conditions	Product	Ref.
1		allyl bromide (50 eq), LiCl, CuCN, THF, −35 °C, 40 min		[22]
2		MeI, THF, 0–20 °C		[23]
3		MgCl (2.5 eq), PhMe, THF, 60 °C, 12 h		[24]
4		MgBr (30 eq), CuBr•Me₂S (3 eq), THF, −10 °C, 6 h		[25]
5		Pd(dba)₂, dppf, 65 °C, 16 h		[26]
6		Pd(dba)₂, PR₃, THF, 25 °C, 20 h		[26]

5.1.3.3 Coupling Reactions with Boranes

Polystyrene-bound trialkylboranes, which can be prepared by hydroboration of support-bound alkenes with 9-BBN, undergo palladium-mediated coupling with alkyl, vinyl, and aryl iodides (Suzuki coupling; Entries 1 and 2, Table 5.3; for vinylations see Section 5.2.4). Because boranes are compatible with many functional groups and do not react with water, these coupling reactions could become a powerful tool for solid-phase synthesis. So far, however, few examples have been reported.

Alternatively, boranes can be prepared in solution and then coupled with support-bound carbon electrophiles. The Suzuki coupling of alkylboranes, generated in situ from 9-BBN and alkenes, with brominated cross-linked polystyrene has been used to link substituted alkyl chains directly to the polymer (Entry **4**, Table 5.3). Alkylboranes have also been used to alkylate polystyrene-bound aryl iodides (Entries **3** and **5**, Table 5.3).

Table 5.3. Alkylations with boranes.

Entry	Starting resin	Conditions	Product	Ref.
1		THPO-(̵)₅-I (4 eq), PdCl₂(dppf) (0.3 eq), K₂CO₃ (3 eq), Triton B (1.5 eq), DMF, 85 °C, 14 h		[27]
2		(4 eq), Pd(OAc)₂ (0.3 eq), PPh₃ (0.9 eq), NaOH (3 eq), Triton B (1.5 eq), DMF, 85 °C, 14 h		[27]
3		BBu₃, Pd(PPh₃)₄ or PdCl₂(dppf), K₂CO₃, DMF, 20 °C, 20 h		[28]
4		Pd(PPh₃)₄, Na₂CO₃, THF Ar: 4-(MeO)C₆H₄		[29] see also [30]
5		PdCl₂(MeCN)₂ (3 eq), dppf (0.2 eq), K₃PO₄ (6 eq), dioxane/DMF 1:6, 50 °C, 30 h		[31]

5.1.3.4 Coupling Reactions with Arylpalladium Compounds

The arylation sketched in Figure 5.3 is a rare example of a palladium-mediated hydroarylation of an alkene. Because of the polycyclic structure of the alkene, the intermediate formed by insertion of the alkene into the Pd–Ar bond does not undergo β-elimination (to yield the product of a normal Heck reaction) but remains unchanged. Reduction of this stable alkylpalladium intermediate with formic acid leads to the formation of the formally hydrogenated Heck product [32].

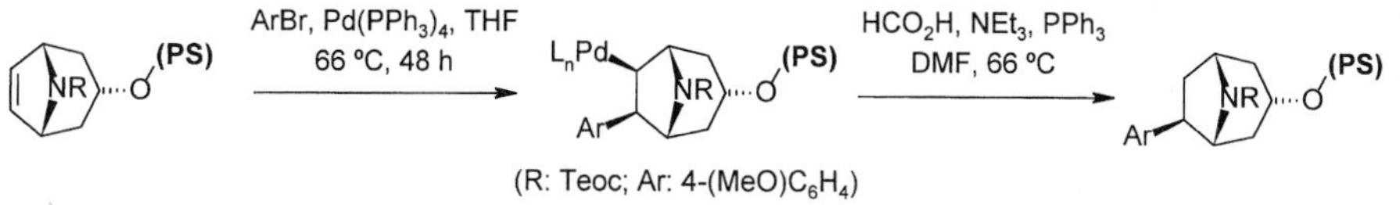

Fig. 5.3. Palladium-mediated hydroarylation of polystyrene-bound alkenes [32].

5.1.3.5 Alkylations with Alkyl Radicals

Alkanes can be prepared by addition of carbon radicals to C–C double bonds (Figure 5.4). The highest yields are usually obtained when electron-rich radicals (e.g. alkyl radicals or heteroatom-substituted radicals) add to acceptor-substituted alkenes or when electron-poor radicals add to electron-rich double bonds. These reactions have also been performed on solid phase, and polystyrene-based supports seem to be particularly suitable for radical-mediated processes [33,34].

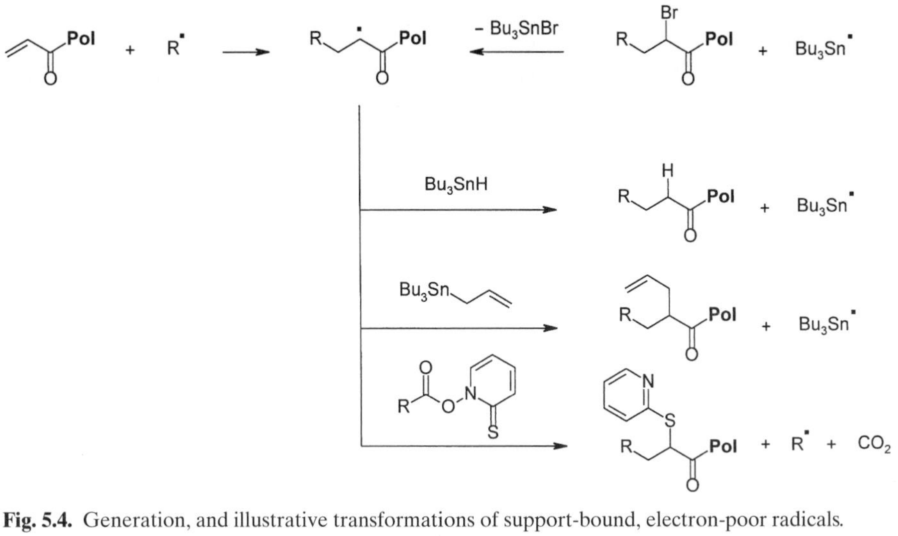

Fig. 5.4. Generation, and illustrative transformations of support-bound, electron-poor radicals.

Examples of radical-mediated C-alkylations are listed in Table 5.4. In these examples radicals are formed by halogen abstraction with tin radicals (Entries **1** and **2**), by photolysis of Barton esters (Entry **3**), and by reduction of organomercury compounds (Entry **4**). Cleavage from supports by homolytic cleavage with simultaneous formation of C–H or C–C bonds is treated in Section 3.16.

Table 5.4. Alkylations with carbon-centered radicals.

Entry	Starting resin	Conditions	Product	Ref.
1		AIBN, C$_6$H$_6$, 80 °C, 14 h		[35]
2		AIBN, C$_6$H$_6$, 80 °C, 14 h		[35]
3		(10 eq), hv, DCM, 0 °C, 4 h		[33]
4		(3 eq), NaBH$_4$ (8 eq), DCM, H$_2$O, 2 h		[36]

5.1.3.6 Preparation of Cycloalkanes

In addition to the alkylations discussed above, some special reactions have been re-ported which enable the solid-phase synthesis of cycloalkanes. These include the intra-molecular ene reaction and the cyclopropanation of alkenes substituted with electron-withdrawing groups (Figure 5.5).

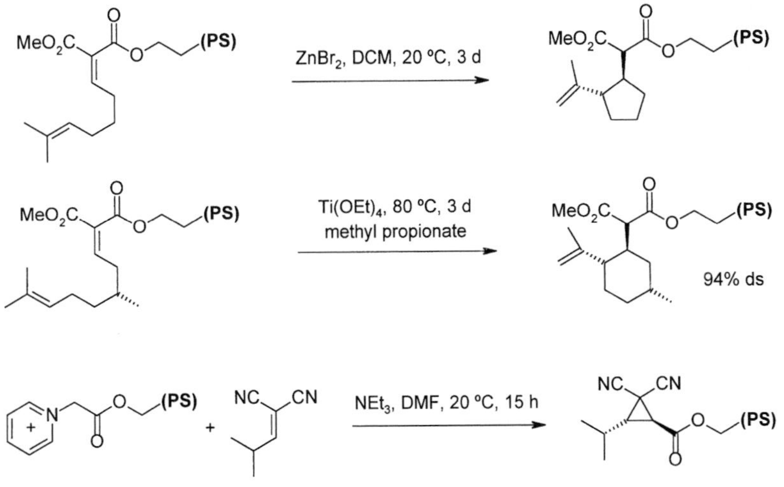

Fig. 5.5. Preparation of cycloalkanes on cross-linked polystyrene [37,38].

5.2 Preparation of Alkenes

5.2.1 Preparation of Alkenes by β-Elimination and Reduction

Linkers have been developed which enable the release of amines or other compounds by base-induced β-elimination; these linkers are thereby converted into alkenes (Chapter 3). Similarly, the β-elimination of resin-bound leaving groups has been used as cleavage strategy to release alkenes from supports (Section 3.15.1).

Two examples of the generation of alkenes by β-elimination are sketched in Figure 5.6. These reactions generally proceed under conditions similar to those used in solution. Further examples have been reported [39,40].

Internal alkynes attached to cross-linked polystyrene can be reduced stereoselectively to Z-olefins by treatment with di(1,2-dimethylpropyl)borane (Figure 5.6 [41]). Other reducing agents, such as 9-BBN, catechol borane, or diisobutylaluminum hydride required higher reaction temperatures and led to poor stereoselectivity.

Vinylpolystyrene, a useful intermediate for the preparation of various functionalized supports for solid-phase synthesis [7,44–46], has been prepared by polymerization of divinylbenzene [7], by Wittig reaction of a Merrifield resin derived phosphonium salt with formaldehyde [46–49], or, most conveniently, by treatment of Merrifield resin with trimethylsulfonium iodide and a base [50] (Figure 5.7).

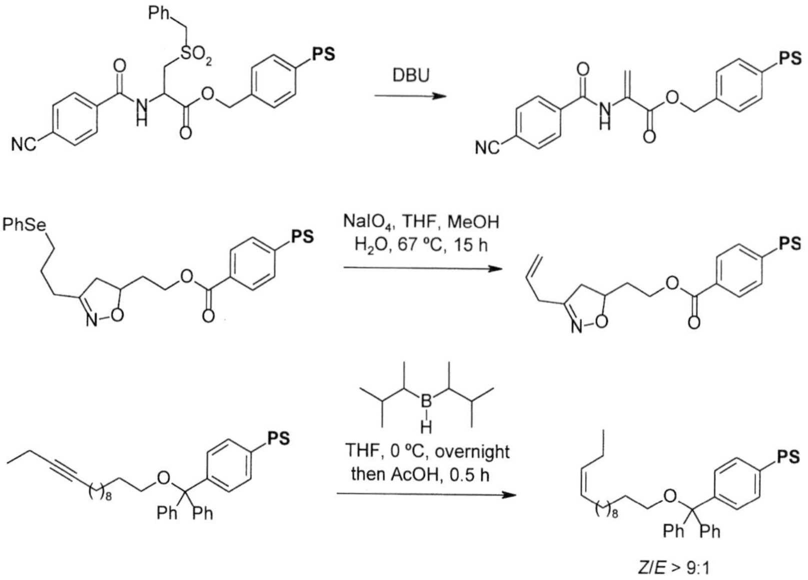

Fig. 5.6. Preparation of resin-bound alkenes by β-elimination and reduction [41–43].

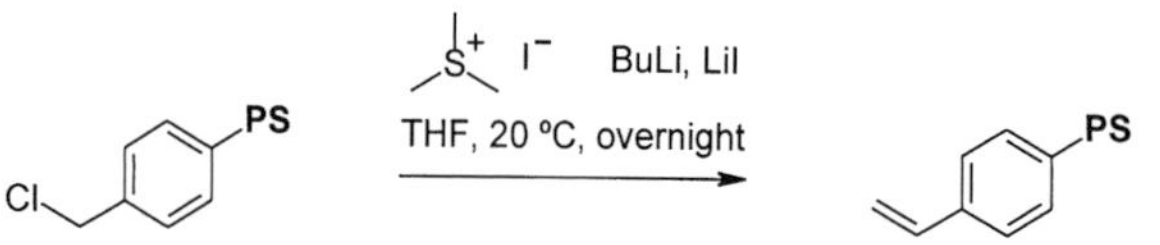

Fig. 5.7. One-step preparation of vinylpolystyrene from Merrifield resin [50].

5.2.2 Preparation of Alkenes by Carbonyl Olefination

Most carbonyl olefinations known to proceed in solution have also been realized with success on insoluble supports. Those Wittig reactions which lead to release of the alkene into solution (because phosphorus is irreversibly bound to the support) are treated in Section 3.15.

5.2.2.1 By Wittig Reaction

Support-bound aldehydes or ketones can be readily olefinated with different types of phosphorus ylide (Table 5.5). Stabilized (acceptor-substituted) ylides, which can already be generated with weak bases, generally react more slowly than non-stabilized ylides, which in turn are only formed upon treatment with strong bases. For Wittig reactions which require strong bases, hydrophobic supports, such as cross-linked polystyrene, are generally more convenient than hydrophilic supports, because the latter are more difficult to dry.

Polystyrene-bound benzaldehydes can be smoothly olefinated with benzyl- or cinnamylphosphonium salts in DMF or THF using sodium methoxide as base (Entry **1**, Table 5.5 [51–54]). Alkylphosphonium salts, however, only react with resin-bound aldehydes upon deprotonation with stronger bases, such as butyllithium [25,55–57]. The more acidic acceptor-substituted phosphonium salts, on the other hand, even react with resin-bound aldeydes and ketones upon treatment with tertiary amines, DBU, sodium ethoxide, or lithium hydroxyde [58–62], but stronger bases are also used occasionally [63].

Table 5.5. Wittig olefination of support-bound aldehydes and ketones.

Entry	Starting resin	Conditions	Product	Ref.
1		BnPPh₃Br (0.15 mol/L), NaOMe (0.77 mol/L), DMF, 20 °C, 24 h		[64] see also [25]
2		MeO₂C⌒PPh₃ THF, 65 °C, 2 h		[58] see also [65]
3		EtO₂C⌒PO(OEt)₂ DBU, LiBr, MeCN, 50 °C, 10 h		[58]
4		Ph₃PMeBr (5 eq), NaN(SiMe₃)₂, THF, 20 °C, 18 h		[66]

Wittig reactions can also be performed with support-bound phosphorus ylides. Polystyrene-bound alkylphosphonium salts have been prepared from the corresponding alkyl mesylates or halides and trialkyl- or triarylphosphines (Figure 5.8 [47,67]). Because polystyrene is a hydrophobic support, salt-formation does not proceed smoothly, and quaternization of phosphines generally requires forcing conditions. The resulting phosphonium salts generally require strong bases such as butyllithium [55,68,69] or sodium bis(trimethylsilyl)amide [67,70] to react with carbonyl compounds (Figure 5.8).

Highly reactive, polystyrene-bound alkylating agents, such as haloketones [72], haloacetic acid derivatives [73,74], or benzyl halides [49,68,70] can be converted into phosphonium salts under milder conditions than alkyl halides (Figure 5.9). Resin-bound, acceptor-substituted phosphonium salts or phosphonates can undergo Wittig olefination in the presence of weak bases, such as tertiary amines [74–77], DBU [73,78], or sodium hydroxide [72]. Stronger bases (BuLi [73]; LiN(SiMe₃)₂ [79,80]; KOtBu [81]) are often used, however, to achieve quantitative ylide formation and thereby accelerate the reaction (Figure 5.9).

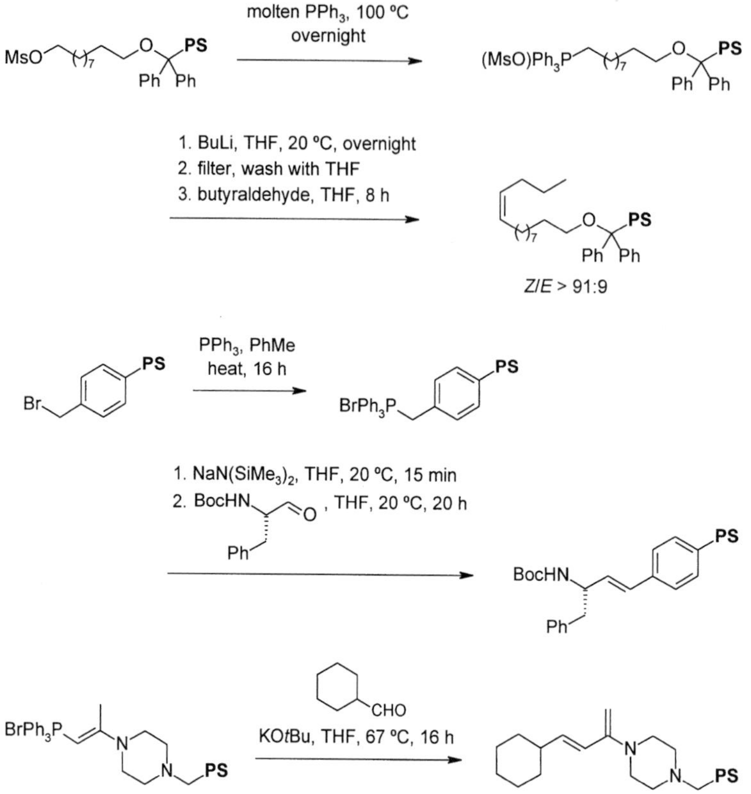

Fig. 5.8. Wittig olefination with polystyrene-bound alkylphosphonium salts [55,70,71].

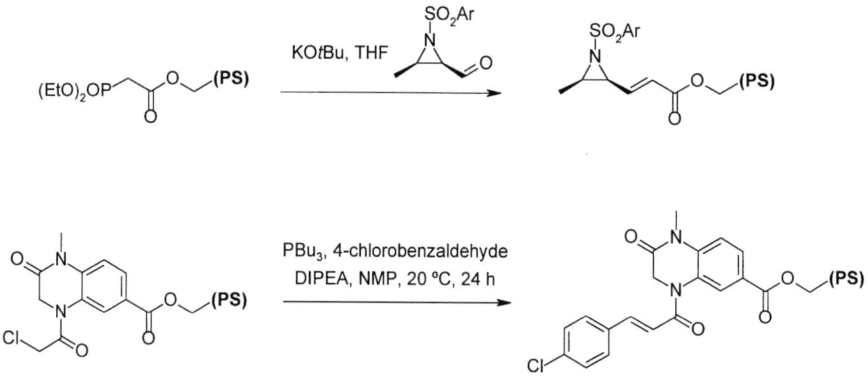

Fig. 5.9. Wittig olefination with acceptor-substituted, polystyrene-bound phosphonates and phosphonium salts [74,81]. Ar: 2-nitrophenyl.

5.2.2.2 By Aldol and Related Condensations

Most C,H-acidic compounds can be condensed with aldehydes or ketones to yield olefins. Some of these reactions have also been realized on solid supports, either with the C,H-acidic (nucleophilic) reactant or the electrophilic reactant linked to the sup-

Table 5.6. Aldol-type condensation reactions on insoluble supports.

Entry	Starting resin	Conditions	Product	Ref.
1		(3 eq), piperidinium acetate (0.2 eq), DCM, 20 °C, 3 h		[37] see also [82-84]
2		piperidine, iPrOH/C$_6$H$_6$, 60 °C		[85] see also [86]
3		1-methylisatin (10 eq), DMF/piperidine 6:1, 20 °C, 20 h		[87]
4		ArCHO (0.22 mol/L, 10 eq), piperidinium acetate (1 eq), C$_6$H$_6$, 80 °C, 24 h Ar: 4-hydroxyphenyl		[8]
5		ArCHO, NaOMe (0.25 mol/L), MeOH/ THF 1:1, 20 °C, 4 d Ar: 4-methoxyphenyl		[88] see also [89]
6		malonic acid (0.22 mol/L), piperidine/pyridine 1:60, 115 °C, 4 h		[68]
7		MeNO$_2$/AcOH 1:1, NH$_4$OAc (0.87 mol/L), 116 °C, 15 h		[90]
8		PhCOMe (0.4 mol/L), DME, LiOH, 20 °C, 16 h		[62] see also [64]

port. Some illustrative examples are listed in Table 5.6. Polystyrene-bound malonic esters or amides, cyanoacetamides, and 3-oxo esters undergo Knoevenagel condensation with aromatic or aliphatic aldehydes. Catalytic amounts of piperidine and heating are generally required, but reactive substrates can react at room temperature.

One problem often encountered with this type of reaction is related to the fact that in solid-phase synthesis reagents generally have to be used in excess. This can occasionally lead to side reactions. If, for instance, a resin-bound aldehyde is condensed under basic conditions with excess ketone, the resulting enone might undergo Michael addition with an additional equivalent of ketone (Figure 5.10). The Wittig reaction might, for some of these enones be a superior synthetic method [62].

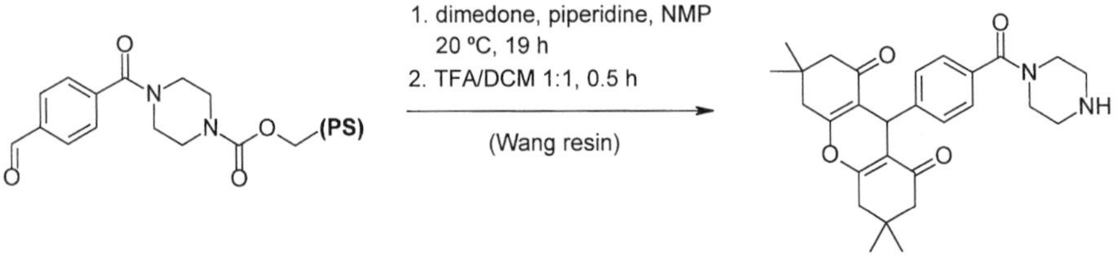

Fig. 5.10. Double condensation of resin-bound aldehydes with 1,3-diketones [91].

5.2.2.3 By Other Carbonyl Olefinations

Phosphorus ylides are generally not suitable for the olefination of esters. This reaction can, however, be accomplished with titanium carbene complexes. The Tebbe reagent, a precursor of $Cp_2Ti=CH_2$, has been used to methylenate polystyrene-bound esters [92] (Figure 5.11). The resulting enol ethers were either cleaved from the support by hydrolysis to yield methyl ketones, or used as starting material for Diels–Alder reactions or the synthesis of thiazoles [92] (Section 15.13).

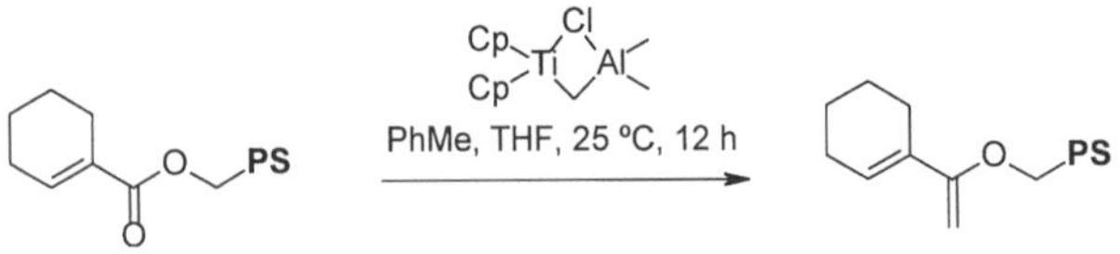

Fig. 5.11. Methylenation of esters using the Tebbe reagent [92].

5.2.3 Preparation of Alkenes by Olefin Metathesis

With the discovery by Grubbs of ruthenium carbene complexes such as $Cl_2(PCy_3)_2Ru=CHR$, which mediate olefin metathesis under mild reaction conditions and which are compatible with a broad range of functional groups [93], the application of olefin metathesis to solid-phase synthesis became a realistic approach to the preparation of alkenes. Both ring-closing metathesis and cross metathesis of alkenes and alkynes bound to insoluble supports have been realized (Figure 5.12).

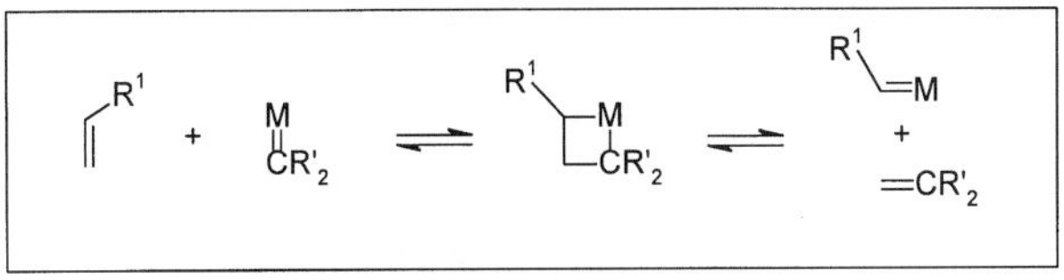

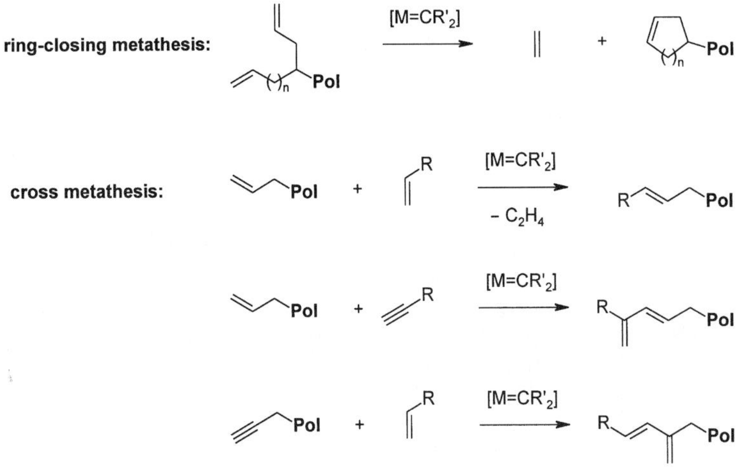

ring-closing metathesis:

$[M=CR'_2]$

cross metathesis:

$[M=CR'_2]$
$- C_2H_4$

$[M=CR'_2]$

$[M=CR'_2]$

Fig. 5.12. Main strategies for the application of olefin metathesis to solid-phase synthesis.

Carbene complex catalyzed olefin metathesis proceeds best in solvents of low nucleophilicity, such as DCM or toluene [93,94]. Tertiary amines, pyridine, or other nucleophiles can poison the (electrophilic) catalyst, and should be carefully removed (e.g. by washing with acetic acid) before addition of the carbene complex. Electron-rich alkenes, such as enol ethers or enamines, can generally not be metathesized, because the intermediate donor-substituted carbene complexes are no longer electrophilic enough to catalyze olefin metathesis.

The reaction rate of metathesis sharply decreases in the series terminal alkene > internal, disubstituted alkene > trisubstituted alkene. Because olefin metathesis is reversible, ring-closing metathesis is usually not well suited to the preparation of strained rings, but works best if five- or six-membered rings are formed.

Five- to eight-membered nitrogen-containing heterocycles [95–97] and macrocyclic peptoids [98] have been prepared by ring-closing metathesis on cross-linked polystyrene and polystyrene–PEG graft polymers. Some illustrative examples of such cyclizations are sketched in Figure 5.13. Ring-closing metathesis has also been used as cleavage reaction, to release into solution non-cyclic olefins or cycloalkenes. These reactions are discussed in Section 3.15.2.

Cross metathesis enables the efficient preparation of acyclic alkenes on insoluble supports (Figure 5.14). Unfortunately, some types of substrate have a high tendency to yield products of self metathesis, i.e. symmetric internal alkenes produced by dimerization of the resin-bound olefin. This is, for instance, the case for allylglycine and homoallylglycine derivatives. Dimerization of the resin-bound alkene can, however, be effectively supressed by reducing the loading of the support [99].

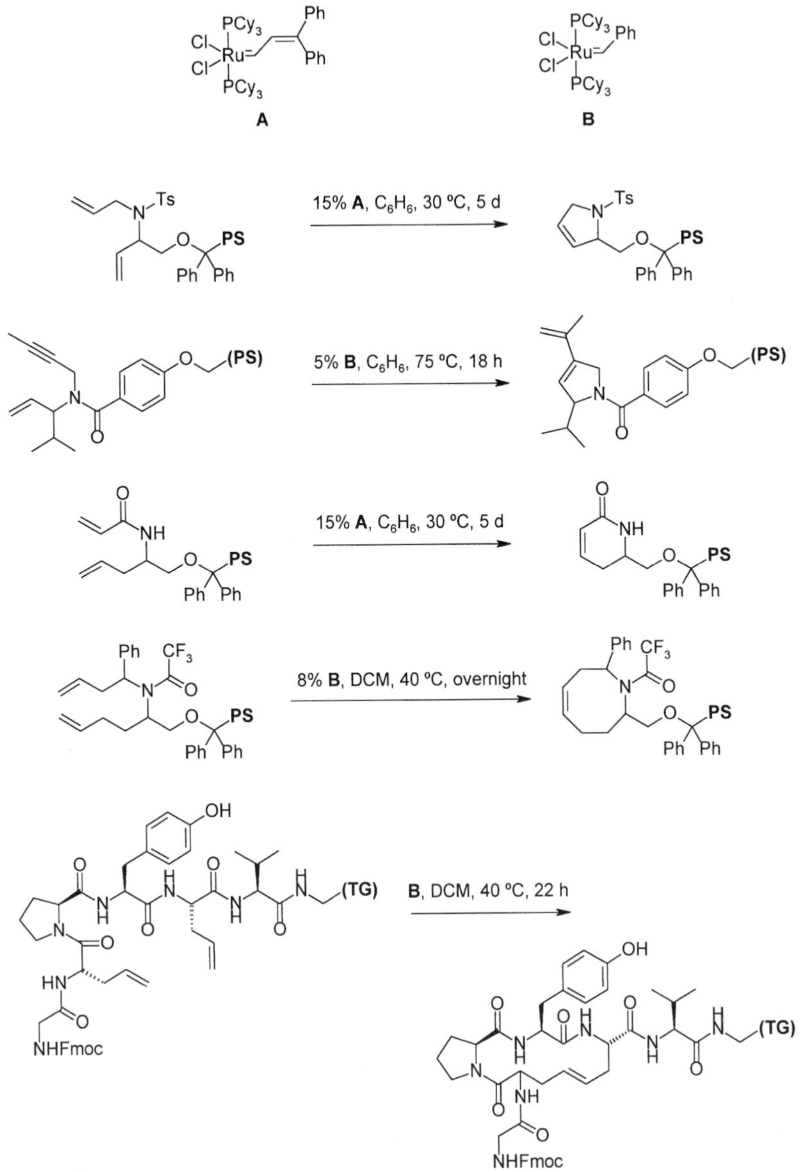

Fig. 5.13. Ring-closing metathesis on insoluble supports [95–98].

An interesting variant of cross metathesis is the so-called ring-opening cross metathesis [100]. When strained, cyclic alkenes (e.g. cyclobutene, norbornene) are treated with metathesis catalysts, rapid and irreversible ring-opening occurs. The newly formed carbene complexes can then react with a second alkene to yield the products of cross metathesis. One example of ring-opening cross metathesis on cross-linked polystyrene [100] is sketched in Figure 5.14.

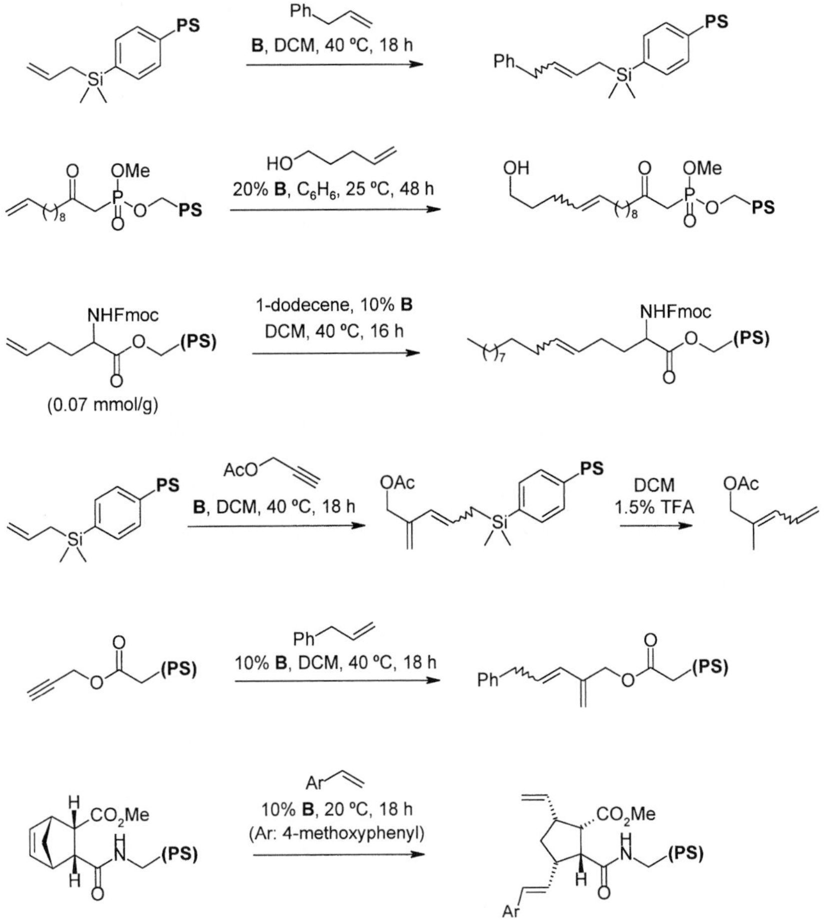

Fig. 5.14. Preparation of alkenes on polystyrene by cross metathesis [99–104]. **B**: Cl$_2$(Cy$_3$P)$_2$Ru=CHPh.

5.2.4 Preparation of Alkenes by C-Vinylation

Some types of alkene can be conveniently prepared by alkylation or arylation of vinyl groups. The Heck, Stille, and Suzuki couplings all belong to this group of reactions. In the Heck reaction an arene or heteroarene with a good leaving group (iodide, bromide, triflate, phenyliodonium, diazonium) is coupled with a (usually electron-poor) alkene in the presence of palladium(0) and a base. This useful reaction has also been realized on solid phase, either with the aryl halide or with the alkene linked to the support (Table 5.7). Further examples have been reported [34,105–110]. Aryl iodides, substituted with either electron-withdrawing or electron-donating groups, have usually been used as resin-bound aryl halide. Typical alkenes are acrylic esters or

amides, acrylonitrile, methacrylates, crotonates, vinylsulfones, α,β-unsaturated aldehydes or ketones, styrene, cinnamates, and 4-vinylbenzoates. The Heck reaction has
also been used for macrocyclizations on insoluble supports [106]. Heck reactions
which lead to heterocycles are treated in Chapter 15.

Table 5.7. C-Vinylation by Heck reaction on insoluble supports.

Entry	Starting resin	Conditions	Product	Ref.
1	(PS)	PhI, Pd(OAc)$_2$, NEt$_3$, DMF, Bu$_4$NCl, 90 °C, 16 h	(PS)	[111]
2	(PS)	2-bromonaphthalene, Pd$_2$(dba)$_3$, NEt$_3$, DMF, P(o-Tol)$_3$, 100 °C, 20 h	(PS)	[111]
3	PS	[Ph$_2$I][BF$_4$] (0.05 mol/L, 1.5 eq), Pd$_2$(dba)$_3$ (0.05 eq), NaHCO$_3$ (2 eq), DMF, P(o-Tol)$_3$ (0.1 eq), 40 °C, 20 h	PS	[112]
4	PEG	4-Tol–I (0.2 mol/L), Pd(OAc)$_2$ (0.01 mol/L), NaHCO$_3$ (0.01 mol/L), DMF, 145 °C, 20 h	PEG	[113]
5	(PS)	Pd(OAc)$_2$, NEt$_3$, DMF, Bu$_4$NCl, 90 °C, 16 h	(PS) (E: CO$_2$Me)	[111]
6	(TG)	methacrylamide (0.3 mol/L), Pd(OAc)$_2$ (0.05 mol/L), PPh$_3$ (0.1 mol/L), Bu$_4$NCl (0.1 mol/L), NEt$_3$/DMF/H$_2$O 1:9:1, 37 °C, 4 h	(TG)	[114]
7	(PS)	styrene (8 eq), Pd(OAc)$_2$ (0.25 eq), NaOAc (3 eq), Bu$_4$NCl (2 eq), DMA, 100 °C, 24 h	(PS)	[115]

Table 5.7. continued.

Entry	Starting resin	Conditions	Product	Ref.
8		(15 eq), Pd(OAc)$_2$ (0.5 eq), NaOAc (5 eq), Bu$_4$NCl (4 eq), DMA, 100 °C, 2 × 24 h		[115]
9		cyclohexene, Pd(OAc)$_2$, PPh$_3$, NEt$_3$, DMF, 80 °C, ultrasound, 24 h		[34]

Experimental Procedure 5.1: Heck reaction with a polystyrene-bound aryl iodide [115]

To Rink amide resin acylated with 3-iodobenzyloxyacetic acid (0.2 g, 0.094 mmol) were added DMA (4.7 mL), sodium acetate (23 mg, 0.28 mmol, 3 eq), Bu$_4$NCl (56 mg, 0.20 mmol, 2 eq), and dimethyl itaconate (119 mg, 0.75 mmol, 8 eq), and the mixture was degassed for 20 min. Palladium(II) acetate (5 mg, 0.022 mmol, 0.2 eq) was then added and the mixture was shaken at 100 °C for 24 h. The support was washed with dioxane, water, ethanol and dioxane, and the coupling reaction was repeated once. After extensive washing the support was dried under reduced pressure and treated with TFA/DCM 2:8 (5 mL, 4 × 1 min). The combined filtrates were concentrated. The residue was coevaporated with toluene (4 ×) and dried under reduced pressure to yield 33 mg of the coupling product, which was slightly contaminated with tetrabutylammonium salts. After purification the yield was 67 %.

In the Suzuki reaction an aryl iodide or synthetic equivalent thereof is coupled, again with palladium(0) as catalyst, with an arylboronic acid or a borane. This reaction is usually used to prepare biaryls, and few examples have been reported of the solid-phase synthesis of olefins via Suzuki coupling (Table 5.8).

Vinylstannanes have also been used with success on solid phase for transition-metal-mediated coupling reactions (Stille coupling). As for the Suzuki reaction, most of the examples reported are biaryl formations. Table 5.9 lists illustrative examples of Stille couplings leading to the formation of alkenes on solid phase. From the results reported it might be concluded that for Stille couplings catalytic systems consisting of

Table 5.8. C-Vinylation by Suzuki coupling on insoluble supports.

Entry	Starting resin	Conditions	Product	Ref.
1		Ph—B(OR)$_2$ Pd(PPh$_3$)$_2$Cl$_2$, KOH (3 mol/L), H$_2$O, DME, 80 °C, 18 h		[116,117]
2		9-hexyl-9-BBN, Pd(PPh$_3$)$_4$, Na$_2$CO$_3$ (2 mol/L), H$_2$O, THF, 65 °C Ar: 4-(MeO)C$_6$H$_4$		[118]
3		9-hexyl-9-BBN, Pd(PPh$_3$)$_4$, Na$_2$CO$_3$ (2 mol/L), H$_2$O, THF, 65 °C Ar: 4-(MeO)C$_6$H$_4$		[118]
4		(1.3 eq), Pd(PPh$_3$)$_4$ or PdCl$_2$(dppf) (0.1 eq), K$_2$CO$_3$ (2 eq), DMF, 20 °C, 20 h		[28]
5		THPO (4 eq), PdCl$_2$(dppf) (0.3 eq), K$_2$CO$_3$ (3 eq), Triton B (1.5 eq), DMF, 85 °C, 14 h		[27]

Pd$_2$(dba)$_3$ and tri(2-furyl)phosphine [119], triphenylarsine [31], or tri(2-tolyl)phosphine are superior to Pd(PPh$_3$)$_4$.

Table 5.9. C-Vinylation with stannanes on insoluble supports.

Entry	Starting resin	Conditions	Product	Ref.
1		EtO̶SnBu₃ (5 eq), Pd₂(dba)₃ (0.05 eq), AsPh₃ (0.2 eq), NMP, 70 °C, 15 h		[105]
2		(3 eq), Pd₂(dba)₃ (0.1 eq), AsPh₃ (0.4 eq), dioxane, 50 °C, 24 h		[31] see also [120]
3		Ph̶SnBu₃ CuI (0.1 eq), NaCl (2 eq), NMP, 100 °C, 20 h		[121]
4		SnBu₃ (3 eq), Pd(PPh₃)₄ (0.05 eq), DMF, 60 °C, 24 h		[119]
5		HO̶SnBu₃ (2 eq), Pd₂(dba)₃ (0.1 eq), AsPh₃ (0.4 eq), NMP, 20 °C, overnight		[76]

5.2.5 Preparation of Cycloalkenes by Cycloaddition

Diels–Alder reactions have been conducted on solid phase, with either the dienophile or the diene linked to the support. The reaction conditions and the regio- and stereoselectivities observed are similar to those in solution [45,122,123]. Illustrative examples of Diels–Alder reactions leading to support-bound cyclohexenes are listed in Table 5.10.

Further cycloadditions used to prepare cycloalkenes on insoluble supports include the cyclopropanation of resin-bound alkynes and of polystyrene [126] (Figure 5.15). The latter reaction has been used to introduce 'tags' on to polystyrene beads, which enable the recognition of a certain bead in compound libraries produced with the mix-and-split method (Section 1.2 [126–128]). The structure of polystyrene tagged in this way has not, however, been rigorously determined.

Table 5.10. Preparation of cyclohexenes by Diels–Alder reaction on insoluble supports.

Entry	Starting resin	Conditions	Product	Ref.
1	(acrylate–O–PS)	diene–Ph (0.52 mol/L), xylene, 140 °C, 18 h	cyclohexene esters, 94:6	[122] see also [124]
2	(acrylate–O–PS)	diene–CO₂Me, PhMe, 110 °C, 24 h	cyclohexene esters, 77:23	[122]
3	(oxazolidinone–O–PS)	cyclopentadiene, DCM, Et₂AlCl, −78 °C to −40 °C (without catalyst no reaction occurred)	86% ee, *endo/exo* 21:1	[123] see also [40]
4	(OtBu cinnamate–triazene–(PS))	cyclopentadiene (neat), 100 °C, 24 h	CO₂tBu norbornene, mixture of *endo/exo*	[34]
5	(morpholine diene–(PS))	N-ethylmaleimide (8 eq), PhMe, 50 °C, 50 h R₂NH: morpholine	fused product	[120]
6	(pyrrolidine diene–(PS))	maleimide, PhMe, 105 °C, 18 h	fused bicyclic product	[97]
7	(Ph enol ether–O–PS)	CHO aldehyde diene, 80–100 °C, PhMe	cyclohexene, mixture of diastereomers 5:3	[92]

Table 5.10. continued.

Entry	Starting resin	Conditions	Product	Ref.
8	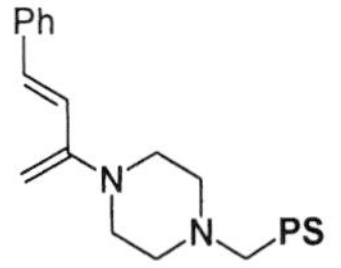	N-methylmaleimide (0.16 mol/L, 10 eq), THF, 4–8 h	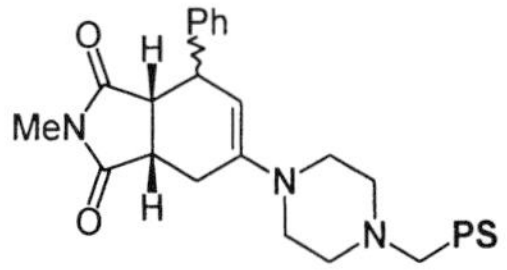	[125]
9	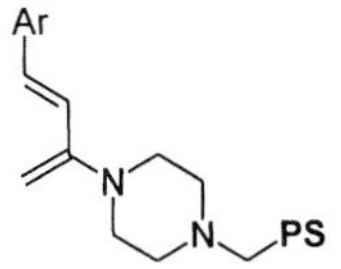	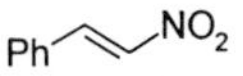 Ph NO$_2$ THF, 2 h	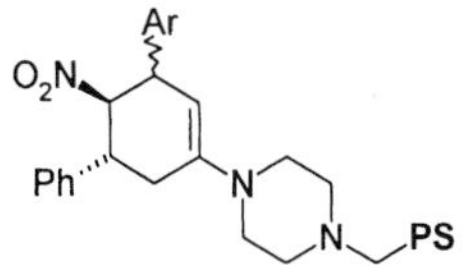	[125]

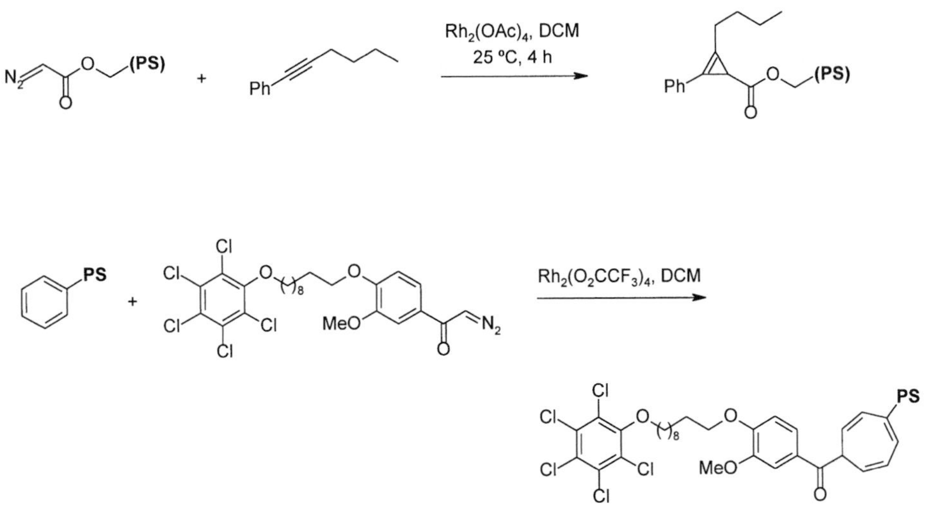

Fig. 5.15. Preparation of cycloalkenes from rhodium carbene complexes [126,129].

5.3 Preparation of Alkynes

The preparations of alkynes on insoluble supports include alkylations, vinylations, arylations, and alkynylations of other alkynes. Most of these reactions can be realized with either the alkyne or the alkylating agent linked to the support.

Terminal alkynes can be deprotonated with butyllithium and then alkylated. Lez-noff and coworkers used this methodology for the solid-phase synthesis of insect phero-mones [41,55]. More amenable to automation are aminoalkylations of alkynes, which can be performed either with the alkyne, the aldehyde, or the amine linked to the sup-port (Table 5.11). The arylation of alkynes can be realized under mild reaction condi-tions using the Sonogashira procedure [130] (Table 5.11). This reaction, in which an aryl or heteroaryl iodide is coupled with a terminal alkyne in the presence of cop-

Table 5.11. Alkylation and arylation of alkynes on insoluble supports.

Entry	Starting resin	Conditions	Product	Ref.
1	MsO–(–)$_4$–O–C(Ph)(Ph)–PS	Li–C≡C–CH=CH$_2$ THF/HMPA 5:3, 20 °C, overnight	(–)$_4$–O–C(Ph)(Ph)–PS	[55]
2	I–C$_6$H$_4$–C(O)–NH–(PS)	Me$_3$SiCCH (7 eq), CuI (0.1 eq), Pd(PPh$_3$)$_4$ (0.1 eq), THF/NEt$_3$ 1:1, 20 °C, 40 h	Me$_3$Si–C≡C–C$_6$H$_4$–C(O)–NH–(PS)	[133]
3	I–C$_6$H$_4$–C(O)–O–(PS)	PhCCH, Pd$_2$(dba)$_3$, P(*o*-Tol)$_3$, NEt$_3$, DMF, 100 °C, 20 h (does not proceed with HC≡C–CO$_2$R)	Ph–C≡C–C$_6$H$_4$–C(O)–O–(PS)	[111]
4	I–C$_6$H$_4$–C(O)–O–(PS)	Bu–C≡C–B(OiPr)$_2$ PdCl$_2$(dppf) (0.1 eq), K$_2$CO$_3$ (2 eq), DMF, 20 °C, 20 h	Bu–C≡C–C$_6$H$_4$–C(O)–O–(PS)	[28]
5	I–C$_6$H$_4$–CH$_2$–O–CH$_2$–O–(PS)	2-propyn-1-ol (4 eq), CuI (0.2 eq), PdCl$_2$(PPh$_3$)$_2$ (0.1 eq), NEt$_3$/dioxane 1:2, 20 °C, 24 h	HO–CH$_2$–C≡C–C$_6$H$_4$–CH$_2$–O–(PS)	[115] see also [134]
6	I–C$_6$H$_4$–C(O)–O–CH$_2$–PS	Ph–C≡C–SnBu$_3$ CuI (0.1 eq), NaCl (2 eq), NMP, 100 °C, 24 h	Ph–C≡C–C$_6$H$_4$–C(O)–O–CH$_2$–PS	[121]
7	Li–C≡C–(–)$_6$–O–C(Ph)(Ph)–PS	butyl bromide, THF/HMPA 1:1, 60–70 °C, 4 h	Bu–C≡C–(–)$_6$–O–C(Ph)(Ph)–PS	[41]
8	HC≡C–C$_6$H$_4$–C(O)–NH–(PS)	ArCHO (4 eq), CuCl (0.1 eq), piperazine (4 eq), dioxane, 90 °C, 36 h Ar: 4-MeC$_6$H$_4$	HN–(piperazine)–N–CH(Ar)–C≡C–C$_6$H$_4$–C(O)–NH–(PS)	[133] see also [135] [136]
9	HC≡C–C(O)–N(H)–O–C(Ph)(Ph(2-Cl))–(PS)	5-iodouridine, CuI, Pd(PPh$_3$)$_4$, NEt$_3$, DMF, 25 °C	uridine–C≡C–C(O)–N(H)–O–C(Ph)(Ph(2-Cl))–(PS)	[137] see also [138]

Table 5.11. continued.

Entry	Starting resin	Conditions	Product	Ref.
10		PhI (0.08 mol/L, 1.5 eq), PdCl$_2$(PPh$_3$)$_2$ (0.05 eq), CuI (0.05 eq), NEt$_3$/DCM 1:6, 20 °C, 18 h		[139]
11		Pd$_2$(dba)$_3$ (2.5 mmol/L), CuI (4 mmol/L), PPh$_3$ (0.02 mol/L), NEt$_3$, 65 °C, 12 h	(Ar: 3-(Me$_3$SiCC)-5-(NC)C$_6$H$_3$)	[140] see also [141]
12		[Ph$_2$I][BF$_4$] (1.5 eq), CuI (0.1 eq), Pd(PPh$_3$)$_4$ (0.05 eq), NaHCO$_3$ (2 eq), DMF, 40 °C, 24 h		[112]
13		1-octyne, CuCl, HONH$_3$Cl, EtOH, PrNH$_2$, 20 °C, 2 d		[142]

per(I), palladium(0), and an amine, tolerates a number of functional groups, such as alcohols, phenols, sulfonamides as RSO$_2$NHR, and Boc-protected primary amines [115]. Oligonucleotides containing 5-iodouridine have been alkynylated on CPG by use of Sonogashira conditions (Table 15.26). In addition to the examples given in Table 5.11 further examples have been reported [131,132]. Conjugated diacetylenes have been prepared on cross-linked polystyrene from terminal alkynes and 1-bromo- or 1-iodoalkynes under standard conditions of Cadiot–Chodkiewicz coupling (Entry **13**, Table 5.11).

Isolable aliphatic organopalladium compounds can also be coupled with alkynes in the presence of copper(I). One example of such a process is sketched in Figure 5.16.

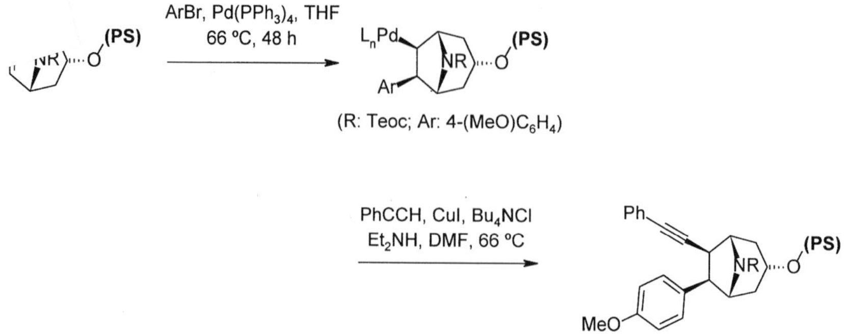

Fig. 5.16. Coupling of alkylpalladium compounds with alkynes [32].

5.4 Preparation of Biaryls

The arylation of support-bound arenes has mainly been performed using the Suzuki and Stille coupling reactions. Both reactions proceed smoothly with arenes and heteroarenes such as furans, thiophenes, or pyridines. Examples of the arylation of heteroarenes are presented in Chapter 15.

In the Suzuki coupling an aryl- or heteroarylboronic acid or ester is coupled with an arene bearing a good leaving group in the presence of palladium(0) and a base. Although examples based on resin-bound boronic acids and resin-bound aryl halides have been reported, the latter variant is generally preferred. This might, however, be because fewer bifunctional boronic acid derivatives are available than functionalized aryl halides.

The Suzuki coupling tolerates additional functional groups in the reacting arenes, such as ethers, aldehydes [143], trifluoromethyl groups, and carbamates. Difficulties have been reported with arenes bearing hydroxymethyl, carboxyl, or amino groups [31]. Further problematic substrates are 2,5-dialkylphenylboronic acids, which react only slowly [144]. Such difficult couplings can occasionally be driven to completion by using higher concentrations of reagents and catalyst, by increasing the reaction temperature and time, by choosing a different catalyst or base, or by repeating the coupling several times.

The examples reported to date do not enable clear ranking of palladium catalysts with regard to their ability to catalyze the Suzuki reaction. It should also be kept in mind that palladium(0) complexes are air-sensitive, and the quality of commercially available catalysts of this type can vary substantially. Some successful Suzuki couplings are listed in Table 5.12. Further examples have been reported [143,145–153].

Table 5.12. Preparation of biaryls by the Suzuki reaction on insoluble supports.

Entry	Starting resin	Conditions	Product	Ref.
1		PhB(OH)$_2$ (4 eq), K$_2$CO$_3$ (9 eq), Pd(OAc)$_2$ (0.1 eq), dioxane/H$_2$O 6:1, 100 °C, 6 h		[31]
2		PhB(OH)$_2$ (0.25 mol/L, 2 eq), Na$_2$CO$_3$ (2.5 eq), Pd(PPh$_3$)$_4$ (0.05 eq), DME/H$_2$O 7:1, 85 °C, overnight		[154]
3		PhB(OH)$_2$ (0.1 mol/L, 4 eq), PdCl$_2$(dppf) (0.2 eq), NEt$_3$/DMF 10:18, 65 °C, overnight		[109]
4		(HO)$_2$B OCONEt$_2$ (3 eq), Pd(PPh$_3$)$_4$ (0.05 eq), Na$_2$CO$_3$ (8 eq), DME, 85 °C, 48 h		[144]
5		PhB(OH)$_2$, Pd$_2$(dba)$_3$ (0.1 eq), K$_2$CO$_3$ (2 eq), DMF, 20 °C, 18 h		[28]
6		(0.1 mol/L, 5 eq), Pd(PPh$_3$)$_4$ (0.02 eq), K$_3$PO$_4$ (5 eq), DMF/H$_2$O 20:1, 80 °C, 20 h		[155] see also [156]
7		[Ph$_2$I][BF$_4$] (0.05 mol/L, 1.5 eq), Pd(PPh$_3$)$_4$ (0.05 eq), Na$_2$CO$_3$ (3 eq), DMF, 20 °C, 20 h		[112] see also [28]

Experimental Procedure 5.2: Suzuki coupling with a polystyrene-bound aryl bromide [144]

To a suspension of polystyrene-bound aryl bromide (0.15 g, approx. 0.09 mmol) in DME (5 mL) under argon were added Pd(PPh$_3$)$_4$ (5.2 mg, 4.5 µmol, 0.05 eq) and a degassed aqueous solution of sodium carbonate (0.36 mL, 2 mol/L, 0.72 mmol, 8 eq). The boronate (86 mg, 0.27 mmol, 3 eq) was added and the mixture was heated to reflux under argon for 48 h. Extensive washing with DME, DME/water, aqueous hydrochloric acid (0.3 mol/L), water, DMF, ethyl acetate, and methanol, followed by drying under reduced pressure yielded the polystyrene-bound biaryl ready for cleavage.

The Stille coupling is closely related to the Suzuki coupling. The only difference is that a stannane is used instead of a boronic acid derivative, and that no base is generally required. Additional functional groups tolerated in the reactants include hydroxyl groups, azides, nitro groups, oxiranes, carbamates, but not primary aliphatic amines [157]. Further substrates which cause problems are (2-alkoxyaryl)stannanes [31]. Most of the reported examples of solid-phase Stille reactions are couplings of a resin-bound aryl halide; few examples have been reported of the 'inverse' variant, in which the organostannane is linked to the support. Illustrative examples of Stille couplings on insoluble supports are listed in Table 5.13. Further examples have been reported [120,147,158].

Biaryls have also been prepared by coupling support-bound aryl halides with arylzinc compounds (Figure 5.17). As in Suzuki or Stille couplings, this reaction also requires transition metal catalysis to proceed. An additional strategy for coupling arenes on solid phase is the oxidative dimerization of phenols (Figure 5.17).

Table 5.13. Preparation of biaryls by the Stille reaction on insoluble supports.

Entry	Starting resin	Conditions	Product	Ref.
1	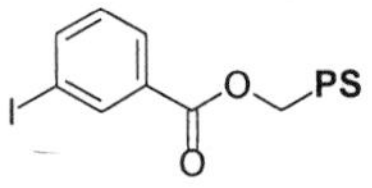	PhSnBu$_3$ (3 eq), Pd(PPh$_3$)$_4$ (0.05 eq), DMF, 60 °C, 24 h	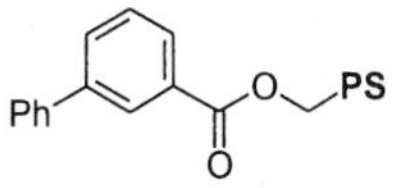	[119]
2	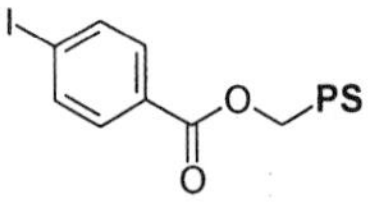	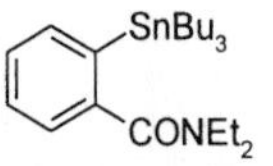 (3 eq), Pd(PPh$_3$)$_4$ (0.05 eq), DMF, 60 °C, 24 h	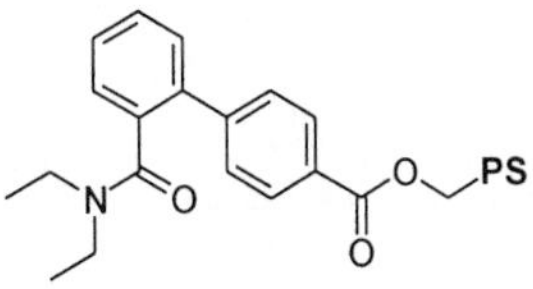	[119]
3	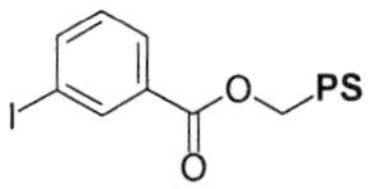	PhSnBu$_3$, CuI (0.1 eq), NaCl (2 eq), NMP, 100 °C, 24 h	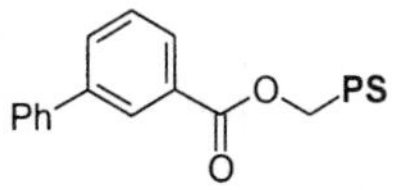	[121]
4	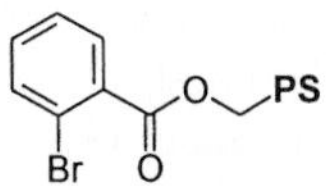	PhSnBu$_3$ (3 eq), Pd(PPh$_3$)$_4$ (0.05 eq), DMF, 60 °C, 24 h	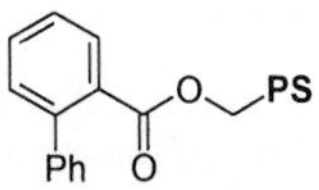	[119]
5	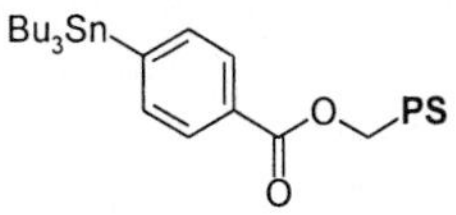	3-bromoaniline, NMP, Pd$_2$(OAc)$_2$[P(o-Tol)$_3$]$_2$ ('palladacycle'), LiCl, 90 °C, 18 h	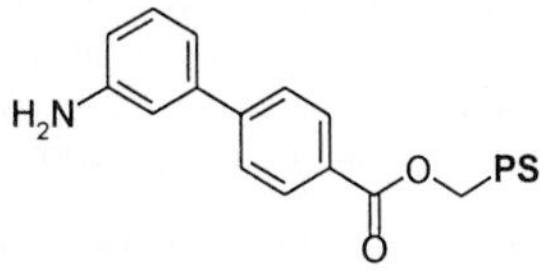	[157]

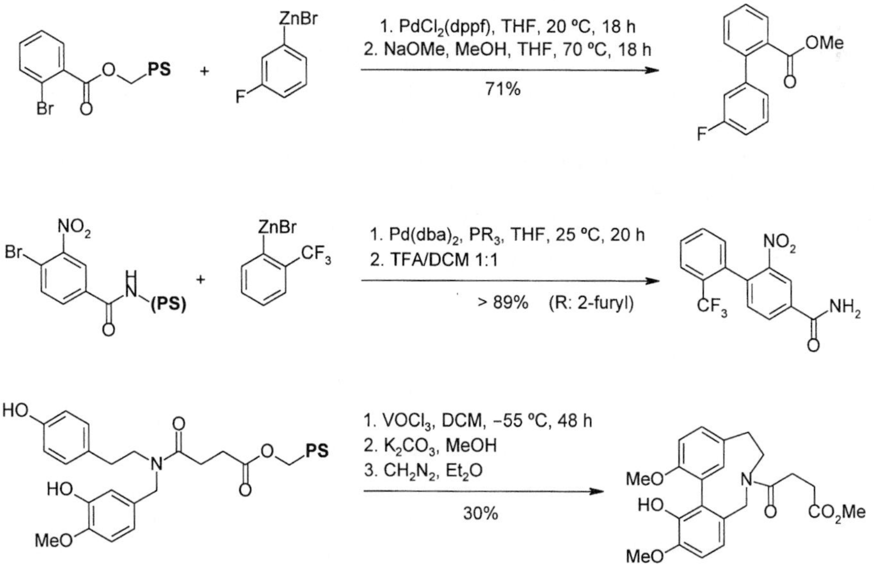

Fig. 5.17. Biaryl formation from resin-bound aryl bromides and arylzinc compounds [26,159], and by oxidative coupling of phenols [160].

172   *5 Preparation of Hydrocarbons*

References for Chapter 5

[1] O'Brien, R. A.; Chen, T.; Rieke, R. D. *J. Org. Chem.* **1992**, *57*, 2667–2677.
[2] Lochmann, L.; Fréchet, J. M. J. *Macromolecules* **1996**, *29*, 1767–1771.
[3] Itsuno, S.; Darling, G. D.; Stöver, H. D. H.; Fréchet, J. M. J. *J. Org. Chem.* **1987**, *52*, 4644–4645.
[4] Colombo, R. *J. Chem. Soc. Chem. Commun.* **1981**, 1012–1013.
[5] Schlatter, J. M.; Mazur, R. H. *Tetrahedron Lett.* **1977**, 2851–2852.
[6] Lacombe, P.; Castagner, B.; Gareau, Y.; Ruel, R. *Tetrahedron Lett.* **1998**, *39*, 6785–6786.
[7] Stranix, B. R.; Gao, J. P.; Barghi, R.; Salha, J.; Darling, G. D. *J. Org. Chem.* **1997**, *62*, 8987–8993.
[8] Brummond, K. M.; Lu, J. *J. Org. Chem.* **1999**, *64*, 1723–1726.
[9] Ojima, I.; Tsai, C. Y.; Zhang, Z. D. *Tetrahedron Lett.* **1994**, *35*, 5785–5788.
[10] Kobayashi, S.; Moriwaki, M. *Tetrahedron Lett.* **1997**, *38*, 4251–4254.
[11] Park, B. D.; Lee, H. I.; Ryoo, S. J.; Lee, Y. S. *Tetrahedron Lett.* **1997**, *38*, 591–594.
[12] Wang, S.; Merrifield, R. B. *J. Am. Chem. Soc.* **1969**, *91*, 6488–6491.
[13] Tomoi, M.; Kori, N.; Kakiuchi, H. *Reactive Polymers* **1985**, *5*, 341–349.
[14] Schiemann, K.; Showalter, H. D. H. *J. Org. Chem.* **1999**, *64*, 4972–4975.
[15] Wolters, E. T. M.; Tesser, G. I.; Nivard, R. J. F. *J. Org. Chem.* **1974**, *39*, 3388–3392.
[16] Scott, R. H.; Barnes, C.; Gerhard, U.; Balasubramanian, S. *Chem. Commun.* **1999**, 1331–1332.
[17] Barrett, A. G. M.; de Miguel, Y. R. *Chem. Commun.* **1998**, 2079–2080.
[18] Ruel, G.; The, N. K.; Dumartin, G.; Delmond, B.; Pereyre, M. *J. Organomet. Chem.* **1993**, *444*, C18–C20.
[19] Li, Z.; Ganesan, A. *Synlett* **1998**, 405–406.
[20] Dygutsch, D. P.; Eilbracht, P. *Tetrahedron* **1996**, *52*, 5461–5468.
[21] Bonds, W. D.; Brubaker, C. H.; Chandrasekaran, E. S.; Gibbons, C.; Grubbs, R. H.; Kroll, L. C. *J. Am. Chem. Soc.* **1975**, *97*, 2128–2132.
[22] Boymond, L.; Rottländer, M.; Cahiez, G.; Knochel, P. *Angew. Chem. Int. Ed. Engl.* **1998**, *37*, 1701–1703.
[23] Kondo, Y.; Komine, T.; Fujinami, M.; Uchiyama, M.; Sakamoto, T. *J. Comb. Chem.* **1999**, *1*, 123–126.
[24] Hu, Y.; Porco, J. A.; Labadie, J. W.; Gooding, O. W.; Trost, B. M. *J. Org. Chem.* **1998**, *63*, 4518–4521.
[25] Doi, T.; Hijikuro, I.; Takahashi, T. *J. Am. Chem. Soc.* **1999**, *121*, 6749–6750.
[26] Rottländer, M.; Knochel, P. *Synlett* **1997**, 1084–1086.
[27] Vanier, C.; Wagner, A.; Mioskowski, C. *Tetrahedron Lett.* **1999**, *40*, 4335–4338.
[28] Guiles, J. W.; Johnson, S. G.; Murray, W. V. *J. Org. Chem.* **1996**, *61*, 5169–5171.
[29] Woolard, F. X.; Paetsch, J.; Ellman, J. A. *J. Org. Chem.* **1997**, *62*, 6102–6103.
[30] Backes, B. J.; Ellman, J. A. *J. Am. Chem. Soc.* **1994**, *116*, 11171–11172.
[31] Wendeborn, S.; Berteina, S.; Brill, W. K. D.; De Mesmaeker, A. *Synlett* **1998**, 671–675.
[32] Koh, J. S.; Ellman, J. A. *J. Org. Chem.* **1996**, *61*, 4494–4495.
[33] Zhu, X.; Ganesan, A. *J. Comb. Chem.* **1999**, *1*, 157–162.
[34] Bräse, S.; Enders, D.; Köbberling, J.; Avemaria, F. *Angew. Chem. Int. Ed. Engl.* **1998**, *37*, 3413–3415.
[35] Sibi, M. P.; Chandramouli, S. V. *Tetrahedron Lett.* **1997**, *38*, 8929–8932.
[36] Yim, A. M.; Vidal, Y.; Viallefont, P.; Martinez, J. *Tetrahedron Lett.* **1999**, *40*, 4535–4538.
[37] Tietze, L. F.; Steinmetz, A. *Angew. Chem. Int. Ed. Engl.* **1996**, *35*, 651–652.
[38] Vo, N. H.; Eyermann, C. J.; Hodge, C. N. *Tetrahedron Lett.* **1997**, *38*, 7951–7954.
[39] Nieuwstad, T. J.; Kieboom, A. P. G.; Breijer, A. J.; van der Linden, J.; van Bekkum, H. *Recl. Trav. Chim. Pays-Bas* **1976**, *95*, 225–254.
[40] Burkett, B. A.; Chai, C. L. L. *Tetrahedron Lett.* **1999**, *40*, 7035–7038.
[41] Fyles, T. M.; Leznoff, C. C.; Weatherston, J. *Can. J. Chem.* **1977**, *55*, 4135–4143.
[42] Gosselin, F.; Di Renzo, M.; Ellis, T. H.; Lubell, W. D. *J. Org. Chem.* **1996**, *61*, 7980–7981.
[43] Kurth, M. J.; Randall, L. A. A.; Takenouchi, K. *J. Org. Chem.* **1996**, *61*, 8755–8761.
[44] Stranix, B. R.; Liu, H. Q.; Darling, G. D. *J. Org. Chem.* **1997**, *62*, 6183–6186.
[45] Stranix, B. R.; Darling, G. D. *J. Org. Chem.* **1997**, *62*, 9001–9004.
[46] Neumann, W. P.; Peterseim, M. *Reactive Polymers* **1993**, *20*, 189–205.
[47] Gerigk, U.; Gerlach, M.; Neumann, W. P.; Vieler, R.; Weintritt, V. *Synthesis* **1990**, 448–452.
[48] Moberg, C.; Rákos, L. *Reactive Polymers* **1991**, *15*, 25–35.
[49] Fréchet, J. M. J.; Eichler, E. *Polym. Bull.* **1982**, *7*, 345–351.
[50] Sylvain, C.; Wagner, A.; Mioskowski, C. *Tetrahedron Lett.* **1998**, *39*, 9679–9680.
[51] Leznoff, C. C.; Greenberg, S. *Can. J. Chem.* **1976**, *54*, 3824–3829.

[52] Ren, Q.; Huang, W.; Ho, P. *Reactive Polymers* **1989**, *11*, 237–244.
[53] Wong, J. Y.; Manning, C.; Leznoff, C. C. *Angew. Chem. Int. Ed. Engl.* **1974**, *13*, 666–667.
[54] Adams, J. H.; Cook, R. M.; Hudson, D.; Jammalamadaka, V.; Lyttle, M. H.; Songster, M. F. *J. Org. Chem.* **1998**, *63*, 3706–3716.
[55] Leznoff, C. C.; Fyles, T. M.; Weatherston, J. *Can. J. Chem.* **1977**, *55*, 1143–1153.
[56] Leznoff, C. C.; Sywanyk, W. *J. Org. Chem.* **1977**, *42*, 3203–3205.
[57] Chan, T.; Huang, W. *J. Chem. Soc. Chem. Commun.* **1985**, 909–911.
[58] Vagner, J.; Krchnák, V.; Lebl, M.; Barany, G. *Collect. Czech. Chem. Commun.* **1996**, *61*, 1697–1702.
[59] Veerman, J. J. N.; van Maarseveen, J. H.; Visser, G. M.; Kruse, C. G.; Schoemaker, H. E.; Hiemstra, H.; Rutjes, F. P. J. T. *Eur. J. Org. Chem.* **1998**, 2583–2589.
[60] Chen, C.; Randall, L. A. A.; Miller, R. B.; Jones, A. D.; Kurth, M. J. *J. Am. Chem. Soc.* **1994**, *116*, 2661–2662.
[61] Chin, J.; Fell, B.; Shapiro, M. J.; Tomesch, J.; Wareing, J. R.; Bray, A. M. *J. Org. Chem.* **1997**, *62*, 538–539.
[62] Marzinzik, A. L.; Felder, E. R. *J. Org. Chem.* **1998**, *63*, 723–727.
[63] Lyngsø, L. O.; Nielsen, J. *Tetrahedron Lett.* **1998**, *39*, 5845–5848.
[64] Leznoff, C. C.; Wong, J. Y. *Can. J. Chem.* **1973**, *51*, 3756–3764.
[65] Chen, C.; Randall, L. A. A.; Miller, R. B.; Jones, A. D.; Kurth, M. J. *Tetrahedron* **1997**, *53*, 6595–6609.
[66] Rotella, D. P. *J. Am. Chem. Soc.* **1996**, *118*, 12246–12247.
[67] Nicolaou, K. C.; Winssinger, N.; Pastor, J.; Ninkovic, S.; Sarabia, F.; He, Y.; Vourloumis, D.; Yang, Z.; Li, T.; Giannakakou, P.; Hamel, E. *Nature* **1997**, *387*, 268–272.
[68] Fréchet, J. M.; Schuerch, C. *J. Am. Chem. Soc.* **1971**, *93*, 492–496.
[69] Piscopio, A. D.; Miller, J. F.; Koch, K. *Tetrahedron Lett.* **1997**, *38*, 7143–7146.
[70] Hall, B. J.; Sutherland, J. D. *Tetrahedron Lett.* **1998**, *39*, 6593–6596.
[71] Hird, N. W.; Irie, K.; Nagai, K. *Tetrahedron Lett.* **1997**, *38*, 7111–7114.
[72] Barco, A.; Benetti, S.; De Risi, C.; Marchetti, P.; Pollini, G. P.; Zanirato, V. *Tetrahedron Lett.* **1998**, *39*, 7591–7594.
[73] Paris, M.; Heitz, A.; Guerlavais, V.; Cristau, M.; Fehrentz, J. A.; Martinez, J. *Tetrahedron Lett.* **1998**, *39*, 7287–7290.
[74] Zaragoza, F.; Stephensen, H. *J. Org. Chem.* **1999**, *64*, 2555–2557.
[75] Johnson, C. R.; Zhang, B. R. *Tetrahedron Lett.* **1995**, *36*, 9253–9256.
[76] Blaskovich, M. A.; Kahn, M. *J. Org. Chem.* **1998**, *63*, 1119–1125.
[77] Sun, S.; Murray, W. V. *J. Org. Chem.* **1999**, *64*, 5941–5945.
[78] Boldi, A. M.; Johnson, C. R.; Eissa, H. O. *Tetrahedron Lett.* **1999**, *40*, 619–622.
[79] Burns, C. J.; Groneberg, R. D.; Salvino, J. M.; McGeehan, G.; Condon, S. M.; Morris, R.; Morrissette, M.; Mathew, R.; Darnbrough, S.; Neuenschwander, K.; Scotese, A.; Djuric, S. W.; Ullrich, J.; Labaudiniere, R. *Angew. Chem. Int. Ed. Engl.* **1998**, *37*, 2848–2850.
[80] Salvino, J. M.; Kiesow, T. J.; Darnbrough, S.; Labaudiniere, R. *J. Comb. Chem.* **1999**, *1*, 134–139.
[81] Wipf, P.; Henninger, T. C. *J. Org. Chem.* **1997**, *62*, 1586–1587.
[82] Hamper, B. C.; Kolodziej, S. A.; Scates, A. M. *Tetrahedron Lett.* **1998**, *39*, 2047–2050.
[83] Hamper, B. C.; Snyderman, D. M.; Owen, T. J.; Scates, A. M.; Owsley, D. C.; Kesselring, A. S.; Chott, R. C. *J. Comb. Chem.* **1999**, *1*, 140–150.
[84] Hamper, B. C.; Gan, K. Z.; Owen, T. J. *Tetrahedron Lett.* **1999**, *40*, 4973–4976.
[85] Gordeev, M. F.; Patel, D. V.; Wu, J.; Gordon, E. M. *Tetrahedron Lett.* **1996**, *37*, 4643–4646.
[86] Tietze, L. F.; Hippe, T.; Steinmetz, A. *Synlett* **1996**, 1043–1044.
[87] Zaragoza, F. *Tetrahedron Lett.* **1995**, *36*, 8677–8678.
[88] Hollinshead, S. P. *Tetrahedron Lett.* **1996**, *37*, 9157–9160.
[89] Chiu, C.; Tang, Z.; Ellingboe, J. W. *J. Comb. Chem.* **1999**, *1*, 73–77.
[90] Beebe, X.; Chiappari, C. L.; Olmstead, M. M.; Kurth, M. J.; Schore, N. E. *J. Org. Chem.* **1995**, *60*, 4204–4212.
[91] Zaragoza, F. unpublished results.
[92] Ball, C. P.; Barrett, A. G. M.; Commercon, A.; Compère, D.; Kuhn, C.; Roberts, R. S.; Smith, M. L.; Venier, O. *Chem. Commun.* **1998**, 2019–2020.
[93] Zaragoza, F. *Metal Carbenes in Organic Synthesis*; Wiley-VCH: Weinheim, New York, **1999**.
[94] Ivin, K. J.; Mol, J. C. *Olefin Metathesis and Metathesis Polymerization*; Academic Press: London, **1997**.
[95] Schuster, M.; Pernerstorfer, J.; Blechert, S. *Angew. Chem. Int. Ed. Engl.* **1996**, *35*, 1979–1980.
[96] Pernerstorfer, J.; Schuster, M.; Blechert, S. *Synthesis* **1999**, 138–144.
[97] Heerding, D. A.; Takata, D. T.; Kwon, C.; Huffman, W. F.; Samanen, J. *Tetrahedron Lett.* **1998**, *39*, 6815–6818.

[98] Miller, S. J.; Blackwell, H. E.; Grubbs, R. H. *J. Am. Chem. Soc.* **1996**, *118*, 9606–9614.
[99] Biagini, S. C. G.; Gibson, S. E.; Keen, S. P. *J. Chem. Soc. Perkin Trans. 1* **1998**, 2485–2499.
[100] Cuny, G. D.; Cao, J.; Hauske, J. R. *Tetrahedron Lett.* **1997**, *38*, 5237–5240.
[101] Schuster, M.; Lucas, N.; Blechert, S. *Chem. Commun.* **1997**, 823–824.
[102] Nicolaou, K. C.; Pastor, J.; Winssinger, N.; Murphy, F. *J. Am. Chem. Soc.* **1998**, *120*, 5132–5133.
[103] Schuster, M.; Blechert, S. *Tetrahedron Lett.* **1998**, *39*, 2295–2298.
[104] Schürer, S. C.; Blechert, S. *Synlett* **1998**, 166–168.
[105] Beaver, K. A.; Siegmund, A. C.; Spear, K. L. *Tetrahedron Lett.* **1996**, *37*, 1145–1148.
[106] Hiroshige, M.; Hauske, J. R.; Zhou, P. *J. Am. Chem. Soc.* **1995**, *117*, 11590–11591.
[107] Pop, I. E.; Dhalluin, C. F.; Déprez, B. P.; Melnyk, P. C.; Lippens, G. M.; Tartar, A. L. *Tetrahedron* **1996**, *52*, 12209–12222.
[108] Hanessian, S.; Xie, F. *Tetrahedron Lett.* **1998**, *39*, 737–740.
[109] Ruhland, B.; Bombrun, A.; Gallop, M. A. *J. Org. Chem.* **1997**, *62*, 7820–7826.
[110] Shaughnessy, K. H.; Kim, P.; Hartwig, J. F. *J. Am. Chem. Soc.* **1999**, *121*, 2123–2132.
[111] Yu, K. L.; Deshpande, M. S.; Vyas, D. M. *Tetrahedron Lett.* **1994**, *35*, 8919–8922.
[112] Kang, S. K.; Yoon, S. K.; Lim, K. H.; Son, H. J.; Baik, T. G. *Synthetic Commun.* **1998**, *28*, 3645–3655.
[113] Blettner, C. G.; König, W. A.; Stenzel, W.; Schotten, T. *Tetrahedron Lett.* **1999**, *40*, 2101–2102.
[114] Hiroshige, M.; Hauske, J. R.; Zhou, P. *Tetrahedron Lett.* **1995**, *36*, 4567–4570.
[115] Berteina, S.; Wendeborn, S.; Brill, W. K. D.; De Mesmaeker, A. *Synlett* **1998**, 676–678.
[116] Brown, S. D.; Armstrong, R. W. *J. Am. Chem. Soc.* **1996**, *118*, 6331–6332.
[117] Brown, S. D.; Armstrong, R. W. *J. Org. Chem.* **1997**, *62*, 7076–7077.
[118] Thompson, L. A.; Moore, F. L.; Moon, Y. C.; Ellman, J. A. *J. Org. Chem.* **1998**, *63*, 2066–2067.
[119] Chamoin, S.; Houldsworth, S.; Snieckus, V. *Tetrahedron Lett.* **1998**, *39*, 4175–4178.
[120] Wendeborn, S.; De Mesmaeker, A.; Brill, W. K. D. *Synlett* **1998**, 865–868.
[121] Kang, S. K.; Kim, J. S.; Yoon, S. K.; Lim, K. H.; Yoon, S. S. *Tetrahedron Lett.* **1998**, *39*, 3011–3012.
[122] Yedidia, V.; Leznoff, C. C. *Can. J. Chem.* **1980**, *58*, 1144–1150.
[123] Winkler, J. D.; McCoull, W. *Tetrahedron Lett.* **1998**, *39*, 4935–4936.
[124] Winkler, J. D.; Kwak, Y. S. *J. Org. Chem.* **1998**, *63*, 8634–8635.
[125] Crawshaw, M.; Hird, N. W.; Irie, K.; Nagai, K. *Tetrahedron Lett.* **1997**, *38*, 7115–7118.
[126] Nestler, H. P.; Bartlett, P. A.; Still, W. C. *J. Org. Chem.* **1994**, *59*, 4723–4724.
[127] Baldwin, J. J.; Burbaum, J. J.; Henderson, I.; Ohlmeyer, M. H. J. *J. Am. Chem. Soc.* **1995**, *117*, 5588–5589.
[128] Burbaum, J. J.; Ohlmeyer, M. H. J.; Reader, J. C.; Henderson, I.; Dillard, L. W.; Li, G.; Randle, T. L.; Sigal, N. H.; Chelsky, D.; Baldwin, J. J. *Proc. Natl. Acad. Sci. USA* **1995**, *92*, 6027–6031.
[129] Cano, M.; Camps, F.; Joglar, J. *Tetrahedron Lett.* **1998**, *39*, 9819–9822.
[130] Sonogashira, K.; Tohda, Y.; Hagihara, N. *Tetrahedron Lett.* **1975**, 4467–4470.
[131] Tan, D. S.; Foley, M. A.; Shair, M. D.; Schreiber, S. L. *J. Am. Chem. Soc.* **1998**, *120*, 8565–8566.
[132] Collini, M. D.; Ellingboe, J. W. *Tetrahedron Lett.* **1997**, *38*, 7963–7966.
[133] Dyatkin, A. B.; Rivero, R. A. *Tetrahedron Lett.* **1998**, *39*, 3647–3650.
[134] Stieber, F.; Grether, U.; Waldmann, H. *Angew. Chem. Int. Ed. Engl.* **1999**, *38*, 1073–1077.
[135] McNally, J. J.; Youngman, M. A.; Dax, S. L. *Tetrahedron Lett.* **1998**, *39*, 967–970.
[136] Youngman, M. A.; Dax, S. L. *Tetrahedron Lett.* **1997**, *38*, 6347–6350.
[137] Khan, S. I.; Grinstaff, M. W. *Tetrahedron Lett.* **1998**, *39*, 8031–8034.
[138] Khan, S. I.; Grinstaff, M. W. *J. Am. Chem. Soc.* **1999**, *121*, 4704–4705.
[139] Bolton, G. L.; Hodges, J. C.; Rubin, J. R. *Tetrahedron* **1997**, *53*, 6611–6634.
[140] Nelson, J. C.; Young, J. K.; Moore, J. S. *J. Org. Chem.* **1996**, *61*, 8160–8168.
[141] Huang, S.; Tour, J. M. *J. Am. Chem. Soc.* **1999**, *121*, 4908–4909.
[142] Montierth, J. M.; DeMario, D. R.; Kurth, M. J.; Schore, N. E. *Tetrahedron* **1998**, *54*, 11741–11748.
[143] Blettner, C. G.; König, W. A.; Rühter, G.; Stenzel, W.; Schotten, T. *Synlett* **1999**, 307–310.
[144] Chamoin, S.; Houldsworth, S.; Kruse, C. G.; Bakker, W. I.; Snieckus, V. *Tetrahedron Lett.* **1998**, *39*, 4179–4182.
[145] Chenera, B.; Finkelstein, J. A.; Veber, D. F. *J. Am. Chem. Soc.* **1995**, *117*, 11999–12000.
[146] Han, Y. X.; Walker, S. D.; Young, R. N. *Tetrahedron Lett.* **1996**, *37*, 2703–2706.
[147] Larhed, M.; Lindeberg, G.; Hallberg, A. *Tetrahedron Lett.* **1996**, *37*, 8219–8222.
[148] Yoo, S. E.; Seo, J. S.; Yi, K. Y.; Gong, Y. D. *Tetrahedron Lett.* **1997**, *38*, 1203–1206.
[149] Raju, B.; Kogan, T. P. *Tetrahedron Lett.* **1997**, *38*, 3373–3376.
[150] Lago, M. A.; Nguyen, T. T.; Bhatnagar, P. *Tetrahedron Lett.* **1998**, *39*, 3885–3888.
[151] Huwe, C. M.; Künzer, H. *Tetrahedron Lett.* **1999**, *40*, 683–686.
[152] Gosselin, F.; Van Betsbrugge, J.; Hatam, M.; Lubell, W. D. *J. Org. Chem.* **1999**, *64*, 2486–2493.
[153] Blettner, C. G.; König, W. A.; Stenzel, W.; Schotten, T. *J. Org. Chem.* **1999**, *64*, 3885–3890.

[154] Frenette, R.; Friesen, R. W. *Tetrahedron Lett.* **1994**, *35*, 9177–9180.
[155] Piettre, S. R.; Baltzer, S. *Tetrahedron Lett.* **1997**, *38*, 1197–1200.
[156] Tempest, P. A.; Armstrong, R. W. *J. Am. Chem. Soc.* **1997**, *119*, 7607–7608.
[157] Brody, M. S.; Finn, M. G. *Tetrahedron Lett.* **1999**, *40*, 415–418.
[158] Forman, F. W.; Sucholeiki, I. *J. Org. Chem.* **1995**, *60*, 523–528.
[159] Marquais, S.; Arlt, M. *Tetrahedron Lett.* **1996**, *37*, 5491–5494.
[160] ApSimon, J. W.; Dixit, D. M. *Can. J. Chem.* **1982**, *60*, 368–370.

6 Preparation of Alkyl and Aryl Halides

The introduction of halogens into organic substrates can be accompanied by a formal oxidation of the substrate or not. Oxidative halogenations include additions to C–C double bonds, aromatic electrophilic substitutions, halomethylations, and the reaction of organometallic compounds with halogens or synthetic equivalents thereof. Because polystyrene is also susceptible to oxidative halogenation, only reactive substrates will undergo clean reaction with halogenating agents before significant halogenation of the support. Acid-labile linkers, which are usually based on electron-rich arenes (e.g. alkoxybenzyl alcohols or benzhydryl derivatives), are generally not compatible with strong oxidizing agents. Halogenation can further be realized by non-oxidative processes, such as aliphatic or aromatic nucleophilic substitutions.

Strategies for the generation of alkyl, vinyl and aryl halides upon cleavage of intermediates from insoluble supports are discussed in Chapter 3.13.

6.1 Preparation of Alkyl Halides

The support originally used for solid-phase synthesis was partially chloromethylated, cross-linked polystyrene, which was prepared by chloromethylation of cross-linked polystyrene with chloromethyl methyl ether and tin tetrachloride [1–3] or zinc chloride [4] (Figure 6.1). Haloalkylations of this type are usually used for the functionalization of supports only, and not for selective transformation of support-bound intermediates. Because of the mutagenicity of α-haloethers, other methods have been developed for the preparation of chloromethyl polystyrene. These include the chlorination of methoxymethyl polystyrene (Figure 6.1 [5]), the use of a mixture of dimethoxymethane, sulfuryl chloride and chlorosulfonic acid instead of chloromethyl

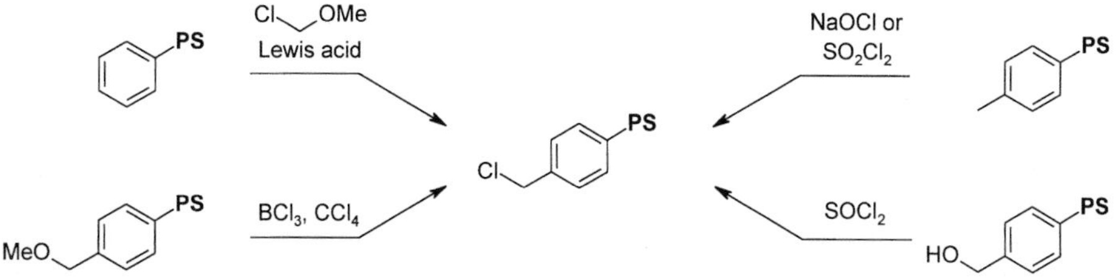

Fig. 6.1. Methods for the preparation of chloromethyl polystyrene.

methyl ether [6], the chlorination of hydroxymethyl polystyrene [7], and the chlorination of cross-linked 4-methylstyrene–styrene copolymer with sodium hypochlorite [8], sulfuryl chloride [8], or cobalt(III) acetate [9] (Figure 6.1, Table 6.1).

Resin-bound organometallic compounds, such as vinylstannanes [10] or organozinc derivatives [11] react cleanly with iodine to yield the corresponding vinyl or alkyl iodides (see also Chapter 3.13).

Additions of halogens or their synthetic equivalents to C–C double bonds on cross-linked polystyrene are limited to reactive substrates, such as styrenes [12,13] or enol ethers (Table 6.1). Iodolactonizations with simultaneous cleavage from an insoluble supports are discussed in Chapter 3.5.2.

Table 6.1. Oxidative halogenation of alkenes and alkanes on polystyrene.

Entry	Starting resin	Conditions	Product	Ref.
1	(4-methyl-PS)	LiCl, AcOH/PhH 1:1, Co(OAc)₃, 50 °C, 2 h; or aq NaOCl, DCE, BnEt₃NCl, 25 °C, 16 h	(ClCH₂-PS)	[8,9]
2	(sulfone resin)	PyHBrBr₂, AcOH, 60 °C, 16 h	(α-bromoketone sulfone resin)	[14]
3	(vinyl-PS)	LiBr, Me₃SiCl, PhMe, H₂O, AIBN, 70 °C, 24 h	(BrCH₂CH₂-PS)	[15]
4	Ph-O (enol ether resin)	Br₂ (1 mol/L, 2 eq), DCM, 25 °C, 1 h	(dibromide)	[16]
5	(allyl ether (PS))	Ph⟶OH (30 eq), NBS (5 eq), DCM, 0 °C, 6 d	(bromo ether (PS))	[17]
6	(glycal resin, OBn, OBn, O-Si-PS)	I(coll)₂ClO₄, PhSO₂NH₂, DCM, 0 °C	(iodo sulfonamide, OBn, OBn, HN-SO₂Ph, O-Si-PS)	[18] see also [19]

Nucleophilic substitutions are a more versatile means of access to alkyl halides than are oxidative halogenations. The most common starting materials for this purpose are resin-bound alcohols, which can be converted into halides either directly or via intermediate formation of sulfonates [20].

Table 6.2. Conversion of resin-bound benzylic and allylic alcohols into halides.

Entry	Starting resin	Conditions	Product	Ref.
1		HBr, DCM, 0 °C, 4 h		[31]
2		CBr$_4$, PPh$_3$, DCM, 20 °C, 16 h		[32]
3	Wang resin	MsCl (0.32 mol/L, 3 eq), DIPEA (4 eq), DMF, 25 °C, 3 d		[33]
4		(3 eq), DCM, 20 °C, 5 h		[34]
5		CBr$_4$ (0.1 mol/L, 2 eq), PPh$_3$ (2 eq), THF, 16 h		[35]
6		1% HCl in THF/DCM; or PPh$_3$ (0.18 mol/L, 5.5 eq), C$_2$Cl$_6$ (5.5 eq), THF, 20 °C, 6 h		[36,37]
7	Ar: 3,4-dialkoxyphenyl	Ph$_3$PBr$_2$ (0.12 mol/L, 1.5 eq), DCM, 20 °C, 1 h		[38]
8		Ph$_3$PBr$_2$ (0.27 mol/L, 3 eq), DCM, 20 °C, 17 h		[39]

Polystyrene-bound allylic or benzylic alcohols react smoothly with hydrogen chloride or hydrogen bromide to yield the corresponding halides. The more stable the intermediate carbocation, the easier will solvolysis proceed. Alternatively, thionyl chloride can be used to convert benzyl alcohols into chlorides [7,21]. A milder alternative for preparing bromides or iodides, which is suitable also for non-benzylic alcohols, is the treatment of alcohols with phosphines and halogens or the preformed adducts thereof (Table 6.2, Experimental Procedure 6.1 [22–26]).

Benzhydryl and trityl alcohols, bound to cross-linked or non-cross-linked polystyrene, are particularly prone to solvolysis, and can be converted into the corresponding chlorides by treatment with acetyl chloride in toluene or similar solvents (Table 6.2 [27–30]).

Experimental Procedure 6.1: Conversion of Wang resin into *p*-benzyloxybenzyl bromide resin [40]

HO–⟨benzyl⟩–⟨C₆H₄⟩–O–CH₂–PS → (PPh₃, CBr₄, DMF) → Br–⟨benzyl⟩–⟨C₆H₄⟩–O–CH₂–PS

A mixture of Wang resin (50.0 g, 36.5 mmol), DMF (500 mL), triphenylphosphine (47.9 g, 183 mmol, 5 eq), and carbon tetrabromide (60.5 g, 182 mmol, 5 eq) was shaken at room temperature for 2.5 h. The resin was filtered and washed with DMF, DCM, *i*PrOH, DMF, DCM, and *i*PrOH (2 × 300 mL of each solvent).

Aliphatic alcohols do not undergo solvolysis as readily as benzylic alcohols, and are generally converted into halides under basic reaction conditions via an intermediate sulfonate. Because of the hydrophobicity of polystyrene, however, nucleophilic substitutions with halides on this support do not always proceed as readily as in solution (Table 6.3). Alternatively, phosphorus-based reagents can also be used to convert aliphatic alcohols into halides.

Table 6.3. Conversion of non-benzylic alcohols into halides.

Entry	Starting resin	Conditions	Product	Ref.
1	HO–CH₂–⟨C₆H₄⟩–PS	SOCl₂ (3.3 eq), pyridine (0.16 eq), CCl₄, 76 °C, 16 h	Cl–CH₂–⟨C₆H₄⟩–PS	[41]
2	HO–CH₂CH₂–⟨C₆H₄⟩–PS	1. TsCl (1.3 eq), (*i*Pr)₂NH (2 eq), CCl₄, 76 °C, 6 h 2. MgBr₂, Et₂O, 35 °C, 17 h	Br–CH₂CH₂–⟨C₆H₄⟩–PS	[41]
3	HO–CH₂CH₂–⟨C₆H₄⟩–PS	NaI (2 eq), Me₃SiCl (2 eq), MeCN, 81 °C, 8 h	I–CH₂CH₂–⟨C₆H₄⟩–PS	[41]
4	HO–(CH₂)₇–O–C(Ph)(Ph)–⟨C₆H₄⟩–PS	1. MsCl (12 eq), pyridine, 20 °C, 48 h 2. NaI (0.2 mol/L), HMPA, 70 °C, 48 h	I–(CH₂)₇–O–C(Ph)(Ph)–⟨C₆H₄⟩–PS	[42]
5	HO–(CH₂)₃–O–CH₂–⟨C₆H₄⟩–PS	I₂, PPh₃, imidazole (4 eq of each), DCM, 0 °C, 4 h	I–(CH₂)₃–O–CH₂–⟨C₆H₄⟩–PS	[23] see also [43]

6.2 Preparation of Aryl and Heteroaryl Halides

Cross-linked polystyrene can be directly brominated in carbon tetrachloride with bromine in the presence of Lewis acids (Experimental Procedure 6.2 [44–47]). Thallium(III) acetate is a particularly suitable catalyst for this reaction [48]. Harsher bromination conditions should be avoided, because these can lead to the decomposition of the polymer. Considering that isopropylbenzene is dealkylated when treated with bromine to yield hexabromobenzene [49], the expected outcome of the extensive bromination of cross-linked polystyrene would be poly(vinyl bromide). In fact, if the bromination of cross-linked polystyrene is attempted using bromine in acetic acid the polymer dissolves and apparently depolymerizes [50].

Iodination of cross-linked polystyrene has been achieved with iodine under strongly acidic reaction conditions [44] or in the presence of thallium(III) acetate [50], but this reaction does not proceed as smoothly as the bromination.

More electron-rich arenes, such as thiophenes [38,51] or phenols [52,53], are readily halogenated, even in the presence of oxidant-labile linkers (Figure 6.2).

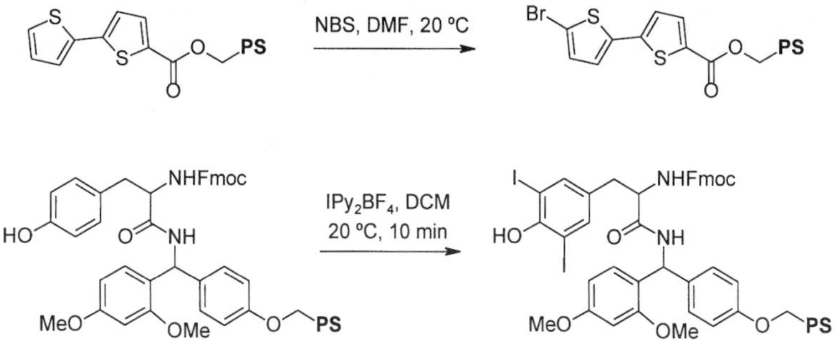

Fig. 6.2. Halogenations of polystyrene-bound arenes [51,52].

Experimental Procedure 6.2: Bromination of cross-linked polystyrene [50]

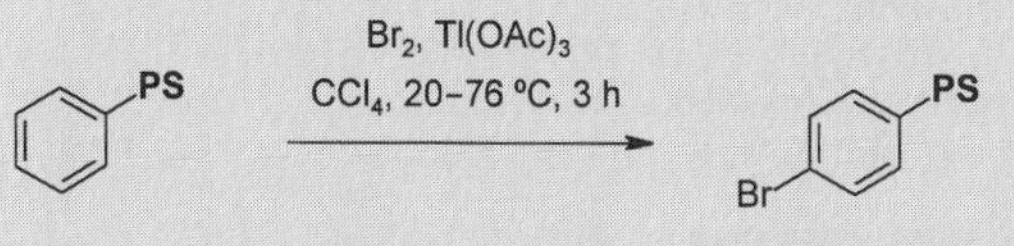

Thallium(III) acetate (1.18 g, 3.09 mmol) was added to a suspension of polystyrene (20 g, 1 % cross-linked) in carbon tetrachloride (300 mL), and the mixture was stirred in the dark for 0.5 h. A solution of bromine (13.6 g, 85.1 mmol) in carbon tetrachloride (20 mL) was added, and the mixture was stirred at room temperature in the dark for 1 h and at 76 °C for 1.5 h. The mixture was filtered and the resin was washed with carbon tetrachloride, acetone, aceton/water 2:1, acetone, benzene*, and methanol. Drying of the resin under reduced pressure yielded 26.3 g of brominated polystyrene with a bromine content of 3.1 mmol/g, what corresponds to a bromination of 43 % of all available phenyl groups.

* Benzene should be replaced by a less toxic solvent, e.g. toluene.

References for Chapter 6

[1] Merrifield, R. B. *J. Am. Chem. Soc.* **1963**, *85*, 2149–2154.
[2] Pepper, K. W.; Paisley, H. M.; Young, M. A. *J. Chem. Soc.* **1953**, 4097–4105.
[3] Pinnell, R. P.; Khune, G. D.; Khatri, N. A.; Manatt, S. L. *Tetrahedron Lett.* **1984**, *25*, 3511–3514.
[4] Ford, W. T.; Yacoub, S. A. *J. Org. Chem.* **1981**, *46*, 819–821.
[5] Arshady, R.; Kenner, G. W.; Ledwith, A. *Makromol. Chem.* **1976**, *177*, 2911–2918.
[6] Neumann, W. P.; Peterseim, M. *Reactive Polymers* **1993**, *20*, 189–205.
[7] Bui, C. T.; Maeji, N. J.; Rasoul, F.; Bray, A. M. *Tetrahedron Lett.* **1999**, *40*, 5383–5386.
[8] Mohanraj, S.; Ford, W. T. *Macromolecules* **1986**, *19*, 2470–2472.
[9] Sheng, Q.; Stöver, H. D. H. *Macromolecules* **1997**, *30*, 6712–6714.
[10] Nicolaou, K. C.; Winssinger, N.; Pastor, J.; Murphy, F. *Angew. Chem. Int. Ed. Engl.* **1998**, *37*, 2534–2537.
[11] Karoyan, P.; Triolo, A.; Nannicini, R.; Giannotti, D.; Altamura, M.; Chassaing, G.; Perrotta, E. *Tetrahedron Lett.* **1999**, *40*, 71–74.
[12] Gerigk, U.; Gerlach, M.; Neumann, W. P.; Vieler, R.; Weintritt, V. *Synthesis* **1990**, 448–452.
[13] Tortolani, D. R.; Biller, S. A. *Tetrahedron Lett.* **1996**, *37*, 5687–5690.
[14] Barco, A.; Benetti, S.; De Risi, C.; Marchetti, P.; Pollini, G. P.; Zanirato, V. *Tetrahedron Lett.* **1998**, *39*, 7591–7594.
[15] Stranix, B. R.; Gao, J. P.; Barghi, R.; Salha, J.; Darling, G. D. *J. Org. Chem.* **1997**, *62*, 8987–8993.
[16] Ball, C. P.; Barrett, A. G. M.; Commercon, A.; Compère, D.; Kuhn, C.; Roberts, R. S.; Smith, M. L.; Venier, O. *Chem. Commun.* **1998**, 2019–2020.
[17] Watanabe, Y.; Ishikawa, S.; Takao, G.; Toru, T. *Tetrahedron Lett.* **1999**, *40*, 3411–3414.
[18] Zheng, C. S.; Seeberger, P. H.; Danishefsky, S. J. *Angew. Chem. Int. Ed. Engl.* **1998**, *37*, 786–789.
[19] Savin, K. A.; Woo, J. C. G.; Danishefsky, S. J. *J. Org. Chem.* **1999**, *64*, 4183–4186.
[20] Guibé, F.; Dangles, O.; Balavoine, G.; Loffet, A. *Tetrahedron Lett.* **1989**, *30*, 2641–2644.
[21] Wong, J. Y.; Manning, C.; Leznoff, C. C. *Angew. Chem. Int. Ed. Engl.* **1974**, *13*, 666–667.
[22] Bhandari, A.; Jones, D. G.; Schullek, J. R.; Vo, K.; Schunk, C. A.; Tamanaha, L. L.; Chen, D.; Yuan, Z. Y.; Needels, M. C.; Gallop, M. A. *Bioorg. Med. Chem. Lett.* **1998**, *8*, 2303–2308.
[23] Nicolaou, K. C.; Winssinger, N.; Vourloumis, D.; Ohshima, T.; Kim, S.; Pfefferkorn, J.; Xu, J. Y.; Li, T. *J. Am. Chem. Soc.* **1998**, *120*, 10814–10826.
[24] van Maarseveen, J. H.; den Hartog, J. A. J.; Engelen, V.; Finner, E.; Visser, G.; Kruse, C. G. *Tetrahedron Lett.* **1996**, *37*, 8249–8252.
[25] Ngu, K.; Patel, D. V. *Tetrahedron Lett.* **1997**, *38*, 973–976.
[26] Raju, B.; Kogan, T. P. *Tetrahedron Lett.* **1997**, *38*, 4965–4968.
[27] Brown, D. S.; Revill, J. M.; Shute, R. E. *Tetrahedron Lett.* **1998**, *39*, 8533–8536.
[28] Fréchet, J. M. J.; Haque, K. E. *Tetrahedron Lett.* **1975**, 3055–3056.
[29] Hayatsu, H.; Khorana, H. G. *J. Am. Chem. Soc.* **1966**, *88*, 3182–3183.
[30] Hayatsu, H.; Khorana, H. G. *J. Am. Chem. Soc.* **1967**, *89*, 3880–3887.
[31] Ajayaghosh, A.; Pillai, V. N. R. *Tetrahedron* **1988**, *44*, 6661–6666.
[32] Hall, B. J.; Sutherland, J. D. *Tetrahedron Lett.* **1998**, *39*, 6593–6596.
[33] Nugiel, D. A.; Wacker, D. A.; Nemeth, G. A. *Tetrahedron Lett.* **1997**, *38*, 5789–5790.
[34] Berteina, S.; Wendeborn, S.; De Mesmaeker, A. *Synlett* **1998**, 1231–1233.
[35] Newlander, K. A.; Chenera, B.; Veber, D. F.; Yim, N. C. F.; Moore, M. L. *J. Org. Chem.* **1997**, *62*, 6726–6732.
[36] Mellor, S. L.; Chan, W. C. *Chem. Commun.* **1997**, 2005–2006.
[37] Garigipati, R. S. *Tetrahedron Lett.* **1997**, *38*, 6807–6810.
[38] Han, Y.; Giroux, A.; Lépine, C.; Laliberté, F.; Huang, Z.; Perrier, H.; Bayly, C. I.; Young, R. N. *Tetrahedron* **1999**, *55*, 11669–11685.
[39] Veerman, J. J. N.; van Maarseveen, J. H.; Visser, G. M.; Kruse, C. G.; Schoemaker, H. E.; Hiemstra, H.; Rutjes, F. P. J. T. *Eur. J. Org. Chem.* **1998**, 2583–2589.
[40] Morales, G. A.; Corbett, J. W.; DeGrado, W. F. *J. Org. Chem.* **1998**, *63*, 1172–1177.
[41] Darling, G. D.; Fréchet, J. M. J. *J. Org. Chem.* **1986**, *51*, 2270–2276.
[42] Leznoff, C. C.; Fyles, T. M.; Weatherston, J. *Can. J. Chem.* **1977**, *55*, 1143–1153.
[43] Nicolaou, K. C.; Winssinger, N.; Pastor, J.; Ninkovic, S.; Sarabia, F.; He, Y.; Vourloumis, D.; Yang, Z.; Li, T.; Giannakakou, P.; Hamel, E. *Nature* **1997**, *387*, 268–272.
[44] Heitz, W.; Michels, R. *Makromol. Chem.* **1971**, *148*, 9–17.
[45] Bernard, M.; Ford, W. T. *J. Org. Chem.* **1983**, *48*, 326–332.
[46] Weinshenker, N. M.; Crosby, G. A.; Wong, J. Y. *J. Org. Chem.* **1975**, *40*, 1966–1971.
[47] Crosby, G. A.; Weinshenker, N. M.; Uh, H. S. *J. Am. Chem. Soc.* **1975**, *97*, 2232–2235.

[48] Camps, F.; Castells, J.; Ferrando, M. J.; Font, J. *Tetrahedron Lett.* **1971**, 1713–1714.
[49] Hennion, G. F.; Anderson, J. G. *J. Am. Chem. Soc.* **1946**, *68*, 424–426.
[50] Farrall, M. J.; Fréchet, J. M. J. *J. Org. Chem.* **1976**, *41*, 3877–3882.
[51] Malenfant, P. R. L.; Fréchet, J. M. J. *Chem. Commun.* **1998**, 2657–2658.
[52] Arsequell, G.; Espuña, G.; Valencia, G.; Barluenga, J.; Carlón, R. P.; González, J. M. *Tetrahedron Lett.* **1998**, *39*, 7393–7396.
[53] Deleuze, H.; Sherrington, D. C. *J. Chem. Soc. Perkin Trans. 2* **1995**, 2217–2221.

7 Preparation of Alcohols and Ethers

7.1 Preparation of Alcohols

Most efforts directed towards the development of solid-phase preparations of alcohols had been limited for many years to the synthesis of biopolymers, such as oligonucleotides and oligosaccharides. Interest in the preparation and chemical transformation of all types of alcohol on insoluble supports began, however, to grow rapidly in the early 1990s, when chemists realized the potential of parallel solid-phase synthesis for high-throughput compound production.

Alcohols are generally prepared on insoluble supports by reduction of carbonyl compounds or by addition of carbon nucleophiles to carbonyl compounds, but other strategies have also been used (Figure 7.1).

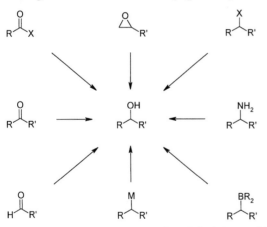

Fig. 7.1. Strategies for the preparation of alcohols on solid phase. M: metal, X: leaving group.

7.1.1 Reduction of Carbonyl Compounds

Support-bound aldehydes or ketones can be reduced to alcohols under mild reaction conditions, which are compatible with most supports and linkers. Typical reducing agents are sodium borohydride or diisobutylaluminum hydride, which can penetrate cross-linked polystyrene, provided a solvent with sufficient swelling ability is chosen.

Table 7.1. Preparation of alcohols by reduction of carbonyl compounds.

Entry	Starting resin	Conditions	Product	Ref.
1		NaBH$_4$, (0.33 mol/L, 4 eq), THF/NMM/EtOH 16:7:7, 20 °C, 24 h		[1]
2		LiBH$_4$, THF, 60 °C		[2]
3		NaBH$_4$ (0.03 mol/L), THF/EtOH 1:1, 4 h		[3] see also [4]
4		L-Selectride (0.5 mol/L), THF, −75 °C to −65 °C, 20 h	(87:13)	[5] see also [6,7]
5		(HCHO)$_n$, KOH, DMF, 15 °C, 48 h		[8]
6		DIBAH (3 eq), THF, 20 °C, 4 h		[9] see also [10,11]
7		LiAlH$_4$ (2 eq), Et$_2$O, 30 °C, 6 h		[12]
8		LiAlH$_4$, Et$_2$O, 6 h		[13]
9		LiBH$_4$, THF, 20 °C, 12 h		[14]
10		1. *i*BuOCOCl, NMM, THF 2. NaBH$_4$, H$_2$O, THF		[15] see also [16]
11		1. EtOCOCl, NEt$_3$, THF, 0 °C 2. Bu$_4$NBH$_4$, MeOH		[17]

Reductions on polystyrene with sodium borohydride can be conducted in DMF or THF, or in mixtures of these solvents with alcohols (Table 7.1). Diastereoselective reductions have been performed on insoluble supports at low temperatures using Selectride (Entry **4**, Table 7.1). Formaldehyde can be used as alternative to hydrides to reduce support-bound ketones (Cannizzaro reaction; Entry **5**, Table 7.1).

Carboxylic esters have been reduced on polystyrene with diisobutylaluminum hydride or lithium aluminum hydride. The latter reagent can, however, lead to the formation of insoluble precipitates, and might readily cause problems if reactions are performed in fritted reactors. An alternative procedure for reducing carboxylic esters to alcohols consists in saponification, followed by activation (e.g. as the mixed anhydride) and reduction with sodium borohydride (Entries **10** and **11**, Table 7.1). Care must be taken during the activation of carboxylic acids with alkyl chloroformates, because the resulting anhydrides decompose at room temperature to yield carboxylic esters.

7.1.2 Addition of Carbon Nucleophiles to C–O Double Bonds

Support-bound carbonyl compounds can be converted into alcohols by treatment with suitable carbon nucleophiles. Aldehydes react readily with ketones or other C,H-acidic compounds upon acid- or base-catalysis to yield the products of aldol addition (Table 7.2). Some types of C,H-acidic compound, such as 1,3-dicarbonyl compounds, will, nevertheless, yield products of aldol condensation directly (Section 5.2.2.2).

Catalyzed aldol additions do not generally proceed with high diastereoselectivity at ambient temperature. Improved stereoselectivity can be achieved with preformed, diastereomerically pure enolates at low temperatures (Entry **5**, Table 7.2). This strategy enables the solid-phase preparation of stereochemically defined polyketides. On cross-linked polystyrene the same diastereoselectivity as in homogeneous phase is observed in the addition of boron enolates to aldehydes [14,18].

Support-bound carbonyl compounds have been converted into alcohols by treatment with Grignard reagents [25–28], organolithium compounds, allylindium compounds, allylsilanes, and organomanganese compounds (Table 7.3). These reactions, of course, require supports and linkers which do not react with the organometallic reagent. The support should, furthermore, swell sufficiently in the solvent chosen. Because cross-linked polystyrene swells substantially in THF and is chemically stable towards strong nucleophiles, polystyrene is well suited for reactions involving Grignard or similar reagents. Careful drying is, however, often necessary to obtain good results. If the support-bound carbonyl compound can enolize, small amounts of water or alcohols can lead to substantial amounts of enolate on the support, which will not react with the organometallic reagent, even if excess of the latter is used. Substrates with a high tendency to enolize are, e.g., 1,3-dicarbonyl compounds and benzylketones. Enolate formation can be avoided either by using a less basic organometallic reagent (e.g. organotitanium, organozirconium, or organoindium reagents). If enolate formation leads to low conversions, the reaction can also be repeated several times, what usually leads to improved results.

Table 7.2. Preparation of alcohols from support-bound carbonyl compounds by aldol addition.

Entry	Starting resin	Conditions	Product	Ref.
1	(PS-bound aldehyde)	MeNO$_2$/EtOH/NEt$_3$/ THF 4:4:1:12, 20 °C, overnight	(PS-bound product)	[19,20]
2	(PEG-bound aldehyde)	MeNO$_2$/NEt$_3$ 20:1, 20 °C, overnight (cross-linked PEG)	(PEG-bound product)	[21]
3	(PS-bound aldehyde, O$_2$N)	PhCOMe, K$_2$CO$_3$, THF, 67 °C, 48 h	(PS-bound product, O$_2$N)	[22]
4	(PS-bound)	LDA, ZnCl$_2$, THF, −78 °C to −40 °C, 2 h	(PS-bound product)	[23]
5	(PS-bound)	BnO..., (C$_6$H$_{11}$)$_2$BO, Et$_2$O, −78 °C, 1 h, 0 °C, 16 h	BnO...(PS) (> 97% ds)	[18] see also [14,24]

PEG-grafted polystyrene is also well suited for reactions with highly reactive organometallic reagents, provided the support has been dried. PEG-containing polymers are generally more difficult to dry than pure polystyrene. Cross-linked PEG (see Section 2.3) is stable towards Lewis acids, and can be used for SnCl$_4$-mediated allylations of aldehydes with allyl silanes [21].

The automated parallel synthesis of compounds in arrays of fritted reactors using organolithium or Grignard reagents is a special challenge. The transfer of these reagents to the reactor can readily lead to clogging of needles and tubing, and during the reaction or upon quenching precipitates might form and clogg fritts. All these potential problems should be carefully considered before starting library production.

Alcohols can also be prepared from support-bound carbon nucleophiles and carbonyl compounds (Table 7.4). Few examples have been reported of the α-alkylation of resin-bound esters with aldehydes or ketones (Table 7.4). This reaction is complicated by the thermal instability of some ester enolates, which can undergo elimination of alkoxide to yield ketenes. Traces of water or alcohols can, furthermore, lead to saponification or transesterification and release of the substrate into solution. Less prone to base-induced cleavage are support-bound imides (Entry **2**, Table 7.4, see also Table 13.8 [36]). Alternatively, support-bound thiol esters can be converted into stable

Table 7.3. Preparation of alcohols from support-bound carbonyl compounds and organometallic reagents.

Entry	Starting resin	Conditions	Product	Ref.
1		allyl bromide (14 eq), In (9 eq), THF/H$_2$O 1:1, ultrasound, 25 °C, 4 h (allyl boronates can also be used)		[29]
2		Me$_3$Si SnCl$_4$, DCM, 2 h (cross-linked PEG)		[21]
3		Ph BuLi (3 eq), THF, −78 °C, 4 h		[30]
4		PhMgBr, THF, −78 °C to −10 °C, 3 h (non-cross-linked PS)		[31]
5		Me$_3$Si≡Li (5 eq), THF, 0–23 °C, 2 h		[32]
6		PhMgBr, CeCl$_3$, THF, 3 h		[33]
7		PhMnI (0.25 mol/L), C$_6$H$_6$, 20 °C, 12 h		[34]
8		PhMgBr (large excess)		[35] see also [10]

silyl ketene acetals, which react with aldehydes upon Lewis-acid catalysis (Entries **3** and **4**, Table 7.4).

Substituted allyl alcohols can be prepared on insoluble supports under mild conditions using the Baylis–Hillman reaction (Figure 7.2). In this reaction an acrylate is treated with a nucleophilic, tertiary amine (typically DABCO) in the presence of an aldehyde. Reversible Michael addition of the amine to the acrylate leads to an ester enolate, which reacts with the aldehyde. The resulting allyl alcohols are valuable intermediates for the preparation of substituted carboxylic acids [37,38].

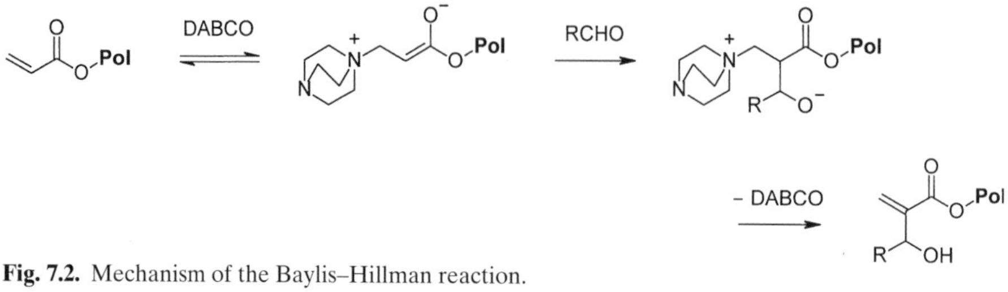

Fig. 7.2. Mechanism of the Baylis–Hillman reaction.

Table 7.4. Preparation of alcohols from support-bound ester enolates and related carbon nucleophiles.

Entry	Starting resin	Conditions	Product	Ref.
1	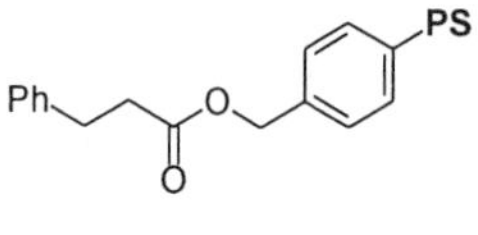	1. LDA, THF, −78 °C 2. ZnCl₂, 0 °C 3. 4-(MeO)C₆H₄CHO	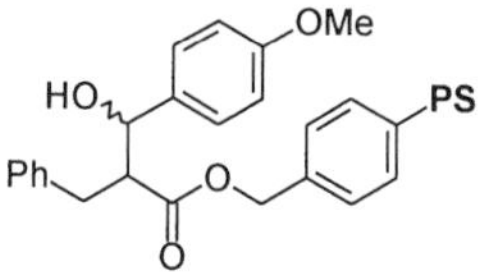	[39]
2	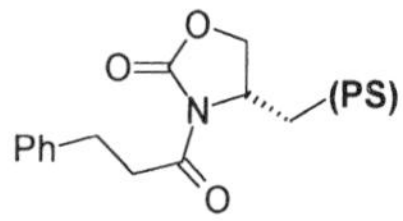	1. Bu₂BOTf (2 × 2 eq), DIPEA, DCM, −20 °C, 2 × 45 min 2. (iPr)CH₂CHO	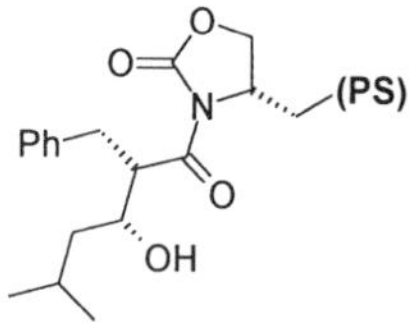	[40] see also [41]
3	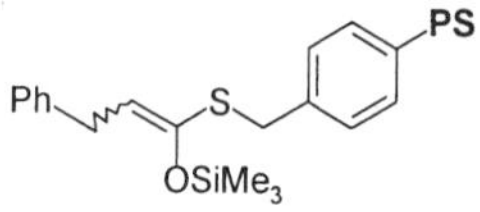	4-(MeO)C₆H₄CHO, Sc(OTf)₃, DCM, −78 °C, 20 h	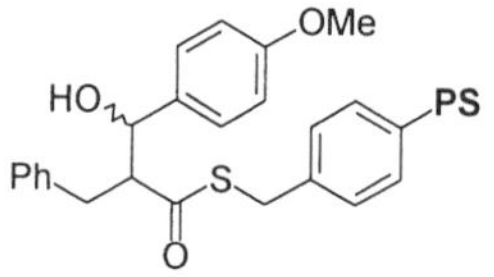	[42]
4	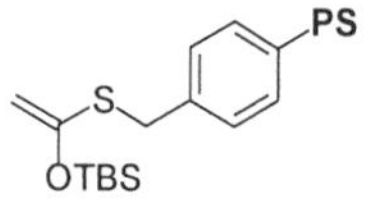	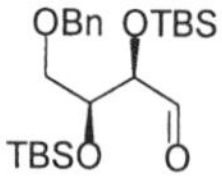 TBSO BF₃OEt₂, DCM, −78 °C, 20 h	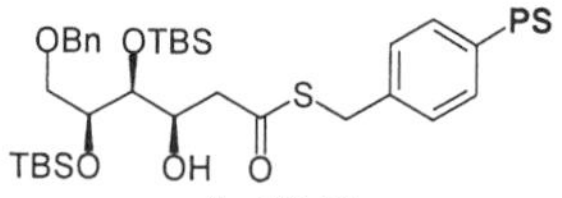 (> 98:2)	[43]
5	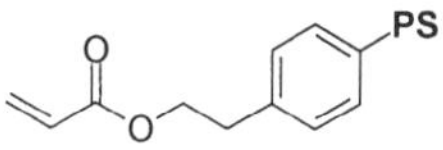	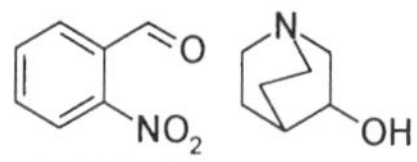 NO₂ DMF, 22 °C, 18 h	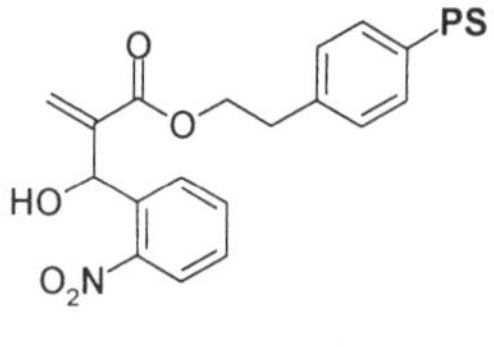	[37]
6	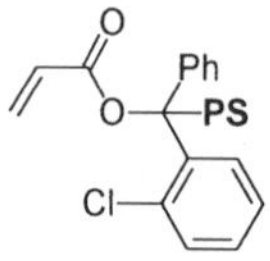	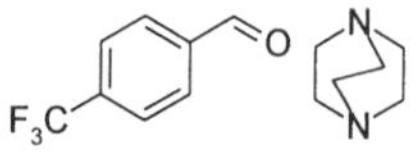 F₃C DMSO/CHCl₃ 1:1, 20 °C, 2 d	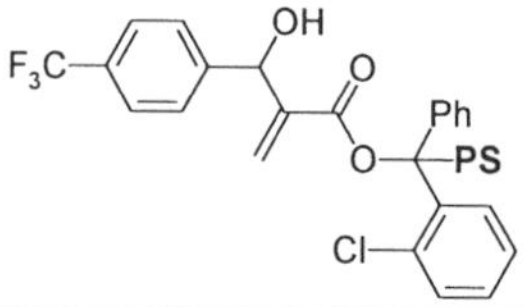	[38]

More reactive carbon nucleophiles than enolates can also be prepared on insoluble supports (see Chapter 4) and used to convert aldehydes or ketones into alcohols. Organolithium compounds have been generated on cross-linked polystyrene by deprotonation of formamidines and by metallation of aryl iodides (Table 7.5). Similarly, support-bound organomagnesium compounds can be prepared by metallation of aryl and vinyl iodides with Grignard reagents. The resulting organometallic compounds react with aldehydes or ketones to yield the expected alcohols (Table 7.5).

Table 7.5. Preparation of alcohols from support-bound organolithium and organomagnesium compounds.

Entry	Starting resin	Conditions	Product	Ref.
1		*tert*-BuLi, THF, PhCHO, −78 °C to 20 °C, overnight		[44]
2		BuLi, THF, −78 °C, 15 min, then		[45]
3		*i*PrMgBr (0.19 mol/L, 7.3 eq), THF, −35 °C, 0.5 h, then PhCHO		[46]
4		1. *i*PrMgBr (10 eq), THF/NMP 40:1, −40 °C, 1.5 h 2. PhCHO (15 eq), −40 °C, 2 h		[47]
5		BuLi, THF, −30 °C, 4 h; add TolCHO, −30 °C to 20 °C, 2 h		[48]

7.1.3 Miscellaneous Preparations of Alcohols

Hydroxymethyl polystyrene has been prepared from chloromethyl polystyrene either by conversion into the acetate by nucleophilic substitution followed by saponification, or directly by treatment with a mixture of potassium acetate and tetrabutyl-ammonium hydroxide in 1,2-dichlorobenzene/water (85 °C, 2 d [49]).

A suitable means of access to functionalized alcohols, also applicable to solid-phase synthesis, is the ring-opening of epoxides with nucleophiles. Hydride (Entry **1**, Table 7.6), organolithium compounds (Entry **2**, Table 7.6), amines (Table 10.1), azide (Table 10.18), alcohols (Table 7.9), and thiols (Table 8.3) have been used with success to convert (mostly support-bound) oxiranes into the corresponding alcohols.

Table 7.6. Miscellaneous preparations of alcohols and phenols.

Entry	Starting resin	Conditions	Product	Ref.
1		LiBH$_4$, THF, 20 °C, 6 h		[51]
2		(190 eq), THF, −50 °C to 20 °C, 18 h		[52]
3		KN(SiMe$_3$)$_2$,		[11]
4		H$_2$O$_2$, NaOH, H$_2$O, THF, 20 °C, 4 h		[53,54]
5		1. O$_3$, DCM, −78 °C, 10 min 2. NaBH$_4$, iPrOH, ultrasound, 20 °C, overnight		[55]
6		K$_3$Fe(CN)$_6$, K$_2$CO$_3$, K$_2$OsO$_4$, (DHQD)$_2$PHAL (AD-mix-β), THF/H$_2$O 1:1, 20 °C, 12 h	(73% yield, 3% ee)	[56]
7		K$_3$Fe(CN)$_6$, K$_2$CO$_3$, K$_2$OsO$_4$, (DHQD)$_2$PHAL (AD-mix-β), THF/H$_2$O 1:1, 20 °C, 18 h	(41% yield, 97% ee)	[56]
8		NaNO$_2$, HCl (30% in H$_2$O), KCl, ultrasound, 3 h (PS-grafted crowns)		[57]
9		1. BuLi, C$_6$H$_{12}$, 65 °C, 4 h 2. O$_2$, C$_6$H$_{12}$, 25 °C, 2 h		[58] see also [49]
10		BnMe$_3$N$^+$OH$^-$, H$_2$O, dioxane, 90 °C, 8 h (X: F, Cl, OMe)		[59]

Table 7.6. continued.

Entry	Starting resin	Conditions	Product	Ref.
11		110 °C, PhMe		[45]
12		NaOH, EtOH, 78 °C, 1 h		[60]

The oxidation of polystyrene-bound boranes or organolithium compounds has been used to prepare aliphatic alcohols and phenols (Table 7.6). Further oxidative protocols applicable to the solid-phase synthesis of alcohols include the α-hydroxylation of ester enolates with *N*-sulfonyloxaziridines, the ozonolysis of alkenes followed by reduction of the resulting ozonides, and the dihydroxylation of alkenes (Table 7.6). Attempts to perform asymmetric dihydroxylation of olefins either on cross-linked polystyrene or on PEG-grafted polystyrene proceeded with lower enantioselectivity than in solution (Entries **6** and **7**, Table 7.6).

Phenols have been prepared on solid phase by aromatic nucleophilic substitution with hydroxide, by thermal rearrangement of vinylcyclobutenones, by oxidative coupling of phenols (Figure 5.17 [50]), and by cyclocondensation reactions with simultaneous release of the phenols into solution (Entry **12**, Table 7.6).

7.1.4 Protective Groups for Alcohols

Some of the strategies used for the protection of alcohols in solution can also be used on insoluble supports. The choice of protective groups is mainly limited by the type of linker chosen. The following section covers those protective groups for alcohols which can be either introduced or removed on insoluble supports. Orthogonal protection strategies for alcohols have mainly been developed for the solid-phase synthesis of oligonucleotides and oligosaccharides (Chapter 16).

7.1.4.1 Protective Groups Cleavable by Acids

Benzyl ethers bearing electron-donating groups can be cleaved by treatment with acids. The rate of solvolysis increases with the number of electron-donating groups, and in the series benzyl < benzhydryl < trityl. For the solid-phase synthesis of oligonu-

cleotides, the 5′-hydroxyl group is usually protected as 4,4′-dimethoxytrityl ether
(DMT-OR) (Entry **1**, Table 7.7), which can readily be cleaved by weak acids, such as
dichloroacetic acid (3 % in DCM, 2 min [61–64]). Monomethoxytrityl ethers (MMT-
OR) are only slightly more resistant towards acidolysis, and can, e.g., be cleaved with
3 % trichloroacetic acid in DCM [65] or 1 % TsOH in DCM [5]. *Tert*-Butyl ethers or
tert-butyl carbamates (Boc-protected amines) are usually stable under these mildly
acidic conditions.

As side-chain protection of serine or threonine in solid-phase peptide synthesis are
used unsubstituted benzyl ethers (Boc methodology), *tert*-butyl ethers [66], or trityl
ethers (Fmoc methodology). Tyrosine is usually protected as the *tert*-butyl ether, benzyl ether, 2,6-dichlorobenzyl ether, or 2-bromobenzyl carbonate. Deprotection takes
place during cleavage of the peptide from the support [67,68].

As an alternative to protection as ethers, alcohols can also be protected as acetals,
the most common being tetrahydropyranyl ethers (THP-OR) and 1-ethoxyethyl

Table 7.7. Acid-labile protective groups for alcohols.

Entry	Starting resin	Conditions	Product	Ref.
1		3% Cl$_3$CCO$_2$H in DCM, 8 s–50 min; or 80% AcOH, 0.5 h; or ZnBr$_2$, MeOH/MeNO$_2$		[70-73]
2		TsOH (0.03 mol/L), DCM/MeOH 97:3, 20 °C, 2 h		[74]
3		AcOH, H$_2$O, THF		[69]
4		MeOH, dioxane, PPTS		[75]
5		TsOH (2% in DCM), 5 × 5 min (TsOH was dissolved in MeOH and diluted with DCM)		[5]

ethers (EE-OR) (Table 7.7). Support-bound secondary, aliphatic alcohols have been converted into THP ethers by treatment with dihydropyran and PPTS [69]. Both THP and EE-ethers can be cleaved under mildly acidic conditions (e.g. with dilute acetic acid [69]), under which no cleavage of *tert*-butyl ethers, *tert*-butyl carbamates, glycosides, or esters of Wang resin occurs.

7.1.4.2 Protective Groups Cleavable by Nucleophiles

The solid-phase synthesis of oligosaccharides is usually performed using acid-resistant linkers and protective groups, because of the slightly acidic reaction conditions required for glycosylations (Section 16.3). Hydroxyl group protection is conveniently ensured by conversion into carboxylic esters, such as acetates, benzoates, or nitrobenzoates. Support-bound esters of primary or secondary aliphatic alcohols can be cleaved by treatment with alcoholates [75–77] (Table 7.8), with DBU in methanol, with hydrazine in DMF [78] or dioxane [79], or with ethylenediamine [80], provided a suitable linker has been chosen.

Esters of 4-oxo carboxylic acids, such as levulinic acid (Entry **5**, Table 7.8) or 3-benzoylpropionic acid [81], can be cleaved under essentially neutral reaction conditions with hydrazinium acetate [82]. The cleavage proceeds via hydrazone formation and ring closure to yield the alcohol and a 2,3,4,5-tetrahydro-3-pyridazinone. Acetates might, however, also be hydrolyzed under these conditions [81].

Silyl ethers which have been used in solid-phase synthesis include TES, TIPS, TBS, and TBDPS ethers. Polystyrene-bound phenols can be converted into TIPS ethers by treatment with TIPS-OTf/imidazole in DMF for 5 min [83]. These silyl ethers are stable towards bases and weak acids, but can be selectively removed by treatment with TBAF (Entries **6** and **8**, Table 7.8) or pyridinium hydrofluoride (THF, 25 °C, 15 h [23,58,84]).

Allyl carbonates can be cleaved by nucleophiles upon catalysis by palladium(0). Allyl carbonates have been proposed for side-chain protection of serine and threonine, and their stability under conditions of *N*-Fmoc or *N*-Boc deprotection has been demonstrated [85]. Tyrosine derivatives and other phenols have been protected as allyl ethers, and deprotection could be achieved by palladium-mediated allylic substitution [85] (Entry **9**, Table 7.8).

4-Methoxyphenyl ethers can be cleaved by mild oxidants (Entry **10**, Table 7.8). Because many acid-labile linkers are also readily oxidized, care must be taken while applying this deprotection strategy.

Table 7.8. Protective groups for alcohols cleavable by nucleophiles and oxidants.

Entry	Starting resin	Conditions	Product	Ref.
1		NaOEt (0.1 mol/L), DMF/pyridine/EtOH 5:4:1, 20 °C, 1 h (macroporous PS)		[86]
2		NH$_3$ (30% in H$_2$O)/ THF/MeOH 2:6:2, 20 °C, 16 h		[87] see also [88]
3		K$_2$CO$_3$ (1.3 eq), MeOH/DCM 6:20, overnight		[89]
4		guanidine (0.1 mol/L, 2 eq), DMF, 20 °C, 16 h (repeat once)		[90] see also [84]
5		N$_2$H$_4$•AcOH, EtOH, 20 °C, 1 d		[91]
6		TBAF (10 eq), THF, 20 °C, 16 h		[75] see also [92]
7		HF•pyridine, THF, 23 °C		[84]
8		TBAF (0.3 mol/L), AcOH (0.3 mol/L), 20 °C, overnight R: alkyl		[88]

Table 7.8. continued.

Entry	Starting resin	Conditions	Product	Ref.
9		Pd(PPh₃)₄, TolSO₂Na, MeOH, THF		[93]
10		CAN (0.1 mol/L), MeCN/H₂O 4:1, 25 °C, 5 min		[65]

7.2 Preparation of Ethers

Ethers are widely used in solid-phase synthesis either as linkers for alcohols or as target compounds. Almost all reported solid-phase syntheses of ethers are O-alkylations or O-arylations of alcohols, which differ only in the type of alkylating agent used and in the precise reaction conditions.

This section covers mainly syntheses of acyclic ethers. Preparations of cyclic ethers are treated in Chapter 15.

7.2.1 Preparation of Dialkyl Ethers

Dialkyl ethers can be prepared either from support-bound electrophiles or from support-bound alcohols. Table 7.9 lists illustrative examples for the preparation of dialkyl ethers from support-bound electrophiles.

Cross-linked chloromethyl polystyrene (Merrifield resin) has frequently been used to O-alkylate aliphatic alcoholates, mainly for attaching linkers, ligands, or various synthetic auxiliaries to the support (Table 7.9). Examples other than those listed in Table 7.9 have been reported [23,92,94–98]. Strong bases, such as alkali metal hydrides, are required to deprotonate aliphatic alcohols. Suitable solvents for the conversion of Merrifield and related resins into alkyl benzyl ethers are THF, DMF, or diglyme, and generally high temperatures and long reaction times are necessary to achieve complete conversions. Support-bound halides with β-hydrogen are generally difficult to convert into dialkyl ethers, because the strongly basic reaction conditions will usually lead to the formation of alkenes by dehydrohalogenation.

Wang resin can also be etherified with aliphatic alcohols under acidic reaction conditions, by conversion of the polystyrene-bound alcohol into a trichloroacetimidate or a thiocarbonate (Entries **5** and **6**, Table 7.9). Support-bound aryldiazomethanes, which can be prepared from hydrazones or tosylhydrazones, also react with alcohols upon Lewis acid catalysis to yield benzyl ethers (Entry **7**, Table 7.9).

Table 7.9. Solid-phase synthesis of dialkyl ethers from support-bound electrophiles.

Entry	Starting resin	Conditions	Product	Ref.
1		KH, 18-crown-6, DMF, 80 °C, 5 d		[99] see also [100-102]
2		(3 eq), KO*t*Bu (1 mol/L, 3 eq), THF, 20 °C, 3.5 d		[103]
3		NaH (0.1 mol/L), DMF, 0 °C, 48 h		[104]
4		NaH, Bu$_4$NI, 18-crown-6, THF, 45 °C, 2 h		[5]
5		DCM/C$_6$H$_{12}$ 1:1, BF$_3$OEt$_2$ (13 mmol/L), 20 °C, 10 min		[105] see also [9,106]
6		AgOTf, DCM, 5 h		[107] see also [11]
7		(2 eq), BF$_3$OEt$_2$ (0.1 eq), DCM, 20 °C, 2 h		[108]
8		1. TFA/DCM 7:3, 20 °C, 5 h 2. ArCH$_2$CH$_2$OH (0.5 mol/L), DCM, 1 h Ar: 2-naphthyl		[109]
9		dodecanol (0.1 mol/L, 8 eq), KO*t*Bu (0.3 mol/L), diglyme, 100 °C, 30 h		[110]

Table 7.9. continued.

Entry	Starting resin	Conditions	Product	Ref.
10		HO (structure) R NaH, THF, 20 °C, 20 h	(structure)	[20]
11		iPrOH, Rh$_2$(OAc)$_4$, DCM, 20 °C, 20 h	(structure)	[111]
12		MeO$_2$C (structure) PhMe$_2$Si Me$_3$SiOMe, Me$_3$SiOTf, DCM, −78 °C to −55 °C, 72 h	(structure)	[112]
13		DIBAH (3 eq), DCM, −70 °C to −20 °C, 4 h	(structure)	[113]
14		DIBAH (3 eq), DCM, −70 °C to −20 °C, 4 h	(structure)	[113]

Other support-bound electrophiles, which react with aliphatic alcohols to yield ethers are epoxides, nitrostyrenes, and rhodium carbenoids generated from diazocarbonyl compounds and rhodium(II) acetate (Entries **9–11**, Table 7.9). The latter strategy enables the preparation of dialkyl ethers under essentially neutral reaction conditions, and thus the etherification of base- or acid-labile alcohols. Methyl ethers have been prepared by C-allylation of support-bound aldehydes or dimethylacetals with allylsilanes (Entry **12**, Table 7.9).

An elegant method for the preparation of monoethers of Wang resin with diols is the reduction of cyclic acetals (Entries **13** and **14**, Table 7.9). A more detailed exploration of the scope of this reaction, and the extension of this methodology to monohydroxy compounds remains to be performed.

Because Wang resin derived ethers are stable towards bases or nucleophiles, but can readily be cleaved with dilute trifluoroacetic acid, etherification represents a convenient method for linking alcohols to insoluble supports [10] (Section 3.11.1).

Polystyrene-bound aliphatic alcohols can be etherified under strongly basic conditions with alkyl halides (Table 7.10). Competing elimination is usually no concern, because a large excess of halide can be used, and olefins can be readily removed by filtration and washing of the support. Alternatively, the addition of resin-bound alcohols to acceptor-substituted alkenes can be used to prepare dialkyl ethers on insoluble supports (Entry **2**, Table 7.10). Treatment of polystyrene-bound alcohols with styrenes and *N*-iodosuccinimide under acidic conditions leads to the formation of 2-iodoalkyl ethers (Entry **3**, Table 7.10). These intermediates have been used to *N*-alkylate imidazole to prepare analogs of the antifungal agent miconazole [114]. Support-bound allylsilanes react with acetals under acidic conditions to yield homoallyl ethers (Entry **4**, Table 7.10).

Table 7.10. Solid-phase synthesis of dialkyl ethers from support-bound nucleophiles.

Entry	Starting resin	Conditions	Product	Ref.
1		NaH, THF/MeCN 5:1, 15-crown-5, ArCH$_2$Cl (0.3 mol/L), 67 °C, overnight Ar: 3,5-(MeO)$_2$C$_6$H$_3$		[44]
2		(0.7 mol/L), DBU (0.2 mol/L), DCM, 20 °C, overnight		[115]
3		(0.9 mol/L), NIS (0.9 mol/L), DME, TfOH (0.018 mol/L), 20 °C, 16 h; repeat once		[114]
4		PhCH(OMe)$_2$, Me$_3$SiOTf, DCM, −78 °C to −55 °C, 72 h	(*syn/anti* 20:1)	[112]

7.2.2 Preparation of Alkyl Aryl Ethers

Phenols attached to insoluble supports can be etherified either by treatment with alkyl halides and a base (Williamson ether synthesis) or by treatment with primary or secondary aliphatic alcohols, a phosphine, and an oxidant (typically DEAD; Mitsunobu reaction). The second methodology is generally preferred, because more alcohols than alkyl halides are commercially available, and because Mitsunobu etherifications proceed quickly at room temperature with high chemoselectivity, as illustrated by Entry **3**, Table 7.11. Thus, neither amines nor C,H-acidic compounds are generally alkylated under Mitsunobu conditions as efficiently as phenols. The reaction proceeds smoothly both with electron-rich and electron-poor phenols. Both primary and secondary aliphatic alcohols can be used to O-alkylate phenols, but variable results have been reported with 2-(Boc-amino)ethanols [116,117].

The Mitsunobu-etherification of polystyrene-bound phenols is usually conducted in THF or NMP, by simply adding the alcohol, the phosphine, and DEAD. Some authors claim that the addition of tertiary amines is beneficial [118], but this seems not always to be the case [116]. It is of course important that neither the support nor any of the reagents is contaminated with traces of methanol (e.g. from previous washings of the resin) or ethanol (from the decomposition of DEAD), which would lead to the formation of product mixtures. It has been suggested [116] that the heat evolved upon mixing a phosphine with DEAD can contribute to the decomposition of the latter, and lead to the formation of ethyl ethers. This can be avoided by using another oxidant (e.g. DIAD or TMAD) or a preformed betaine (e.g. Entry **4**, Table 7.11).

Table 7.11. Preparation of alkyl aryl ethers from support-bound phenols.

Entry	Starting resin	Conditions	Product	Ref.
1		1,3-propanediol, PPh₃, DEAD, THF, 1 h		[116] see also [120]
2		iPrOH, PPh₃, DEAD, THF, 1 h		[116]
3		ArCH₂OH (0.1 mol/L, 5 eq), TMAD (5 eq), PBu₃ (5 eq), THF/DCM 1:1, 1 h		[121] see also [122]
4		2-(4-fluorophenyl)-ethanol, DCM/PhMe 1:1, 3 d		[88] see also [117,123]
5		EtOH, DEAD, PPh₃, THF, −15 °C, 1 h, then 20 °C, 24 h		[93]
6		BuI (0.25 mol/L), DBU (0.25 mol/L), DMSO/NMP 1:1, 20 h		[124] see also [119,125]
7		ArO~~~~I, Cs₂CO₃, DMF, 25 °C, 30 h		[58]

The etherification of support-bound phenols with alkyl halides is usually performed in dipolar aprotic solvents (DMF, NMP, DMSO) in the presence of bases such as DBU, KN(SiMe₃)₂, phosphazenes [119], or cesium carbonate (Entries **6** and **7**, Table 7.11).

Polystyrene-bound alkylating agents can react with phenols under basic conditions to yield aryl ethers. The reaction conditions must be carefully adjusted for halides with β-hydrogen, to prevent dehydrohalogenation. This is of course no problem when

Table 7.12. Preparation of alkyl aryl ethers from support-bound alkylating agents.

Entry	Starting resin	Conditions	Product	Ref.
1		OHC—〈〉—OH, MeO, KOtBu, DMA, 90–50 °C, 14 h	('Sasrin-aldehyde')	[1,134] see also [26,135]
2		3-(HO)C₆H₄COMe, Cs₂CO₃, NaI, DMF		[136] see also [137]
3		2-cyanophenol, Cs₂CO₃, NMP, 70 °C, overnight		[138]
4		O₂N—〈〉—OH (0.5 mol/L), NMP/ 2,6-lutidine 4:1, 20 h		[139] see also [109]
5		Br—〈〉—CHO, OH (0.22 mol/L, 5 eq), DIPEA (5 eq), DMAP, DCM, 25 °C, 18 h		[140]
6		2-iodophenol (0.13 mol/L, 6 eq), BEMP (6 eq), dioxane, 100 °C, 90 h		[141]
7		K₂CO₃, KI, DMF, 80 °C, 16 h		[142]

using benzylic halides, which are in fact the most common support-bound alkylating agents used to O-alkylate phenols (Table 7.12). These reactions are usually conducted in DMF or DMA, and bases such as sodium hydride [60,95,123,126,127], alkali metal carbonates (Entries 2, 3, and 7; Table 7.12), sodium methanolate [28,128], or sodium hydroxide [129,130] can be used. Functionalized phenols, such as hydroxybenzaldehydes [130], hydroxybenzamides [129], or phenols containing keto [28] or hydroxy-alkyl groups do usually not undergo side reactions under the conditions of aryl ether formation, and can be selectively O-alkylated with support-bound benzyl halides (Table 7.12). PEG-bound mesylates and benzyl halides have been used to etherify phenols under conditions similar to those used in solution [131–133].

Alternatively, alkyl aryl ethers can be prepared from support-bound aliphatic alcohols by Mitsunobu etherification with phenols (Table 7.13). In this variant of the Mitsunobu reaction the presence of residual methanol or ethanol is less critical than in the etherification of support-bound phenols, because no dialkyl ethers can be formed by the Mitsunobu reaction. For this reason good results will also be obtained if the reaction mixture is left to warm upon mixing DEAD and the phosphine. Both triphenyl- or tributylphosphine can be used as phosphines. Tributylphosphine is a liquid, and generally does not give rise to insoluble precipitates. This reagent must, however, be handled with care because it readily ignites in air.

Table 7.13. Preparation of alkyl aryl ethers from support-bound aliphatic alcohols.

Entry	Starting resin	Conditions	Product	Ref.
1		PPh₃, DEAD, THF, 20 °C, 24 h; repeat once		[145]
2		Boc-Tyr-OMe (0.3 mol/L, 3 eq), PPh₃ (3 eq), DEAD (3 eq), NMM, 25 °C, 16 h		[143,146] see also [147]
3		phenol, PPh₃, DEAD, NMM		[15]
4		2-fluorophenol, PPh₃, DIAD, NMM, 20 °C, 12 h		[148]
5		4-(HO)C₆H₄CHO, DCM		[149]

It has been reported that tertiary amines (as additive or as solvent) lead to increased yields when etherifying tyrosine derivatives with polystyrene-bound benzyl alcohols [143]. Nevertheless, other phenols react smoothly without addition of bases [41,144]. When only a slight excess of phenol is used for the etherification of support-bound alcohols *N,N'*-bis(ethoxycarbonyl)hydrazine (the byproduct of the Mitsunobu reaction) can compete to a significant extent with the phenol and become attached to the support. This reaction can be supressed by use of a larger excess of phenol [137].

The most common resin-bound substrates for Mitsunobu etherification are primary benzylic alcohols, but a few non-benzylic alcohols have also been converted into aryl ethers (Table 7.13). Support-bound secondary alcohols are less suitable alkylating agents because elimination often predominates.

Experimental Procedure 7.1: Etherification of Fmoc-tyrosine methyl ester with a dialkoxybenzyl alcohol linker [147]

A solution of Fmoc-Tyr-OMe (96 mg, 0.23 mmol, 5.2 eq) and PPh$_3$ (72 mg, 0.28 mmol, 6.2 eq) in DCM (2 mL) was added to the DCM-swollen support (100 mg, 0.044 mmol). To this suspension a solution of DEAD (0.042 mL, 0.27 mmol, 6.1 eq) in DCM (0.2 mL) was added and the resulting mixture was shaken at room temperature overnight. Filtration, washing with DCM, DMF, MeOH, and Et$_2$O, and drying under reduced pressure yielded a resin with a loading of 0.17 mmol/g. The residual hydroxyl groups were acetylated by treatment of the support with Ac$_2$O (6 eq) and DIPEA (6 eq) in DCM (0.5 mL) for 20 min.

7.2.3 Preparation of Diaryl Ethers

Acceptor-substituted haloarenes have been used with success to O-arylate phenols (Table 7.14). The most common arylating agents are 2-fluoro-1-nitroarenes, 2-halopyridines, 2-halopyrimidines, and 2-halotriazines. When sufficiently reactive haloarenes are used the reaction proceeds smoothly, either with the arylating agent or with the phenol linked to the support.

Burgess and coworkers have developed a solid-phase synthesis of β-turn mimetics based on ring closure by aromatic nucleophilic substitution (Entry **4**, Table 7.14; see also Table 10.5). Phenols, alkylamines, and thiols have been used with success as nucleophiles for this type of macrocyclization [150].

Table 7.14. Preparation of diaryl ethers.

Entry	Starting resin	Conditions	Product	Ref.
1	NO₂, F, benzoate O—(PS)	salicylaldehyde, K₂CO₃, DMF, 40 °C, 16 h	NO₂ diaryl ether O—(PS)	[151]
2	HO— PS phenol	1. NaOH, H₂O 2. cyanuric chloride, Me₂CO, 0 °C, 8 h, then 20 °C, 2 d	triazinyl ether PS	[79]
3	HO— amidine N(TG)	pentafluoropyridine (0.16 mol/L), NaH, DMF, 2 h	pyridyl ether amidine (TG)	[152]
4	TIPSO, NO₂, F, peptide (PS)	1. TBAF, THF 2. K₂CO₃, DMF (cyclization also occurs without K₂CO₃ treatment)	NO₂ macrocyclic ether (PS)	[83] see also [150,153]

References for Chapter 7

[1] Katritzky, A. R.; Toader, D.; Watson, K.; Kiely, J. S. *Tetrahedron Lett.* **1997**, *38*, 7849–7850.
[2] Brown, D. S.; Revill, J. M.; Shute, R. E. *Tetrahedron Lett.* **1998**, *39*, 8533–8536.
[3] Purandare, A. V.; Poss, M. A. *Tetrahedron Lett.* **1998**, *39*, 935–938.
[4] Chambers, S. L.; Ronald, R.; Hanesworth, J. M.; Kinder, D. H.; Harding, J. W. *Peptides* **1997**, *18*, 505–512.
[5] Lee, C. E.; Kick, E. K.; Ellman, J. A. *J. Am. Chem. Soc.* **1998**, *120*, 9735–9747.
[6] Thompson, L. A.; Moore, F. L.; Moon, Y. C.; Ellman, J. A. *J. Org. Chem.* **1998**, *63*, 2066–2067.
[7] Chen, S.; Janda, K. D. *Tetrahedron Lett.* **1998**, *39*, 3943–3946.
[8] Ren, Q.; Huang, W.; Ho, P. *Reactive Polymers* **1989**, *11*, 237–244.
[9] Furman, B.; Thürmer, R.; Kaluza, Z.; Lysek, R.; Voelter, W.; Chmielewski, M. *Angew. Chem. Int. Ed. Engl.* **1999**, *38*, 1121–1123.
[10] Hanessian, S.; Xie, F. *Tetrahedron Lett.* **1998**, *39*, 737–740.
[11] Hanessian, S.; Ma, J.; Wang, W. *Tetrahedron Lett.* **1999**, *40*, 4631–4634.
[12] Chandrasekhar, S.; Padmaja, M. B. *Synthetic Commun.* **1998**, *28*, 3715–3720.
[13] Wang, S. *J. Am. Chem. Soc.* **1973**, *95*, 1328–1333.
[14] Reggelin, M.; Brenig, V.; Welcker, R. *Tetrahedron Lett.* **1998**, *39*, 4801–4804.
[15] Gayo, L. M.; Suto, M. J. *Tetrahedron Lett.* **1997**, *38*, 211–214.
[16] Bhandari, A.; Jones, D. G.; Schullek, J. R.; Vo, K.; Schunk, C. A.; Tamanaha, L. L.; Chen, D.; Yuan, Z. Y.; Needels, M. C.; Gallop, M. A. *Bioorg. Med. Chem. Lett.* **1998**, *8*, 2303–2308.
[17] Moran, E. J.; Wilson, T. E.; Cho, C. Y.; Cherry, S. R.; Schultz, P. G. *Biopolymers* **1995**, *37*, 213–219.
[18] Gennari, C.; Ceccarelli, S.; Piarulli, U.; Aboutayab, K.; Donghi, M.; Paterson, I. *Tetrahedron* **1998**, *54*, 14999–15016.
[19] Beebe, X.; Schore, N. E.; Kurth, M. J. *J. Org. Chem.* **1995**, *60*, 4196–4203.

[20] Beebe, X.; Chiappari, C. L.; Olmstead, M. M.; Kurth, M. J.; Schore, N. E. *J. Org. Chem.* **1995**, *60*, 4204–4212.
[21] Rademann, J.; Meldal, M.; Bock, K. *Chem. Eur. J.* **1999**, *5*, 1218–1225.
[22] Ruhland, T.; Künzer, H. *Tetrahedron Lett.* **1996**, *37*, 2757–2760.
[23] Nicolaou, K. C.; Winssinger, N.; Pastor, J.; Ninkovic, S.; Sarabia, F.; He, Y.; Vourloumis, D.; Yang, Z.; Li, T.; Giannakakou, P.; Hamel, E. *Nature* **1997**, *387*, 268–272.
[24] Reggelin, M.; Brenig, V. *Tetrahedron Lett.* **1996**, *37*, 6851–6852.
[25] Leznoff, C. C.; Wong, J. Y. *Can. J. Chem.* **1973**, *51*, 3756–3764.
[26] Routledge, A.; Stock, H. T.; Flitsch, S. L.; Turner, N. J. *Tetrahedron Lett.* **1997**, *38*, 8287–8290.
[27] Xu, Z. H.; McArthur, C. R.; Leznoff, C. C. *Can. J. Chem.* **1983**, *61*, 1405–1409.
[28] Léger, R.; Yen, R.; She, M. W.; Lee, V. J.; Hecker, S. J. *Tetrahedron Lett.* **1998**, *39*, 4171–4174.
[29] Cavallaro, C. L.; Herpin, T.; McGuinness, B. F.; Shimshock, Y. C.; Dolle, R. E. *Tetrahedron Lett.* **1999**, *40*, 2711–2714.
[30] Routledge, A.; Abell, C.; Balasubramanian, S. *Tetrahedron Lett.* **1997**, *38*, 1227–1230.
[31] Gosselin, F.; Van Betsbrugge, J.; Hatam, M.; Lubell, W. D. *J. Org. Chem.* **1999**, *64*, 2486–2493.
[32] Fraley, M. E.; Rubino, R. S. *Tetrahedron Lett.* **1997**, *38*, 3365–3368.
[33] Chen, C.; Munoz, B. *Tetrahedron Lett.* **1998**, *39*, 3401–3404.
[34] Leznoff, C. C.; Yedidia, V. *Can. J. Chem.* **1980**, *58*, 287–290.
[35] Liu, G. C.; Ellman, J. A. *J. Org. Chem.* **1995**, *60*, 7712–7713.
[36] Backes, B. J.; Ellman, J. A. *J. Am. Chem. Soc.* **1994**, *116*, 11171–11172.
[37] Prien, O.; Rölfing, K.; Thiel, M.; Künzer, H. *Synlett* **1997**, 325–326.
[38] Richter, H.; Walk, T.; Höltzel, A.; Jung, G. *J. Org. Chem.* **1999**, *64*, 1362–1365.
[39] Kurth, M. L.; Randall, L. A. A.; Chen, C.; Melander, C.; Miller, R. B.; McAlister, K.; Reitz, G.; Kang, R.; Nakatsu, T.; Green, C. *J. Org. Chem.* **1994**, *59*, 5862–5864.
[40] Purandare, A. V.; Natarajan, S. *Tetrahedron Lett.* **1997**, *38*, 8777–8780.
[41] Phoon, C. W.; Abell, C. *Tetrahedron Lett.* **1998**, *39*, 2655–2658.
[42] Kobayashi, S.; Hachiya, I.; Yasuda, M. *Tetrahedron Lett.* **1996**, *37*, 5569–5572.
[43] Kobayashi, S.; Wakabayashi, T.; Yasuda, M. *J. Org. Chem.* **1998**, *63*, 4868–4869.
[44] Furth, P. S.; Reitman, M. S.; Gentles, R.; Cook, A. F. *Tetrahedron Lett.* **1997**, *38*, 6643–6646.
[45] Tempest, P. A.; Armstrong, R. W. *J. Am. Chem. Soc.* **1997**, *119*, 7607–7608.
[46] Boymond, L.; Rottländer, M.; Cahiez, G.; Knochel, P. *Angew. Chem. Int. Ed. Engl.* **1998**, *37*, 1701–1703.
[47] Rottländer, M.; Knochel, P. *J. Comb. Chem.* **1999**, *1*, 181–183.
[48] Li, Z.; Ganesan, A. *Synlett* **1998**, 405–406.
[49] Fréchet, J. M. J.; de Smet, M. D.; Farrall, M. J. *Polymer* **1979**, *20*, 675–680.
[50] ApSimon, J. W.; Dixit, D. M. *Can. J. Chem.* **1982**, *60*, 368–370.
[51] Rotella, D. P. *J. Am. Chem. Soc.* **1996**, *118*, 12246–12247.
[52] Darling, G. D.; Fréchet, J. M. J. *J. Org. Chem.* **1986**, *51*, 2270–2276.
[53] Stranix, B. R.; Gao, J. P.; Barghi, R.; Salha, J.; Darling, G. D. *J. Org. Chem.* **1997**, *62*, 8987–8993.
[54] Sylvain, C.; Wagner, A.; Mioskowski, C. *Tetrahedron Lett.* **1998**, *39*, 9679–9680.
[55] Sylvain, C.; Wagner, A.; Mioskowski, C. *Tetrahedron Lett.* **1997**, *38*, 1043–1044.
[56] Riedl, R.; Tappe, R.; Berkessel, A. *J. Am. Chem. Soc.* **1998**, *120*, 8994–9000.
[57] Bui, C. T.; Maeji, N. J.; Rasoul, F.; Bray, A. M. *Tetrahedron Lett.* **1999**, *40*, 5383–5386.
[58] Nicolaou, K. C.; Winssinger, N.; Pastor, J.; DeRoose, F. *J. Am. Chem. Soc.* **1997**, *119*, 449–450.
[59] Cohen, B. J.; Karoly-Hafeli, H.; Patchornik, A. *J. Org. Chem.* **1984**, *49*, 922–924.
[60] Katritzky, A. R.; Belyakov, S. A.; Fang, Y. F.; Kiely, J. S. *Tetrahedron Lett.* **1998**, *39*, 8051–8054.
[61] Yang, X. B.; Sierzchala, A.; Misiura, K.; Niewiarowski, W.; Sochacki, M.; Stec, W. J.; Wieczorek, M. W. *J. Org. Chem.* **1998**, *63*, 7097–7100.
[62] Wright, P.; Lloyd, D.; Rapp, W.; Andrus, A. *Tetrahedron Lett.* **1993**, *34*, 3373–3376.
[63] Adinolfi, M.; Barone, G.; De Napoli, L.; Iadonisi, A.; Piccialli, G. *Tetrahedron Lett.* **1998**, *39*, 1953–1956.
[64] Swayze, E. E. *Tetrahedron Lett.* **1997**, *38*, 8643–8646.
[65] Peyman, A.; Weiser, C.; Uhlmann, E. *Bioorg. Med. Chem. Lett.* **1995**, *5*, 2469–2472.
[66] Chang, C. D.; Waki, M.; Ahmad, M.; Meienhofer, J.; Lundell, E. O.; Haug, J. D. *Int. J. Pept. Prot. Res.* **1980**, *15*, 59–66.
[67] Fields, G. B.; Noble, R. L. *Int. J. Pept. Prot. Res.* **1990**, *35*, 161–214.
[68] Novabiochem Catalog and Peptide Synthesis Handbook, Läufelfingen, **1999**.
[69] Zhu, T.; Boons, G. J. *Angew. Chem. Int. Ed. Engl.* **1998**, *37*, 1898–1900.
[70] Davis, P. W.; Vickers, T. A.; Wilson-Lingardo, L.; Wyatt, J. R.; Guinosso, C. J.; Sanghvi, Y. S.; De Baets, E. A.; Acevedo, O. L.; Cook, P. D.; Ecker, D. J. *J. Med. Chem.* **1995**, *38*, 4363–4366.
[71] Bayer, E.; Bleicher, K.; Maier, M. *Z. Naturforschung Sect. B* **1995**, *50*, 1096–1100.

[72] Adams, S. P.; Kavka, K. S.; Wykes, E. J.; Holder, S. B.; Galluppi, G. R. *J. Am. Chem. Soc.* **1983**, *105*, 661–663.

[73] Eckstein, F. *Oligonucleotides and Analogues; A Practical Approach*; Oxford University Press: Oxford, **1991**.

[74] Kuisle, O.; Quiñoá, E.; Riguera, R. *Tetrahedron Lett.* **1999**, *40*, 1203–1206.

[75] Wunberg, T.; Kallus, C.; Opatz, T.; Henke, S.; Schmidt, W.; Kunz, H. *Angew. Chem. Int. Ed. Engl.* **1998**, *37*, 2503–2505.

[76] Zehavi, U.; Patchornik, A. *J. Am. Chem. Soc.* **1973**, *95*, 5673–5677.

[77] Martin, G. E.; Shambhu, M. B.; Shakhshir, S. R.; Digenis, G. A. *J. Org. Chem.* **1978**, *43*, 4571–4574.

[78] Liang, R.; Yan, L.; Loebach, J.; Ge, M.; Uozumi, Y.; Sekanina, K.; Horan, N.; Gildersleeve, J.; Thompson, C.; Smith, A.; Biswas, K.; Still, W. C.; Kahne, D. *Science* **1996**, *274*, 1520–1522.

[79] Deleuze, H.; Sherrington, D. C. *J. Chem. Soc. Perkin Trans. 2* **1995**, 2217–2221.

[80] Rodebaugh, R.; Joshi, S.; Fraser-Reid, B.; Geysen, H. M. *J. Org. Chem.* **1997**, *62*, 5660–5661.

[81] Belorizky, N.; Excoffier, G.; Gagnaire, D.; Utille, J. P.; Vignon, M.; Vottero, P. *Bull. Soc. Chim. Fr.* **1972**, 4749–4753.

[82] Leikauf, E.; Barnekow, F.; Koster, H. *Tetrahedron* **1996**, *52*, 6913–6930.

[83] Burgess, K.; Lim, D.; Bois-Choussy, M.; Zhu, J. P. *Tetrahedron Lett.* **1997**, *38*, 3345–3348.

[84] Hunt, J. A.; Roush, W. R. *J. Am. Chem. Soc.* **1996**, *118*, 9998–9999.

[85] Loffet, A.; Zhang, H. X. *Int. J. Pept. Prot. Res.* **1993**, *42*, 346–351.

[86] Köster, H.; Cramer, F. *Liebigs Ann. Chem.* **1974**, 946–958.

[87] Fancelli, D.; Fagnola, M. C.; Severino, D.; Bedeschi, A. *Tetrahedron Lett.* **1997**, *38*, 2311–2314.

[88] Pavia, M. R.; Cohen, M. P.; Dilley, G. J.; Dubuc, G. R.; Durgin, T. L.; Forman, F. W.; Hediger, M. E.; Milot, G.; Powers, T. S.; Sucholeiki, I.; Zhou, S.; Hangauer, D. G. *Bioorg. Med. Chem.* **1996**, *4*, 659–666.

[89] Singh, R.; Nuss, J. M. *Tetrahedron Lett.* **1999**, *40*, 1249–1252.

[90] Heckel, A.; Mross, E.; Jung, K. H.; Rademann, J.; Schmidt, R. R. *Synlett* **1998**, 171–173.

[91] Shimizu, H.; Ito, Y.; Kanie, O.; Ogawa, T. *Bioorg. Med. Chem. Lett.* **1996**, *6*, 2841–2846.

[92] Nicolaou, K. C.; Winssinger, N.; Vourloumis, D.; Ohshima, T.; Kim, S.; Pfefferkorn, J.; Xu, J. Y.; Li, T. *J. Am. Chem. Soc.* **1998**, *120*, 10814–10826.

[93] Barber, A. M.; Hardcastle, I. R.; Rowlands, M. G.; Nutley, B. P.; Marriott, J. H.; Jarman, M. *Bioorg. Med. Chem. Lett.* **1999**, *9*, 623–626.

[94] Moberg, C.; Rákos, L. *Reactive Polymers* **1991**, *15*, 25–35.

[95] Furth, P. S.; Reitman, M. S.; Cook, A. F. *Tetrahedron Lett.* **1997**, *38*, 5403–5406.

[96] Worster, P. M.; McArthur, C. R.; Leznoff, C. C. *Angew. Chem. Int. Ed. Engl.* **1979**, *18*, 221–222.

[97] McArthur, C. R.; Worster, P. M.; Jiang, J.; Leznoff, C. C. *Can. J. Chem.* **1982**, *60*, 1836–1841.

[98] Rademann, J.; Schmidt, R. R. *J. Org. Chem.* **1997**, *62*, 3650–3653.

[99] Allin, S. M.; Shuttleworth, S. J. *Tetrahedron Lett.* **1996**, *37*, 8023–8026.

[100] Tietze, L. F.; Steinmetz, A. *Angew. Chem. Int. Ed. Engl.* **1996**, *35*, 651–652.

[101] Burgess, K.; Lim, D. *Chem. Commun.* **1997**, 785–786.

[102] Colwell, A. R.; Duckwall, L. R.; Brooks, R.; McManus, S. P. *J. Org. Chem.* **1981**, *46*, 3097–3102.

[103] Hernández, A. S.; Hodges, J. C. *J. Org. Chem.* **1997**, *62*, 3153–3157.

[104] Vidal-Ferran, A.; Bampos, N.; Moyano, A.; Pericàs, M. A.; Riera, A.; Sanders, J. K. M. *J. Org. Chem.* **1998**, *63*, 6309–6318.

[105] Hanessian, S.; Xie, F. *Tetrahedron Lett.* **1998**, *39*, 733–736.

[106] Craig, D.; Robson, M. J.; Shaw, S. J. *Synlett* **1998**, 1381–1383.

[107] Hanessian, S.; Huynh, H. K. *Tetrahedron Lett.* **1999**, *40*, 671–674.

[108] Mergler, M.; Dick, F.; Gosteli, J.; Nyfeler, R. *Tetrahedron Lett.* **1999**, *40*, 4663–4664.

[109] Tommasi, R. A.; Nantermet, P. G.; Shapiro, M. J.; Chin, J.; Brill, W. K. D.; Ang, K. *Tetrahedron Lett.* **1998**, *39*, 5477–5480.

[110] Brill, W. K. D.; De Mesmaeker, A.; Wendeborn, S. *Synlett* **1998**, 1085–1090.

[111] Zaragoza, F.; Petersen, S. V. *Tetrahedron* **1996**, *52*, 5999–6002.

[112] Panek, J. S.; Zhu, B. *J. Am. Chem. Soc.* **1997**, *119*, 12022–12023.

[113] Furman, B.; Thürmer, R.; Kaluza, Z.; Voelter, W.; Chmielewski, M. *Tetrahedron Lett.* **1999**, *40*, 5909–5912.

[114] Tortolani, D. R.; Biller, S. A. *Tetrahedron Lett.* **1996**, *37*, 5687–5690.

[115] Heinonen, P.; Lönnberg, H. *Tetrahedron Lett.* **1997**, *38*, 8569–8572.

[116] Krchnák, V.; Flegelová, Z.; Weichsel, A. S.; Lebl, M. *Tetrahedron Lett.* **1995**, *36*, 6193–6196.

[117] Johnson, M. G.; Bronson, D. D.; Gillespie, J. E.; Gifford-Moore, D. S.; Kalter, K.; Lynch, M. P.; McCowan, J. R.; Redick, C. C.; Sall, D. J.; Smith, G. F.; Foglesong, R. J. *Tetrahedron* **1999**, *55*, 11641–11652.

[118] Valerio, R. M.; Bray, A. M.; Patsiouras, H. *Tetrahedron Lett.* **1996**, *37*, 3019–3022.

[119] Du, X.; Armstrong, R. W. *J. Org. Chem.* **1997**, *62*, 5678–5679.

[120] Krchnák, V.; Weichsel, A. S.; Lebl, M.; Felder, S. *Bioorg. Med. Chem. Lett.* **1997**, *7*, 1013–1016.

[121] Rano, T. A.; Chapman, K. T. *Tetrahedron Lett.* **1995**, *36*, 3789–3792.

[122] Newlander, K. A.; Chenera, B.; Veber, D. F.; Yim, N. C. F.; Moore, M. L. *J. Org. Chem.* **1997**, *62*, 6726–6732.

[123] Brummond, K. M.; Lu, J. *J. Org. Chem.* **1999**, *64*, 1723–1726.

[124] Dankwardt, S. M.; Phan, T. M.; Krstenansky, J. L. *Mol. Diversity* **1995**, *1*, 113–120.

[125] Leznoff, C. C.; Dixit, D. M. *Can. J. Chem.* **1977**, *55*, 3351–3355.

[126] van Maarseveen, J. H.; den Hartog, J. A. J.; Engelen, V.; Finner, E.; Visser, G.; Kruse, C. G. *Tetrahedron Lett.* **1996**, *37*, 8249–8252.

[127] Bräse, S.; Köbberling, J.; Enders, D.; Lazny, R.; Wang, M.; Brandtner, S. *Tetrahedron Lett.* **1999**, *40*, 2105–2108.

[128] Kohlbau, H. J.; Tschakert, J.; Al-Qawasmeh, R. A.; Nizami, T. A.; Malik, A.; Voelter, W. *Z. Naturforschung Sect. B* **1998**, *53*, 753–764.

[129] Kobayashi, S.; Aoki, Y. *Tetrahedron Lett.* **1998**, *39*, 7345–7348.

[130] Kobayashi, S.; Akiyama, R. *Tetrahedron Lett.* **1998**, *39*, 9211–9214.

[131] Zhao, X. Y.; Janda, K. D. *Tetrahedron Lett.* **1997**, *38*, 5437–5440.

[132] Chen, S.; Janda, K. D. *J. Am. Chem. Soc.* **1997**, *119*, 8724–8725.

[133] Benaglia, M.; Annunziata, R.; Cinquini, M.; Cozzi, F.; Ressel, S. *J. Org. Chem.* **1998**, *63*, 8628–8629.

[134] Sarantakis, D.; Bicksler, J. J. *Tetrahedron Lett.* **1997**, *38*, 7325–7328.

[135] Zhong, H. M.; Greco, M. N.; Maryanoff, B. E. *J. Org. Chem.* **1997**, *62*, 9326–9330.

[136] Hollinshead, S. P. *Tetrahedron Lett.* **1996**, *37*, 9157–9160.

[137] Heerding, D. A.; Takata, D. T.; Kwon, C.; Huffman, W. F.; Samanen, J. *Tetrahedron Lett.* **1998**, *39*, 6815–6818.

[138] Goff, D.; Fernandez, J. *Tetrahedron Lett.* **1999**, *40*, 423–426.

[139] Brill, W. K. D.; Schmidt, E.; Tommasi, R. A. *Synlett* **1998**, 906–908.

[140] Haap, W. J.; Kaiser, D.; Walk, T. B.; Jung, G. *Tetrahedron* **1998**, *54*, 3705–3724.

[141] Berteina, S.; De Mesmaeker, A.; Wendeborn, S. *Synlett* **1999**, 1121–1123.

[142] Rölfing, K.; Thiel, M.; Künzer, H. *Synlett* **1996**, 1036–1038.

[143] Richter, L. S.; Gadek, T. R. *Tetrahedron Lett.* **1994**, *35*, 4705–4706.

[144] McNally, J. J.; Youngman, M. A.; Dax, S. L. *Tetrahedron Lett.* **1998**, *39*, 967–970.

[145] Devraj, R.; Cushman, M. *J. Org. Chem.* **1996**, *61*, 9368–9373.

[146] Winkler, J. D.; McCoull, W. *Tetrahedron Lett.* **1998**, *39*, 4935–4936.

[147] Cabrele, C.; Langer, M.; Beck-Sickinger, A. G. *J. Org. Chem.* **1999**, *64*, 4353–4361.

[148] Ruhland, T.; Andersen, K.; Pedersen, H. *J. Org. Chem.* **1998**, *63*, 9204–9211.

[149] Swayze, E. E. *Tetrahedron Lett.* **1997**, *38*, 8465–8468.

[150] Feng, Y.; Wang, Z.; Jin, S.; Burgess, K. *J. Am. Chem. Soc.* **1998**, *120*, 10768–10769.

[151] Wijkmans, J. C. H. M.; Culshaw, A. J.; Baxter, A. D. *Mol. Diversity* **1998**, *3*, 117–120.

[152] Mohan, R.; Yun, W. Y.; Buckman, B. O.; Liang, A.; Trinh, L.; Morrissey, M. M. *Bioorg. Med. Chem. Lett.* **1998**, *8*, 1877–1882.

[153] Kiselyov, A. S.; Eisenberg, S.; Luo, Y. *Tetrahedron Lett.* **1999**, *40*, 2465–2468.

8 Preparation of Sulfur Compounds

8.1 Preparation of Thiols

Thiols bound irreversibly to supports have mainly been used as linkers for carboxylic acids (Chapters 3.1.2 and 3.3.3), linkers for pyrimidines and triazines (Chapter 3.8), or linkers for other thiols (Chapter 3.12). When working with these resins, special care must be taken to prevent atmospheric oxidation of the thiols to the corresponding disulfides.

Some strategies used for the preparation of support-bound thiols are listed in Table 8.1. Oxidative thiolation of lithiated polystyrene has been used to prepare polymeric thiophenol (Entry **1**, Table 8.1). Polystyrene functionalized with 2-mercaptoethyl groups has been prepared by radical addition of thioacetic acid to cross-linked vinylpolystyrene, followed by hydrolysis of the intermediate thiol ester (Entry **2**, Table 8.1). A more controllable introduction of thiol groups, suitable also for the selective transformation of support-bound substrates, is based on nucleopilic substitution with thiourea or potassium thioacetate. The resulting isothiuronium salts and acetic acid thiol esters can be saponified, preferably under reductive conditions, to yield thiols (Entries **5–7**, Table 8.1). Thiolacetates have been saponified on insoluble supports with mercaptoethanol [1], propylamine [2], sodium or lithium borohydride, alcoholates, or hydrochloric acid (Table 8.1).

Additional functional groups, which can be transformed into thiols on insoluble supports are methoxytrityl and trityl thioethers and mixed disulfides (Table 8.2). These functionalities can serve as protective groups for support-bound thiols, cleavable without simultaneous cleavage of the thiol from the support. Most other protective groups for thiols, which have mainly been developed as side-chain protection of cysteine for solid-phase peptide synthesis, are removed during or after cleavage of the peptide from the support. Conventional side-chain protective groups for cysteine include trityl, *tert*-butyl, and acetamidomethyl thioethers (AcNH–CH$_2$–SR; Acm-SR), or unsymmetric disulfides. Deprotection is effected by treatment with TFA, mercury(II) salts, or silver(I) salts [16–19]. Alkoxycarbonyl groups are not well suited as protective groups for thiols because of their lability towards nucleophiles. For instance, allyloxycarbonyl thiols are completely deprotected within 3 h upon treatment with 30 % piperidine in DMF [20].

Table 8.1. Preparation of thiols.

Entry	Starting resin	Conditions	Product	Ref.
1		1. S_8, THF, 20 °C, 1 h 2. LiAlH$_4$, THF		[3,4]
2		1. AcSH, AIBN, PhMe, 70 °C, 2 d 2. HCl (37% in H$_2$O)/ dioxane 1:2, 70 °C, 2 d		[5]
3		1. AcSH, PPh$_3$, DIAD, THF, 0 °C 2. NaBH$_4$, THF, MeOH, 65 °C		[6]
4		1. (H$_2$N)$_2$CS (0.9 mol/L), DMA, 85 °C, 20 h 2. pyrrolidine/dioxane 1:4, 110 °C, 2 h		[7] see also [8,9]
5		1. AcSK, DMF 2. LiBH$_4$, Et$_2$O (saponification under basic conditions failed)		[10,11] see also [2,12,13]
6		1. AcSH (5 eq), BF$_3$OEt$_2$ (0.8 eq), DME, 23 °C, 16 h 2. LiOBn, THF, 23 °C, 2 h		[14]
7		NaOMe (0.2 mol/L), THF/MeOH 3:1, 1 h		[15]

Table 8.2. Deprotection of support-bound, protected thiols.

Entry	Starting resin	Conditions	Product	Ref.
1		TFA/(*i*Bu)₃SiH/DCM 5:5:90		[21] see also [22–24]
2		3–5% TFA in DCM, 45 min		[25,26]
3		PBu₃, DMF, *i*PrOH, H₂O		[27]
4		2-mercaptoethanol, DIPEA, DMF		[28]

8.2 Preparation of Thioethers

Thioethers are usually prepared on solid phase by S-alkylation of thiols. Thiols are powerful nucleophiles and, because of their high reactivity and ready availability, the most suitable starting materials for the solid-phase synthesis of thioethers. Numerous examples have been reported for both the reaction of support-bound electrophiles with thiols and the reaction of support-bound thiols with alkylating agents.

Table 8.3 lists illustrative examples of the preparation of thioethers from support-bound alkylating agents. Alkyl halides and trichloroacetimidates have been used as alkylating agents for aromatic or aliphatic thiols. Additional suitable support-bound electrophiles are 1-fluoro-2-nitroarenes, α,β-unsaturated carbonyl compounds, and epoxides. Nucleophilic substitutions with thiols do not require the use of strong bases and generally proceed smoothly and cleanly at room temperature to yield thioethers. Some S-alkylations or S-arylations can even be performed at room temperature in dilute acetic acid. Because thiols are sensitive to oxidation by air (disulfide formation), excess thiol should be used to afford complete conversion of the substrate.

As in other reactions of support-bound reagents with bifunctional compounds, the reaction of Merrifield resin with α,ω-dithiols can lead to significant cross-linking and thereby to a severe reduction of the swelling capacity of the support. By careful optimization of the reaction conditions, however, almost exclusive monoalkylation of α,ω-dithiols can be achieved (Entry **2**, Table 8.3).

Table 8.3. Preparation of thioethers from support-bound electrophiles.

Entry	Starting resin	Conditions	Product	Ref.
1		4-mercaptophenol (0.6 mol/L), KOH (0.4 mol/L), DMF, 90 °C, 16 h		[29] see also [30]
2		HS⁓⁓⁓SH (1 mol/L), DBU (0.3 mol/L), PhMe, 20 °C, 24 h		[25] see also [31,32]
3	(1.35 mmol/g)	cysteine (0.07 mol/L, 2 eq), DMF/H₂O/DBU 200:1:4, overnight	(0.38 mmol/g)	[33] see also [34]
4		3,4-dichlorothiophenol, DIPEA, DMSO		[35] see also [36-38]
5		BnSH (5 eq), BF₃OEt₂ (0.7 eq), DCM, 23 °C, 16 h		[14]
6		MeO₂C⁓⁓SH NMM, DMA, 20 °C, 16 h		[39] see also [40]
7		PhSH (1.25 mol/L), NaOMe, THF, 20 °C, 3 d		[41] see also [42,43]
8		PhSNa, DMF, 20 °C, 12 h		[44]

Support-bound thiols can be alkylated under mild reaction conditions with alkyl halides, α,β-unsaturated carbonyl compounds, and reactive aryl halides (Table 8.4). These reactions should be conducted under inert gas or in the presence of a reducing additive to prevent disulfide formation. PEG-bound thiophenol has been S-alkylated under conditions similar to those used for low-molecular-weight thiophenols (DMF, Cs₂CO₃, 20 °C, 15 h [45–47]).

Table 8.4. Preparation of thioethers from support-bound thiols.

Entry	Starting resin	Conditions	Product	Ref.
1		TMG, DMF, 23 °C		[6] see also [21,48]
2		(0.13 mol/L), 3% pyridine in DMF, 20 °C, 24 h		[9] see also [49]
3		AcOH, EtOH, 20 °C, 3 h		[11]
4		(0.13 mol/L), NEt₃ (0.13 mol/L), DMF, 1 h		[50]
5		(2 eq), 15-crown-5, THF, 20 °C, 16 h		[51]
6		(10 eq), KO*t*Bu (12 eq), THF/DMSO 2:1, 0 °C to 20 °C, 18 h		[52]
7		1. Hg(O₂CCF₃)₂, BnNH₂, THF, 20 °C 2. BnBr, THF, 20 °C		[53]

Because thioether formation is a rapid, irreversible reaction, this reaction has been widely used for the macrocyclization of peptides or peptoids on supports (Entries **1–4**, Table 8.5). For this purpose either S-protected cysteine or S-protected ω-aminomercaptans are linked to a support and then elongated by standard solid-phase peptide/peptoid chemistry. When a suitable length has been reached, the terminal amine is acylated with an acid bearing a reactive functionality. After deprotection of the thiol, cyclization is induced by treatment with a base. Although this strategy gives good results, even when

Table 8.5. Miscellaneous preparations of thioethers.

Entry	Starting resin	Conditions	Product	Ref.
1		TMG (2.5% in THF), 17 h		[27] see also [15,22] [56]
2		5% DBU in DMF, 20 °C, 24 h		[26] see also [57,58]
3		1. TFA/TES/DCM 1:2:97, 2 × 10 min, 1 × 20 min 2. DIPEA, NMP		[59]
4		TBAF (4 eq), DCM, 20 °C, 3 h		[60]
5		1. H-Cys-OEt•HCl, AIBN, EtOH, PhMe, PEG, 70 °C, 24 h		[5]
6		(10 eq), hv, DCM, 0 °C, 4 h		[61]

Table 8.5. continued.

Entry	Starting resin	Conditions	Product	Ref.
7		MeSSMe, THF, 20 °C, 15 min, then 65 °C, 0.5 h		[3] see also [62,63]
8		iPrMgBr (0.19 mol/L, 7.3 eq), THF, −35 °C, 15 min, then CuCN•2 LiCl (10 eq), 15 min, then PhSSPh		[64]
9		1. BuLi (0.14 mol/L, 3 eq), THF, −60 °C, 20 min 2. MeSSMe (10 eq), −60 °C to 20 °C, 2 h		[65]

strained rings are prepared, it has certain limitations. As in most other macrocyclizations, low yields are mainly because of competing intermolecular bond formation, and for macrocyclization to occur efficiently some preorganization of the precursor is usually required. A similar strategy for preparing macrocyclic peptides is based on the intramolecular reaction of N-terminal cysteine with an aldehyde to yield thiazolidines [54].

Thioethers have also been prepared on cross-linked polystyrene by radical addition of thiols to support-bound alkenes and by reaction of support-bound carbon radicals (generated by addition of carbon radicals to resin-bound acrylates) with esters of 1-hydroxy-1,2-dihydro-2-pyridinethione ('Barton esters'; Entry **6**, Table 8.5). Additional methods include the reaction of metallated supports with symmetric disulfides (Entries **7–9**, Table 8.5) and the alkylation of polystyrene-bound, α-lithiated thioanisol [55].

Thioglycosides have been prepared on solid phase by glycosylation of thiols with various types of glycosyl donor. Carbohydrate-derived thioethers have been used either to link carbohydrates to thiol-functionalized supports [8,25,66,67] or as glycosyl donors for the preparation of glycosides on solid phase (see Chapter 16.3.3).

8.3 Preparation of Sulfoxides and Sulfones

Sulfoxides and sulfones can be prepared on cross-linked polystyrene by oxidation of thioethers. The most common reagent for this purpose is MCPBA in DCM [7,11,50,68–70] or dioxane [71] (Table 8.6), but other oxidants such as H_2O_2 in acetic acid [30] or Oxone (Entry **7**, Table 8.6) have also been used. PEG-bound thioethers have been converted into sulfones by oxidation with MCPBA in DCM [45,47] or with OsO_4/NMO [72]. The oxidation of thioethers to sulfoxides requires careful control of the reaction conditions to prevent the formation of sulfones. Sulfones have, furthermore, been prepared by S-alkylation of polystyrene-bound sulfinates (Entry **8**, Table 8.6), and by α-alkylation of sulfones (BuLi, THF, alkyl halide [73]).

Table 8.6. Preparation of sulfoxides and sulfones.

Entry	Starting resin	Conditions	Product	Ref.
1		MCPBA (0.05 mol/L, 1.4 eq), DCM, 0 °C, 20 h (Wang resin)	(94:6)	[74]
2		MCPBA (0.12 mol/L, 1.2 eq), DCM, 20 °C, 18 h		[75]
3		MCPBA (0.2 mol/L, 5 eq), DCM, 20 °C, 96 h		[74]
4		MCPBA (1 mol/L, 3 eq), DCM, 20 °C, 15 h		[76]
5		MCPBA (0.15 mol/L, 3 eq), DCM, 25 °C, 1 h		[34]
6		MCPBA (0.6 mol/L, 10 eq), DCM, 20 °C, 16 h		[52]
7		Oxone, DMF, H_2O, 20 °C, 12 h		[32]
8		1. SO_2, THF 2. allyl bromide, THF, 67 °C, overnight		[73]

8.4 Preparation of Sulfonamides

Sulfonamides have usually been prepared on solid supports by sulfonylation of aliphatic or aromatic amines with sulfonyl chlorides or by N-alkylation of other sulfonamides (Figure 8.1).

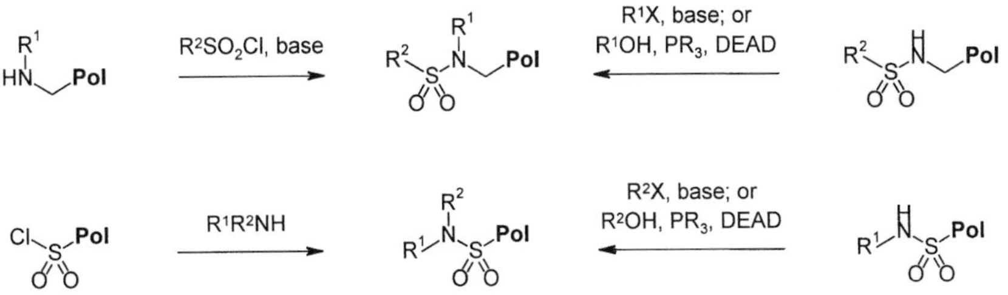

Fig. 8.1. Strategies for the preparation of sulfonamides on solid phase.

The clean conversion of support-bound, primary amines into sulfonamides is more difficult to perform than the acylation of amines with carboxylic acid derivatives, probably because primary amines can easily be sulfonylated twice. Weak bases (pyridine, NMM, collidine), short reaction times and only a slight excess of sulfonyl chloride should therefore be used to convert primary amines into sulfonamides (Table 8.7).

Sulfonyl chlorides with α-hydrogen are unstable under basic reaction conditions and can give variable results [77,78]. For base-labile sulfonyl chlorides the use of *O*-silyl keteneacetals as scavenger for HCl has been recommended [77]. Table 8.7 lists some illustrative procedures for the preparation of sulfonamides from primary amines on solid phase. Further examples have been reported [79–81].

Support-bound secondary amines are more readily sulfonylated than primary amines, as shown by the many examples reported in the literature (Table 8.8 [95–98]). Typically the sulfonylation of polystyrene-bound amines is conducted in DCM, but THF [99], DMF [78,100], pyridine [101], or mixtures thereof give sometimes better results [85]. NMM, DIPEA, and pyridine have mainly been used as bases.

Alternatively, sulfonamides can also be prepared (in low yield and purity) by oxidation of sulfinamides (Entry **3**, Table 8.8). Polystyrene-bound sulfonyl chlorides, which can be prepared from polystyrene-bound sulfonic acids (e.g. ion-exchange resins) by treatment with PCl_5, $SOCl_2$ [102–104], or SO_2Cl_2/PPh_3 [105], react smoothly with amines to yield the corresponding sulfonamides (Entry **4**, Table 8.8). Support-bound carbamates of primary aliphatic or aromatic amines can be N-sulfonylated in the presence of strong bases, and can therefore be used as backbone amide linkers for sulfonamides (Entries **5** and **6**, Table 8.8).

Sulfonamides of primary amines are readily deprotonated (pK_a 9–11) and N-alkylated. Because of their high nucleophilicity and low basicity, deprotonated sulfonamides also react smoothly with less reactive electrophiles, such as *n*-alkyl bromides [110] (Table 8.9). Sulfonamides can also be N-alkylated with aliphatic alcohols under Mitsunobu conditions. Suitable solvents for the N-alkylation of sulfonamides on polystyrene by Mitsunobu reaction are DCM, toluene, or THF. Because many more alcohols than alkyl halides are commercially availble, the Mitsunobu reaction enables the synthesis of larger and more diverse compound arrays than the alkylations with alkyl halides.

2-Nitrobenzene- or 2,4-dinitrobenzenesulfonamides of primary or secondary amines can be hydrolyzed under mildly basic conditions, and are increasingly being used for amine protection [100,111,112]. *N*-(2-Nitrobenzenesulfonyl)amino acids can

Table 8.7. Preparation of sulfonamides from support-bound primary amines and sulfonyl chlorides.

Entry	Starting resin	Conditions	Product	Ref.
1		TsCl (0.15 mol/L, 2 eq), NEt$_3$ (3 eq), DCM, 7 h		[82]
2		ArSO$_2$Cl (3.5 eq), NEt$_3$ (3.5 eq), DCM Ar: 4-PrC$_6$H$_4$		[83] see also [84]
3		2-(O$_2$N)C$_6$H$_4$SO$_2$Cl (4 eq), DIPEA (6 eq), THF/DCM 2:1, 20 °C, 3 h		[85] see also [86-88]
4		TsCl (5 eq), pyridine/DCM 1:1, 20 °C, 15 h		[89]
5		Ph BocHN (3 eq), NMM (6 eq), DCM, 2 × 2.5 h		[90] see also [77,91]
6		MsCl, pyridine, DCM, 28 h		[92]
7		PhSO$_2$Cl (3 eq), pyridine (1 eq), DMAP, DCM, 25 °C, 18 h		[93] see also [94]

Table 8.8. Miscellaneous preparations of sulfonamides from sulfonyl chlorides and from sulfinamides.

Entry	Starting resin	Conditions	Product	Ref.
1		(0.5 mol/L), DMF, pyridine, 2–18 h		[78]
2		TsCl (0.4 mol/L, 5 eq), DMAP, pyridine, 60 °C, 14 h		[106]
3		NMM, DCM, overnight; then NaIO$_4$, RuCl$_3$, DCM, MeCN, H$_2$O, 1 h		[107]
4		1. PCl$_5$, DMF, 23 °C, 4 h 2. 1,3-diamino-propane, DMF, pyridine		[108]
5		LiN(SiMe$_3$)$_2$, THF, –78 °C, 45 min, then TsCl		[109]
6		NaH, DMA, 20 °C, 8 h, then add TsCl		[109]

be used as alternative to *N*-Fmoc amino acids for the solid-phase synthesis of peptides [113]. Deprotection is achieved by treatment of the polystyrene-bound sulfonamide with a solution of PhSH (0.5 mol/L) and K$_2$CO$_3$ (2 mol/L) in DMF for 10 min at room temperature [113], whereby neither cleavage of esters (e.g. of the Wang linker) nor racemization occurs.

Sulfonamides can also be alkylated by support-bound electrophiles (Table 8.10). Poly-styrene-bound allylic alcohols have been used to N-alkylate sulfonamides under Mitsu-nobu reaction conditions. Oxidative iodo-sulfonylamidation of support-bound enol ethers (e.g. glycals; Entry **3**, Table 8.10) has been used to prepare *N*-sulfonyl aminals. Jung and coworkers have reported an interesting variant of the Baylis–Hillman reaction in which tosylamide and an aromatic aldehyde were condensed with polystyrene-bound acrylic acid to yield 2-(sulfonamidomethyl)acrylates (Entry **4**, Table 8.10).

Table 8.9. N-Alkylation of support-bound sulfonamides.

Entry	Starting resin	Conditions	Product	Ref.
1		KOtBu (5 eq), BnBr (10 eq), THF, 50 °C, 2 h		[89]
2		phenacyl bromide (6 eq), DBU (6 eq), DMSO/ NMP 1:1, overnight Ar: 4-(MeO)C₆H₄		[110]
3		Br—⟨cyclohexyl⟩ (6 eq), DBU (6 eq), DMSO/NMP 1:1, overnight		[110]
4		propargyl bromide (0.2 mol/L, 4 eq), Cs₂CO₃ (2 eq), DMF, 20 h		[82]
5		iodoacetonitrile (1.2 mol/L), DIPEA, NMP, 24 h		[114,115]
6		CH₂N₂, Me₂CO, Et₂O		[116]
7		⟨allyl⟩–O–CO₂Me (15 eq), Pd₂(dba)₃, PPh₃, THF, 2 h		[113]
8		EtOH (5 eq), PPh₃ (5 eq), DEAD (5 eq), DCM, 0.5 h		[100] see also [85,87] [110,117] [118]
9		Ph⟨⟩OH PPh₃, DEAD, THF, 20 h		[119]

Table 8.10. N-Alkylation of sulfonamides on insoluble supports.

Entry	Starting resin	Conditions	Product	Ref.
1	HO~~/~~PS	MeO$_2$C ~ NHSO$_2$Ar; PPh$_3$, DEAD, THF, 20 °C, 16 h; Ar: 2,4-(O$_2$N)$_2$C$_6$H$_3$	MeO$_2$C–N(SO$_2$Ar)~~/~~PS	[120] see also [111,121]
2	HO~~/~~PS	O$_2$N / NO$_2$; BocHN–S(O$_2$); PPh$_3$, DIAD (3 eq of each), THF, 12 h	tBuO$_2$C–N(SO$_2$Ar)~~/~~PS	[122]
3	OBn, OBn (glycal), O–Si(iPr)$_2$–PS	PhSO$_2$NH$_2$ (8 eq), I(coll)$_2$ClO$_4$ (5.5 eq), DCM, 0 °C, 6 h	I, OBn, OBn; PhO$_2$S–NH; O–Si–PS	[123]
4	acryloyl–O–CH(Ph)–PS, Cl–C$_6$H$_4$	4-(F$_3$C)C$_6$H$_4$CHO (16 eq), TsNH$_2$ (20 eq), DABCO (1.6 eq), dioxane, 70 °C, 20 h	Tol–SO$_2$–NH–C(Ar)... O–CH(Ph)–PS, Cl–C$_6$H$_4$	[124]

8.5 Preparation of Sulfonic Esters

Sulfonates of primary and secondary alcohols are strong alkylating agents, and are usually prepared on solid phase as synthetic intermediates only. Sulfonic esters of phenols are, however, sufficiently stable to serve as potential lead structures for various types of drug [125].

The most common synthesis of sulfonic esters, which can also be conducted on insoluble supports, is the sulfonylation of alcohols with sulfonyl chlorides under basic reaction conditions. Several examples of the sulfonylation of support-bound alcohols and of the reaction of support-bound sulfonyl chlorides with alcohols have been reported (Table 8.11). For the preparation of highly reactive sulfonates, bases of low nucleophilicity, such as DIPEA or 2,6-lutidine, should be used to prevent alkylation of the base by the newly formed sulfonate.

Treatment of Wang resin with methanesulfonyl chloride in the presence of a weak base only yields the mesylate if the reaction is conducted at low temperature and for a short time (Entry 1, Table 8.11). Longer reaction times lead, in both DCM or DMF as

Table 8.11. Preparation of sulfonic esters.

Entry	Starting resin	Conditions	Product	Ref.
1		MsCl, DIPEA, DCM, 0–25 °C, 75 min		[127]
2		MsCl (5 eq), pyridine, DCM, 0 °C, 24 h		[128] see also [129,130]
3		TsCl (20 eq), NEt$_3$ (20 eq), DCM, 20 °C, overnight		[2]
4		O$_2$N—⟨ ⟩—SO$_2$Cl pyridine, DCM, 20 °C, 9 h		[131]
5		Tf$_2$O (5 eq), 2,6-lutidine (7 eq), DCM, 0 °C, 6 h		[132]
6		MsCl, pyridine, DCM, 20 °C, 7 h		[133]
7		Tf$_2$O (8 eq), pyridine, DCM, 20 °C, 5–10 h		[134]
8		PhSO$_2$Cl (0.09 mol/L), DIPEA (0.09 mol/L), DCM, 16 h		[125]
9		(1.1 eq), NEt$_3$ (1.0 eq), DCM, 20 °C, overnight		[102] see also [103,135]
10		tBuO$_2$C—⟨ ⟩—OH NEt$_3$, DCM		[104]
11		(20 eq), THF, −78 °C to −25 °C, 20 h		[105]

solvents, to the formation of the corresponding benzyl chloride [106,126]. Attempts to convert Wang resin or Sasrin into the even more reactive tosylates or nosylates have been unsuccessful [95].

The preparation of non-benzylic sulfonates of primary or secondary alcohols does not usually cause problems (Table 8.11). Secondary alcohols do not react smoothly with aromatic sulfonyl chlorides but only with alkylsulfonyl chlorides. Tertiary alcohols are unreactive towards sulfonylation (Entry **6**, Table 8.11).

Support-bound sulfonyl chlorides, which can be prepared from sulfonic acids by treatment with PCl$_5$ [108], SOCl$_2$ [102–104], or SO$_2$Cl$_2$/PPh$_3$ [105], react with primary aliphatic alcohols or phenols to yield the expected sulfonates. Sulfonates have, moreover, been prepared by treating Merrifield resin with lithiated isopropyl methanesulfonate (Entry **11**, Table 8.11).

References for Chapter 8

[1] Camarero, J. A.; Cotton, G. J.; Adeva, A.; Muir, T. W. *J. Pept. Res.* **1998**, *51*, 303–316.
[2] Bettinger, T.; Remy, J. S.; Erbacher, P.; Behr, J. P. *Bioconjugate Chem.* **1998**, *9*, 842–846.
[3] Farrall, M. J.; Fréchet, J. M. J. *J. Org. Chem.* **1976**, *41*, 3877–3882.
[4] Fréchet, J. M. J.; de Smet, M. D.; Farrall, M. J. *Polymer* **1979**, *20*, 675–680.
[5] Stranix, B. R.; Gao, J. P.; Barghi, R.; Salha, J.; Darling, G. D. *J. Org. Chem.* **1997**, *62*, 8987–8993.
[6] Moran, E. J.; Wilson, T. E.; Cho, C. Y.; Cherry, S. R.; Schultz, P. G. *Biopolymers* **1995**, *37*, 213–219.
[7] Masquelin, T.; Meunier, N.; Gerber, F.; Rossé, G. *Heterocycles* **1998**, *48*, 2489–2505.
[8] Chiu, S. H. L.; Anderson, L. *Carbohyd. Res.* **1976**, *50*, 227–238.
[9] Adamczyk, M.; Fishpaugh, J. R.; Mattingly, P. G. *Tetrahedron Lett.* **1999**, *40*, 463–466.
[10] Kobayashi, S.; Hachiya, I.; Suzuki, S.; Moriwaki, M. *Tetrahedron Lett.* **1996**, *37*, 2809–2812.
[11] Barco, A.; Benetti, S.; De Risi, C.; Marchetti, P.; Pollini, G. P.; Zanirato, V. *Tetrahedron Lett.* **1998**, *39*, 7591–7594.
[12] Bertini, V.; Lucchesini, F.; Pocci, M.; De Munno, A. *Tetrahedron Lett.* **1998**, *39*, 9263–9266.
[13] Kobayashi, S.; Wakabayashi, T.; Yasuda, M. *J. Org. Chem.* **1998**, *63*, 4868–4869.
[14] Phoon, C. W.; Oliver, S. F.; Abell, C. *Tetrahedron Lett.* **1998**, *39*, 7959–7962.
[15] Souers, A. J.; Virgilio, A. A.; Rosenquist, A.; Fenuik, W.; Ellman, J. A. *J. Am. Chem. Soc.* **1999**, *121*, 1817–1825.
[16] Yoshida, M.; Tatsumi, T.; Fujiwara, Y.; Iinuma, S.; Kimura, T.; Akaji, K.; Kiso, Y. *Chem. Pharm. Bull.* **1990**, *38*, 1551–1557.
[17] Houseman, B. T.; Mrksich, M. *J. Org. Chem.* **1998**, *63*, 7552–7555.
[18] Pearson, D. A.; Blanchette, M.; Baker, M. L.; Guindon, C. A. *Tetrahedron Lett.* **1989**, *30*, 2739–2742.
[19] Novabiochem Catalog and Peptide Synthesis Handbook, Läufelfingen, **1999**.
[20] Loffet, A.; Zhang, H. X. *Int. J. Pept. Prot. Res.* **1993**, *42*, 346–351.
[21] Nefzi, A.; Giulianotti, M.; Houghten, R. A. *Tetrahedron Lett.* **1998**, *39*, 3671–3674.
[22] Cho, C. Y.; Youngquist, R. S.; Paikoff, S. J.; Beresini, M. H.; Hébert, A. R.; Berleau, L. T.; Liu, C. W.; Wemmer, D. E.; Keough, T.; Schultz, P. G. *J. Am. Chem. Soc.* **1998**, *120*, 7706–7718.
[23] Marsh, I. R.; Bradley, M. *Tetrahedron* **1997**, *53*, 17317–17334.
[24] Pátek, M.; Drake, B.; Lebl, M. *Tetrahedron Lett.* **1995**, *36*, 2227–2230.
[25] Rademann, J.; Schmidt, R. R. *J. Org. Chem.* **1997**, *62*, 3650–3653.
[26] Kiselyov, A. S.; Eisenberg, S.; Luo, Y. *Tetrahedron* **1998**, *54*, 10635–10640.
[27] Virgilio, A. A.; Bray, A. A.; Zhang, W.; Trinh, L.; Snyder, M.; Morrissey, M. M.; Ellman, J. A. *Tetrahedron* **1997**, *53*, 6635–6644.
[28] Sucholeiki, I. *Tetrahedron Lett.* **1994**, *35*, 7307–7310.
[29] Dressman, B. A.; Singh, U.; Kaldor, S. W. *Tetrahedron Lett.* **1998**, *39*, 3631–3634.
[30] Marshall, D. L.; Liener, I. E. *J. Org. Chem.* **1970**, *35*, 867–868.
[31] Farrall, M. J.; Fréchet, J. M. J. *J. Am. Chem. Soc.* **1978**, *100*, 7998–7999.
[32] Kroll, F. E. K.; Morphy, R.; Rees, D.; Gani, D. *Tetrahedron Lett.* **1997**, *38*, 8573–8576.
[33] Delaet, N. G. J.; Tsuchida, T. *Lett. Pept. Sci.* **1995**, *2*, 325–331.

[34] Yamada, M.; Miyajima, T.; Horikawa, H. *Tetrahedron Lett.* **1998**, *39*, 289–292.
[35] Bhandari, A.; Jones, D. G.; Schullek, J. R.; Vo, K.; Schunk, C. A.; Tamanaha, L. L.; Chen, D.; Yuan, Z. Y.; Needels, M. C.; Gallop, M. A. *Bioorg. Med. Chem. Lett.* **1998**, *8*, 2303–2308.
[36] Tumelty, D.; Schwarz, M. K.; Needels, M. C. *Tetrahedron Lett.* **1998**, *39*, 7467–7470.
[37] Karoyan, P.; Triolo, A.; Nannicini, R.; Giannotti, D.; Altamura, M.; Chassaing, G.; Perrotta, E. *Tetrahedron Lett.* **1999**, *40*, 71–74.
[38] Barn, D. R.; Morphy, J. R.; Rees, D. C. *Tetrahedron Lett.* **1996**, *37*, 3213–3216.
[39] Yan, B.; Kumaravel, G. *Tetrahedron* **1996**, *52*, 843–848.
[40] Schwarz, M. K.; Tumelty, D.; Gallop, M. A. *J. Org. Chem.* **1999**, *64*, 2219–2231.
[41] Chen, C.; Randall, L. A. A.; Miller, R. B.; Jones, A. D.; Kurth, M. J. *Tetrahedron* **1997**, *53*, 6595–6609.
[42] Burns, C. J.; Groneberg, R. D.; Salvino, J. M.; McGeehan, G.; Condon, S. M.; Morris, R.; Morrissette, M.; Mathew, R.; Darnbrough, S.; Neuenschwander, K.; Scotese, A.; Djuric, S. W.; Ullrich, J.; Labaudiniere, R. *Angew. Chem. Int. Ed. Engl.* **1998**, *37*, 2848–2850.
[43] Chen, C.; Randall, L. A. A.; Miller, R. B.; Jones, A. D.; Kurth, M. J. *J. Am. Chem. Soc.* **1994**, *116*, 2661–2662.
[44] Le Hetet, C.; David, M.; Carreaux, F.; Carboni, B.; Sauleau, A. *Tetrahedron Lett.* **1997**, *38*, 5153–5156.
[45] Zhao, X. Y.; Jung, K. W.; Janda, K. D. *Tetrahedron Lett.* **1997**, *38*, 977–980.
[46] Jung, K. W.; Zhao, X. Y.; Janda, K. D. *Tetrahedron* **1997**, *53*, 6645–6652.
[47] Zhao, X. Y.; Janda, K. D. *Tetrahedron Lett.* **1997**, *38*, 5437–5440.
[48] Forman, F. W.; Sucholeiki, I. *J. Org. Chem.* **1995**, *60*, 523–528.
[49] Katoh, M.; Sodeoka, M. *Bioorg. Med. Chem. Lett.* **1999**, *9*, 881–884.
[50] Gayo, L. M.; Suto, M. J. *Tetrahedron Lett.* **1997**, *38*, 211–214.
[51] Hummel, G.; Hindsgaul, O. *Angew. Chem. Int. Ed. Engl.* **1999**, *38*, 1782–1784.
[52] Kulkarni, B. A.; Ganesan, A. *Tetrahedron Lett.* **1999**, *40*, 5633–5636.
[53] Girdwood, J. A.; Shute, R. E. *Chem. Commun.* **1997**, 2307–2308.
[54] Botti, P.; Pallin, T. D.; Tam, J. P. *J. Am. Chem. Soc.* **1996**, *118*, 10018–10024.
[55] Crosby, G. A.; Kato, M. *J. Am. Chem. Soc.* **1977**, *99*, 278–280.
[56] Roberts, K. D.; Lambert, J. N.; Ede, N. J.; Bray, A. M. *Tetrahedron Lett.* **1998**, *39*, 8357–8360.
[57] Feng, Y.; Wang, Z.; Jin, S.; Burgess, K. *J. Am. Chem. Soc.* **1998**, *120*, 10768–10769.
[58] Fotsch, C.; Kumaravel, G.; Sharma, S. K.; Wu, A. D.; Gounarides, J. S.; Nirmala, N. R.; Petter, R. C. *Bioorg. Med. Chem. Lett.* **1999**, *9*, 2125–2130.
[59] Sharma, S. K.; Wu, A. D.; Chandramouli, N. *Tetrahedron Lett.* **1996**, *37*, 5665–5668.
[60] Ueki, M.; Ikeo, T.; Iwadate, M.; Asakura, T.; Williamson, M. P.; Slaninová, J. *Bioorg. Med. Chem. Lett.* **1999**, *9*, 1767–1772.
[61] Zhu, X.; Ganesan, A. *J. Comb. Chem.* **1999**, *1*, 157–162.
[62] Farrall, M. J.; Durst, T.; Fréchet, J. M. *Tetrahedron Lett.* **1979**, 203–206.
[63] Crosby, G. A.; Weinshenker, N. M.; Uh, H. S. *J. Am. Chem. Soc.* **1975**, *97*, 2232–2235.
[64] Boymond, L.; Rottländer, M.; Cahiez, G.; Knochel, P. *Angew. Chem. Int. Ed. Engl.* **1998**, *37*, 1701–1703.
[65] Havez, S.; Begtrup, M.; Vedsø, P.; Andersen, K.; Ruhland, T. *J. Org. Chem.* **1998**, *63*, 7418–7420.
[66] Heckel, A.; Mross, E.; Jung, K. H.; Rademann, J.; Schmidt, R. R. *Synlett* **1998**, 171–173.
[67] Rademann, J.; Schmidt, R. R. *Tetrahedron Lett.* **1996**, *37*, 3989–3990.
[68] Panek, J. S.; Zhu, B. *Tetrahedron Lett.* **1996**, *37*, 8151–8154.
[69] Gosselin, F.; Di Renzo, M.; Ellis, T. H.; Lubell, W. D. *J. Org. Chem.* **1996**, *61*, 7980–7981.
[70] García Echeverría, C. *Tetrahedron Lett.* **1997**, *38*, 8933–8934.
[71] Flanigan, E.; Marshall, G. R. *Tetrahedron Lett.* **1970**, 2403–2406.
[72] Han, H. S.; Janda, K. D. *Tetrahedron Lett.* **1997**, *38*, 1527–1530.
[73] Halm, C.; Evarts, J.; Kurth, M. J. *Tetrahedron Lett.* **1997**, *38*, 7709–7712.
[74] Mata, E. G. *Tetrahedron Lett.* **1997**, *38*, 6335–6338.
[75] Masquelin, T.; Sprenger, D.; Baer, R.; Gerber, F.; Mercadal, Y. *Helv. Chim. Acta* **1998**, *81*, 646–660.
[76] Obrecht, D.; Abrecht, C.; Grieder, A.; Villalgordo, J. M. *Helv. Chim. Acta* **1997**, *80*, 65–72.
[77] Gude, M.; Piarulli, U.; Potenza, D.; Salom, B.; Gennari, C. *Tetrahedron Lett.* **1996**, *37*, 8589–8592.
[78] Floyd, C. D.; Lewis, C. N.; Patel, S. R.; Whittaker, M. *Tetrahedron Lett.* **1996**, *37*, 8045–8048.
[79] Tomasi, S.; Le Roch, M.; Renault, J.; Corbel, J. C.; Uriac, P.; Carboni, B.; Moncoq, D.; Martin, B.; Delcros, J. G. *Bioorg. Med. Chem. Lett.* **1998**, *8*, 635–640.
[80] Pop, I. E.; Déprez, B. P.; Tartar, A. L. *J. Org. Chem.* **1997**, *62*, 2594–2603.
[81] Gordeev, M. F.; Luehr, G. W.; Hui, H. C.; Gordon, E. M.; Patel, D. V. *Tetrahedron* **1998**, *54*, 15879–15890.

[82] Bolton, G. L.; Hodges, J. C.; Rubin, J. R. *Tetrahedron* **1997**, *53*, 6611–6634.
[83] Kim, S. W.; Hong, C. Y.; Koh, J. S.; Lee, E. J.; Lee, K. *Mol. Diversity* **1998**, *3*, 133–136.
[84] Kim, S. W.; Hong, C. Y.; Lee, K.; Lee, E. J.; Koh, J. S. *Bioorg. Med. Chem. Lett.* **1998**, *8*, 735–738.
[85] Yang, L.; Chiu, K. *Tetrahedron Lett.* **1997**, *38*, 7307–7310.
[86] Raman, P.; Stokes, S. S.; Angell, Y. M.; Flentke, G. R.; Rich, D. H. *J. Org. Chem.* **1998**, *63*, 5734–5735.
[87] Scicinski, J. J.; Barker, R. D.; Murray, P. J.; Jarvie, E. M. *Bioorg. Med. Chem. Lett.* **1998**, *8*, 3609–3614.
[88] Mohamed, N.; Bhatt, U.; Just, G. *Tetrahedron Lett.* **1998**, *39*, 8213–8216.
[89] Beaver, K. A.; Siegmund, A. C.; Spear, K. L. *Tetrahedron Lett.* **1996**, *37*, 1145–1148.
[90] de Bont, D. B. A.; Dijkstra, G. D. H.; den Hartog, J. A. J.; Liskamp, R. M. J. *Bioorg. Med. Chem. Lett.* **1996**, *6*, 3035–3040.
[91] Gennari, C.; Longari, C.; Ressel, S.; Salom, B.; Piarulli, U.; Ceccarelli, S.; Mielgo, A. *Eur. J. Org. Chem.* **1998**, 2437–2449.
[92] Zhang, H. C.; Brumfield, K. K.; Jaroskova, L.; Maryanoff, B. E. *Tetrahedron Lett.* **1998**, *39*, 4449–4452.
[93] Meyers, H. V.; Dilley, G. J.; Durgin, T. L.; Powers, T. S.; Winssinger, N. A.; Zhu, H.; Pavia, M. R. *Mol. Diversity* **1995**, *1*, 13–20.
[94] Wang, Y.; Huang, T. N. *Tetrahedron Lett.* **1998**, *39*, 9605–9608.
[95] Ngu, K.; Patel, D. V. *Tetrahedron Lett.* **1997**, *38*, 973–976.
[96] Fivush, A. M.; Willson, T. M. *Tetrahedron Lett.* **1997**, *38*, 7151–7154.
[97] Swayze, E. E. *Tetrahedron Lett.* **1997**, *38*, 8643–8646.
[98] Swayze, E. E. *Tetrahedron Lett.* **1997**, *38*, 8465–8468.
[99] Dixit, D. M.; Leznoff, C. C. *Israel J. Chem.* **1978**, *17*, 248–252.
[100] Reichwein, J. F.; Liskamp, R. M. J. *Tetrahedron Lett.* **1998**, *39*, 1243–1246.
[101] Chenera, B.; Finkelstein, J. A.; Veber, D. F. *J. Am. Chem. Soc.* **1995**, *117*, 11999–12000.
[102] ten Holte, P.; Thijs, L.; Zwanenburg, B. *Tetrahedron Lett.* **1998**, *39*, 7407–7410.
[103] Bicak, N.; Senkal, B. F. *Reactive and Functional Polymers* **1996**, *29*, 123–128.
[104] Jin, S.; Holub, D. P.; Wustrow, D. J. *Tetrahedron Lett.* **1998**, *39*, 3651–3654.
[105] Hunt, J. A.; Roush, W. R. *J. Am. Chem. Soc.* **1996**, *118*, 9998–9999.
[106] Raju, B.; Kogan, T. P. *Tetrahedron Lett.* **1997**, *38*, 4965–4968.
[107] de Bont, D. B. A.; Moree, W. J.; Liskamp, R. M. J. *Bioorg. Med. Chem.* **1996**, *4*, 667–672.
[108] Zhong, H. M.; Greco, M. N.; Maryanoff, B. E. *J. Org. Chem.* **1997**, *62*, 9326–9330.
[109] Raju, B.; Kogan, T. P. *Tetrahedron Lett.* **1997**, *38*, 3373–3376.
[110] Dankwardt, S. M.; Smith, D. B.; Porco, J. A.; Nguyen, C. H. *Synlett* **1997**, 854–856.
[111] Piscopio, A. D.; Miller, J. F.; Koch, K. *Tetrahedron Lett.* **1998**, *39*, 2667–2670.
[112] Wipf, P.; Henninger, T. C. *J. Org. Chem.* **1997**, *62*, 1586–1587.
[113] Miller, S. C.; Scanlan, T. S. *J. Am. Chem. Soc.* **1998**, *120*, 2690–2691.
[114] Backes, B. J.; Virgilio, A. A.; Ellman, J. A. *J. Am. Chem. Soc.* **1996**, *118*, 3055–3056.
[115] Backes, B. J.; Ellman, J. A. *J. Org. Chem.* **1999**, *64*, 2322–2330.
[116] Kenner, G. W.; McDermott, J. R.; Sheppard, R. C. *J. Chem. Soc. Chem. Commun.* **1971**, 636–637.
[117] Ngu, K.; Patel, D. V. *J. Org. Chem.* **1997**, *62*, 7088–7089.
[118] Boeijen, A.; Kruijtzer, J. A. W.; Liskamp, R. M. J. *Bioorg. Med. Chem. Lett.* **1998**, *8*, 2375–2380.
[119] Kay, C.; Murray, P. J.; Sandow, L.; Holmes, A. B. *Tetrahedron Lett.* **1997**, *38*, 6941–6944.
[120] Piscopio, A. D.; Miller, J. F.; Koch, K. *Tetrahedron Lett.* **1997**, *38*, 7143–7146.
[121] Veerman, J. J. N.; van Maarseveen, J. H.; Visser, G. M.; Kruse, C. G.; Schoemaker, H. E.; Hiemstra, H.; Rutjes, F. P. J. T. *Eur. J. Org. Chem.* **1998**, 2583–2589.
[122] Piscopio, A. D.; Miller, J. F.; Koch, K. *Tetrahedron* **1999**, *55*, 8189–8198.
[123] Zheng, C. S.; Seeberger, P. H.; Danishefsky, S. J. *Angew. Chem. Int. Ed. Engl.* **1998**, *37*, 786–789.
[124] Richter, H.; Jung, G. *Tetrahedron Lett.* **1998**, *39*, 2729–2730.
[125] Dankwardt, S. M.; Phan, T. M.; Krstenansky, J. L. *Mol. Diversity* **1995**, *1*, 113–120.
[126] Nugiel, D. A.; Wacker, D. A.; Nemeth, G. A. *Tetrahedron Lett.* **1997**, *38*, 5789–5790.
[127] Richter, L. S.; Desai, M. C. *Tetrahedron Lett.* **1997**, *38*, 321–322.
[128] Rölfing, K.; Thiel, M.; Künzer, H. *Synlett* **1996**, 1036–1038.
[129] Fyles, T. M.; Leznoff, C. C.; Weatherston, J. *Can. J. Chem.* **1977**, *55*, 4135–4143.
[130] Leznoff, C. C.; Fyles, T. M.; Weatherston, J. *Can. J. Chem.* **1977**, *55*, 1143–1153.
[131] Lee, C. E.; Kick, E. K.; Ellman, J. A. *J. Am. Chem. Soc.* **1998**, *120*, 9735–9747.
[132] Uozumi, Y.; Danjo, H.; Hayashi, T. *J. Org. Chem.* **1999**, *64*, 3384–3388.
[133] Hanessian, S.; Huynh, H. K. *Synlett* **1999**, 102–104.
[134] Rottländer, M.; Knochel, P. *Synlett* **1997**, 1084–1086.
[135] Rueter, J. K.; Nortey, S. O.; Baxter, E. W.; Leo, G. C.; Reitz, A. B. *Tetrahedron Lett.* **1998**, *39*, 975–978.

9 Preparation of Organoselenium Compounds

Support-bound organoselenium compounds have mainly been used as synthetic intermediates and linkers. The use of organoselenium compounds as linkers for solid-phase synthesis has been investigated by several groups [1–3]. The selenium–carbon bond is stable under a broad variety of reaction conditions, but can be selectively cleaved by tin radicals or by oxidants (see Chapters 3.15.1 and 3.16.4).

Polystyrene-bound arylselenium derivatives have been prepared from lithiated polystyrene and metallic selenium [2] or dimethyldiselenide [1] (Figure 9.1). These derivatives are suitable starting materials for the preparation of other selenium reagents, which enable the facile attachment of organic substrates to insoluble supports. Alkali metal selenides or selenoborate complexes are strong nucleophiles and react swiftly with organic halides (including DCM!) to yield selenides. Electrophilic selenium derivatives, such as the bromide or phthalimide, undergo addition to alkenes under mild reaction conditions. Some representative synthetic transformations of organoselenium compounds on solid phase are outlined in Figure 9.1.

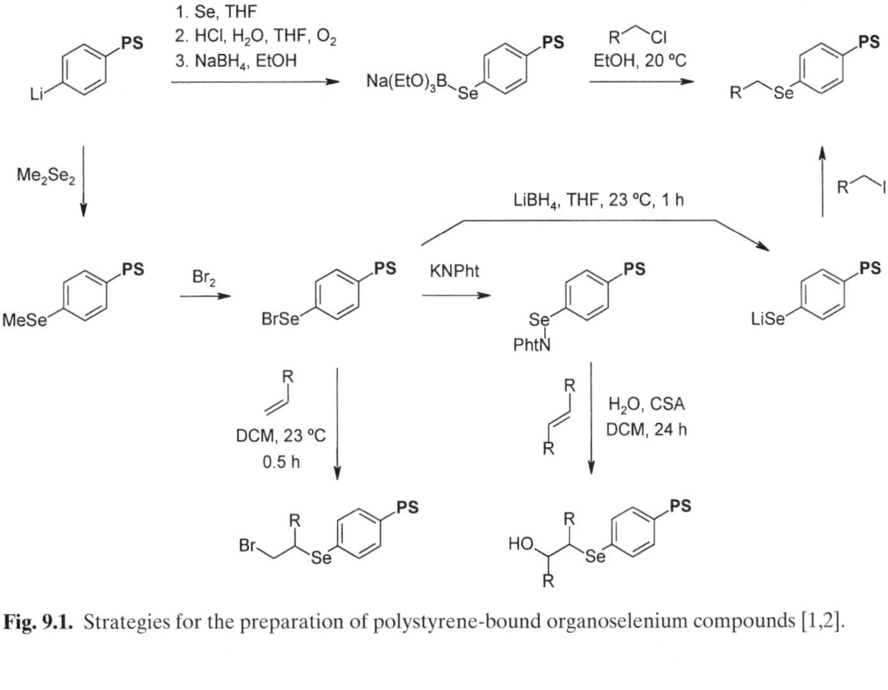

Fig. 9.1. Strategies for the preparation of polystyrene-bound organoselenium compounds [1,2].

References for Chapter 9

[1] Nicolaou, K. C.; Pastor, J.; Barluenga, S.; Winssinger, N. *Chem. Commun.* **1998**, 1947–1948.
[2] Ruhland, T.; Andersen, K.; Pedersen, H. *J. Org. Chem.* **1998**, *63*, 9204–9211.
[3] Michels, R.; Kato, M.; Heitz, W. *Makromol. Chem.* **1976**, *177*, 2311–2320.

10 Preparation of Nitrogen Compounds

10.1 Preparation of Amines

Amines are both valuable synthetic intermediates and interesting target molecules. A number of solid-phase syntheses have been developed, most of which are also suitable for the automated, parallel synthesis of this class of compound.

The most common preparations of amines on insoluble supports include nucleophilic aliphatic and aromatic substitutions, Michael-type additions, and the reduction of imines, amides, nitro groups, and azides (Figure 10.1). Further methods include the addition of carbon nucleophiles to imines (e.g. the Mannich reaction) and oxidative degradation of carboxylic acids or amides. Linkers for primary, secondary, and tertiary amines are discussed in Sections 3.6, 3.7, and 3.8.

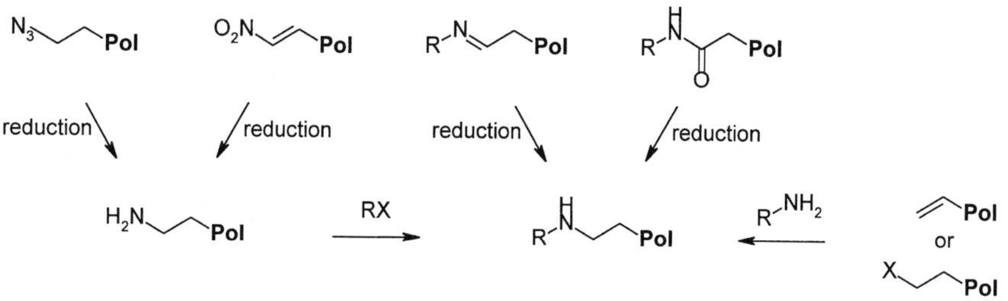

N₃–Pol   O₂N–Pol   R–N=–Pol   R–N(H)–C(O)–Pol

reduction   reduction   reduction   reduction

H₂N–Pol   —RX→   R–N(H)–Pol   ←   R–NH₂   CH₂=–Pol or X–Pol

Fig. 10.1. Strategies for the synthesis of amines on insoluble supports. X: leaving group; R: alkyl, aryl.

10.1.1 Preparation of Amines by Aliphatic Nucleophilic Substitution

10.1.1.1 With Support-Bound Alkylating Agents

Support-bound alkyl halides, sulfonates, or epoxides can be used to alkylate amines. Alkylating agents of low reactivity, such as non-benzylic alkyl halides or sulfonates, generally require high amine concentrations, dipolar, aprotic solvents (e.g. DMSO), and high reaction temperatures to undergo substitution (Table 10.1). Forcing conditions can, however, also promote several other processes, such as elimination [1], cleavage of the linker [2], oxidation, or substitutions with competing nucleophiles (e.g. the solvent or water; see also Entry 6, Table 10.1). This is probably why few examples for N-alkylations with non-activated alkylating agents have been reported. The reactivity of the amine is also decisive for the success of N-alkylations, and best results are usually obtained with primary or secondary alkylamines. Particularly poor nucleophiles are amino acid esters and aromatic or heteroaromatic amines. Table 10.1 lists illustrative examples of the alkylation of amines with support-bound alkylating agents of low reactivity.

Table 10.1. Preparation of amines from support-bound weak alkylating agents.

Entry	Starting resin	Conditions	Product	Ref.
1	NosO...O(PS), N$_3$, Ph	Ar$\sim$NH$_2$ (1 mol/L), NMP, 80 °C, 36 h, Ar: 2,4-Cl$_2$C$_6$H$_3$	Ar...N(H)...O(PS), N$_3$, Ph	[3] see also [4,5]
2	Br$\sim\sim$O(PS)	piperidine (0.7 mol/L), DMF, 18 h	piperidine-N$\sim\sim$O(PS)	[6]
3	MsO$\sim$S-S$\sim$N(H) PS	isobutylamine (1 mol/L), NMP, 50 °C, 16 h	$\sim$N(H)$\sim$S-S$\sim$N(H) PS	[7]
4	MsO$\sim$...O(PS)	benzylamine, MeCN, 81 °C, 20 h	Ph$\sim$N(H)$\sim$...O(PS)	[8] see also [1]
5	Br$\sim$OEt$\sim$O$\sim$TG	butylamine (2 mol/L), DMSO, 60 °C, 15 h	$\sim$N(H)$\sim$OEt$\sim$O$\sim$TG	[9]
6	Br$\sim\sim$C(O)N(H)(PS)	ArNH$_2$, DMSO, Ar: halophenyl	Ar-HN$\sim\sim$C(O)NH(PS) + pyrrolidinone-N(PS)	[10]
7	Ph...(epoxide)$\sim$O$\sim$PS	piperidine, LiClO$_4$ (6 mol/L each), MeCN, 55 °C, 48 h	Ph...OH$\sim$O$\sim$PS, N-piperidine	[11] see also [12]
8	Cl...(bicyclic dioxolane)(PS)	BnNH$_2$ (8 eq), dioxane, 80 °C, 68 h	Cl...(bicyclic dioxolane)(PS), Ph$\sim$N(H), OH	[13] see also [14-16]

More reactive electrophiles, such as benzyl and allyl halides, and α- or β-halocarbonyl compounds react smoothly with amines, often even at room temperature. Support-bound chloro- and bromoacetamides, for instance, react cleanly with a broad range of aliphatic and aromatic amines to yield glycine derivatives (Entries **1–4**, Table 10.2 [17–27]). This reaction is usually conducted in DMSO at room temperature (2–12 h), but for sensitive

Table 10.2. Alkylation of amines with support-bound strong alkylating agents.

Entry	Starting resin	Conditions	Product	Ref.
1	Br–CH₂C(O)NH–(PS)	Ph₂CH–CH₂–NH₂ (2.5 mol/L), DMSO, 20 °C, 2 h	Ph₂CH–CH₂–NH–CH₂C(O)NH–(PS)	[35]
2	Br–CH₂C(O)NH–(PS)	allylamine (0.5 mol/L), DMF, 20 °C, 12 h	allyl–NH–CH₂C(O)NH–(PS)	[36]
3	Br–CH₂C(O)NH–(PS)	glycoside–O–CH₂CH₂–NH₂ (1 mol/L), DMF, DBU, 50 °C	glycoside–O–CH₂CH₂–NH–CH₂C(O)NH–(PS)	[37]
4	Cl–CH₂C(O)–N(CH₂-thienyl)(PS)	BuNH₂ (2 mol/L), NMP, NaI, 80 °C, overnight	Bu–NH–CH₂C(O)–N(CH₂-thienyl)(PS)	[38]
5	Cl–CH₂–C₆H₄–PS	BnNH₂, DMF, 60 °C, 48 h	Ph–CH₂–NH–CH₂–C₆H₄–PS	[39]
6	ClCH₂–/I–C₆H₃–CH₂–O–C(O)–(PS)	BuNH₂ (10 eq), Bu₄NBr (2 eq), NEt₃ (3 eq), 20 °C, 96 h	Bu–NH–CH₂–/I–C₆H₃–CH₂–O–C(O)–(PS)	[40]
7	Cl–CH₂–C₆H₄–C(O)NH–(PS)	allylamine (1 mol/L) DMF, NaI, 75 °C, overnight	allyl–NH–CH₂–C₆H₄–C(O)NH–(PS)	[41] see also [42]
8	ClCH₂–/I–C₆H₃–CH₂–O–C(O)–(PS)	PhNH₂ (10 eq), DIPEA (3 eq), Bu₄NI (2 eq), dioxane, 100 °C, 36 h	Ph–NH–CH₂–/I–C₆H₃–CH₂–O–C(O)–(PS)	[40]
9	Br–CH₂–C₆H₄–N(C(O)CH=CH–C(O)O–CH₂–(PS))(CH₂Ph)	ArCH₂NH₂ (0.2 mol/L), DMF, 20 °C, 10 h; Ar: 4-methoxyphenyl	Ar–CH₂–NH–CH₂–C₆H₄–N(C(O)CH=CH–C(O)O–CH₂–(PS))(CH₂Ph)	[43] see also [44]
10	Cl–CH₂–C₆H₄–O–CH₂–PS	4-nitroaniline, proton sponge, NaI, DMF, 90 °C, 24 h	O₂N–C₆H₄–NH–CH₂–C₆H₄–O–CH₂–PS	[45]
11	Ar–C(O)–O–CH₂–C₆H₄–O–CH₂–PS	piperazine (neat), 130 °C, 20 h; Ar: 4-FC₆H₄	piperazinyl–CH₂–C₆H₄–O–CH₂–PS	[46]

Table 10.2. continued.

Entry	Starting resin	Conditions	Product	Ref.
12		(0.5 mol/L), NMP, 20 °C, 20 h Ar: 2-ClC$_6$H$_4$		[47]
13		(1.5 mol/L), THF, 20 °C, 48 h		[38]
14		1. TFA/DCM 7:3, 20 °C, 5 h 2. ArNH$_2$ (0.5 mol/L), DCM, 1 h Ar: 4-(PhO)C$_6$H$_4$		[48]
15		2-iodoaniline (0.2 mol/L), DMF, DIPEA, 80 °C, 18 h		[49] see also [50,51]
16		Bn$_2$NH, Pd(dba)$_2$, PPh$_3$, THF, 24 h		[39]
17		cyclopropylamine, BEMP, DMF, 20 °C, 6 h		[52]

amines DMF or NMP might furnish milder reaction conditions (Entry **3**, Table 10.2). Higher yields can often be obtained by increasing the reaction temperature and the concentration of the amine.

Merrifield resin [28–32] and other support-bound benzyl halides [33] have also been used to alkylate amines (Entries **5–10**, Table 10.2). Similarly, resin-bound allyl bromides react cleanly with aliphatic or aromatic amines (Entry **15**, Table 10.2). Allylic esters or carbonates undergo nucleophilic substitution upon catalysis by palladium(0). Illustrative examples of these reactions are listed in Table 10.2. When using ammonia, e.g. to convert cross-linked chloromethyl polystyrene into aminomethyl polystyrene, good results are obtained only with anhydrous ammonia in DCM [34]. Other solvents or aqueous ammonia generally lead to extensive cross-linking.

10.1.1.2 By Alkylation of Support-Bound Amines

Amines can also be prepared on insoluble supports by treating support-bound primary or secondary amines with alkylating agents. The reaction seems to be limited to strong alkylating agents, such as benzyl or allyl halides and α-halo carbonyl compounds. Because quaternization does not proceed readily on hydrophobic supports, the alkylation of polystyrene-bound amines can be conducted with excess alkylating agent to provide products of acceptable purity (Table 10.3). Primary amines will, however, usually be alkylated twice under these conditions. Diarylmethyl halides can be used to alkylate primary amines only once, because steric hindrance usually prevents twofold diarylmethylation (Entry **2**, Table 10.3). The resulting secondary benzhydrylamines can, however, be alkylated by sterically less demanding alkylating agents (Entries **4** and **5**, Table 10.3). Because alkoxy-substituted diarylmethyl groups can be cleaved from aliphatic amines by treatment with TFA (see Section 10.1.10.6), diarylmethylation can be used to protect primary amines from twofold alkylation by sterically less demanding alkylating agents.

Polystyrene-bound secondary aliphatic amines or *N*-alkyl amino acids can be allylated by treatment with a diene and an aryl iodide or bromide in the presence of palladium(II) acetate (Entry **10**, Table 10.3). As dienes can be used 1,3-, 1,4-, and 1,5-dienes, and besides aryl halides also heteroaryl bromides have been used with success [53]. This remarkable reaction proceeds probably via formation of an aryl palladium complex, followed by insertion of an alkene into the C–Pd bond. The resulting organopalladium compound does not undergo β-elimination (as in the Heck reaction), but isomerizes to an allyl palladium complex, which reacts with the amine to give the observed allyl amines.

10.1.2 Preparation of Amines by Aromatic Nucleophilic Substitution

10.1.2.1 With Support-Bound Arylating Agents

Aryl- and heteroaryl halides can undergo thermal or transition metal catalyzed substitution reactions with amines. These reactions proceed on insoluble supports under conditions similar to those used in solution.

Support-bound 4-fluoro-3-nitrobenzoic acid has become a widely used intermediate for the preparation of a variety of heterocycles (see Chapter 15). Aromatic nucleophilic substitution of fluoride in this reagent proceeds smoothly with a broad range of nucleophiles, including aliphatic and aromatic amines (Table 10.4). Aliphatic amines even react at room temperature (Entry **1**, Table 10.4 [42,62]), whereas anilines generally require heating (Entry **3**, Table 10.4). Unprotected β-amino acids can be N-arylated directly by 4-fluoro-3-nitrobenzamides linked to PEG-grafted polystyrene (Entry **2**, Table 10.4). Less strongly activated fluorobenzenes have also been used as resin-bound arylating agents for amines, but higher reaction temperatures are often required (Entry **6**, Table 10.4).

Table 10.3. Alkylation of support-bound amines.

Entry	Starting resin	Conditions	Product	Ref.
1		Ph$_2$CHBr (0.7 mol/L), NMP, DIPEA, 80 °C, 18 h		[54] see also [55]
2		Ar$_2$CHCl (0.02 mol/L), DCM, DIPEA, 20 min Ar: 4-(MeO)C$_6$H$_4$		[56]
3		allyl bromide (4 eq), BEMP, dioxane, 20 °C, 70 h		[40] see also [52]
4		BnBr (3 eq), DIPEA, DMF/Me$_2$CO 1:1, 20 °C Ar: 4-fluorophenyl		[57]
5		phenacyl chloride (0.05 mol/L), DMF/Me$_2$CO 1:1, 45 °C, 10 h		[57]
6		BocHN [structure] (2.5 eq), DIPEA, DMF, 3–6 h		[58] see also [59]
7		HO [structure] OCO$_2$Me Pd(PPh$_3$)$_4$, AcOH, PhMe, 20 °C, 3 h		[60]
8		2-methoxybenzyl bromide (0.4 mol/L), DMSO, DBU, 3 h Ar: 2,5-(MeO)$_2$C$_6$H$_3$		[61]

Table 10.3. continued.

Entry	Starting resin	Conditions	Product	Ref.
9		ArCH₂Br (0.3 mol/L, 10 eq), DMF, DIPEA, 80 °C, 22 h Ar: 4-(F₃C)C₆H₄		[49]
10		(0.75 mol/L, 10 eq), PhI (2 eq), DIPEA (10 eq), LiCl (2 eq), Pd(OAc)₂ (0.1 eq), DMF, 100 °C, 2 d		[53]

Non-activated aryl bromides (but not fluorides) can be used as substrates for palladium(0)-catalyzed aromatic nucleophilic substitutions with aliphatic or aromatic amines. These reactions require sodium alcoholates or cesium carbonate as base, and sterically demanding phosphines as ligands. High reaction temperatures are, furthermore, often necessary to attain complete conversion (Entries **7** and **8**, Table 10.4; Experimental Procedure 10.1). Unfortunately, the choice of substituents on the amine or aryl bromide is currently limited to non-protic functional groups [63]. As byproduct of these N-arylations the dehalogenated arenes are often observed [63,64].

N-Arylations have also been realized with support-bound heteroaromatic halides (Entries **9–11**, Table 10.4). The reactivity of these arylating agents depends strongly on their precise substitution pattern, and generally increases with decreasing electron density of the heteroarene. Illustrative examples are given in Table 10.4. The arylation of amines with simultaneous cleavage of the substrate from the support are discussed in Section 3.8.

Experimental Procedure 10.1: Arylation of benzylamine with a support-bound aryl bromide [64]

To the 3-bromobenzyl urea (bound to Tentagel via a Rink amide linker, 0.15 g) in dioxane (2 mL) under nitrogen were added benzylamine (15 eq) and a solution of Pd₂(dba)₃ (0.2 eq), BINAP (0.8 eq), and sodium *tert*-butoxide (18 eq) in dioxane (1 mL). The mixture was stirred at 80 °C for 16–20 h, filtered, and the support was washed with water, DMF, THF, DCM, *i*PrOH, DCM, and AcOH. Treatment of the resin with TFA/water 9:1 (2 × 20 min) and concentration of the filtrates yielded 3-(benzylamino)benzylurea (92 % yield; 89 % pure by HPLC, 220 nm). The product contained 3 % of benzylurea.

Table 10.4. Preparation of aromatic amines from resin-bound arylating agents.

Entry	Starting resin	Conditions	Product	Ref.
1		EtO$_2$C$\sim$NH$_2$ (0.22 mol/L, 10 eq), DMF, 20 °C, 2 h		[65]
2		HO$_2$C$\sim$NH$_2$ NaHCO$_3$ (0.5 mol/L), H$_2$O/Me$_2$CO 1:1, 75 °C, 24 h		[66]
3		PhNH$_2$ (2 mol/L), DMSO, 50 °C, 12 h		[18]
4		piperazine, NMP, 110 °C, 4 h		[67]
5		BuNH$_2$ (5 mol/L), DMSO, 20 °C, 24 h		[68]
6		piperazine, NMP, 110 °C, 10 h		[46]
7		morpholine, dioxane, Pd$_2$(dba)$_3$, P(o-Tol)$_3$, NaOtBu, 80 °C, 20 h		[64]
8		2,6-dimethylaniline, PhMe, Pd$_2$(dba)$_3$, P(o-Tol)$_3$, NaOtBu, 100 °C, 20 h		[63]
9		(1.9 mol/L), NMP, 24 h		[69]

Table 10.4. continued.

Entry	Starting resin	Conditions	Product	Ref.
10		RNH$_2$ (0.25 mol/L), DMSO, 70 °C, 12 h		[70] see also [71,72]
11		3-bromoaniline, HCl, *i*PrOH/DMF 1:1, 20 °C, overnight Ar: 3-bromophenyl		[73]

10.1.2.2 By Arylation of Support-Bound Amines

Few examples have been reported of the arylation of support-bound amines. Unlike alkylations, the arylation of a primary amine usually leads to a strong reduction of the nucleophilicty of the nitrogen atom, so bis-arylation of resin-bound primary amines cannot be expected. Illustrative examples are listed in Table 10.5.

Table 10.5. Preparation of aromatic amines by N-arylation of support-bound amines.

Entry	Starting resin	Conditions	Product	Ref.
1		(0.5 mol/L, 10 eq), DIPEA (5 eq), DMSO, 60 °C, 12 h		[74]
2		(0.6 mol/L, 10 eq), NMP, DTBP, 60 °C, 11 h		[75]
3		2-fluoropyridine (neat), NBu$_3$, 110 °C, 30 h		[76]
4		K$_2$CO$_3$, DMF, 25 °C, 30 h		[77] see also [78]

Table 10.6. Addition of amines to support-bound, acceptor-substituted alkenes.

Entry	Starting resin	Conditions	Product	Ref.
1		tetrahydroisoquinoline (0.6 mol/L), DMF, 20 °C, 18 h		[54] see also [80,81]
2		Ph$_2$CHNH$_2$, (1.7 mol/L), DMSO, 20 °C, 48 h (no reaction with H-Phe-OMe or PhNH$_2$)	(55% conversion only)	[79]
3		EtNH$_2$ (0.9 mol/L), DMSO, 70 °C, 9 h		[82]
4		BnNH$_2$ (2.3 mol/L), DMSO, 50 °C, 24 h		[79]
5		(2.3 mol/L), DMSO, 50 °C, 72 h	(56% conversion only)	[79]
6		pyrrolidine (30 eq), DMF, 22 °C, 18 h		[83]
7		morpholine (10 eq), DMF, 20 °C, 16 h		[84] see also [85]
8		indoline (2 mol/L, 20 eq), DMF, 20 °C, overnight		[86]

10.1.3 Preparation of Amines by Addition of Amines to C–C Double Bonds

Support-bound alkenes substituted with at least one electron-withdrawing group can react with primary or secondary amines (Table 10.6). If acrylates are used as Michael acceptors, the products (β-alanine derivatives) are generally stable, and do not undergo β-elimination either upon N-acylation or on treatment with TFA (for a longer synthetic sequence with a support-bound 2-(1-piperazinyl)propionate, see [36]). Suitable solvents for the addition of amines to electron-poor alkenes are THF, NMP, or DMSO. Less suitable is DMF, because this solvent often contains dimethyl-amine, which also adds to Michael acceptors [79].

Acrylic acid esterified with cross-linked hydroxymethyl polystyrene or Wang resin reacts smoothly with primary or secondary aliphatic amines at room temperature (Entries 1 and 2, Table 10.6). Only sterically demanding amines or amines of low nucleophilicity (anilines, α-amino acid esters) fail to add to polystyrene-bound acrylate. Support-bound acrylamides are less reactive than acrylic esters, and generally require heating to undergo addition reaction with amines (Entries 4 and 5, Table 10.6). Also α,β-unsaturated esters with substituents in position 3 (e.g. crotonates, Entry 3, Table 10.6) react significantly more slowly with nucleophiles than do acrylates. The examples in Table 10.6 also show that polystyrene-bound esters are rather stable towards aminolysis, and provide for robust attachment even in the presence of high concentrations of amines.

10.1.4 Preparation of Amines by Reduction of Imines

One of the most convenient preparations of alkylamines on insoluble supports is the reduction of imines, which can be readily prepared either from support-bound amines or from support-bound carbonyl compounds (Figure 10.2).

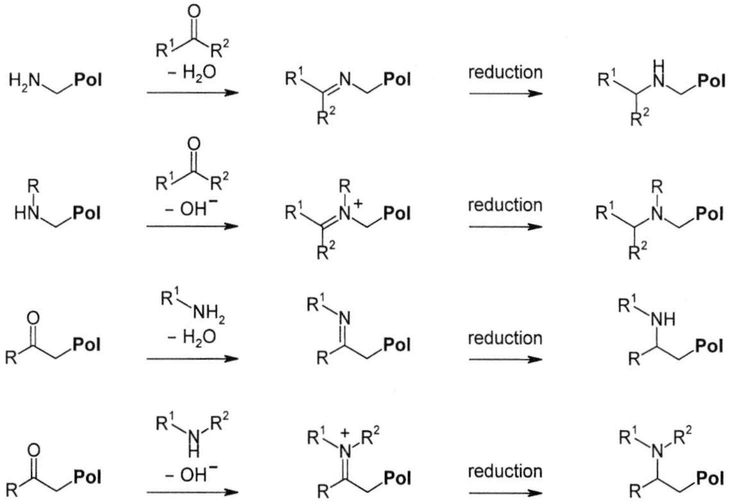

Fig. 10.2. Reductive amination on solid phase.

Many aldehydes react spontaneously with primary aliphatic amines to yield the corresponding imines (Schiff bases). These are often sufficiently stable to enable filtration and washing of the support. Ketones are generally less readily converted into imines than are aldehydes, and the stoichiometric conversion of support-bound amines into ketone-derived imines is more conveniently achieved by treatment of the amine with an imine (Entry **10**, Table 10.7). The formation of imines can be promoted by acids [87–89], by dehydrating agents, such as orthoformates [89–92], or by higher reaction temperatures.

Reduction of resin-bound imines can be performed with $LiBH_4$ [93] or $NaBH_4$ under basic reaction conditions, or with $NaCNBH_3$ in the presence of acetic acid. Other reagents are given in Table 10.7. The two-step-methodology (imine-formation, washing, reduction) enables the clean monoalkylation of support-bound primary amines. Alternatively, amines can be alkylated on solid phase by one-pot imine formation and reduction. $NaCNBH_3$ is a particularly convenient reagent for this purpose, because it does not usually reduce aldehydes, ketones, or (non-protonated) imines, but solely iminium salts. The main advantage of one-pot reductive aminations is that carbonyl compounds which do not readily form imines, e.g. ketones, can also be used. The most serious inconvenience of the one-pot method is that support-bound primary amines might be alkylated twice.

Monoalkylation of support-bound amines can also be realized by a three-step protocol involving initial protection/activation of the amine (e.g. as 2-nitrobenzenesulfonamide [94,95], trifluoroacetamide [70], or 4,4′-dimethoxybenzhydrylamine [56]) followed by N-alkylation and deprotection. In particular, if monomethylations (see, e.g., Entry **7**, Table 10.7) or allylations are to be performed this three-step strategy will generally give the best results.

Illustrative examples of the reductive alkylation of support-bound amines are listed in Table 10.7. Further examples have been reported [96–100].

Examples of the conversion of support-bound carbonyl compounds into amines are listed in Table 10.8. Additional examples have been reported [111,112]. The loading of backbone amide linkers (Section 3.3.1) with amines is the most important application of this reaction (Experimental Procedure 10.2). Because the reduction of imines with $NaCNBH_3$ or similar reducing agents generally requires slightly acidic reaction conditions, more acid than amine must always be used.

Entry **7** in Table 10.8 is an interesting example of a preparation of MBHA resin, in which the initial product of reductive amination is a formamide (Leuckart reaction). This formamide can be hydrolyzed by treatment with aqueous hydrochloric acid, without cleavage of the benzylic C–N bond.

As in the Leuckart reaction, reductive aminations with $NaCNBH_3/AcOH$ under anhydrous conditions can lead to the formation of acetamides as side products [82]. This unwanted acetylation can be avoided by performing the reductive amination in the presence of small amounts of water (e.g. Entry **2**, Table 10.7).

Table 10.7. Reductive alkylation of support-bound amines.

Entry	Starting resin	Conditions	Product	Ref.
1		1. Me$_2$CHCHO (0.25 mol/L, 5 eq), 1% AcOH in DMF, 1.5 h 2. NaBH$_4$ (8 eq), DMF/MeOH 10:4, 2 h		[101] see also [93]
2		cyclohexanone (1.2 eq), THF/AcOH/ H$_2$O 90:5:5, NaCNBH$_3$, 23 °C, 3 h		[102] see also [57]
3		ArCHO (0.3 mol/L), 1% AcOH in DMA, NaCNBH$_3$ (1.7 mol/L), overnight		[61] see also [103]
4		C$_6$H$_{11}$CHO, NaBH(OAc)$_3$ (30 eq of each), DCM, 25 °C, 16 h		[104]
5		FmocHN $\overset{R^1}{\diagdown}$ 0.5% AcOH in DMF, NaCNBH$_3$, 3 h		[105] see also [90]
6		Ph CO$_2$Et (25 eq), 1% AcOH in DMF, NaCNBH$_3$ (40 eq)		[106]
7		HCHO, 1% AcOH in DMF, NaCNBH$_3$, 1 h Ar: 4-(MeO)C$_6$H$_4$		[56]
8		2-butanone, pyridine•BH$_3$ (10 eq of each), DMF/EtOH 3:1, 25 °C, 4 d (does not proceed with acetophenone)		[107] see also [108]

Table 10.7. continued.

Entry	Starting resin	Conditions	Product	Ref.
9		PhCHO (1.5 mol/L, 20 eq), DMF, AcOH (0.5 eq), NaBH(OAc)$_3$ (20 eq), 18 h		[81]
10		DCM, 20 °C, 16 h; then Ac$_2$O, DIPEA, DCM, 20 °C, 0.5 h; then NaCNBH$_3$, AcOH, DMA		[109] see also [110]

Experimental Procedure 10.2: Racemization-free loading of amino acid esters on to a backbone amide linker [113]

The resin-bound aldehyde (0.50 g, approx. 0.3 mmol) was suspended in DMF (20 mL) containing 1 % acetic acid, and NaBH(OAc)$_3$ (0.64 g, 3.00 mmol) was added. An α-amino acid methyl ester hydrochloride (3.00 mmol) was added to the suspension, and the mixture was stirred for 1 h. The resin was filtered and washed with methanol, DMF, DCM, and finally with methanol, and dried under reduced pressure.

10.1.5 Preparation of Amines by Reaction of Carbon Nucleophiles with Imines or Aminals

Nucleophiles other than hydride can be added to support-bound imines to yield amines. These include C,H-acidic compounds, alkynes, electron-rich heterocycles, organometallic compounds, and ketene acetals. When basic reaction conditions are used, stoichiometric amounts of the imine must be prepared on the support (Entries **1–3**, Table 10.9).

Alternatively, if the carbon-nucleophile is stable under acidic conditions, imines or iminium salts might be generated in situ, as in Mannich reactions in solution. Few examples have been reported of Mannich reactions on insoluble supports; most are based on alkynes as C-nucleophiles. Illustrative examples are listed in Table 10.9.

Support-bound C-nucleophiles have also been added to imines with success. Polystyrene-bound thiol esters can be converted into ketene acetals by O-silylation, and

Table 10.8. Reductive amination of support-bound carbonyl compounds.

Entry	Starting resin	Conditions	Product	Ref.
1		pyridine•BH$_3$, 1% AcOH in MeOH/HC(OMe)$_3$		[114] see also [115]
2		1. EtNH$_2$ (5 eq), THF, 18 h 2. filtration, washing 3. NaBH$_4$ (10 eq), THF/EtOH 3:1, 8 h		[116]
3		EtNH$_2$, Me$_4$NBH(OAc)$_3$, DCE, 16 h; then NaCNBH$_3$, MeOH, 6 h		[117]
4		BnNH$_2$ (15 eq), NaBH(OAc)$_3$ (5 eq), Na$_2$SO$_4$ (10 eq), 1% AcOH in DCM, ultrasound		[118] see also [119] [120]
5		indoline (15 eq), NaBH(OAc)$_3$ (5 eq), Na$_2$SO$_4$ (10 eq), 1% AcOH in DCM, ultrasound		[118]
6		Ti(OiPr)$_4$ (0.6 mol/L), PhMe, BnNH$_2$ (0.4 mol/L, 2.5 eq), 2 h; then AcOH (10%), NaBH(OAc)$_3$ (2.4 mol/L, 10 eq), THF, 24 h		[121]
7		1. HCO$_2$NH$_4$, HCONH$_2$, HCO$_2$H/PhNO$_2$ 1:5, 165 °C, 22 h 2. HCl (12 mol/L in H$_2$O)/ EtOH 1:1, 78 °C, 1 h		[122]

Table 10.9. Addition of C-nucleophiles to support-bound imines and hemiaminals.

Entry	Starting resin	Conditions	Product	Ref.
1		BuLi, THF, −78 °C to 20 °C, 31 h		[93]
2		(10 eq), PhH, 20 °C, 1 h		[123] see also [124]
3		Ph━━━MgBr Et$_2$O, PhMe, 60 °C, 24 h		[93]
4		Yb(OTf)$_3$, DCM/MeCN 1:1, 20 °C, 20 h		[87]
5		PhCHO, CuCl, dioxane, 85 °C, 3 h		[125]
6		Ph━━━ 1-phenylpiperazine, CuCl, dioxane, 75 °C		[125]
7		HO$_2$C⟍⟍CO$_2$H citric acid buffer (pH 4), 20 °C, 12 h		[126]
8		(2 eq), ArOK (1.3 eq), THF, 20 °C, overnight, then HCl, THF, 4 h		[127]

then alkylated with imines in the presence of Lewis acids. Further examples include Mannich reactions of support-bound alkynes and indoles (Table 10.10).

Some Mannich-type products (e.g. 3-(aminomethyl)indoles, 2-(aminomethyl)phenols, β-amino ketones) are unstable and can decompose upon treatment with acids. 3-(Aminomethyl)indoles have, for instance, been used as acid-labile linkers for amines and amides (Tables 3.9 and 3.24). Hence the choice of a suitable linker can be critical for Mannich reactions on insoluble supports.

Table 10.10. Addition of support-bound C-nucleophiles to imines.

Entry	Starting resin	Conditions	Product	Ref.
1		Ph–N=CH–Ph Sc(OTf)$_3$, DCM, 20 °C, 20 h		[128] see also [129]
2		Ph–N=CH–C$_6$H$_{11}$ Sc(OTf)$_3$, DCM, 20 °C, 20 h		[128]
3		1-naphthaldehyde, 1-cyclohexylpiperazine, CuCl, dioxane, 90 °C, 36 h R: cyclohexyl		[130] see also [131]
4		BnNH$_2$, HCHO, dioxane/AcOH 4:1, 23 °C, 1.5 h		[132]

10.1.6 Preparation of Amines by Reduction of Amides and Carbamates

Polystyrene-bound amides, including peptides, can be reduced to the corresponding amines by treatment with diborane in ethers. Other reagents, such as lithium aluminum hydride, are less convenient for reductions on insoluble supports, because insoluble precipitates can readily form and clogg fritts. Carbamates can also be reduced, e.g., by lithium aluminum hydride (Entry 2, Table 3.21 [133]) or by Red-Al (Entry 4, Table 10.11). Illustrative examples of the reduction of amides and carbamates on solid phase are listed in Table 10.11.

Table 10.11. Reduction of support-bound amides and carbamates.

Entry	Starting resin	Conditions	Product	Ref.
1		B₂H₆ (0.5 mol/L), THF, 50 °C, 1 h; quenching with DBU (0.06 mol/L in NMP/MeOH 9:1)		[134] see also [135]
2		B(OH)₃, B(OMe)₃, B₂H₆ (0.5 mol/L), THF, 65 °C, 72 h		[136] see also [137–139]
3		BH₃•SMe₂ (0.26 mol/L), DME, 85 °C, 20 h		[87]
4		MeO–, MeO–AlH₂Na (70 eq), PhMe, 20 °C, overnight		[140]

10.1.7 Preparation of Amines by Reduction of Nitro Compounds

Anilines can be readily prepared on insoluble supports by reduction of nitroarenes (Table 10.12). The most common reagent for this purpose is tin(II) chloride dihydrate, which is highly soluble in NMP or DMF, and does not generally lead to insoluble precipitates in these solvents [141–145]. The reduction is usually performed with 1–5 molar solutions of the reducing agent at 20–100 °C for 5–16 h. Other reagents which have been used to reduce support-bound aromatic nitro compounds include phenylhydrazine at high temperatures (Entry **5**, Table 10.12), sodium borohydride in the presence of copper(II) acetylacetonate [68], chromium(II) chloride [146], lithium aluminum hydride (Entry **3**, Table 10.12), and sodium dithionite (Entry **4**, Table 10.12). The reduction of nitroarenes with tin(II) chloride proceeds via intermediate formation of a hydroxylamine, which can be trapped inter- or intramolecularly with different electrophiles [147,148]. For instance, treatment of Wang resin bound benzonitriles with nitrobenzene and tin(II) chloride leads to resin-bound *N*-hydroxyamidines [82].

Nitroalkanes are generally more difficult to reduce than nitroarenes, and as far as I am aware no examples have been reported of the reduction of nitroalkanes on insoluble supports.

Table 10.12. Reduction of support-bound nitro compounds.

Entry	Starting resin	Conditions	Product	Ref.
1		SnCl$_2$•2 H$_2$O (1.6 mol/L), NMP, 20 °C, 16 h		[145]
2		SnCl$_2$•2 H$_2$O (8 eq), DMF, 25 °C, 16 h (does not proceed with Na$_2$S$_2$O$_4$ or FeSO$_4$•7 H$_2$O)		[104]
3		LiAlH$_4$, Et$_2$O, 35 °C, 48 h		[149]
4		Na$_2$S$_2$O$_4$, EtOH, 78 °C, 1.5 h		[150]
5		PhNHNH$_2$ (neat), 200 °C, 1 h		[151]

10.1.8 Preparation of Amines by Reduction of Azides

The azido group is another functional group which can serve as a precursor to an amine. The preparation of azides on solid phase is discussed in Section 10.4. Azides can be reduced to amines by several types of reagent, but few of these are compatible with insoluble supports. The most convenient reducing agents, which are also effective on cross-linked polystyrene, are phosphines in the presence of water, and thiols (Table 10.13). Phosphines react with azides to yield iminophosphoranes, which can be hydrolyzed to yield amines. The reduction can also be performed in the presence of an activated carboxylic acid derivative, whereby amides are formed directly. The latter strategy was used in Entry **5** (Table 10.13), because the intermediate amine would have undergone rapid intramolecular reaction with the ester linkage to yield a lactam. Reduction in the presence of a HOBt ester efficiently prevented intramolecular nucleophilic cleavage.

Table 10.13. Reduction of support-bound azides.

Entry	Starting resin	Conditions	Product	Ref.
1	N₃ ...(PEGA)	dithiothreitol (2 mol/L), DIPEA (1 mol/L), DMF, 50 °C, 0.5 h	H₂N ...(PEGA)	[152] see also [153]
2	N₃ ...O(PS), Ar, N, R	SnCl₂ (0.2 mol/L), PhSH (0.8 mol/L), NEt₃ (1 mol/L), THF, 20 °C, 4 h	NH₂ ...O(PS), Ar, N, R	[3,4] see also [5,154] [155]
3	N₃ ...O(TG), peptide	SnCl₂, PhSH, NEt₃, THF (acyl migration only in *cis* isomer)	H₂N ...O(TG), peptide	[156]
4	TIPSO, OAc, N₃, O(PS), O=S–N, Ar	HS⌒⌒SH DIPEA 1:1, 24 h Ar: 9-anthracenyl	TIPSO, OAc, H₂N, O(PS), HN, O	[157]
5	N₃ ...O(PS)	PBu₃, EDC, PhCO₂H, HOBt, dioxane, 20 °C, 18 h	Ph–C(O)–NH ...O(PS) (no lactam formation)	[158] see also [159]
6	OCOR ...O(PS), N₃	PPh₃ (10 eq), H₂O (30 eq), THF, 25 °C, 8 h	OCOR ...O(PS), NH₂	[160] see also [161] [162]

10.1.9 Miscellaneous Preparations of Amines

Amines have been prepared on insoluble supports by Hofmann degradation of amides [163] followed by hydrolysis of the intermediate isocyanates (Figure 10.3). One reagent suitable for this purpose is [bis(trifluoroacetoxy)iodo]benzene, which can be used both on cross-linked polystyrene [164] and on more hydrophilic supports, such as polyacrylamides (Figure 10.4).

Support-bound carboxylic acids can also be degraded via the acyl azides (Curtius degradation [165,166]) to yield isocyanates. These can be converted into amines by direct hydrolysis, or into Fmoc-protected amines by reaction with 9-fluorenylmethanol [165].

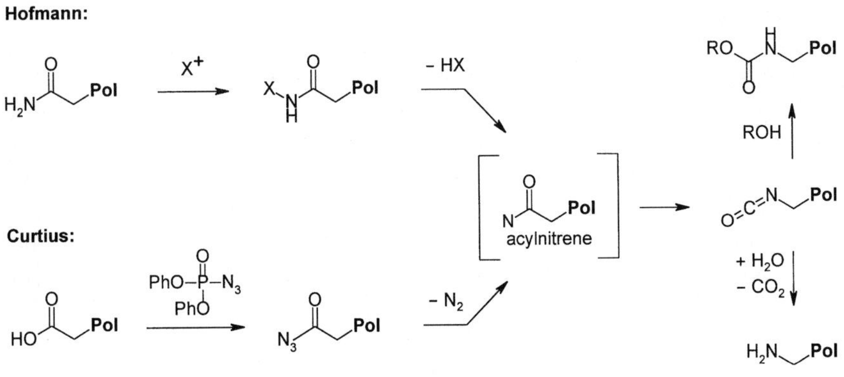

Hofmann:

Curtius:

Fig. 10.3. Hofmann and Curtius degradation of support-bound amides and acids.

The degradation of support-bound α-amino acid amides has been used to prepare retro-inverso peptide mimetics ([167], second equation in Figure 10.4). These compounds are of interest because of their potentially improved metabolic stability, selectivity, and biological activity, compared with peptides [167]. Although retro-inverso peptides are aminals susceptible to acid-catalyzed hydrolysis, N,N'-diacylated aminals can be sufficiently stable to withstand the conditions of Boc-group removal [168] or of acidolytic cleavage of peptides from Wang resin [167].

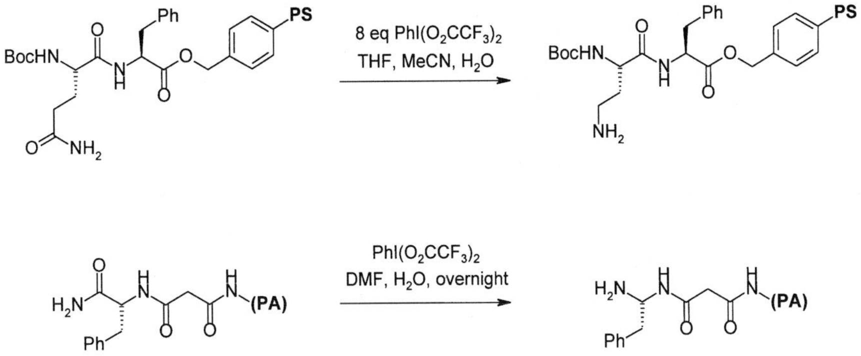

Fig. 10.4. Conversion of amides into amines by oxidative degradation [163,169].

10.1.10 Protective Groups for Amines

Amines are important synthetic intermediates, and numerous protective groups have been developed for temporarily preventing amines from being acylated or alkylated. The following sections cover protective groups for amines which can be introduced or removed on insoluble supports. The development of such protective groups was mainly driven by solid-phase peptide synthesis. A more detailed collection of protective groups can be found in [170].

10.1.10.1 Carbamates

Carbamates are by far the most common type of amine protection used in solid-phase synthesis. Various types of carbamate have been developed which can be cleaved under mild reaction conditions on solid phase. Less well developed, however, are techniques which enable the protection of support-bound amines as carbamates. Protection of amino acids as carbamates (Boc or Fmoc) is usually performed in solution using aqueous base (Schotten–Baumann conditions). These conditions enable the selective protection of amines, without simultaneous formation of imides or acylation of hydroxyl groups. Unfortunately, Schotten–Baumann conditions are not compatible with insoluble, hydrophobic supports. Other bases and solvents have to be used for preparing carbamates on, e.g., cross-linked polystyrene, and more side reactions are generally observed than in aqueous solution.

Most carbamates which are used as protective groups for amines are either acid-labile or base-labile. Deprotection proceeds by the mechanism outlined in Figure 10.5. During the deprotection of acid-labile carbamates carbocations are formed, which can alkylate electron-rich structural elements in a given substrate (e.g. phenols, thiols, indoles, pyrroles) but not the deprotected (protonated) amine. Some base-labile carbamates, however, can lead to the formation of acceptor-substituted alkenes which will alkylate the deprotected amine unless scavengers are added. As scavengers for such Michael acceptors can be used nucleophilic primary or secondary amines, such as pyrrolidine or piperidine.

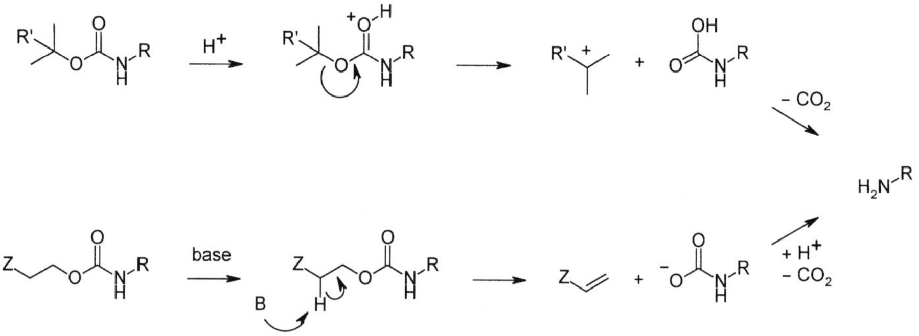

Fig. 10.5. Mechanism of deprotection of acid-labile and base-labile carbamates. R: alkyl, aryl; R′: alkyl, aryl; Z: electron-withdrawing group; B: base.

tert-Butyl Carbamates (Boc Protection)

Tert-Butyl carbamates (Boc amines) were developed as acid-labile protective groups for α-amino acids (see Section 16.1.2). Boc amines are stable towards bases, nucleophiles, weak acids, oxidants, and weak reducing agents, but can be selectively converted into amine trifluoroacetates by treatment with TFA/DCM 1:1.

The most common reagent for converting amines into *tert*-butyl carbamates is di-*tert*-butyl dicarbonate ('Boc anhydride'). This reagent has almost completely replaced the more hazardous *tert*-butyloxycarbonyl azide, which had been the standard reagent for many years [171].

Few examples have been reported of the conversion of support-bound amines into *tert*-butyl carbamates (Table 10.14). Most substrates are secondary amines devoid of further acylable groups. It has been shown that polystyrene-bound anilides [172] or *N*-alkylamides in solution [173] can be N-alkoxycarbonylated with di-*tert*-butyl dicarbonate (NEt$_3$, DMAP, DCM, 6–72 h). These observations suggest that the selective Boc-protection of primary amines or amide-containing substrates on hydrophobic supports might require a careful optimization.

Table 10.14. Conversion of support-bound amines into *tert*-butyl carbamates.

Entry	Starting resin	Conditions	Product	Ref.
1		Boc$_2$O, DIPEA		[174]
2		Boc$_2$O, DIPEA, DCM, 20 °C, overnight		[25]
3		Boc$_2$O, THF, 25 °C, overnight		[175]
4		KF, NEt$_3$, DMF, 20 °C, 24 h		[176]
5		Boc$_2$O (4 eq), DMAP, THF, 20 °C, 16 h		[177]

Tert-Butyl carbamates are stable towards bases and weak acids (half-life in 80 % AcOH: 58 d [178]). Stronger acids, however, lead to dealkylation via formation of a *tert*-butyl cation and decarboxylation of the intermediate carbamic acid. The removal of Boc groups from support-bound amines has been extensively studied. For the synthesis of peptides on cross-linked polystyrene using Boc methodology (Section 16.1.2) the standard deprotection protocol consists in treatment of the resin with TFA/DCM 1:1 for 15–30 min at room temperature. Additional reagents useful for the acidolysis of Boc groups are 10 % sulfuric acid in dioxane [179], aqueous hydrochloric acid

(6 mol/L, only suitable for polyacrylamide-based supports [180]), hydrochloric acid (1.2 mol/L) in acetic acid [181], and hydrochloric acid (4–5 mol/L) in dioxane [182] or DCM [183]. Mixtures of chlorotrimethylsilane (1 mol/L) and phenol (3 mol/L) in DCM (22 °C, 20 min) have been claimed superior to TFA/DCM [184]. Figure 10.6 depicts an interesting procedure, which enables the hydrolysis of Boc groups from peptides bound to (TFA-labile) Rink amide resin [185].

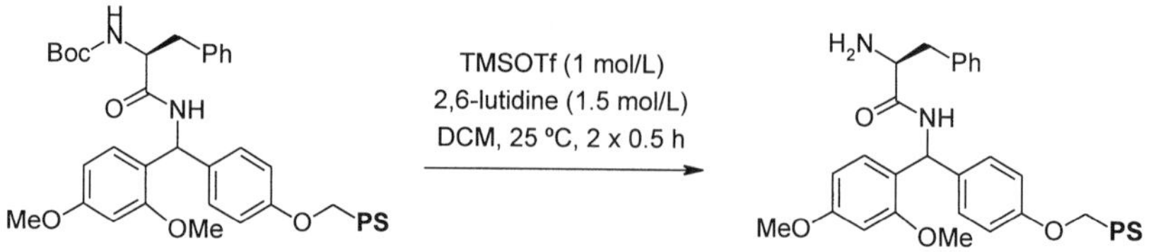

Fig. 10.6. Boc group removal from Rink amide resin [185].

Other Acid-Labile Carbamates

Several other acid-labile carbamates have been developed, but most have found only little application in solid-phase synthesis. A selection of these protective groups is listed in Table 10.15. Highly acid-labile carbamates, such as Bpoc, Mpc, or Ddz can be cleaved from substrates esterified with Wang resin, without cleavage of the linker.

Some of the protective groups listed in Table 10.15, e.g. Bpoc or Azoc, contain a chromophore which enables photometric monitoring of the deprotection reaction. Such monitoring is important for automated solid-phase peptide synthesis, for quickly assessing the quality of a peptide (before cleavage from the support) and the location of those positions in the peptide where peptide bond formation is difficult (see Section 16.1).

Benzyl carbamate protection (Cbz or Z group; see Table 10.15) was initially chosen by Merrifield for solid-phase peptide synthesis [191]. The strongly acidic conditions required for its solvolysis (30 % HBr in AcOH, 25 °C, 5 h) demanded the use of an acid-resistant nitrobenzyl alcohol linker. Z protection of the α-amino group in solid-phase peptide synthesis was, however, quickly abandoned and replaced by the more acid-labile Boc protection.

Benzyl carbamates can be cleaved by strongly ionizing acids (HF, HBr, TfOH, HBF₄, etc.) or by hydrogenolysis, and today this group is only used in peptide synthesis as permanent side-chain protection of lysine. Deprotection is effected during cleavage of the peptide from the support or afterwards in solution.

Secondary aliphatic amines bound to cross-linked polystyrene have been converted into benzyl carbamates by treatment with benzyl chloroformate (10 eq) and triethylamine (15 eq) in DCM (0 °C, 1 h, then 20 °C, 18 h) [192].

Table 10.15. Selection of acid-labile carbamates.

Entry	Starting resin	Conditions	Product	Ref.
1-(4-biphenylyl)-1-methylethoxycarbonyl **Bpoc**		80% AcOH, 20 °C, 2.5 h; or 3% TFA in DCM, 5 min; or 0.5% TFA in DCM, 20 min		[178] [186] [187]
1-(4-methylphenyl)-1-methylethoxycarbonyl **Mpc**		80% AcOH, 20 °C, 35 min; or 0.2% TFA in DCM, 1 min; or 5% Cl$_2$CHCO$_2$H in DCM, 5 min; or B-chlorocatecholborane (0.05 mol/L, 8 eq), DIPEA (4 eq), DCM, 2 min		[178] [186]
1-(3,5-dimethoxyphenyl)-1-methylethoxycarbonyl **Ddz**		AcOH/H$_2$O 8:2, 20 °C, 3 h; or 1% TFA in DCM, 20 °C, 1 h; or photolysis (280 nm) in THF		[188]
1-(1-adamantyl)-1-methylethoxycarbonyl **Adpoc**		3% TFA in DCM, 20 °C, 3 min		[189]
1-(4-phenylazophenyl)-1-methylethoxycarbonyl **Azoc**		1.5% TFA in DCM, 20 °C, 5 min; or AcOH/H$_2$O 8:2, 20 °C, 6 h		[190]
benzyloxycarbonyl **Cbz or Z**		30% HBr in AcOH, 25 °C, 5 h; or Me$_3$SiI (1.2 eq), CHCl$_3$, 25 °C, 10 min		[171] [191]

Fluorenylmethyl Carbamates (Fmoc Protection)

The 9-fluorenylmethoxycarbonyl group, developed by Carpino and coworkers in 1972 [193], has become one of the most common protective groups for aliphatic or aromatic amines in solid-phase synthesis. In particular for solid-phase peptide synthesis this protective group plays an important role [194] (Section 16.1). The Fmoc group is not well suited for liquid phase synthesis because non-volatile side products are formed during deprotection.

As for Boc protection, the Fmoc group is usually not introduced on solid phase but in solution, by use of an activated Fmoc derivative (e.g. the chloroformate Fmoc-Cl or O-Fmoc-N-hydroxysuccinimide, Fmoc-OSu) and aqueous base (Experimental Procedure 10.3). N-Alkylamino acids bound to cross-linked polystyrene have been Fmoc-protected by treatment with Fmoc-Cl (4 eq) and DIPEA (6 eq) in DCM for 2 h [94,195]. Also primary amines can be converted into Fmoc derivatives on insoluble supports under these conditions [196].

The deprotection of Fmoc amines proceeds by base-induced β-elimination (Figure 10.7). The most common reagent for this purpose is piperidine, which serves both as base and as scavenger of dibenzofulvene, which would otherwise react irreversibly with the deprotected amine.

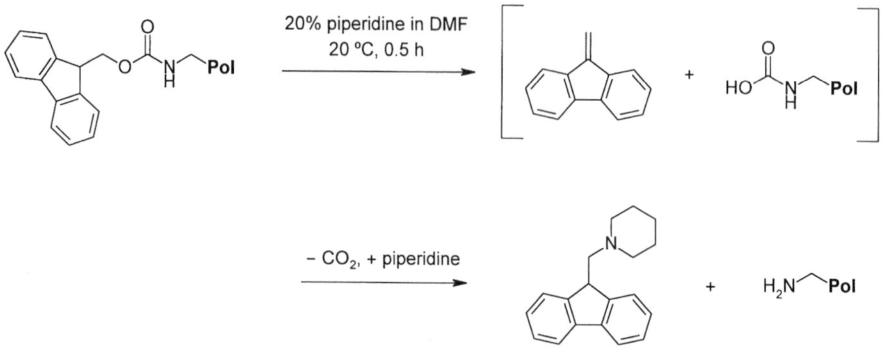

Fig. 10.7. Mechanism of deprotection of Fmoc amines.

The stability of Fmoc amines towards different bases has been investigated in detail [194,197,198]. Ammonia and primary or secondary aliphatic amines in polar aprotic solvents lead to swift deprotection. Further reagents claimed to be useful for Fmoc deprotection are 2 % DBU in DMF [199], KF/NEt$_3$ in DMF [176], 40 % Et$_2$NH in DCM (3 h [200]), and TBAF in DMF (20 °C, 2 min; this reagent also cleaves esters [201,202]). Tertiary amines do not deprotect Fmoc amines as quickly as secondary amines, and can therefore be used for Fmoc introduction. For instance, the half-life of Fmoc-Val-OH in DMF/DIPEA 1:1 is approximately 10 h [197]. No deprotection is generally observed upon treatment of Fmoc amines with pyridine, but in DMAP/DMF 1:9 cleavage occurs with a half-life of about 1.5 h [197]. Deprotection of Fmoc amines has also been observed upon treatment with phosphines under the conditions of Mitsunobu reaction [203].

Experimental Procedure 10.3: Fmoc protection of α-amino acids [204,205]

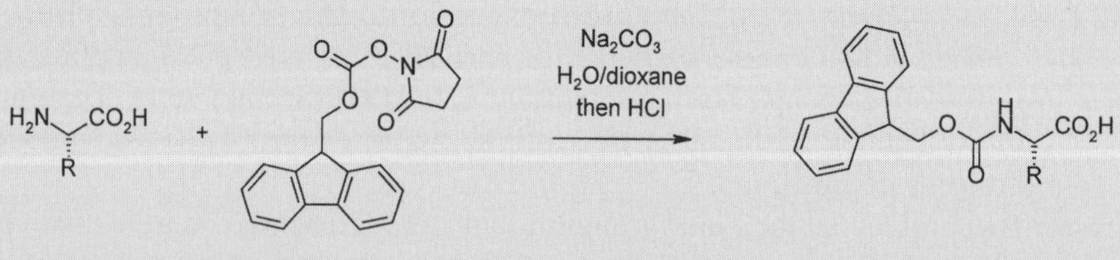

A solution of Fmoc-OSu (11.8 g, 35 mmol) in dioxane (100 mL) is added in one portion to an ice-cooled, stirred mixture of the α-amino acid (42 mmol), sodium carbonate (9.0 g, 85 mmol), and water (100 mL). The mixture is stirred at room temperature for 10 min, diluted with water (1.2 L), and washed with ether (80 mL) and ethyl acetate (2 × 100 mL). The aqueous layer is cooled and acidified to pH 2 with concentrated aqueous hydrochloric acid, whereby a precipitate forms. The mixture is extracted with ethyl acetate (6 × 100 mL), and the combined extracts are washed with an aqueous saturated sodium chloride solution (3 × 60 mL), with water (2 × 60 mL), dried with sodium sulfate, and concentrated under reduced pressure. The Fmoc-amino acid precipitates upon addition of petroleum ether.

This procedure is also suitable for the protection of proline (86 % yield) and 4-hydroxyproline (78 % yield).

Other Base-Labile Carbamates

Further carbamates susceptible to base-induced cleavage have been investigated, but none of these has been extensively used in solid-phase synthesis. 2,2-Bis(4-nitro-phenyl)ethyl carbamates (Bnpeoc, Figure 10.8 [206]), (1,1-dioxobenzothiophen-2-yl)methyl carbamates (Bsmoc, Figure 10.8 [207]), and 2-(4-nitrophenyl)sulfonylethyl carbamates (Nsc [193,208]) show chemical stability similar to that of Fmoc amines. More stable towards bases than Fmoc amines are 2-(4-nitrophenyl)ethyl carbamates (Npeoc). This group has, for instance, been used in the solid-phase synthesis of pep-tidyl phosphonates [203,209]. The more base-labile Fmoc protection could not be used, because deprotection occurred under the conditions of P–O bond formation (PAr$_3$, DIPEA, DIAD, THF) [203].

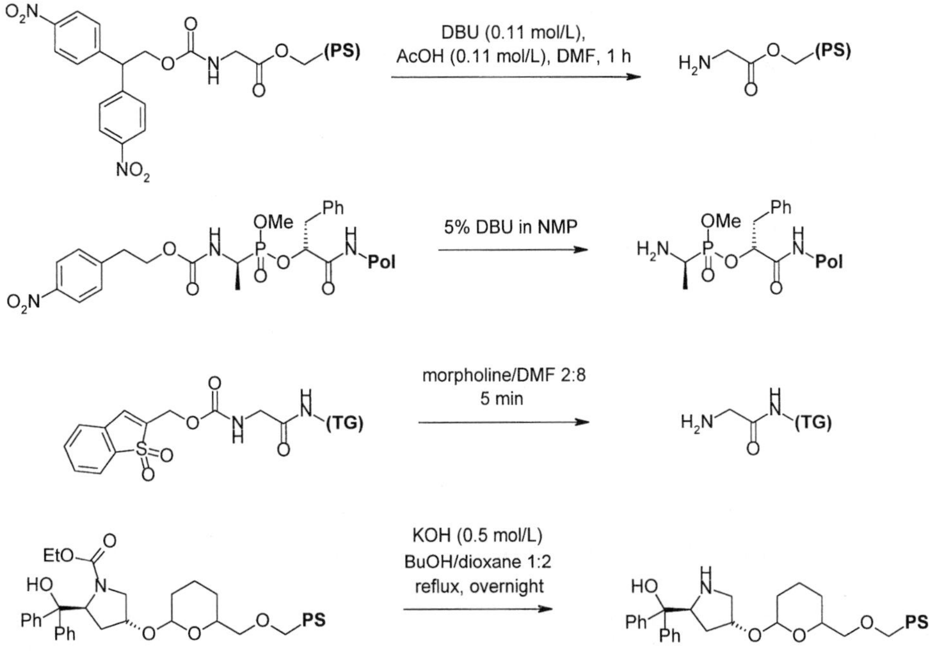

Fig. 10.8. Base-induced cleavage of support-bound carbamates [140,203,206,207].

Alkyl carbamates are generally more stable towards nucleophiles than amides, and therefore of limited utility as protective groups. Amines lacking other base-sensitive functionalities can, however, be protected as alkyl carbamates. An illustrative example for the use of ethyl carbamate as a protective group is sketched in Figure 10.8 [140].

Miscellaneous Carbamates

Carbamate-based protective groups which can be cleaved by other agents than acids or bases have been developed; these supplement the protective groups presented above.

Allyl carbamates (Alloc) have been used for side-chain protection of peptides (at lysine, arginine, or histidine), and can be selectively removed by soft nucleophiles in the presence of palladium(0) complexes [210–212]. Polystyrene-bound anilines have been converted into allyl carbamates by treatment with allyl chloroformate and $NaHCO_3$ in THF at 80 °C overnight [144]. A typical deprotection, in which a stannane is used as hydride source to reduce the intermediate palladium allyl complex, consists in treating the allyl carbamate with a solution of $PdCl_2(PPh_3)_2$ (8 mmol/L), AcOH (0.5 mol/L), and Bu_3SnH (0.4 mol/L) in DCM for 10 min at room temperature [210]. Another example is sketched in Figure 10.9.

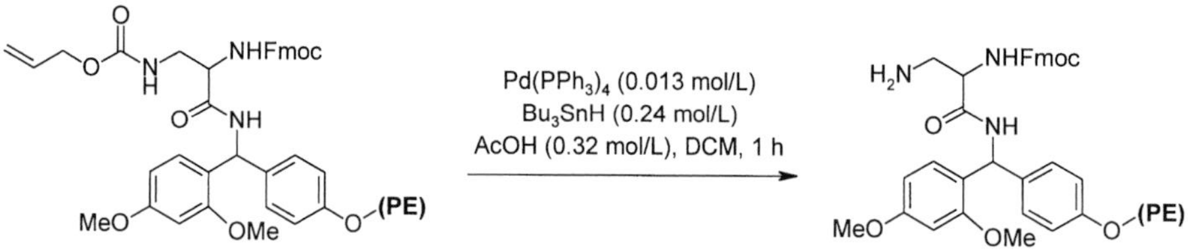

Fig. 10.9. Deprotection of Alloc amines [213].

2-(Trimethylsilyl)ethyl carbamate (Teoc) can be selectively removed with fluoride (e.g. TBAF, THF, 50 °C, 5–20 h [108,214,215]), but also by acidolysis with TFA [114].

10.1.10.2 Amides

Amides with electron-withdrawing substituents can be sufficiently labile towards nucleophilic attack to enable their use as protective groups. This is, e.g., true for trifluoro- [70,216] or trichloroacetamides [123], which are readily hydrolyzed under mild conditions (Figure 10.10). Suitable nucleophiles are hydrazine [217], aliphatic amines, or hydroxide, but if a hydrophobic support has been chosen, it must be taken into account that the reactivity of alkali metal hydroxides will be reduced because of poor diffusion into the support. Amides of electron-poor amines (e.g. anilides) can also be readily cleaved by nucleophiles [218].

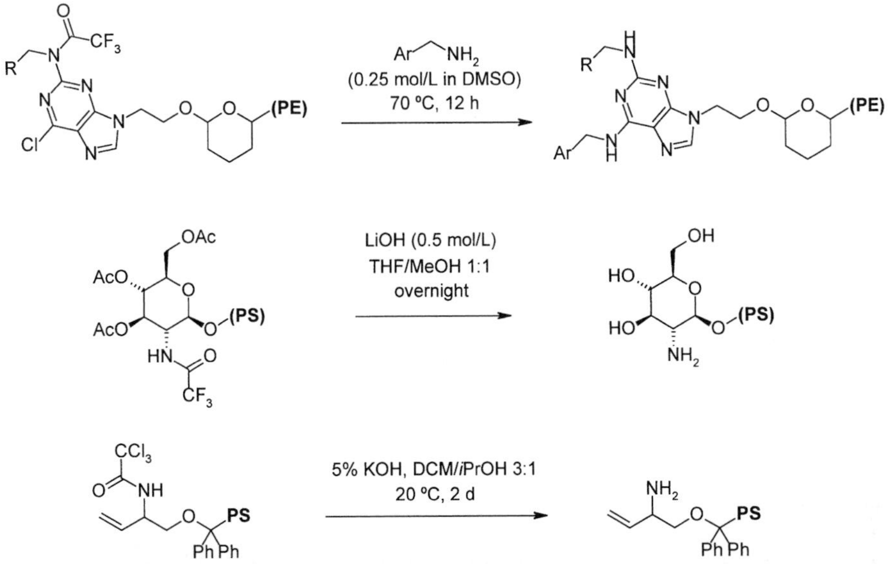

Fig. 10.10. Nucleophilic cleavage of support-bound amides [70,123,216].

10.1.10.3 Cyclic Imides

Phthalimide protection is stable towards acids and bases, but can be cleaved with strong nucleophiles, such as hydrazines or sulfides, or by reduction with sodium borohydride [170]. More sensitive towards nucleophilic attack than unsubstituted phthalimide is tetrachlorophthalimide [28].

Few examples of the conversion of support-bound amines into phthalimides have been reported. An example of the conversion of polystyrene-bound amines into the corresponding phthalimides by heating with other phthalimides (BuOH, 85 °C, 18 h) has been reported [28]. Support-bound phthalimides have also been prepared by N-alkylation of phthalimide derivatives [219] or by amidomethylation of cross-linked polystyrene with N-hydroxymethyl or N-chloromethylphthalimide [220,221].

Typical conditions for the removal of phthaloyl protection on cross-linked polystyrene include treatment of the resin with hydrazine hydrate [219,221], with methyl hydrazine [222], or with primary aliphatic amines [223] in DMF, EtOH, or solvent mixtures for several hours at room temperature or higher temperature [220,221,224]. Illustrative examples are sketched in Figure 10.11. It has been claimed that the hydrazinolysis of polystyrene-bound phthalimides proceeds more readily in DCM or DCE than in DMF [225].

Another cyclic imide which has been used as a protective group for primary amines is the dithiasuccinoyl group (Dts) [170,227]. This group is stable towards acids (e.g. during deprotection of Boc amines) but can be cleaved with thiols under basic conditions [2-mercaptoethanol (0.2 mol/L), NEt$_3$ (0.5 mol/L), DCM, 25 °C, 5 min [227]].

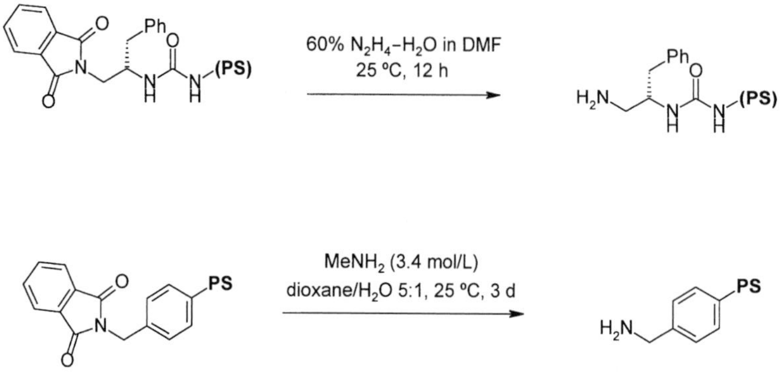

Fig. 10.11. Examples of nucleophilic cleavage of phthaloyl groups [223,226].

10.1.10.4 Enamines

Enamines derived from simple ketones and aliphatic amines are too acid-labile and nucleophilic for being useful as protective group for amines. Triacylmethanes, however, form less basic enamines, which are sufficiently stable to be suitable as amine protection.

Primary aliphatic and aromatic amines react reversibly with 2-acetyldimedone to yield non-nucleophilic enamines (1-[(4,4-dimethyl-2,6-dioxocyclohexylidene)ethyl] derivatives, 'Dde', Figure 10.12). Aliphatic amines react even at room temperature [200], whereas anilines generally require heating [228]. The resulting enamines are sufficiently stable towards TFA or piperidine to enable their use for efficient side-chain protection of lysine in solid-phase peptide synthesis [229,230]. A similar protective group, which is more stable under conditions of Fmoc group removal, is 2-(3-methylbutyryl)dimedone ('isovaleroyl'dimedone, ivDde [231]).

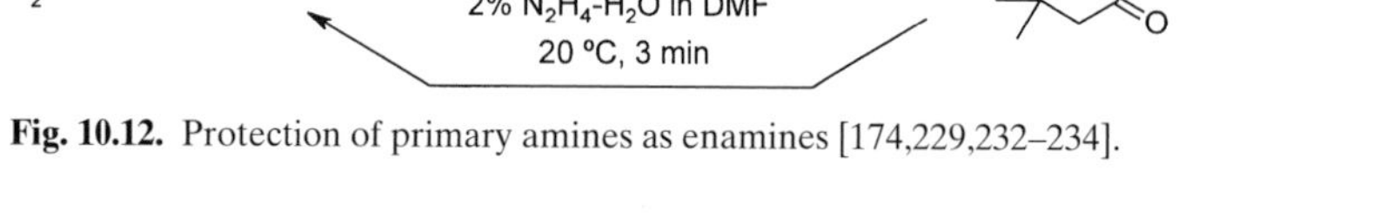

Fig. 10.12. Protection of primary amines as enamines [174,229,232–234].

Secondary amines do not generally form enamines with 2-acetyldimedone, and this protective group can therefore be used for selective protection of primary amines in the presence of secondary amines [174,232]. Support-bound 2-acetyldimedone has also been used as linker for amines [235] (see Section 3.6).

10.1.10.5 Imines

Several types of imine have been used as protective groups for amines in solution [170]. Most are stable towards bases, but can be hydrolyzed by acids.

Benzophenone-derived imines can be prepared by treating support-bound aliphatic primary amines with benzophenone imine [109,196], but not by treatment with benzophenone. Polystyrene-bound benzophenone imines of glycine are sufficiently C,H-acidic to enable C-alkylation with alkyl halides [196,236] or Michael acceptors [237], and have mainly been used for this purpose.

Deprotection of polystyrene-bound ketimines has been achieved by treatment with aqueous HCl or TFA in THF [196,236], or by benzophenone oxime formation with hydroxylamine hydrochloride [237,238] (Figure 10.13).

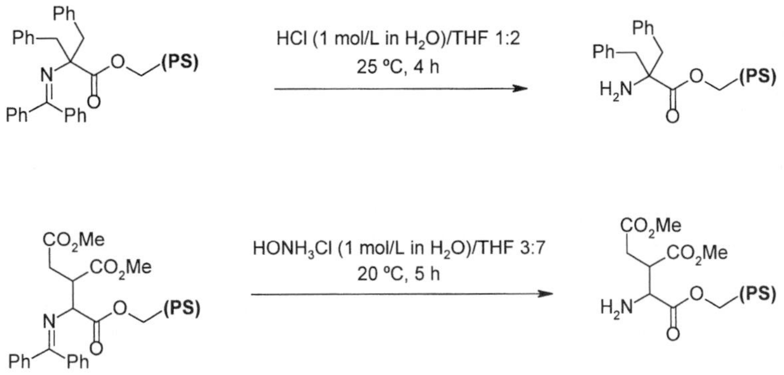

Fig. 10.13. Deprotection of polystyrene-bound ketimines [236,237]. (PS): Wang resin.

10.1.10.6 *N*-Alkyl and *N*-Aryl Derivatives

Amines can also be protected from unwanted acylation or from twofold alkylation by alkylation with a sterically demanding group, susceptible to selective removal after completion of the synthesis. The most common groups for this purpose are triphenylmethyl derivatives. Amines are readily tritylated by trityl chloride in the presence of a tertiary amine. Deprotection is achieved by treatment with dilute trifluoroacetic acid [137,239] (Figure 10.14).

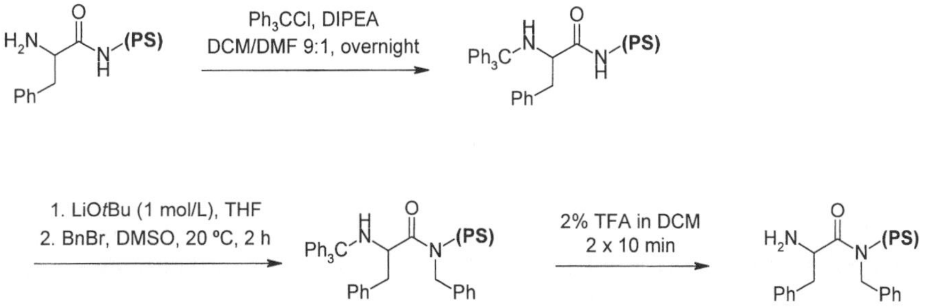

Fig. 10.14. Protection of primary amines by tritylation [136,137]. (PS): MBHA polystyrene.

Barlos and coworkers investigated the synthesis of peptides on cross-linked polystyrene with TFA-labile linkers, using N-trityl amino acid HOBt esters as building blocks [240]. These authors found that trityl amino acids, esterified with Wang resin, could be selectively detritylated with 2 % TFA in DCM (2 × 2 min), without significant cleavage of the Wang linker. More sensitive towards acidolytic cleavage than tritylamines are (4-methyltrityl)amines [233], cleavable in 2 min by 1 % TFA in DCM, and (4-methoxytrityl)amines, which can even be hydrolyzed with trichloroacetic acid (Figure 10.15 [241,242]).

The 4,4′-dimethoxybenzhydryl group has been used to protect primary amines from twofold methylation during reductive alkylation with formaldehyde (Figure 10.15 [56]). Removal of this protective group requires 50 % TFA and will therefore only be compatible with linkers of higher stability towards acids than the Wang linker.

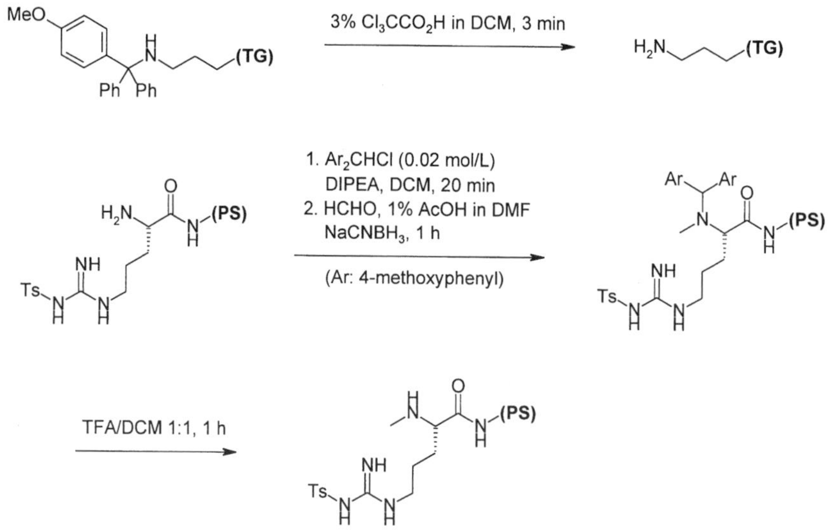

Fig. 10.15. Methoxytrityl and 4,4′-dimethoxybenzhydryl protection of amines [56,241]. (PS): MBHA polystyrene.

10.1.10.7 N-Sulfenyl and N-Sulfonyl Derivatives

The 2-nitrobenzenesulfenyl protective group (2-$(O_2N)C_6H_4$-S-NH-R), developed as amino group protection for α-amino acids [243,244], has been completely superseded by the Fmoc group. N-Sulfenylamines are prepared from sulfenyl chlorides ArSCl, and can be cleaved by acids, phosphines, or various nucleophiles [170,243].

Sulfonamides have found wider application as protective groups than sulfenamides (for the preparation of sulfonamides, see Section 8.4). Arenesulfonamides are generally very stable towards acidolytic or nucleophilic cleavage, unless substituted with electron-withdrawing groups. Thus, 2-nitrobenzene- and 2,4-dinitrobenzenesulfonamides derived from primary or secondary aliphatic amines can be selectively hydrolyzed under mild conditions with nucleophiles such as thiols or primary amines [245], even in the presence of carboxylic esters (Figure 10.16 [94,246–248]). *N*-(2-Nitroben-zenesulfonyl)amino acids have been used to prepare *N*-alkylated peptides on insoluble supports [246,249]. 4-Nitrobenzenesulfonamides are not well suited as protective group for amines, because cleavage with thiols can lead to the formation of 4-thioben-zenesulfonamides, which are no longer easy to hydrolyze [250].

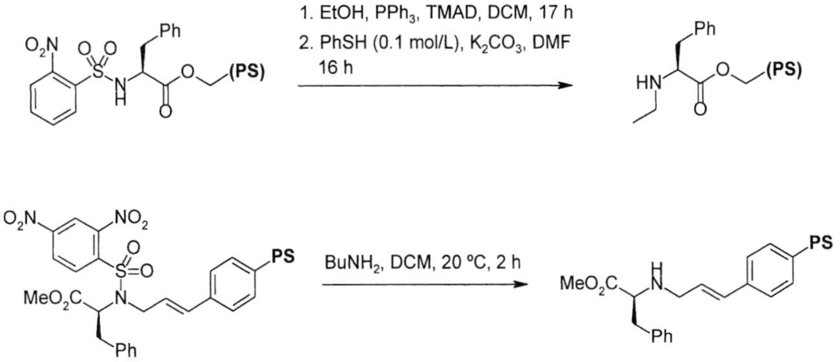

Fig. 10.16. Nucleophilic cleavage of polystyrene-bound sulfonamides [95,245,251]. (PS): Wang resin.

Sulfonamides prepared from 9-(chlorosulfonyl)anthracene and polystyrene-bound primary amines can be converted into amides by N-acylation of the sulfonamide (carboxylic acid anhydride, DMAP, pyridine, THF, 24 h) followed by nucleophilic desulfo-nylation with neat 1,3-propanedithiol/DIPEA [157] (Entry **4**, Table 10.13). An example of the use of sulfonamides as linkers for amines is given in Table 3.22.

10.2 Preparation of Quaternary Ammonium Salts

The quaternization of tertiary amines on cross-linked polystyrene has been investigated in detail. The most common substrates in these studies were N-dialkylated β-alanine derivatives, because after quaternization pure tertiary amines can be released from the support by treatment with a base (see Section 3.7).

The formation of charged molecules within hydrophobic supports does not proceed as smoothly as in polar solvents. High reaction temperatures are, for instance, required to quaternize polystyrene-bound phosphines [252,253] and the N-benzylation of pyridines with Merrifield resin also proceeds sluggishly [32]. Quaternization of tertiary amines on cross-linked polystyrene only proceeds well with reactive alkylating agents, such as methyl, allyl, propargyl, phenacyl, or benzyl halides or sulfonates (Table 10.16). Most reported examples were conducted at room temperature, to avoid premature Hoffmann elimination and cleavage from the support [54].

Table 10.16. Quaternization of support-bound tertiary amines.

Entry	Starting resin	Conditions	Product	Ref.
1		O₂N— benzyl-Br (0.29 mol/L), DMF, 20 °C, 18 h		[54] see also [84,85]
2		MeI (10 eq), DMF, 20 °C, 18 h		[54]
3		phenacyl bromide (2.4 mol/L), DMF, 20 °C, 18 h		[81]

10.3 Preparation of Hydrazines and Hydroxylamines

Few examples of the preparation of hydrazines or hydroxylamines on insoluble supports have been reported (Table 10.17). Hydrazines have been prepared by reduction of aromatic diazonium salts and by N-amination of support-bound amines (Entries **1** and **2**, Table 10.17). The second approach does not proceed cleanly when using 3-aryl-oxaziridines as the aminating reagents, because benzaldehydes are formed as byproducts, which undergo condensation with the support-bound amine to yield imines [254]. These were hydrolyzed by treatment with a hydrazine and the resulting amines were treated again with the oxaziridine to yield, after three repetitions, a sufficiently pure, Boc-protected hydrazine (Entry **2**, Table 10.17). Hydrazines have further been prepared by N-alkylation of free or protected hydrazines with resin-bound haloacetamides [255,256].

Support-bound hydroxylamines have been prepared by reaction of hydroxylamines with support-bound alkylating agents (Entries **4** and **5**, Table 10.17). Derivatives of hydroxylamine protected at nitrogen (e.g. *N*-hydroxyphthalimide [24,257,258] or *N*-Fmoc hydroxylamine [55]) have been O-alkylated with Wang resin or with trityl chloride resins to yield supports suitable for the preparation of hydroxamic acids (Section 3.4). The deprotection of support-bound *N*-phthaloyl hydroxylamines can be accomplished under conditions similar to those used for the hydrolysis of phthalimides, e.g. by treatment with hydrazine or methylhydrazine [259]. Polystyrene-bound hydroxylamines have also been prepared by addition of alkyl radicals to support-bound *O*-alkyloximes (Entry **6**, Table 10.17) and by reduction of oximes with borane (Entry **7**, Table 10.17).

Table 10.17. Solid-phase synthesis of hydrazines and hydroxylamines.

Entry	Starting resin	Conditions	Product	Ref.
1		SnCl$_2$, conc HCl, 20 °C, 2 h, then 60 °C, 1 h		[260]
2		(1 eq), DCM, 20 °C, 0.5 h; 3 repetitions		[254]
3		DIPEA, DMSO		[255]
4		HONH$_3$Cl (0.23 mol/L, 10 eq), DIPEA (10 eq), DMF, 25 °C, 3 h		[261]
5		BnONH$_2$, 2,6-lutidine, DCM, 0 °C		[262]
6		*i*PrI (7.1 eq), Bu$_3$SnH (2.1 eq), Et$_3$B (1.1 eq), DCM, 20 °C, 1 h		[263]
7		pyridine•BH$_3$ (10 eq), THF, 20 °C, 10 h		[264]

10.4 Preparation of Azides

Azides are convenient intermediates for the solid-phase preparation of primary amines (Section 10.1.8). Their preparation and handling on insoluble supports is significantly less hazardous than the handling of isolated organic azides. The most common strategy for preparing azides on insoluble supports is the nucleophilic substitution of halides or sulfonates (Table 10.18). Further suitable electrophiles are support-bound oxiranes, which can undergo ring-opening upon treatment with sodium azide to yield azido alcohols. This reaction can be catalyzed by enantiomerically pure chromium(III) complexes, and enantiomerically enriched azido alcohols have been prepared by use of this strategy (Entry **4**, Table 10.18). Resin-bound primary aliphatic

Table 10.18. Solid-phase synthesis of azides.

Entry	Starting resin	Conditions	Product	Ref.
1		NaN$_3$ (1 mol/L), DMSO, 60 °C, overnight		[265] see also [155] [266]
2		NaN$_3$ (1 mol/L), DMF, 45 °C, 24 h Ar: 4-(MeO)C$_6$H$_4$		[3]
3		NaN$_3$ (0.2 mol/L), NH$_4$Cl (0.2 mol/L), DMF, 100 °C, 2 h		[267]
4		Me$_3$SiN$_3$ (20 eq), [(salen*)CrN$_3$] (0.2 eq), Et$_2$O, 24 h salen*H$_2$: Ar:3,5-(*t*Bu)$_2$-2-(OH)C$_6$H$_2$	94% ee	[156]
5		PPh$_3$, DEAD, THF, (PhO)$_2$PON$_3$, overnight		[268] see also [160]

alcohols can be converted directly into azides by treatment with DPPA, DEAD, and a phosphine (Entry **5**, Table 10.18).

10.5 Preparation of Diazo Compounds

Diazocarbonyl compounds can be prepared on insoluble supports by diazo group transfer with sulfony azides or by diazotation of primary amines. Diazo group transfer with sulfonyl azides to 1,3-dicarbonyl compounds proceeds on cross-linked polystyrene as smoothly as in solution (Table 10.19). When 3-keto esters or amides are used as substrates for diazo group transfer, the resulting diazoketones can readily be deacylated with secondary amines to yield diazoacetic esters or amides (Entry **3**, Table 10.19). A less convenient method for the preparation of support-bound diazoacetic acid derivatives is the direct diazotation of glycine, because this reaction generally gives good yields only when conducted in a biphasic solvent mixture (dilute aqueous HCl or H$_2$SO$_4$/DCM [269]). Aryldiazomethanes have been prepared on cross-linked polystyrene by heating tosylhydrazones with aqueous base and by oxidation of hydra-

Table 10.19. Solid-phase synthesis of diazo compounds.

Entry	Starting resin	Conditions	Product	Ref.
1		TsN$_3$ (0.5 mol/L), DIPEA (1.4 mol/L), DMF, 20 °C, 1 h		[273] see also [274,275]
2		NaNO$_2$ (4 eq), HCl, DCM, H$_2$O, 25 °C, 4 h		[276]
3		pyrrolidine (3 mol/L), DMF, 20 °C, 2 h		[273]
4		KOH (2 eq), MeOH/THF 1:4, 90 °C, 7 min Ar: 2,4,6-(iPr)$_3$C$_6$H$_2$		[270]
5		AcOH, AcO$_2$H, TMG, I$_2$, DCM, 0 °C		[271,272]

zones (Table 10.19). These diazo compounds react quantitatively under mild conditions with carboxylic acids to yield the corresponding benzylic esters [270,271], and with alcohols in the presence of BF$_3$OEt$_2$ to yield benzyl ethers [272].

10.6 Preparation of Nitro Compounds

Cross-linked polystyrene can be nitrated directly with fuming nitric acid at low temperatures (−25 °C to 0 °C) [141,151,277]; polymers with up to one nitro group per arene result [151]. Partial nitration can be achieved with milder nitrating agents, such as acetyl nitrate [151]. Because direct nitrations are not compatible with most linkers (which are often acid- or oxidant-labile), nitro compounds are generally not prepared on supports but in solution, and then linked to the support.

References for Chapter 10

[1] Darling, G. D.; Fréchet, J. M. J. *J. Org. Chem.* **1986**, *51*, 2270–2276.
[2] Karoyan, P.; Triolo, A.; Nannicini, R.; Giannotti, D.; Altamura, M.; Chassaing, G.; Perrotta, E. *Tetrahedron Lett.* **1999**, *40*, 71–74.
[3] Lee, C. E.; Kick, E. K.; Ellman, J. A. *J. Am. Chem. Soc.* **1998**, *120*, 9735–9747.
[4] Kick, E. K.; Ellman, J. A. *J. Med. Chem.* **1995**, *38*, 1427–1430.
[5] Zhou, J.; Termin, A.; Wayland, M.; Tarby, C. M. *Tetrahedron Lett.* **1999**, *40*, 2729–2732.
[6] Barn, D. R.; Morphy, J. R.; Rees, D. C. *Tetrahedron Lett.* **1996**, *37*, 3213–3216.
[7] Virgilio, A. A.; Schürer, S. C.; Ellman, J. A. *Tetrahedron Lett.* **1996**, *37*, 6961–6964.
[8] Rölfing, K.; Thiel, M.; Künzer, H. *Synlett* **1996**, 1036–1038.
[9] Vojkovsky, T.; Weichsel, A.; Pátek, M. *J. Org. Chem.* **1998**, *63*, 3162–3163.
[10] Bhandari, A.; Jones, D. G.; Schullek, J. R.; Vo, K.; Schunk, C. A.; Tamanaha, L. L.; Chen, D.; Yuan, Z. Y.; Needels, M. C.; Gallop, M. A. *Bioorg. Med. Chem. Lett.* **1998**, *8*, 2303–2308.
[11] Vidal-Ferran, A.; Bampos, N.; Moyano, A.; Pericàs, M. A.; Riera, A.; Sanders, J. K. M. *J. Org. Chem.* **1998**, *63*, 6309–6318.
[12] Moberg, C.; Rákos, L. *Reactive Polymers* **1991**, *15*, 25–35.
[13] Wendeborn, S.; De Mesmaeker, A.; Brill, W. K. D. *Synlett* **1998**, 865–868.
[14] Brill, W. K. D.; De Mesmaeker, A.; Wendeborn, S. *Synlett* **1998**, 1085–1090.
[15] Fréchet, J. M. J.; Eichler, E. *Polym. Bull.* **1982**, *7*, 345–351.
[16] Berteina, S.; De Mesmaeker, A.; Wendeborn, S. *Synlett* **1999**, 1121–1123.
[17] Pei, Y. H.; Moos, W. H. *Tetrahedron Lett.* **1994**, *35*, 5825–5828.
[18] Tumelty, D.; Schwarz, M. K.; Needels, M. C. *Tetrahedron Lett.* **1998**, *39*, 7467–7470.
[19] Richter, L. S.; Zuckermann, R. N. *Bioorg. Med. Chem. Lett.* **1995**, *5*, 1159–1162.
[20] Goff, D. A.; Zuckermann, R. N. *J. Org. Chem.* **1995**, *60*, 5744–5745.
[21] Rano, T. A.; Chapman, K. T. *Tetrahedron Lett.* **1995**, *36*, 3789–3792.
[22] Goff, D. A.; Zuckermann, R. N. *J. Org. Chem.* **1995**, *60*, 5748–5749.
[23] Scott, B. O.; Siegmund, A. C.; Marlowe, C. K.; Pei, Y.; Spear, K. L. *Mol. Diversity* **1995**, *1*, 125–134.
[24] Floyd, C. D.; Lewis, C. N.; Patel, S. R.; Whittaker, M. *Tetrahedron Lett.* **1996**, *37*, 8045–8048.
[25] Byk, G.; Frederic, M.; Scherman, D. *Tetrahedron Lett.* **1997**, *38*, 3219–3222.
[26] Virgilio, A. A.; Bray, A. A.; Zhang, W.; Trinh, L.; Snyder, M.; Morrissey, M. M.; Ellman, J. A. *Tetrahedron* **1997**, *53*, 6635–6644.
[27] Révész, L.; Bonne, F.; Manning, U.; Zuber, J. F. *Bioorg. Med. Chem. Lett.* **1998**, *8*, 405–408.
[28] Stangier, P.; Hindsgaul, O. *Synlett* **1996**, 179–181.
[29] Hird, N. W.; Irie, K.; Nagai, K. *Tetrahedron Lett.* **1997**, *38*, 7111–7114.
[30] Conti, P.; Demont, D.; Cals, J.; Ottenheijm, H. C. J.; Leysen, D. *Tetrahedron Lett.* **1997**, *38*, 2915–2918.
[31] Adrian, F. M.; Altava, B.; Burguete, M. I.; Luis, S. V.; Salvador, R. V.; García-España, E. *Tetrahedron* **1998**, *54*, 3581–3588.
[32] Obika, S.; Nishiyama, T.; Tatematsu, S.; Nishimoto, M.; Miyashita, K.; Imanishi, T. *Heterocycles* **1998**, *49, P261–267*, 261–267.
[33] Ngu, K.; Patel, D. V. *Tetrahedron Lett.* **1997**, *38*, 973–976.
[34] Rich, D. H.; Gurwara, S. K. *J. Am. Chem. Soc.* **1975**, *97*, 1575–1579.
[35] Zuckermann, R. N.; Kerr, J. M.; Kent, S. B. H.; Moos, W. H. *J. Am. Chem. Soc.* **1992**, *114*, 10646–10647.
[36] Kiselyov, A. S.; Eisenberg, S.; Luo, Y. *Tetrahedron* **1998**, *54*, 10635–10640.
[37] Yuasa, H.; Kamata, Y.; Kurono, S.; Hashimoto, H. *Bioorg. Med. Chem. Lett.* **1998**, *8*, 2139–2144.
[38] Brown, D. S.; Revill, J. M.; Shute, R. E. *Tetrahedron Lett.* **1998**, *39*, 8533–8536.
[39] Bräse, S.; Enders, D.; Köbberling, J.; Avemaria, F. *Angew. Chem. Int. Ed. Engl.* **1998**, *37*, 3413–3415.
[40] Berteina, S.; Wendeborn, S.; De Mesmaeker, A. *Synlett* **1998**, 1231–1233.
[41] Marx, M. A.; Grillot, A. L.; Louer, C. T.; Beaver, K. A.; Bartlett, P. A. *J. Am. Chem. Soc.* **1997**, *119*, 6153–6167.
[42] Dankwardt, S. M.; Newman, S. R.; Krstenansky, J. L. *Tetrahedron Lett.* **1995**, *36*, 4923–4926.
[43] Bhalay, G.; Blaney, P.; Palmer, V. H.; Baxter, A. D. *Tetrahedron Lett.* **1997**, *38*, 8375–8378.
[44] Newlander, K. A.; Chenera, B.; Veber, D. F.; Yim, N. C. F.; Moore, M. L. *J. Org. Chem.* **1997**, *62*, 6726–6732.
[45] Raju, B.; Kogan, T. P. *Tetrahedron Lett.* **1997**, *38*, 4965–4968.
[46] Yamamoto, Y.; Ajito, K.; Ohtsuka, Y. *Chem. Lett.* **1998**, 379–380.

[47] Brill, W. K. D.; Schmidt, E.; Tommasi, R. A. *Synlett* **1998**, 906–908.
[48] Tommasi, R. A.; Nantermet, P. G.; Shapiro, M. J.; Chin, J.; Brill, W. K. D.; Ang, K. *Tetrahedron Lett.* **1998**, *39*, 5477–5480.
[49] Zhang, H. C.; Maryanoff, B. E. *J. Org. Chem.* **1997**, *62*, 1804–1809.
[50] Goff, D. *Tetrahedron Lett.* **1998**, *39*, 1477–1480.
[51] van Maarseveen, J. H.; den Hartog, J. A. J.; Engelen, V.; Finner, E.; Visser, G.; Kruse, C. G. *Tetrahedron Lett.* **1996**, *37*, 8249–8252.
[52] Richter, H.; Walk, T.; Höltzel, A.; Jung, G. *J. Org. Chem.* **1999**, *64*, 1362–1365.
[53] Wang, Y.; Huang, T. N. *Tetrahedron Lett.* **1999**, *40*, 5837–5840.
[54] Brown, A. R.; Rees, D. C.; Rankovic, Z.; Morphy, J. R. *J. Am. Chem. Soc.* **1997**, *119*, 3288–3295.
[55] Mellor, S. L.; McGuire, C.; Chan, W. C. *Tetrahedron Lett.* **1997**, *38*, 3311–3314.
[56] Kaljuste, K.; Undén, A. *Int. J. Pept. Prot. Res.* **1993**, *42*, 118–124.
[57] Purandare, A. V.; Poss, M. A. *Tetrahedron Lett.* **1998**, *39*, 935–938.
[58] Chambers, S. L.; Ronald, R.; Hanesworth, J. M.; Kinder, D. H.; Harding, J. W. *Peptides* **1997**, *18*, 505–512.
[59] Alewood, P. F.; Brinkworth, R. I.; Dancer, R. J.; Garnham, B.; Jones, A.; Kent, S. B. H. *Tetrahedron Lett.* **1992**, *33*, 977–980.
[60] Flegelová, Z.; Pátek, M. P. *J. Org. Chem.* **1996**, *61*, 6735–6738.
[61] Green, J. *J. Org. Chem.* **1995**, *60*, 4287–4290.
[62] Pan, P. C.; Sun, C. M. *Bioorg. Med. Chem. Lett.* **1999**, *9*, 1537–1540.
[63] Ward, Y. D.; Farina, V. *Tetrahedron Lett.* **1996**, *37*, 6993–6996.
[64] Willoughby, C. A.; Chapman, K. T. *Tetrahedron Lett.* **1996**, *37*, 7181–7184.
[65] Shapiro, M. J.; Kumaravel, G.; Petter, R. C.; Beveridge, R. *Tetrahedron Lett.* **1996**, *37*, 4671–4674.
[66] Schwarz, M. K.; Tumelty, D.; Gallop, M. A. *Tetrahedron Lett.* **1998**, *39*, 8397–8400.
[67] MacDonald, A. A.; DeWitt, S. H.; Hogan, E. M.; Ramage, R. *Tetrahedron Lett.* **1996**, *37*, 4815–4818.
[68] Phillips, G. B.; Wei, G. P. *Tetrahedron Lett.* **1996**, *37*, 4887–4890.
[69] Mohan, R.; Yun, W. Y.; Buckman, B. O.; Liang, A.; Trinh, L.; Morrissey, M. M. *Bioorg. Med. Chem. Lett.* **1998**, *8*, 1877–1882.
[70] Norman, T. C.; Gray, N. S.; Koh, J. T.; Schultz, P. G. *J. Am. Chem. Soc.* **1996**, *118*, 7430–7431.
[71] Gray, N. S.; Kwon, S.; Schultz, P. G. *Tetrahedron Lett.* **1997**, *38*, 1161–1164.
[72] Deleuze, H.; Sherrington, D. C. *J. Chem. Soc. Perkin Trans. 2* **1995**, 2217–2221.
[73] Cobb, J. M.; Fiorini, M. T.; Goddard, C. R.; Theoclitou, M. E.; Abell, C. *Tetrahedron Lett.* **1999**, *40*, 1045–1048.
[74] Tumelty, D.; Schwarz, M. K.; Cao, K.; Needels, M. C. *Tetrahedron Lett.* **1999**, *40*, 6185–6188.
[75] Gordeev, M. F.; Luehr, G. W.; Hui, H. C.; Gordon, E. M.; Patel, D. V. *Tetrahedron* **1998**, *54*, 15879–15890.
[76] Brummond, K. M.; Lu, J. *J. Org. Chem.* **1999**, *64*, 1723–1726.
[77] Feng, Y.; Wang, Z.; Jin, S.; Burgess, K. *J. Am. Chem. Soc.* **1998**, *120*, 10768–10769.
[78] Fotsch, C.; Kumaravel, G.; Sharma, S. K.; Wu, A. D.; Gounarides, J. S.; Nirmala, N. R.; Petter, R. C. *Bioorg. Med. Chem. Lett.* **1999**, *9*, 2125–2130.
[79] Hamper, B. C.; Kolodziej, S. A.; Scates, A. M.; Smith, R. G.; Cortez, E. *J. Org. Chem.* **1998**, *63*, 708–718.
[80] Kolodziej, S. A.; Hamper, B. C. *Tetrahedron Lett.* **1996**, *37*, 5277–5280.
[81] Ouyang, X. H.; Armstrong, R. W.; Murphy, M. M. *J. Org. Chem.* **1998**, *63*, 1027–1032.
[82] Zaragoza, F. unpublished results.
[83] Prien, O.; Rölfing, K.; Thiel, M.; Künzer, H. *Synlett* **1997**, 325–326.
[84] Kroll, F. E. K.; Morphy, R.; Rees, D.; Gani, D. *Tetrahedron Lett.* **1997**, *38*, 8573–8576.
[85] Heinonen, P.; Lönnberg, H. *Tetrahedron Lett.* **1997**, *38*, 8569–8572.
[86] Garibay, P.; Nielsen, J.; Høeg-Jensen, T. *Tetrahedron Lett.* **1998**, *39*, 2207–2210.
[87] Kobayashi, S.; Aoki, Y. *Tetrahedron Lett.* **1998**, *39*, 7345–7348.
[88] Boyd, E. A.; Chan, W. C.; Loh, V. M. *Tetrahedron Lett.* **1996**, *37*, 1647–1650.
[89] Gordeev, M. F.; Gordon, E. M.; Patel, D. V. *J. Org. Chem.* **1997**, *62*, 8177–8181.
[90] Szardenings, A. K.; Burkoth, T. S.; Look, G. C.; Campbell, D. A. *J. Org. Chem.* **1996**, *61*, 6720–6722.
[91] Look, G. C.; Murphy, M. M.; Campbell, D. A.; Gallop, M. A. *Tetrahedron Lett.* **1995**, *36*, 2937–2940.
[92] Ruhland, B.; Bhandari, A.; Gordon, E. M.; Gallop, M. A. *J. Am. Chem. Soc.* **1996**, *118*, 253–254.
[93] Katritzky, A. R.; Xie, L.; Zhang, G.; Griffith, M.; Watson, K.; Kiely, J. S. *Tetrahedron Lett.* **1997**, *38*, 7011–7014.
[94] Yang, L.; Chiu, K. *Tetrahedron Lett.* **1997**, *38*, 7307–7310.

[95] Scicinski, J. J.; Barker, R. D.; Murray, P. J.; Jarvie, E. M. *Bioorg. Med. Chem. Lett.* **1998**, *8*, 3609–3614.

[96] Tourwé, D.; Piron, J.; Defreyn, P.; Van Binst, G. *Tetrahedron Lett.* **1993**, *34*, 5499–5502.

[97] Szardenings, A. K.; Burkoth, T. S.; Lu, H. H.; Tien, D. W.; Campbell, D. A. *Tetrahedron* **1997**, *53*, 6573–6593.

[98] Kim, S. W.; Ahn, S. Y.; Koh, J. S.; Lee, J. H.; Ro, S.; Cho, H. Y. *Tetrahedron Lett.* **1997**, *38*, 4603–4606.

[99] Matthews, J.; Rivero, R. A. *J. Org. Chem.* **1997**, *62*, 6090–6092.

[100] Matthews, J.; Rivero, R. A. *J. Org. Chem.* **1998**, *63*, 4808–4810.

[101] Chan, W. C.; Mellor, S. L. *J. Chem. Soc. Chem. Commun.* **1995**, 1475–1477.

[102] Brown, E. G.; Nuss, J. M. *Tetrahedron Lett.* **1997**, *38*, 8457–8460.

[103] Devraj, R.; Cushman, M. *J. Org. Chem.* **1996**, *61*, 9368–9373.

[104] Arumugam, V.; Routledge, A.; Abell, C.; Balasubramanian, S. *Tetrahedron Lett.* **1997**, *38*, 6473–6476.

[105] Meyer, J. P.; Davis, P.; Lee, K. B.; Porreca, F.; Yamamura, H. I.; Hruby, V. J. *J. Med. Chem.* **1995**, *38*, 3462–3468.

[106] Blackburn, C.; Pingali, A.; Kehoe, T.; Herman, L. W.; Wang, H. Q.; Kates, S. A. *Bioorg. Med. Chem. Lett.* **1997**, *7*, 823–826.

[107] Nawaz, M. K.; Arumugam, V.; Balasubramanian, S. *Tetrahedron Lett.* **1996**, *37*, 4819–4822.

[108] Koh, J. S.; Ellman, J. A. *J. Org. Chem.* **1996**, *61*, 4494–4495.

[109] Lee, S. H.; Chung, S. H.; Lee, Y. S. *Tetrahedron Lett.* **1998**, *39*, 9469–9472.

[110] Steele, J.; Gordon, D. W. *Bioorg. Med. Chem. Lett.* **1995**, *5*, 47–50.

[111] Fivush, A. M.; Willson, T. M. *Tetrahedron Lett.* **1997**, *38*, 7151–7154.

[112] Hone, N. D.; Davies, S. G.; Devereux, N. J.; Taylor, S. L.; Baxter, A. D. *Tetrahedron Lett.* **1998**, *39*, 897–900.

[113] Boojamra, C. G.; Burow, K. M.; Thompson, L. A.; Ellman, J. A. *J. Org. Chem.* **1997**, *62*, 1240–1256.

[114] Swayze, E. E. *Tetrahedron Lett.* **1997**, *38*, 8643–8646.

[115] Swayze, E. E. *Tetrahedron Lett.* **1997**, *38*, 8465–8468.

[116] Sarantakis, D.; Bicksler, J. J. *Tetrahedron Lett.* **1997**, *38*, 7325–7328.

[117] Estep, K. G.; Neipp, C. E.; Stramiello, L. M. S.; Adam, M. D.; Allen, M. P.; Robinson, S.; Roskamp, E. J. *J. Org. Chem.* **1998**, *63*, 5300–5301.

[118] Ley, S. V.; Mynett, D. M.; Koot, W. J. *Synlett* **1995**, 1017–1020.

[119] Bray, A. M.; Chiefari, D. S.; Valerio, R. M.; Maeji, N. J. *Tetrahedron Lett.* **1995**, *36*, 5081–5084.

[120] Bui, C. T.; Bray, A. M.; Ercole, F.; Pham, Y.; Rasoul, F. A.; Maeji, N. J. *Tetrahedron Lett.* **1999**, *40*, 3471–3474.

[121] Breitenbucher, J. G.; Hui, H. C. *Tetrahedron Lett.* **1998**, *39*, 8207–8210.

[122] Matsueda, G. R.; Stewart, J. M. *Peptides* **1981**, *2*, 45–50.

[123] Schuster, M.; Pernerstorfer, J.; Blechert, S. *Angew. Chem. Int. Ed. Engl.* **1996**, *35*, 1979–1980.

[124] Chenera, B.; Finkelstein, J. A.; Veber, D. F. *J. Am. Chem. Soc.* **1995**, *117*, 11999–12000.

[125] McNally, J. J.; Youngman, M. A.; Dax, S. L. *Tetrahedron Lett.* **1998**, *39*, 967–970.

[126] Jönsson, D.; Molin, H.; Undén, A. *Tetrahedron Lett.* **1998**, *39*, 1059–1062.

[127] O'Donnell, M. J.; Delgado, F.; Drew, M. D.; Pottorf, R. S.; Zhou, C. Y.; Scott, W. L. *Tetrahedron Lett.* **1999**, *40*, 5831–5835.

[128] Kobayashi, S.; Hachiya, I.; Suzuki, S.; Moriwaki, M. *Tetrahedron Lett.* **1996**, *37*, 2809–2812.

[129] Kobayashi, S.; Moriwaki, M. *Tetrahedron Lett.* **1997**, *38*, 4251–4254.

[130] Dyatkin, A. B.; Rivero, R. A. *Tetrahedron Lett.* **1998**, *39*, 3647–3650.

[131] Youngman, M. A.; Dax, S. L. *Tetrahedron Lett.* **1997**, *38*, 6347–6350.

[132] Zhang, H. C.; Brumfield, K. K.; Jaroskova, L.; Maryanoff, B. E. *Tetrahedron Lett.* **1998**, *39*, 4449–4452.

[133] Ho, C. Y.; Kukla, M. J. *Tetrahedron Lett.* **1997**, *38*, 2799–2802.

[134] Paikoff, S. J.; Wilson, T. E.; Cho, C. Y.; Schultz, P. G. *Tetrahedron Lett.* **1996**, *37*, 5653–5656.

[135] Karigiannis, G.; Mamos, P.; Balayiannis, G.; Katsoulis, I.; Papaioannou, D. *Tetrahedron Lett.* **1998**, *39*, 5117–5120.

[136] Nefzi, A.; Ostresh, J. M.; Houghten, R. A. *Tetrahedron* **1999**, *55*, 335–344.

[137] Nefzi, A.; Ostresh, J. M.; Meyer, J. P.; Houghten, R. A. *Tetrahedron Lett.* **1997**, *38*, 931–934.

[138] Ostresh, J. M.; Schoner, C. C.; Hamashin, V. T.; Nefzi, A.; Meyer, J. P.; Houghten, R. A. *J. Org. Chem.* **1998**, *63*, 8622–8623.

[139] Hall, D. G.; Laplante, C.; Manku, S.; Nagendran, J. *J. Org. Chem.* **1999**, *64*, 698–699.

[140] Liu, G. C.; Ellman, J. A. *J. Org. Chem.* **1995**, *60*, 7712–7713.

[141] Dowling, L. M.; Stark, G. R. *Biochemistry* **1969**, *8*, 4728–4734.

[142] Meyers, H. V.; Dilley, G. J.; Durgin, T. L.; Powers, T. S.; Winssinger, N. A.; Zhu, H.; Pavia, M. R. *Mol. Diversity* **1995**, *1*, 13–20.
[143] Kiselyov, A. S.; Armstrong, R. W. *Tetrahedron Lett.* **1997**, *38*, 6163–6166.
[144] Morales, G. A.; Corbett, J. W.; DeGrado, W. F. *J. Org. Chem.* **1998**, *63*, 1172–1177.
[145] Zaragoza, F.; Stephensen, H. *J. Org. Chem.* **1999**, *64*, 2555–2557.
[146] Hari, A.; Miller, B. L. *Tetrahedron Lett.* **1999**, *40*, 245–248.
[147] Stephensen, H.; Zaragoza, F. *Tetrahedron Lett.* **1999**, *40*, 5799–5802.
[148] Ruhland, T.; Künzer, H. *Tetrahedron Lett.* **1996**, *37*, 2757–2760.
[149] Beebe, X.; Chiappari, C. L.; Olmstead, M. M.; Kurth, M. J.; Schore, N. E. *J. Org. Chem.* **1995**, *60*, 4204–4212.
[150] Hughes, I. *Tetrahedron Lett.* **1996**, *37*, 7595–7598.
[151] Seliger, H. *Makromol. Chem.* **1973**, *169*, 83–93.
[152] Meldal, M.; Juliano, M. A.; Jansson, A. M. *Tetrahedron Lett.* **1997**, *38*, 2531–2534.
[153] Long, D. D.; Smith, M. D.; Marquess, D. G.; Claridge, T. D. W.; Fleet, G. W. J. *Tetrahedron Lett.* **1998**, *39*, 9293–9296.
[154] Kim, J. M.; Bi, Y. Z.; Paikoff, S. J.; Schultz, P. G. *Tetrahedron Lett.* **1996**, *37*, 5305–5308.
[155] Tortolani, D. R.; Biller, S. A. *Tetrahedron Lett.* **1996**, *37*, 5687–5690.
[156] Annis, D. A.; Helluin, O.; Jacobsen, E. N. *Angew. Chem. Int. Ed. Engl.* **1998**, *37*, 1907–1909.
[157] Savin, K. A.; Woo, J. C. G.; Danishefsky, S. J. *J. Org. Chem.* **1999**, *64*, 4183–4186.
[158] Tang, Z. L.; Pelletier, J. C. *Tetrahedron Lett.* **1998**, *39*, 4773–4776.
[159] Tremblay, M. R.; Poirier, D. *Tetrahedron Lett.* **1999**, *40*, 1277–1280.
[160] Nicolaou, K. C.; Winssinger, N.; Vourloumis, D.; Ohshima, T.; Kim, S.; Pfefferkorn, J.; Xu, J. Y.; Li, T. *J. Am. Chem. Soc.* **1998**, *120*, 10814–10826.
[161] Osborn, N. J.; Robinson, J. A. *Tetrahedron* **1993**, *49*, 2873–2884.
[162] Liang, R.; Yan, L.; Loebach, J.; Ge, M.; Uozumi, Y.; Sekanina, K.; Horan, N.; Gildersleeve, J.; Thompson, C.; Smith, A.; Biswas, K.; Still, W. C.; Kahne, D. *Science* **1996**, *274*, 1520–1522.
[163] Kowalski, J.; Lipton, M. A. *Tetrahedron Lett.* **1996**, *37*, 5839–5840.
[164] Blettner, C.; Bradley, M. *Tetrahedron Lett.* **1994**, *35*, 467–470.
[165] Richter, L. S.; Andersen, S. *Tetrahedron Lett.* **1998**, *39*, 8747–8750.
[166] Shao, H.; Colucci, M.; Tong, S.; Zhang, H.; Castelhano, A. L. *Tetrahedron Lett.* **1998**, *39*, 7235–7238.
[167] Fletcher, M. D.; Campbell, M. M. *Chem. Rev.* **1998**, *98*, 763–795.
[168] Rivier, J. E.; Jiang, G. C.; Koerber, S. C.; Porter, J.; Simon, L.; Craig, A. G.; Hoeger, C. A. *Proc. Natl. Acad. Sci. USA* **1996**, *93*, 2031–2036.
[169] Pessi, A.; Pinori, M.; Verdini, A. S.; Viscomi, G. C. *J. Chem. Soc. Chem. Commun.* **1983**, 195–197.
[170] Green, T. W.; Wuts, P. G. M. *Protective Groups in Organic Synthesis*; John Wiley & Sons: New York, **1991**.
[171] Jones, J. *The Chemical Synthesis of Peptides*; Oxford University Press: Oxford, **1994**.
[172] Hulme, C.; Peng, J.; Morton, G.; Salvino, J. M.; Herpin, T.; Labaudiniere, R. *Tetrahedron Lett.* **1998**, *39*, 7227–7230.
[173] Flynn, D. L.; Zelle, R. E.; Grieco, P. A. *J. Org. Chem.* **1983**, *48*, 2424–2426.
[174] Nash, I. A.; Bycroft, B. W.; Chan, W. C. *Tetrahedron Lett.* **1996**, *37*, 2625–2628.
[175] Yamada, M.; Miyajima, T.; Horikawa, H. *Tetrahedron Lett.* **1998**, *39*, 289–292.
[176] Furlán, R. L. E.; Mata, E. G. *Tetrahedron Lett.* **1998**, *39*, 6421–6422.
[177] Panek, J. S.; Zhu, B. *Tetrahedron Lett.* **1996**, *37*, 8151–8154.
[178] Sieber, P.; Iselin, B. *Helv. Chim. Acta* **1968**, *51*, 614–622.
[179] Houghten, R. A.; Beckman, A.; Ostresh, J. M. *Int. J. Pept. Prot. Res.* **1986**, *27*, 653–658.
[180] Naharissoa, H.; Sarrade, V.; Follet, M.; Calas, B. *Pept. Res.* **1992**, *5*, 293–299.
[181] Bayer, E.; Dengler, M.; Hemmasi, B. *Int. J. Pept. Prot. Res.* **1985**, *25*, 178–186.
[182] Andreatta, R. H.; Rink, H. *Helv. Chim. Acta* **1973**, *56*, 1205–1218.
[183] Allin, S. M.; Shuttleworth, S. J. *Tetrahedron Lett.* **1996**, *37*, 8023–8026.
[184] Kaiser, E.; Picart, F.; Kubiak, T.; Tam, J. P.; Merrifield, R. B. *J. Org. Chem.* **1993**, *58*, 5167–5175.
[185] Zhang, A. J.; Russell, D. H.; Zhu, J. P.; Burgess, K. *Tetrahedron Lett.* **1998**, *39*, 7439–7442.
[186] Plunkett, M. J.; Ellman, J. A. *J. Org. Chem.* **1997**, *62*, 2885–2893.
[187] Kuisle, O.; Quiñoá, E.; Riguera, R. *Tetrahedron Lett.* **1999**, *40*, 1203–1206.
[188] Birr, C.; Lochinger, W.; Stahnke, G.; Lang, P. *Liebigs Ann. Chem.* **1972**, *763*, 162–172.
[189] Kalbacher, H.; Voelter, W. *Angew. Chem. Int. Ed. Engl.* **1978**, *17*, 944–945.
[190] Tun-Kyi, A.; Schwyzer, R. *Helv. Chim. Acta* **1976**, *59*, 1642–1646.
[191] Merrifield, R. B. *J. Am. Chem. Soc.* **1963**, *85*, 2149–2154.
[192] Wen, J. J.; Spatola, A. F. *J. Pept. Res.* **1997**, *49*, 3–14.
[193] Carpino, L. A.; Han, G. Y. *J. Org. Chem.* **1972**, *37*, 3404–3409.

[194] Fields, G. B.; Noble, R. L. *Int. J. Pept. Prot. Res.* **1990**, *35*, 161–214.
[195] Ede, N. J.; Ang, K. H.; James, I. W.; Bray, A. M. *Tetrahedron Lett.* **1996**, *37*, 9097–9100.
[196] O'Donnell, M. J.; Zhou, C. Y.; Scott, W. L. *J. Am. Chem. Soc.* **1996**, *118*, 6070–6071.
[197] Atherton, E.; Logan, C. J.; Sheppard, R. C. *J. Chem. Soc. Perkin Trans. 1* **1981**, 538–546.
[198] Chang, C. D.; Waki, M.; Ahmad, M.; Meienhofer, J.; Lundell, E. O.; Haug, J. D. *Int. J. Pept. Prot. Res.* **1980**, *15*, 59–66.
[199] Wade, J. D.; Bedford, J.; Sheppard, R. C.; Tregear, G. W. *Pept. Res.* **1991**, *4*, 194–199.
[200] Page, P.; Bradley, M.; Walters, I.; Teague, S. *J. Org. Chem.* **1999**, *64*, 794–799.
[201] Ueki, M.; Kai, K.; Amemiya, M.; Horino, H.; Oyamada, H. *J. Chem. Soc. Chem. Commun.* **1988**, 414–415.
[202] Ueki, M.; Amemiya, M. *Tetrahedron Lett.* **1987**, *28*, 6617–6620.
[203] Campbell, D. A.; Bermak, J. C. *J. Am. Chem. Soc.* **1994**, *116*, 6039–6040.
[204] Lapatsanis, L.; Milias, G.; Froussios, K.; Kolovos, M. *Synthesis* **1983**, 671–673.
[205] Paquet, A. *Can. J. Chem.* **1982**, *60*, 976–980.
[206] Ramage, R.; Blake, A. J.; Florence, M. R.; Gray, T.; Raphy, G.; Roach, P. L. *Tetrahedron* **1991**, *47*, 8001–8024.
[207] Carpino, L. A.; Ismail, M.; Truran, G. A.; Mansour, E. M. E.; Iguchi, S.; Ionescu, D.; El-Faham, A.; Riemer, C.; Warrass, R. *J. Org. Chem.* **1999**, *64*, 4324–4338.
[208] Samukov, V. V.; Sabirov, A. N.; Pozdnyakov, P. I. *Tetrahedron Lett.* **1994**, *35*, 7821–7824.
[209] Campbell, D. A.; Bermak, J. C.; Burkoth, T. S.; Patel, D. V. *J. Am. Chem. Soc.* **1995**, *117*, 5381–5382.
[210] Loffet, A.; Zhang, H. X. *Int. J. Pept. Prot. Res.* **1993**, *42*, 346–351.
[211] Kunz, H.; Unverzagt, C. *Angew. Chem. Int. Ed. Engl.* **1984**, *23*, 436–437.
[212] Girdwood, J. A.; Shute, R. E. *Chem. Commun.* **1997**, 2307–2308.
[213] Valerio, R. M.; Bray, A. M.; Stewart, K. M. *Int. J. Pept. Prot. Res.* **1996**, *47*, 414–418.
[214] Kim, S. W.; Hong, C. Y.; Lee, K.; Lee, E. J.; Koh, J. S. *Bioorg. Med. Chem. Lett.* **1998**, *8*, 735–738.
[215] Kim, S. W.; Hong, C. Y.; Koh, J. S.; Lee, E. J.; Lee, K. *Mol. Diversity* **1998**, *3*, 133–136.
[216] Silva, D. J.; Wang, H.; Allanson, N. M.; Jain, R. K.; Sofia, M. J. *J. Org. Chem.* **1999**, *64*, 5926–5929.
[217] Sofia, M. J.; Allanson, N.; Hatzenbuhler, N. T.; Jain, R.; Kakarla, R.; Kogan, N.; Liang, R.; Liu, D.; Silva, D. J.; Wang, H.; Gange, D.; Anderson, J.; Chen, A.; Chi, F.; Dulina, R.; Huang, B.; Kamau, M.; Wang, C.; Baizman, E.; Branstrom, A.; Bristol, N.; Goldman, R.; Han, K.; Longley, C.; Midha, S.; Axelrod, H. R. *J. Med. Chem.* **1999**, *42*, 3193–3198.
[218] Collini, M. D.; Ellingboe, J. W. *Tetrahedron Lett.* **1997**, *38*, 7963–7966.
[219] Aronov, A. M.; Gelb, M. H. *Tetrahedron Lett.* **1998**, *39*, 4947–4950.
[220] Mitchell, A. R.; Kent, S. B. H.; Erickson, B. W.; Merrifield, R. B. *Tetrahedron Lett.* **1976**, 3795–3798.
[221] Zikos, C. C.; Ferderigos, N. G. *Tetrahedron Lett.* **1995**, *36*, 3741–3744.
[222] Wess, G.; Bock, K.; Kleine, H.; Kurz, M.; Guba, W.; Hemmerle, H.; Lopez-Calle, E.; Baringhaus, K. H.; Glombik, H.; Enhsen, A.; Kramer, W. *Angew. Chem. Int. Ed. Engl.* **1996**, *35*, 2222–2224.
[223] Adams, J. H.; Cook, R. M.; Hudson, D.; Jammalamadaka, V.; Lyttle, M. H.; Songster, M. F. *J. Org. Chem.* **1998**, *63*, 3706–3716.
[224] Wells, N. J.; Basso, A.; Bradley, M. *Biopolymers* **1998**, *47*, 381–396.
[225] Nielsen, J.; Rasmussen, P. H. *Tetrahedron Lett.* **1996**, *37*, 3351–3354.
[226] Burgess, K.; Ibarzo, J.; Linthicum, D. S.; Russell, D. H.; Shin, H.; Shitangkoon, A.; Totani, R.; Zhang, A. J. *J. Am. Chem. Soc.* **1997**, *119*, 1556–1564.
[227] Barany, G.; Merrifield, R. B. *J. Am. Chem. Soc.* **1977**, *99*, 7363–7365.
[228] Chan, W. C.; Bycroft, B. W.; Evans, D. J.; White, P. D. *J. Chem. Soc. Chem. Commun.* **1995**, 2209–2210.
[229] Lelièvre, D.; Daguet, D.; Brack, A. *Tetrahedron Lett.* **1995**, *36*, 9317–9320.
[230] Bloomberg, G. B.; Askin, D.; Gargaro, A. R.; Tanner, M. J. A. *Tetrahedron Lett.* **1993**, *34*, 4709–4712.
[231] Chhabra, S. R.; Hothi, B.; Evans, D. J.; White, P. D.; Bycroft, B. W.; Chan, W. C. *Tetrahedron Lett.* **1998**, *39*, 1603–1606.
[232] Kellam, B.; Bycroft, B. W.; Chhabra, S. R. *Tetrahedron Lett.* **1997**, *38*, 4849–4852.
[233] Novabiochem Catalog and Peptide Synthesis Handbook, Läufelfingen, **1999**.
[234] Bycroft, B. W.; Chan, W. C.; Chhabra, S. R.; Hone, N. D. *J. Chem. Soc. Chem. Commun.* **1993**, 778–779.
[235] Chhabra, S. R.; Khan, A. N.; Bycroft, B. W. *Tetrahedron Lett.* **1998**, *39*, 3585–3588.
[236] Griffith, D. L.; O'Donnell, M. J.; Pottorf, R. S.; Scott, W. L.; Porco, J. A. *Tetrahedron Lett.* **1997**, *38*, 8821–8824.
[237] Domínguez, E.; O'Donnell, M. J.; Scott, W. L. *Tetrahedron Lett.* **1998**, *39*, 2167–2170.

[238] O'Donnell, M. J.; Lugar, C. W.; Pottorf, R. S.; Zhou, C. Y.; Scott, W. L.; Cwi, C. L. *Tetrahedron Lett.* **1997**, *38*, 7163–7166.
[239] Kaljuste, K.; Undén, A. *Tetrahedron Lett.* **1996**, *37*, 3031–3034.
[240] Barlos, K.; Gatos, D.; Kallitsis, J.; Papaioannou, D.; Sotiriu, P.; Schäfer, W. *Liebigs Ann. Chem.* **1987**, 1031–1035.
[241] Will, D. W.; Langner, D.; Knolle, J.; Uhlmann, E. *Tetrahedron* **1995**, *51*, 12069–12082.
[242] van der Laan, A. C.; Meeuwenoord, N. J.; Kuyl-Yeheskiely, E.; Oosting, R. S.; Brands, R.; van Boom, J. H. *Recl. Trav. Chim. Pays-Bas* **1995**, *114*, 295–297.
[243] Kessler, W.; Iselin, B. *Helv. Chim. Acta* **1966**, *49*, 1330–1344.
[244] Matsueda, R.; Maruyama, H.; Kitazawa, E.; Takahagi, H.; Mukaiyama, T. *J. Am. Chem. Soc.* **1975**, *97*, 2573–2575.
[245] Piscopio, A. D.; Miller, J. F.; Koch, K. *Tetrahedron* **1999**, *55*, 8189–8198.
[246] Miller, S. C.; Scanlan, T. S. *J. Am. Chem. Soc.* **1998**, *120*, 2690–2691.
[247] Hidai, Y.; Kan, T.; Fukuyama, T. *Tetrahedron Lett.* **1999**, *40*, 4711–4714.
[248] Wipf, P.; Henninger, T. C. *J. Org. Chem.* **1997**, *62*, 1586–1587.
[249] Raman, P.; Stokes, S. S.; Angell, Y. M.; Flentke, G. R.; Rich, D. H. *J. Org. Chem.* **1998**, *63*, 5734–5735.
[250] Wuts, P. G. M.; Northuis, J. M. *Tetrahedron Lett.* **1998**, *39*, 3889–3890.
[251] Piscopio, A. D.; Miller, J. F.; Koch, K. *Tetrahedron Lett.* **1997**, *38*, 7143–7146.
[252] Leznoff, C. C.; Fyles, T. M.; Weatherston, J. *Can. J. Chem.* **1977**, *55*, 1143–1153.
[253] Nicolaou, K. C.; Winssinger, N.; Pastor, J.; Ninkovic, S.; Sarabia, F.; He, Y.; Vourloumis, D.; Yang, Z.; Li, T.; Giannakakou, P.; Hamel, E. *Nature* **1997**, *387*, 268–272.
[254] Klinguer, C.; Melnyk, O.; Loing, E.; Gras-Masse, H. *Tetrahedron Lett.* **1996**, *37*, 7259–7262.
[255] Lohse, A.; Jensen, K. B.; Bols, M. *Tetrahedron Lett.* **1999**, *40*, 3033–3036.
[256] Wilson, L. J.; Li, M.; Portlock, D. E. *Tetrahedron Lett.* **1998**, *39*, 5135–5138.
[257] Richter, L. S.; Desai, M. C. *Tetrahedron Lett.* **1997**, *38*, 321–322.
[258] Bauer, U.; Ho, W. B.; Koskinen, A. M. P. *Tetrahedron Lett.* **1997**, *38*, 7233–7236.
[259] Morvan, F.; Sanghvi, Y. S.; Perbost, M.; Vasseur, J. J.; Bellon, L. *J. Am. Chem. Soc.* **1996**, *118*, 255–256.
[260] Semenov, A. N.; Gordeev, K. Y. *Int. J. Pept. Prot. Res.* **1995**, *45*, 303–304.
[261] Haap, W. J.; Kaiser, D.; Walk, T. B.; Jung, G. *Tetrahedron* **1998**, *54*, 3705–3724.
[262] Hanessian, S.; Yang, R. Y. *Tetrahedron Lett.* **1996**, *37*, 5835–5838.
[263] Miyabe H; Fujishima Y; Naito T *J. Org. Chem.* **1999**, *64*, 2174–2175.
[264] Kobayashi, S.; Akiyama, R. *Tetrahedron Lett.* **1998**, *39*, 9211–9214.
[265] Drewry, D. H.; Gerritz, S. W.; Linn, J. A. *Tetrahedron Lett.* **1997**, *38*, 3377–3380.
[266] Schneider, S. E.; Bishop, P. A.; Salazar, M. A.; Bishop, O. A.; Anslyn, E. V. *Tetrahedron* **1998**, *54*, 15063–15086.
[267] Le Hetet, C.; David, M.; Carreaux, F.; Carboni, B.; Sauleau, A. *Tetrahedron Lett.* **1997**, *38*, 5153–5156.
[268] Hanessian, S.; Xie, F. *Tetrahedron Lett.* **1998**, *39*, 737–740.
[269] Searle, N. E. *Org. Synth.* **1963**, *Coll. Vol. IV*, 424–426.
[270] Bhalay, G.; Dunstan, A. R. *Tetrahedron Lett.* **1998**, *39*, 7803–7806.
[271] Chapman, P. H.; Walker, D. *J. Chem. Soc. Chem. Commun.* **1975**, 690–691.
[272] Mergler, M.; Dick, F.; Gosteli, J.; Nyfeler, R. *Tetrahedron Lett.* **1999**, *40*, 4663–4664.
[273] Zaragoza, F.; Petersen, S. V. *Tetrahedron* **1996**, *52*, 5999–6002.
[274] Gowravaram, M. R.; Gallop, M. A. *Tetrahedron Lett.* **1997**, *38*, 6973–6976.
[275] Whitehouse, D. L.; Nelson, K. H.; Savinov, S. N.; Löwe, R. S.; Austin, D. J. *Bioorg. Med. Chem.* **1998**, *6*, 1273–1282.
[276] Cano, M.; Camps, F.; Joglar, J. *Tetrahedron Lett.* **1998**, *39*, 9819–9822.
[277] Kumari, K. A.; Sreekumar, K. *Polymer* **1996**, *37*, 171–176.

11 Preparation of Phosphorus Compounds

Compounds containing phosphorus can both be valuable synthetic intermediates or target compounds of solid-phase synthesis. Important synthetic intermediates include phosphonium salts and phosphorus ylides, which are key intermediates in carbonyl olefinations. Their preparation is discussed in Section 5.2.2.1. The preparation of oligonucleotides, being the most important phosphorus-containing target molecules in solid-phase synthesis, is treated in Section 16.2. In this section the preparation of phosphonic acid derivatives is discussed.

11.1 Phosphonic Acid Derivatives

H-Phosphonates react with aldehydes and imines to yield α-(hydroxyalkyl)- or α-(aminoalkyl)phosphonates, respectively. These reactions can also be conducted on cross-linked polystyrene. In Figure 11.1 a sequence is outlined in which polystyrene-bound *H*-phosphonates are treated with imines and aldehydes. The variant in which a support-bound imine is converted into an α-(aminoalkyl)phosphonate has also been reported [1].

The formation of dialkyl phosphonates by coupling of monoalkyl phosphonates with support-bound alcohols is discussed in Section 16.2.3. Some additional examples, not related to the synthesis of oligonucleotides, are sketched in Figure 11.2.

Support-bound, enantiomerically pure alcohols can be converted into phosphonates by Mitsunobu esterification, which results in complete inversion at the stereogenic center. This strategy has been used to prepare peptidyl phosphonates on solid phase; these are interesting transition state analogs with potential utility as peptidase inhibitors (Figure 11.2 [4,5]).

Serine or threonine derivatives can be converted into phosphonates by direct phosphonylation with an activated monoalkyl phosphonate [6] or by treatment with phosphonamidites $RP(OR)NR_2$ in the presence of tetrazole, followed by oxidation [7].

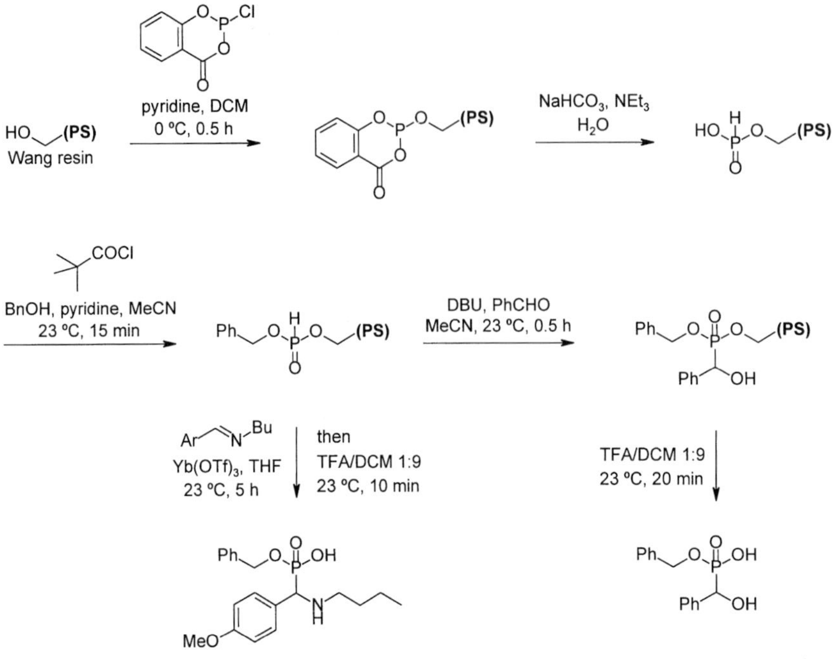

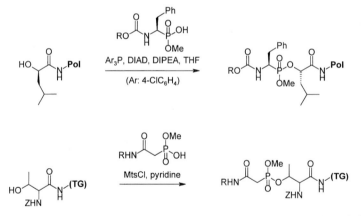

Fig. 11.1. Preparation of α-(hydroxyalkyl)- and α-(aminoalkyl)phosphonates on insoluble supports [2,3].

Fig. 11.2. Preparation of phosphonates by phosphonylation of support-bound alcohols [4–6].

References for Chapter 11

[1] Boyd, E. A.; Chan, W. C.; Loh, V. M. *Tetrahedron Lett.* **1996**, *37*, 1647–1650.
[2] Cao, X. D.; Mjalli, A. M. M. *Tetrahedron Lett.* **1996**, *37*, 6073–6076.
[3] Zhang, C. Z.; Mjalli, A. M. M. *Tetrahedron Lett.* **1996**, *37*, 5457–5460.
[4] Campbell, D. A.; Bermak, J. C. *J. Am. Chem. Soc.* **1994**, *116*, 6039–6040.
[5] Campbell, D. A.; Bermak, J. C.; Burkoth, T. S.; Patel, D. V. *J. Am. Chem. Soc.* **1995**, *117*, 5381–5382.
[6] Johnson, C. R.; Zhang, B. R. *Tetrahedron Lett.* **1995**, *36*, 9253–9256.
[7] Wijkmans, J. C. H. M.; Meeuwenoord, N. J.; Bloemhoff, W.; van der Marel, G. A.; van Boom, J. H. *Tetrahedron* **1996**, *52*, 2103–2112.

12 Preparation of Aldehydes and Ketones

Aldehydes and ketones are usually prepared on insoluble supports by acylation of arenes, C,H-acidic compounds, or organometallic reagents. Alcohols or other substrates can also be converted into carbonyl compounds by oxidation (Figure 12.1). Linkers which enable the generation of aldehydes and ketones upon cleavage from a support are treated in Section 3.14.

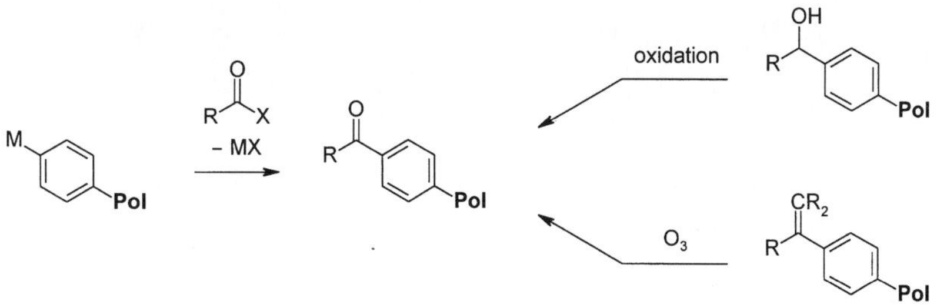

Fig. 12.1. Preparation of aldehydes and ketones on insoluble supports. M: H, metal; X: leaving group.

12.1 Preparation of Aldehydes and Ketones by C-Acylation

C-Acylation is one of the most powerful reactions for the generation of C–C bonds. As starting materials can be combined either highly reactive carbon nucleophiles (e.g. Grignard reagents) with unreactive acylating agents (e.g. Weinreb amides) or unreactive carbon nucleophiles (e.g. polystyrene) with strong acylating agents (e.g. acyl halides + AlCl$_3$). Most of these strategies have also been used with success for the preparation of carbonyl compounds on insoluble supports.

Cross-linked polystyrene can be acylated with aliphatic and aromatic acyl halides in the presence of AlCl$_3$ (Friedel–Crafts acylation, Table 12.1). This reaction has mainly been used for the functionalization of polystyrene-based supports, but not for the modification of support-bound substrates. Suitable solvents for Friedel–Crafts acylations of cross-linked polystyrene are tetrachloroethylene [1], DCE [2], CS$_2$ [3,4], nitrobenzene [5,6], and CCl$_4$ [7]. As in the bromination of polystyrene, Friedel–Crafts acylations at high temperatures (e.g. DCE, 83 °C, 15 min [2]) can lead to partial dealkylation of phenyl groups and yield a soluble polymer.

Table 12.1. Preparation of aldehydes and ketones by C-acylation of support-bound carbon nucleophiles.

Entry	Starting resin	Conditions	Product	Ref.
1		*i*PrCOCl, AlCl$_3$, CS$_2$, 46 °C, 4 h		[9] see also [2]
2		BrCHMeCOBr, AlCl$_3$, DCM, 0–20 °C, 21 h		[5]
3		ArCOCl, Pd$_2$(dba)$_3$, K$_2$CO$_3$, DIPEA, THF, 1 h Ar: 4-(MeO)C$_6$H$_4$		[10]
4		PhCO$_2$Et (1 mol/L, 30 eq), NaH, DMA, 90 °C, 1 h		[11]
5		*t*BuOCH(NMe$_2$)$_2$, THF, 20 °C; then HCl (2 mol/L in H$_2$O)/THF 1:2 (Rink amide linker)		[12] see also [13]
6		PhCO$_2$H (10 eq), NEt$_3$ (20 eq), (EtO)$_2$(O)PCN (10 eq), DMF, 0–20 °C, 15 h		[14]
7		(PhCO)$_2$O, NEt$_3$, DMF, 20 °C, overnight		[15]
8		MgCl$_2$, NEt$_3$, PhMe, 20 °C, 12 h		[16]
9		BuLi (5 eq), THF, −30 °C, 4 h; then DMA (10 eq), 20 °C, 2 h		[17]
10		BuLi (5 eq), THF, −30 °C, 4 h; then DMF (10 eq), 20 °C, 2 h		[17]

Table 12.2. Preparation of aldehydes and ketones from support-bound acylating agents.

Entry	Starting resin	Conditions	Product	Ref.
1		LiAlH$_4$ (1.5 eq), THF, 0 °C, 25 min		[18] see also [19]
2		MeMgCl (5 eq), THF, overnight		[20]
3		BnMgBr (0.2 mol/L, 4.2 eq), THF, 4 °C, 8 h		[21] see also [22]
4		PhCdCl, C$_6$H$_6$, 20 °C, 24 h		[23]

Aryl ketones can also be prepared by C-acylation of support-bound aryl stannanes with acyl halides (Entry **3**, Table 12.1). The reaction conditions are mild and suitable for selective chemical transformations of polystyrene-bound intermediates.

C-Acylations of C,H-acidic compounds have also been realized on insoluble supports. The few examples reported include the C-acylation of support-bound ester enolates with acyl halides [8], Claisen condensations of polystyrene-bound ketones with benzoic acid esters, the C-acylation of nitriles with acyl cyanides or anhydrides, and the C-acylation of phosphonates with acyl halides (Entries **4–8**, Table 12.1). The α-formylation of support-bound arylacetonitriles can be accomplished in two steps by α-aminomethylenation followed by hydrolysis of the intermediate enamine (Entry **5**, Table 12.1).

Polystyrene-bound Weinreb and related amides react with organolithium or Grignard reagents to yield ketones (Table 12.2). Amides have also been used for this purpose, without significant formation of tertiary alcohols. Aldehydes can be prepared on solid phase by reduction of *N,O*-dialkylhydroxamates with LiAlH$_4$ (Entry **1**, Table 12.2) and the successful conversion of polystyrene-bound acyl chlorides into ketones by treatment with organocadmium compounds has also been reported (Entry **4**, Table 12.2).

12.2 Preparation of Aldehydes and Ketones by Oxidation

Substrates suitable for oxidative conversion into carbonyl compounds are alkenes, primary or secondary alcohols, and benzyl halides. Polystyrene-bound alkenes have been converted into aldehydes (with loss of one carbon atom) by ozonolysis, followed by reductive cleavage of the intermediate ozonide (Entry **1**, Table 12.3). Olefins can be transformed into ketones by Wacker oxidation (Entry **2**, Table 12.3), but this reaction does not seem to proceed cleanly on polymeric supports. Janda and coworkers were able to oxidize styrenes bound to macroporous polystyrene to the corresponding acetophenones, but reported that the reaction did not proceed on PEG in homogeneous phase [24].

Polystyrene-bound benzaldehyde can be prepared directly from Merrifield resin by treatment of the latter with DMSO and a base at high temperatures (Entry **3**, Table 12.3). These reaction conditions are, however, not suitable for sensitive intermediates.

The oxidation of alcohols to carbonyl compounds proceeds under mild conditions, which are compatible with a number of linkers and additional functional groups. Suitable oxidants are NMO in the presence of catalytic amounts of TPAP, sulfur trioxide/pyridine, oxalyl chloride/DMSO (Swern oxidation), the Dess–Martin periodinane, $CrO_2(OtBu)_2$, and pyridinium dichromate (Table 12.3). 1,2-Diols and aminoalcohols can be oxidatively cleaved with sodium periodate to yield aldehydes (Entry **13**, Table 12.3).

12.3 Miscellaneous Preparations of Aldehydes and Ketones

Other solid-phase preparations of carbonyl compounds include the hydrolysis of acetals (Table 12.4), inter- [44] and intramolecular Pauson–Khand reactions, and the α-alkylation and α-arylation of other ketones. Tietze reported the generation of acetoacetyl dianions on cross-linked polystyrene and their selective alkylation at C-4 (Entry **4**, Table 12.4). Use of weaker bases resulted in single or twofold alkylation at C-2 [45].

Leznoff and coworkers reported the preparation of 1,2-diketones (in low yield) from polystyrene-bound benzaldehydes via the benzoin condensation and simultaneous oxidation [46]. The 1,4-addition of support-bound organometallic compounds to enones and the oxidative degradation of benzyl cyanides have also been reported (Entries **7** and **10**, Table 12.4).

Table 12.3. Preparation of aldehydes and ketones by oxidation.

Entry	Starting resin	Conditions	Product	Ref.
1		O_3, DCM, −78 °C, then PPh$_3$, 20 °C, 16 h		[25] see also [26]
2		PdCl$_2$ (1 eq), CuCl$_2$ (3 eq), H$_2$O, air, overnight (macroporous PS)		[24]
3		DMSO, NaHCO$_3$, 155 °C, 6 h		[27-29] see also [7]
4		NMO (0.3 mol/L, 10 eq), TPAP (0.2 eq), 20 °C, 1.5 h		[30] see also [31]
5		SO$_3$•pyridine (0.56 mol/L), NEt$_3$/DMSO 1:1.8, 20 °C, 3 h		[32] see also [33]
6		(Dess–Martin periodinane; 1.5 eq), DCM, 25 °C, 6 h		[34]
7		Dess–Martin periodinane (10 eq), DCM, 20 °C, 18 h		[33]
8		(COCl)$_2$ (4 eq), DMSO (8 eq), NEt$_3$ (13 eq), −78 °C to −25 °C		[35] see also [36]
9		CrO$_2$Cl$_2$ (0.17 mol/L, 5 eq), tBuOH (10 eq), pyridine (15 eq), DCM, 20 °C, 28 h (formation of CrO$_2$(OtBu)$_2$) Ar: 4-(MeO)C$_6$H$_4$		[37] see also [38]
10		SO$_3$•pyridine (0.5 mol/L), NEt$_3$/DMSO 1:15, overnight, repeat once (oxidation did not proceed on pure PS)		[39] see also [40]

Table 12.3. continued.

Entry	Starting resin	Conditions	Product	Ref.
11		Dess–Martin periodinane (3 eq), DCM, 40 °C, 2 h		[41]
12		pyridinium dichromate (0.2 mol/L), DMF, 37 °C (Rink amide linker)		[42]
13		NaIO$_4$ (0.23 mol/L, 10 eq), H$_2$O/NaH$_2$PO$_4$-buffer (pH 7), 3 h (cross-linked PEG)		[43]

Table 12.4. Miscellaneous preparations of aldehydes and ketones.

Entry	Starting resin	Conditions	Product	Ref.
1		TsOH, NMP/Me₂CO 2:1, 50 °C, 3 × 17 h		[47]
2		NMO (0.22 mol/L), DCM, 2 h		[48]
3		TBAF (1 mol/L, 10 eq), THF, 20 °C, 2 h, then EtI (44 eq), 20 °C, 2 h		[11]
4		1. LDA (6 eq), THF, 0 °C, 1 h 2. *i*PrI (5 eq), THF, 0–25 °C, 12 h		[49]
5		(3 eq), Pd(PPh₃)₄ (0.1 eq), THF, 20 °C, 1 h		[45]
6		2,4-pentanedione (0.7 mol/L, 17 eq), DMF/DBU 10:1, 20 °C, 16 h		[50]
7		(2-thienyl)Cu(CN)Li, THF, −20 °C, then 2-cyclohexenone		[51]
8		(5 eq), THF, −78 °C to 20 °C		[52] see also [53]
9		MeOTf (10 eq), DCM, 20 °C, 20 min		[54] see also [55]
10		Ph CN (0.8 mol/L, 32 eq), KN(SiMe₃)₂ (4 eq), air, DMF, 20 °C, 18 h		[50]

References for Chapter 12

[1] Chapman, P. H.; Walker, D. *J. Chem. Soc. Chem. Commun.* **1975**, 690–691.
[2] Matsueda, G. R.; Stewart, J. M. *Peptides* **1981**, *2*, 45–50.
[3] Fréchet, J. M. J.; Haque, K. E. *Tetrahedron Lett.* **1975**, 3055–3056.
[4] Kobayashi, S.; Moriwaki, M. *Tetrahedron Lett.* **1997**, *38*, 4251–4254.
[5] Mizoguchi, T.; Shigezane, K.; Takamura, N. *Chem. Pharm. Bull.* **1970**, *18*, 1465–1474.
[6] Ajayaghosh, A.; Pillai, V. N. R. *Tetrahedron* **1988**, *44*, 6661–6666.
[7] Kumari, K. A.; Sreekumar, K. *Polymer* **1996**, *37*, 171–176.
[8] Patchornik, A.; Kraus, M. A. *J. Am. Chem. Soc.* **1970**, *92*, 7587–7589.
[9] Ren, Q.; Huang, W.; Ho, P. *Reactive Polymers* **1989**, *11*, 237–244.
[10] Plunkett, M. J.; Ellman, J. A. *J. Org. Chem.* **1997**, *62*, 2885–2893.
[11] Marzinzik, A. L.; Felder, E. R. *Tetrahedron Lett.* **1996**, *37*, 1003–1006.
[12] Wilson, R. D.; Watson, S. P.; Richards, S. A. *Tetrahedron Lett.* **1998**, *39*, 2827–2830.
[13] MacDonald, A. A.; DeWitt, S. H.; Hogan, E. M.; Ramage, R. *Tetrahedron Lett.* **1996**, *37*, 4815–4818.
[14] Sim, M. M.; Lee, C. L.; Ganesan, A. *Tetrahedron Lett.* **1998**, *39*, 2195–2198.
[15] Sim, M. M.; Lee, C. L.; Ganesan, A. *Tetrahedron Lett.* **1998**, *39*, 6399–6402.
[16] Kim, D. Y.; Suh, K. H. *Synthetic Commun.* **1999**, *29*, 1271–1275.
[17] Li, Z.; Ganesan, A. *Synlett* **1998**, 405–406.
[18] Gosselin, F.; Van Betsbrugge, J.; Hatam, M.; Lubell, W. D. *J. Org. Chem.* **1999**, *64*, 2486–2493.
[19] Paris, M.; Douat, C.; Heitz A; Gibbons, W.; Martinez, J.; Fehrentz, J. A. *Tetrahedron Lett.* **1999**, *40*, 5179–5182.
[20] Kim, S. W.; Bauer, S. M.; Armstrong, R. W. *Tetrahedron Lett.* **1998**, *39*, 6993–6996.
[21] Lee, C. E.; Kick, E. K.; Ellman, J. A. *J. Am. Chem. Soc.* **1998**, *120*, 9735–9747.
[22] Wallace, O. B. *Tetrahedron Lett.* **1997**, *38*, 4939–4942.
[23] Leznoff, C. C.; Yedidia, V. *Can. J. Chem.* **1980**, *58*, 287–290.
[24] Hori, M.; Gravert, D. J.; Wentworth, P.; Janda, K. D. *Bioorg. Med. Chem. Lett.* **1998**, *8*, 2363–2368.
[25] Gennari, C.; Ceccarelli, S.; Piarulli, U.; Aboutayab, K.; Donghi, M.; Paterson, I. *Tetrahedron* **1998**, *54*, 14999–15016.
[26] Sylvain, C.; Wagner, A.; Mioskowski, C. *Tetrahedron Lett.* **1997**, *38*, 1043–1044.
[27] Fréchet, J. M.; Schuerch, C. *J. Am. Chem. Soc.* **1971**, *93*, 492–496.
[28] Sheng, Q.; Stöver, H. D. H. *Macromolecules* **1997**, *30*, 6712–6714.
[29] Beebe, X.; Schore, N. E.; Kurth, M. J. *J. Org. Chem.* **1995**, *60*, 4196–4203.
[30] Yan, B.; Sun, Q.; Wareing, J. R.; Jewell, C. F. *J. Org. Chem.* **1996**, *61*, 8765–8770.
[31] Li, W.; Yan, B. *J. Org. Chem.* **1998**, *63*, 4092–4097.
[32] Chen, C.; Randall, L. A. A.; Miller, R. B.; Jones, A. D.; Kurth, M. J. *Tetrahedron* **1997**, *53*, 6595–6609.
[33] Reggelin, M.; Brenig, V.; Welcker, R. *Tetrahedron Lett.* **1998**, *39*, 4801–4804.
[34] Nicolaou, K. C.; Pastor, J.; Winssinger, N.; Murphy, F. *J. Am. Chem. Soc.* **1998**, *120*, 5132–5133.
[35] Nicolaou, K. C.; Winssinger, N.; Pastor, J.; Ninkovic, S.; Sarabia, F.; He, Y.; Vourloumis, D.; Yang, Z.; Li, T.; Giannakakou, P.; Hamel, E. *Nature* **1997**, *387*, 268–272.
[36] Marx, M. A.; Grillot, A. L.; Louer, C. T.; Beaver, K. A.; Bartlett, P. A. *J. Am. Chem. Soc.* **1997**, *119*, 6153–6167.
[37] Miller, P. C.; Owen, T. J.; Molyneaux, J. M.; Curtis, J. M.; Jones, C. R. *J. Comb. Chem.* **1999**, *1*, 223–234.
[38] Leznoff, C. C.; Fyles, T. M.; Weatherston, J. *Can. J. Chem.* **1977**, *55*, 1143–1153.
[39] Page, P.; Bradley, M.; Walters, I.; Teague, S. *J. Org. Chem.* **1999**, *64*, 794–799.
[40] Furth, P. S.; Reitman, M. S.; Cook, A. F. *Tetrahedron Lett.* **1997**, *38*, 5403–5406.
[41] Thompson, L. A.; Moore, F. L.; Moon, Y. C.; Ellman, J. A. *J. Org. Chem.* **1998**, *63*, 2066–2067.
[42] Bray, A. M.; Chiefari, D. S.; Valerio, R. M.; Maeji, N. J. *Tetrahedron Lett.* **1995**, *36*, 5081–5084.
[43] Rademann, J.; Meldal, M.; Bock, K. *Chem. Eur. J.* **1999**, *5*, 1218–1225.
[44] Spitzer, J. L.; Kurth, M. J.; Schore, N. E. *Tetrahedron* **1997**, *53*, 6791–6808.
[45] Tietze, L. F.; Hippe, T.; Steinmetz, A. *Chem. Commun.* **1998**, 793–794.
[46] Leznoff, C. C.; Wong, J. Y. *Can. J. Chem.* **1973**, *51*, 3756–3764.
[47] Veerman, J. J. N.; van Maarseveen, J. H.; Visser, G. M.; Kruse, C. G.; Schoemaker, H. E.; Hiemstra, H.; Rutjes, F. P. J. T. *Eur. J. Org. Chem.* **1998**, 2583–2589.
[48] Bolton, G. L.; Hodges, J. C.; Rubin, J. R. *Tetrahedron* **1997**, *53*, 6611–6634.
[49] Tietze, L. F.; Steinmetz, A. *Synlett* **1996**, 667–668.
[50] Stephensen, H.; Zaragoza, F. *Tetrahedron Lett.* **1999**, *40*, 5799–5802.

[51] Kondo, Y.; Komine, T.; Fujinami, M.; Uchiyama, M.; Sakamoto, T. *J. Comb. Chem.* **1999**, *1*, 123–126.
[52] Ley, S. V.; Mynett, D. M.; Koot, W. J. *Synlett* **1995**, 1017–1020.
[53] Gutke, H.-J.; Spitzner, D. *Tetrahedron* **1999**, *55*, 3931–3936.
[54] Lee, H. B.; Balasubramanian, S. *J. Org. Chem.* **1999**, *64*, 3454–3460.
[55] Routledge, A.; Abell, C.; Balasubramanian, S. *Tetrahedron Lett.* **1997**, *38*, 1227–1230.

13 Preparation of Carboxylic Acid Derivatives

13.1 Preparation of Amides

The acylation of amines on insoluble supports is one of the most thoroughly investigated reactions in solid-phase synthesis. The continuous optimization of peptide bond formation in recent decades has led to protocols which enable racemization-free, quantitative acylations of support-bound peptides with protected amino acids. In recent years the range of amides available by solid-phase synthesis has expanded significantly to include, e.g., N-alkylated peptides and anilides. New strategies of the preparation of amides, such as C-carbamoylations and the Ugi reaction, have, furthermore, been realized with success on insoluble supports.

13.1.1 Preparation of Amides by Acylation of Amines with Isolated Acylating Agents

Amides are usually prepared by treating an amine with an acylating agent (Figure 13.1). Carboxylic acids are not suitable acylating agents for this purpose, because amines are sufficiently basic to form stable salts with acids. These salts can, however, be converted into amides by treatment with a dehydrating reagent (e.g. a carbodiimide) or by strong heating [1]. Alternatively, carboxylic acids can be converted into acylating agents ('activation'), such as HOBt-esters or symmetric anhydrides, and then treated with an amine (Figure 13.1). Acylating agents such as symmetric anhydrides or acyl halides are sufficiently reactive to acylate ammonium salts, and can therefore be used to acylate support-bound amines in the presence of excess carboxylic acid.

Many different acylating agents have been evaluated for solid-phase peptide synthesis. Ideally, the activated amino acid derivatives should be chemically and stereochemically stable, highly soluble in DCM, and lead to a fast and complete acylation of all types of amine. Optimization of these parameters led to the activated acid derivatives used today in solid-phase synthesis. Figure 13.2 shows a selection of the most common types of acylating agent used for the solid-phase synthesis of amides; these are ordered according to their approximate reactivity towards amines. It should be noted, however, that the rate of N-acylations depends on the solvent, and can be

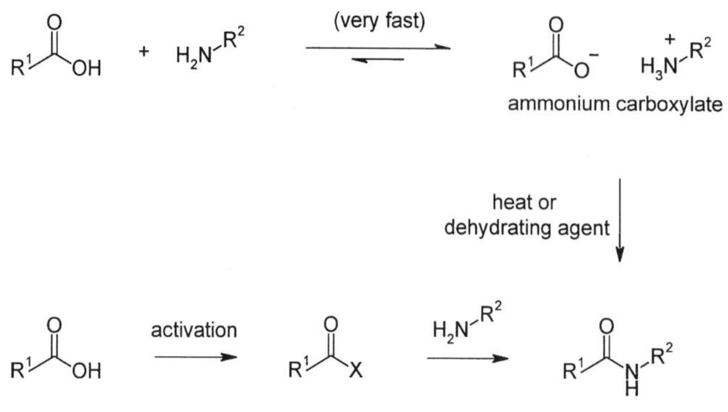

Fig. 13.1. Acylation of amines with carboxylic acids. X: see Fig. 13.2.

effectively increased by various catalysts (e.g. DMAP, HOBt [2], or HOAt [3,4]). For this reason the ranking given in Figure 13.2 should be considered as a rough guide only. Comparative studies of various acylating agents for solid-phase peptide synthesis have been reported [2,5,6].

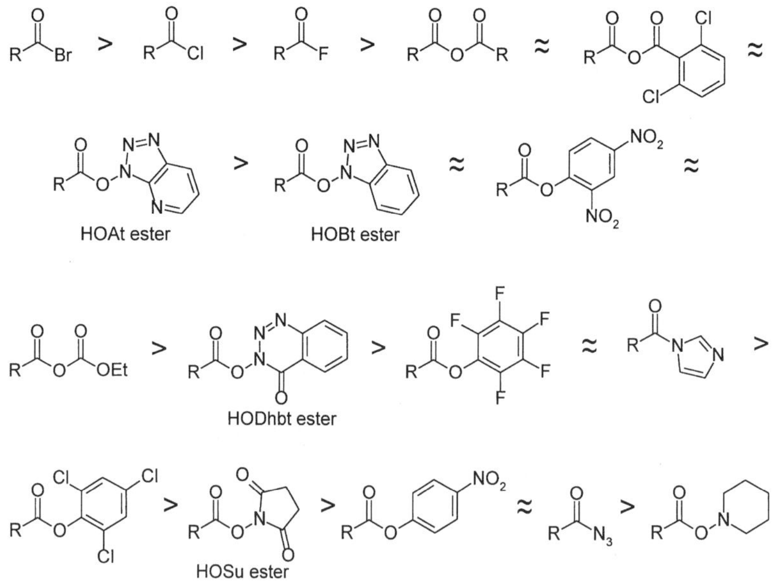

Fig. 13.2. Representative acylating agents for the solid-phase synthesis of amides, ranked according to their approximate reactivity towards amines [3,6–12].

The reagents sketched in Figure 13.2 are stable and can be prepared either in solution or on insoluble supports. Activated Boc- or Fmoc-protected amino acid derivatives, which are sufficiently stable to be isolated, and some of which are commercially available, include acyl chlorides [9,13], fluorides [10,14,15], symmetric anhydrides [16], pentafluorophenyl esters, N-hydroxysuccinimidyl esters, and 4-nitrophenyl esters [17,18].

Difficult N-acylations, such as acylations of *N*-alkylanilines or α-alkylamino acid derivatives, are most conveniently performed with acyl halides in non-nucleophilic solvents (e.g. DCM, DCP) in the presence of pyridine or DIPEA [13]. Acyl halides can be prepared on insoluble supports under conditions similar to those used in solution. Typical reagents for the preparation of acyl chlorides include oxalyl chloride [19–21], thionyl chloride [22,23], and triphosgene [13]. Anhydrous solvents must be used for all washing of the polymer to prevent hydrolysis [24–26]. The acylation of support-bound amines with acyl halides is generally limited to substrates devoid of other acylable functionalities, because of the high reactivity and low selectivity of these acylating agents.

Activated α-amino acids and other enantiomerically pure acids with a center of chirality at the α-position can racemize upon treatment with a base. The main mechanisms of racemization of such acylating agents are elimination with concomitant formation of ketenes, or enolate formation. For instance, acyl halides with α-hydrogen will usually undergo dehydrohalogenation to yield ketenes if a tertiary amine is used as the base. In the case of α-acylamino acids, racemization is facilitated by the formation of oxazolones (see Figure 16.2). The danger of racemization of such activated acid derivatives generally increases with their reactivity. Hence, if amines are to be acylated with peptide fragments or other sensitive, enantiomerically pure acids, only derivatives of low acylating power, such as acyl azides, should be used. Furthermore, only small amounts of weak bases should be added to the reaction mixture, in order to keep racemization minimal. Examples of acylations with α-amino acid derivatives are given below. For additional notes concerning the problem of racemization, see Section 13.4.1.1.

Illustrative examples of the preparation of amides with isolated acyl halides on insoluble supports are listed in Table 13.1.

Less reactive than acyl halides, but still suitable for difficult couplings are symmetric or mixed anhydrides (e.g. with pivalic or 2,6-dichlorobenzoic acid) and HOAt-derived active esters. HOBt esters smoothly acylate primary or secondary aliphatic amines, including amino acid esters or amides, without concomitant esterification of alcohols or phenols [33]. HOBt esters are the most common type of activated esters used in automated solid-phase peptide synthesis. For reasons not fully understood, acylations with HOBt esters or halophenyl esters can be effectively catalyzed by HOBt and HOAt [3], and mixtures of BOP (in situ formation of HOBt esters) and HOBt are among the most efficient coupling agents for solid-phase peptide synthesis [2]. In acylations with activated amino acid derivatives the addition of HOBt or HOAt also retards racemization [4,12,34].

The synthesis of large peptides on insoluble supports requires rapid coupling with coupling yields > 99.95 % [8]. Acylating agents of lower reactivity than HOBt esters are generally not suitable for solid-phase peptide synthesis (unless additives are used), because reaction rates are too low and quantitative acylations can no longer be attained. The lower reactivity of these reagents, however, also implies greater stability towards hydrolysis, a reduced tendency to racemize (in the case of activated α-amino acids), and increased chemoselectivity. These properties turn out to be particularly profitable for *support-bound* acylating agents, and pentafluorophenyl esters, imidazo-

Table 13.1. Preparation of amides by use of isolated acyl halides.

Entry	Starting resin	Conditions	Product	Ref.
1		MeCOCl, NEt$_3$, DCM		[27] see also [28,29]
2		PhCOCl/DCM 1:1, pyridine (1 eq), 20 °C, 16 h		[30]
3		COCl (0.2 mol/L, 2 eq), NEt$_3$ (3 eq), DCM, 20 °C, 2 h		[31]
4	(Ar: 4-(MeO)C$_6$H$_4$)	FmocHN—COF (0.2 mol/L, 10 eq), DTBMP (10 eq), DCM, 24 h, 20 °C		[32] see also [13]
5		isoquinoline (0.37 mol/L, 5 eq), DCM, 20 °C, 20 min, then Me$_3$SiCN (4.5 eq), 48 h		[22]

lides, and acyl azides have proven suitable for the cyclization of peptides on insoluble supports, or as resin-bound acylating agents for intermolecular acylations (Table 13.2).

Simple alkyl esters (excluding formates) do not generally react with amines. The aminolysis of esters can, however, be efficiently catalyzed by trialkylalanes (Entry **9**, Table 13.2).

13.1.2 Acylation of Amines with Acylating Reagents Formed In Situ

Acylating agents as those shown in Figure 13.2 can either be prepared in the pure form, and then used to acylate amines, or, alternatively, be generated in the presence of an amine ('in situ activation'). Because activated acid derivatives can not be stored indefinitely, in situ activation is generally the preferred strategy for preparing amides on solid phase. Several coupling reagents have been developed for this purpose; these react only slowly or not at all with amines, but efficiently convert free acids or carbox-

Table 13.2. Preparation of amides by acylation of amines with isolated anhydrides and esters.

Entry	Starting resin	Conditions	Product	Ref.
1	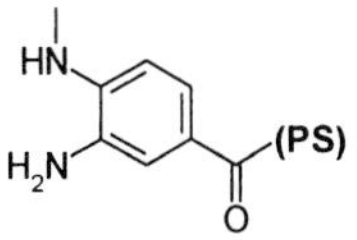	$(ClCH_2CO)_2O$ (0.85 mol/L), DMAP (0.03 mol/L), DCE, 20 °C, 24 h	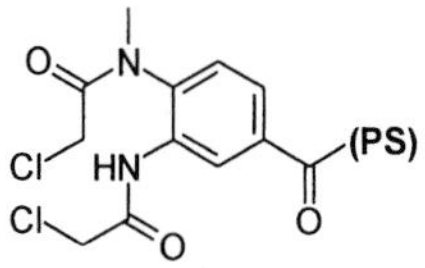	[35] see also [36]
2	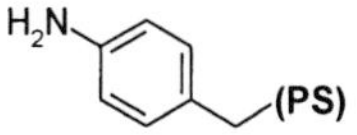	HCO_2H, Ac_2O, DCM	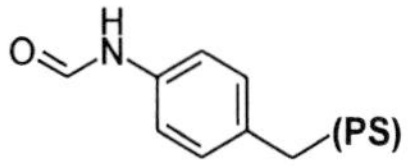	[37]
3	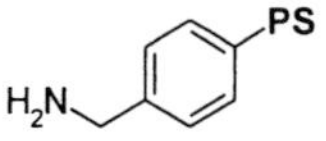	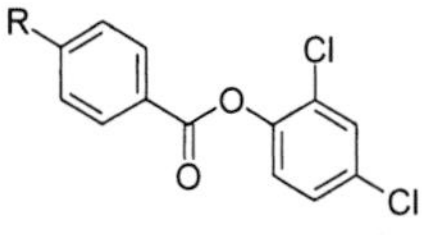 (1.5 eq), HOBt, DIPEA, DMF, 20 °C, 5 h	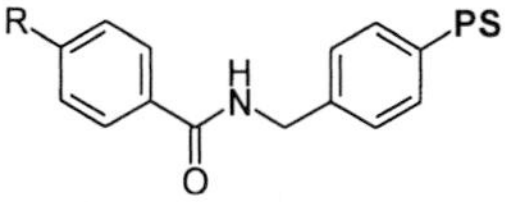	[38] see also [39,40]
4	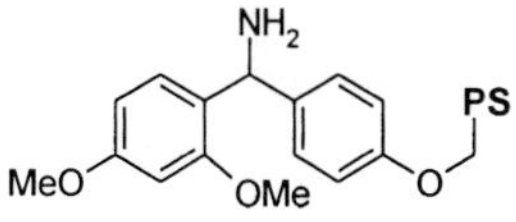	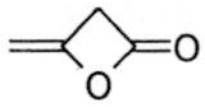 DCM, −15 °C to 20 °C, 2.5 h	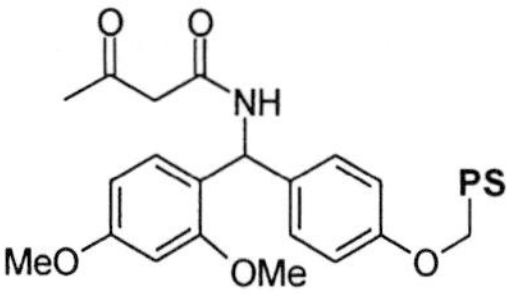	[41]
5	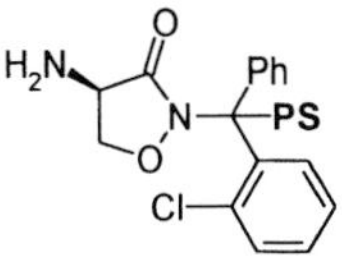	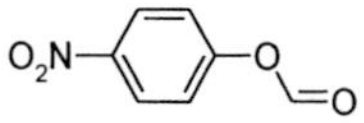 (0.18 mol/L, 6 eq), DMF/pyridine 9:1, 20 °C, 16 h	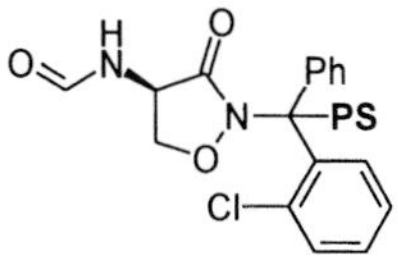	[42]
6	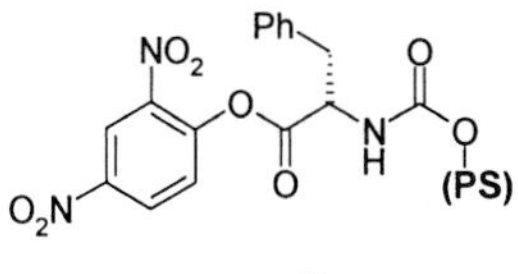	H-His(Tr)-OLi, DMF	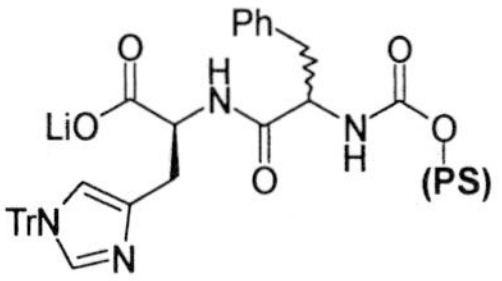	[43]
7	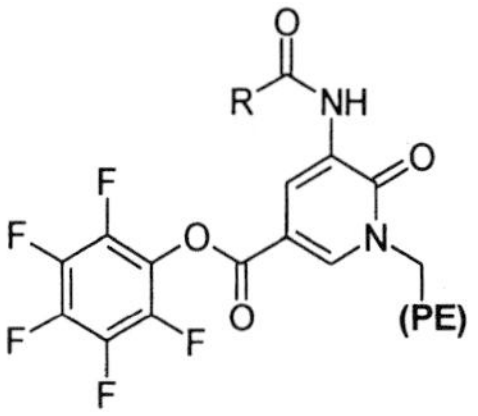	$BnNH_2$ (0.53 mol/L), DMF, 20 °C, 18 h	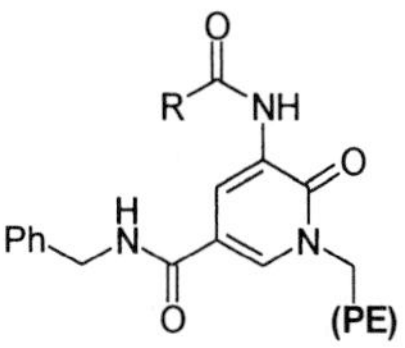	[44] see also [45]
8	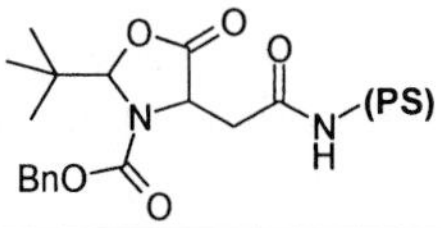	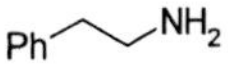 (0.43 mol/L), DMF, 20 °C, 20 h	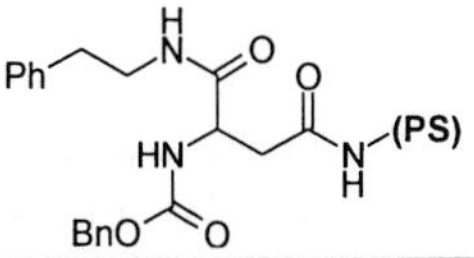	[46,47]

Table 13.2. continued.

Entry	Starting resin	Conditions	Product	Ref.
9	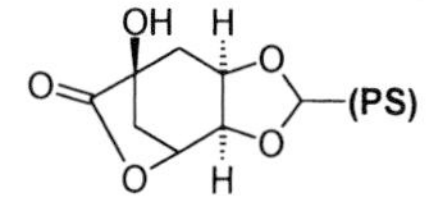	BnNH₂, AlMe₃, DCM, 50 °C, 3 h	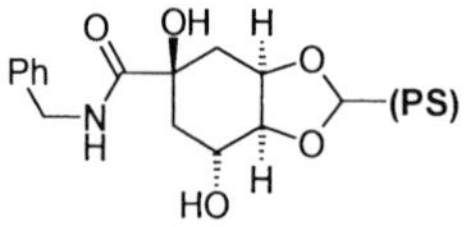	[48] see also [49]
10	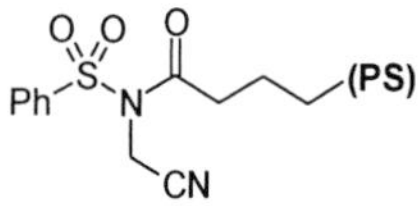	BnNH₂ (1 mol/L), NMP, 20 °C	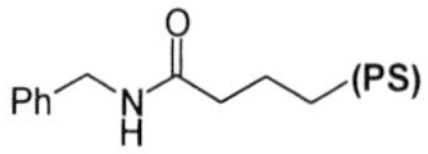	[50]

ylates into acylating agents. The most important reagents of this type are carbodi-imides and phosphonium and uronium salts. Other, less common, reagents include EEDQ (formation of mixed carboxylic carbonic anhydrides), Bop-Cl (formation of mixed carboxylic phosphinic anhydrides [51,52]), DPPA (formation of acyl azides), DECP (formation of acyl cyanides), MSNT (formation of mixed carboxylic sulfonic anhydrides), and benzisoxazolium salts (generation of phenyl esters [53]).

13.1.2.1 Activation of Acids with Carbodiimides

Carbodiimides were introduced as coupling reagents for peptide synthesis by Shee-han and Hess in 1955 [54]. Dicyclohexylcarbodiimide (DCC) has been the standard reagent for many years, but this compound is gradually being replaced by diisopropyl-carbodiimide (DIC) and the water-soluble N-ethyl-N'-(3-dimethylaminopropyl)carbo-diimide (EDC) (Figure 13.3). DIC is an easy-to-handle liquid, whereas DCC (mp 34 °C) is generally sold as solidified mass, difficult to remove from its container. Diiso-propylurea is, moreover, more soluble than dicyclohexylurea in most solvents, and can be washed away more easily during solid-phase synthesis.

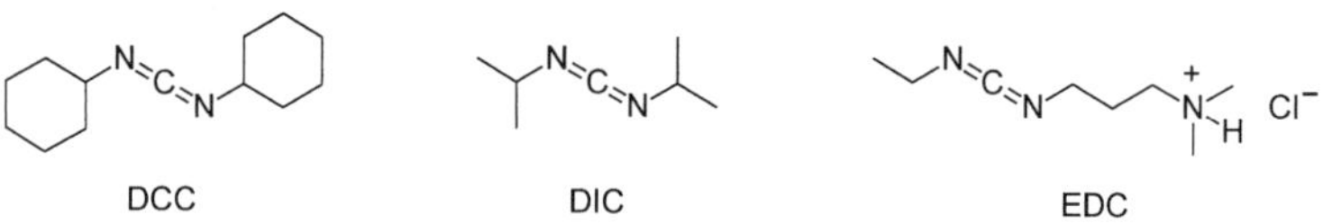

DCC   DIC   EDC

Fig. 13.3. Representative carbodiimides used in solid-phase synthesis.

The mechanism of carbodiimide-mediated amide formation is outlined in Figure 13.4. The first intermediate formed during the reaction of a carbodiimide with a car-boxylic acid (or an ammonium carboxylate [55]) is an O-acylisourea [1]. This inter-mediate reacts under acidic conditions [56,57] with acids to yield anhydrides, and with alcohols, phenols, HOBt, or other compounds containing hydroxyl groups, to yield the corresponding esters. The carbodiimide is thereby transformed into a urea.

Under basic reaction conditions, however, reaction of the O-acylisourea with a car-boxylate is slow [57], and rearrangement of the O-acylisourea to an N-acylurea can efficiently compete with anhydride formation [58]. This can, for instance, occur in the presence of excess carbodiimide [59], a tertiary amine [55,60], or pyridine [58]. *N*-Acylureas are stable products, and will generally not acylate amines. Generation of *O*-acylisoureas in the presence of primary or secondary amines can lead to the formation of amides, but *N*-acylureas can readily be formed as byproducts, in particular if a large excess of amine is used. The formation of *N*-acylureas can be supressed by using excess carboxylic acid or by adding an acidic, hydroxyl group containing compound (e.g. HOBt) to the reaction mixture. Under these conditions the *O*-acylisourea will be mainly converted into a symmetric anhydride or into an active ester, which then acylates the amine. Amines react only slowly with carbodiimides (to yield guanidines), and this side reaction generally does not interfere significantly with amide formation.

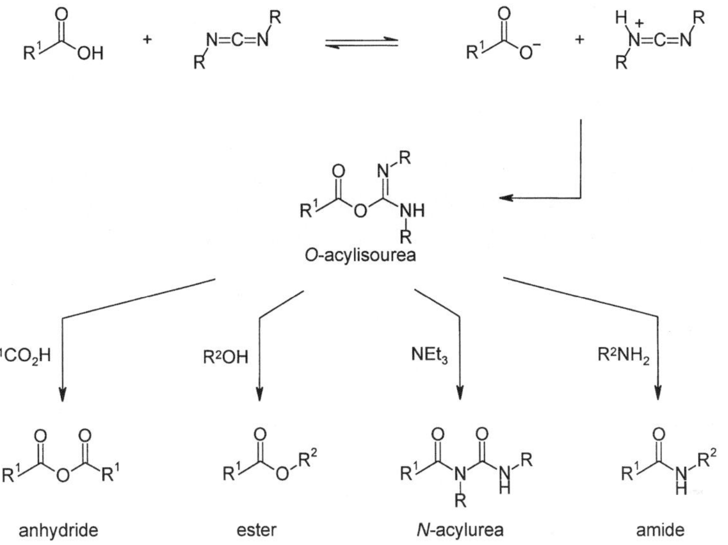

Fig. 13.4. Activation of carboxylic acids with carbodiimides.

For the solid-phase synthesis of amides it makes a significant difference whether the amine or the acid is linked to the support. Resin-bound amines are readily acylated by adding a carboxylic acid and then a carbodiimide to the immobilized amine (Table 13.3). The acid/carbodiimide ratio is not critical, because both the *O*-acylisourea (ratio 1:1) and the symmetric anhydride (ratio 2:1) will lead to N-acylation. It should, however, be kept in mind that the half-lives of *O*-acylisoureas are shorter than those of anhydrides, and for difficult couplings it might be a better choice to acylate with a symmetric anhydride (two equivalents of acid and one of carbodiimide).

Illustrative examples of the acylation of support-bound amines with carbodiimides as coupling agents are listed in Table 13.3. Difficulties are usually encountered during the acylation of α-alkylamino acid derivatives (which are significantly less nucleophilic than simple secondary amines; Entries **3** and **4**) and *N*-alkyl (Entries **5** and **6**) or *N*-aryl anilines. Acylations with haloacetic or related acids containing a leaving group prone to nucleophilic displacement should not be performed with the aid of HOBt and bases, because O-alkylation of HOBt by the product occurs readily.

Table 13.3. Acylation of support-bound amines with carboxylic acids and carbodiimides.

Entry	Starting resin	Conditions	Product	Ref.
1		Boc-Gly-OH (4 eq), DCC (4 eq), DCM, 0.5 h		[61] see also [62,63]
2		RCO$_2$H (0.5 eq), C$_6$F$_5$OH (2.5 eq), DIC (2.5 eq), DMF, 24 h		[64]
3		FmocHN (20 eq), DIC (10 eq), DCM/DMF 4:1, 2 h; repeat once X: H or COR		[65] see also [66]
4		EDC (0.44 mol/L), anthranilic acid (0.37 mol/L), NMP, 12 h		[67]
5		BrCH$_2$CO$_2$H (0.6 mol/L, 12 eq), DIC (13 eq), DMF, 20 °C, 0.5 h; repeat once		[68]
6		(3 eq), HOAt (3 eq), DIC (3 eq), DMF, 20 °C, 8 h		[69]

Experimental Procedure 13.1: Typical acylation of support-bound amines with anhydrides formed in situ [70]

To Wang resin bound piperazine (5.0 g, approx. 5 mmol [71]) was added a solution of 4-(chloromethyl)benzoic acid (8.85 g, 51.9 mmol, 10 eq) in a mixture of NMP (20 mL) and DCM (60 mL). (If the acid used is sufficiently soluble, pure DCM or DCP can be used as solvent instead). To the mixture was added DIC (4.0 mL, 25.7 mmol, 5 eq), and after shaking for 4–8 h at room temperature, the mixture was filtered, and the resin was washed with acetonitrile and DCM (4 × 100 mL of each solvent) and dried at the air. The resulting resin remained unchanged upon storage at room temperature for several months.

Because carbodiimide-mediated N-acylations lead mainly to acid-derived bypro-ducts, the conversion of support-bound acids into amides is less straightforward than the opposite variant. If a support-bound acid is treated with excess amine and a carbo-diimide, large amounts of N-acylureas will generally be formed. This can be avoided by adding HOBt to the mixture and thereby generating an intermediate HOBt ester (Table 13.4). Yields of amides are not as high as for the acylation of support-bound

Table 13.4. Acylation of amines with support-bound carboxylic acids and carbodiimides.

Entry	Starting resin	Conditions	Product	Ref.
1		BnNH$_2$, HOBt, EDC (5 eq of each), DMF, 20 °C, 16 h		[43] see also [72]
2		EDC, HOBt, NEt$_3$ (3.5 eq of each), DMF, 12 h		[73] see also [74]
3		HOBt (0.63 mol/L, 4 eq), N-(2-hydroxy-ethyl)aniline (4 eq), DIC (4 eq), DMF, 20 °C, 12–120 h		[75] see also [76]
4		MeO$_2$C~R~NH$_2$ HOBt, DCC (5 eq of each, 0.12 mol/L), DMF		[77] see also [78]
5		Ph~NH$_2$ (0.17 mol/L, 5 eq), HOBt (4 eq), DIPEA (5 eq), EDC (4 eq), DMF/DCM 1:1, 20 °C, 6 h; repeat once		[79]

amines, however, which makes this strategy unsuitable for the solid-phase synthesis of peptides. An additional problem of peptide synthesis with support-bound active esters is that activated *N*-acyl amino acid derivatives racemize more readily than activated *N*-alkoxycarbonyl amino acid esters (see, e.g., [5] and Figure 16.2). These problems are less critical for the solid-phase synthesis of small non-peptides, and several successful conversions of support-bound acids into amides have been reported (Table 13.4). As discussed above, support-bound acylating agents of low reactivity will often lead to cleaner products than highly activated acid derivatives.

Experimental Procedure 13.2: Synthesis of unsymmetric malonamides [70,76]

To Wang resin bound 1,3-diamino-2,2-dimethylpropane (40 mg, approx. 0.04 mmol [71,80]) was added a solution of HOBt (130 mg, 0.96 mmol, 24 eq) and malonic acid (48 mg, 0.46 mmol, 12 eq) in a mixture of NMP (1 mL) and DCP (1 mL). DIC (0.18 mL, 1.16 mmol, 29 eq) was added, and the resulting mixture was shaken at room temperature for 3–20 h. After filtration the resin was washed with NMP (3 × 2 mL). A solution of HOBt (148 mg, 1.10 mmol, 27 eq) in a mixture of NMP (1 mL) and DCP (1 mL) was added, followed by the addition of DIC (0.16 mL, 1.03 mmol, 26 eq) and piperidine (0.12 mL, 1.21 mmol, 30 eq). The mixture was shaken at room temperature for 6–20 h, filtered, and the resin was washed with NMP and DCM.

13.1.2.2 Activation of Acids with Phosphonium Salts

Triaminophosphonium salts, unlike carbodiimides, react with carboxylic acids under basic conditions only to yield acyloxyphosphonium salts (Figure 13.5). Depending on the counterion X^- and on the precise reaction conditions, the acyloxyphosphonium salt can either acylate the amine directly or be transformed into the symmetric anhydride or an active ester [2,81], which then acylates the amine. At low temperatures direct acylation with the acyloxyphosphonium intermediate seems to be the prevalent mechanism [82].

Phosphonium salts as those shown in Figure 13.5 do not react with amines, and are well suited for preparing amides on insoluble supports, either with the amine or with the acid linked to the support (Table 13.5). A selection of commercially available phosphonium salts suitable for the activation of carboxylic acids in the presence of amines, are sketched in Figure 13.5. Solutions of these reagents in dry DMF are quite stable and can be used even after standing at room temperature for several days [83].

BOP, one of the first phosphonium salts used for peptide synthesis [84], leads to the formation of mutagenic HMPA, and is therefore being replaced by the less hazardous PyBOP [85].

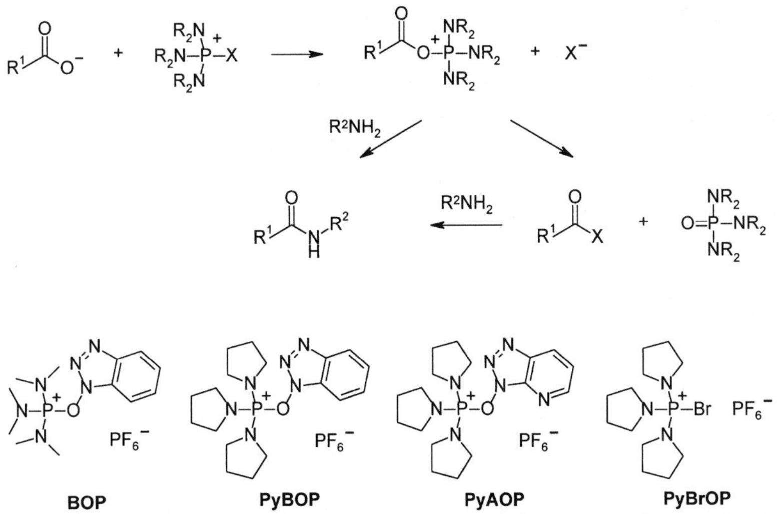

Fig. 13.5. Activation of carboxylates by triaminophosphonium salts [17,81,84–87]. X: leaving group.

Table 13.5. Preparation of amides by use of phosphonium salts as coupling agents.

Entry	Starting resin	Conditions	Product	Ref.
1	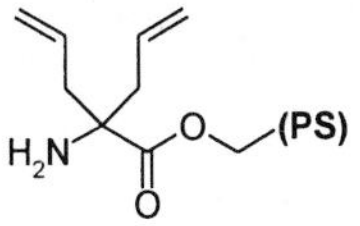	quinaldic acid, PyBrOP, DIPEA (10 eq of each), NMP, 24 h	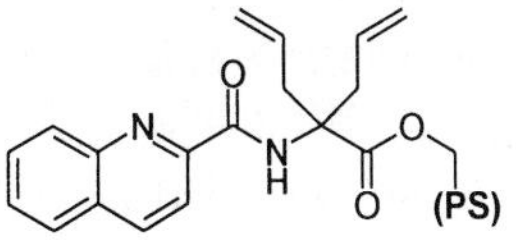	[88]
2	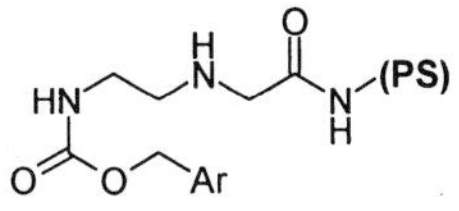	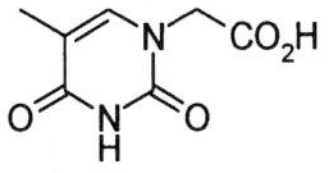 PyBrOP (each 0.5 mol/L), DMSO/NMM 1:1, 0.5 h; repeat once Ar: 4-(MeO)C$_6$H$_4$	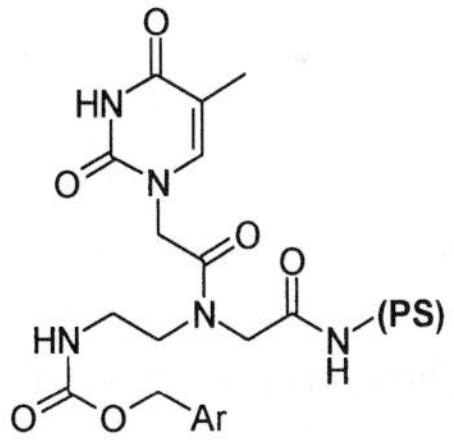	[89]
3	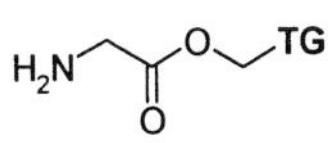	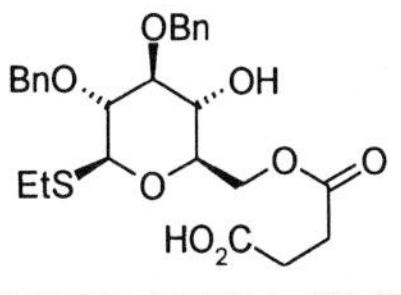 PyBOP, DIPEA, DMF	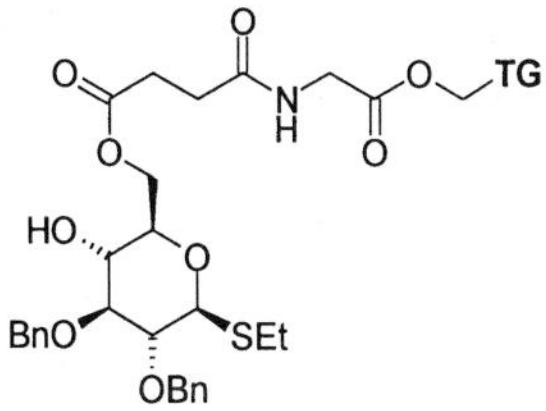	[90] see also [91]
4	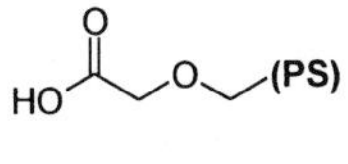	PyBOP, NMM, 2-aminophenol (5 eq of each, 0.38 mol/L), DMF, 20 °C, 17 h	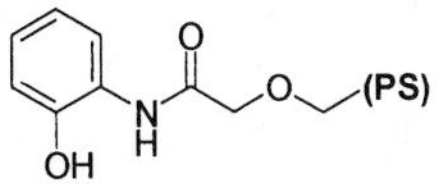	[92]
5	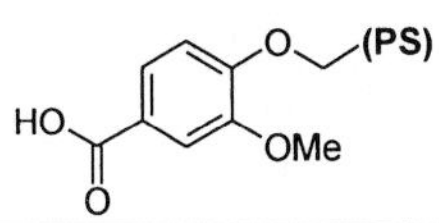	iPr$_2$NH, PyBrOP, DCM (low yield)	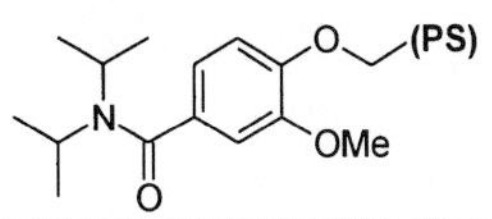	[93]

13.1.2.3 Activation of Acids with Uronium Salts

Uronium salts can be used to convert carboxylates into HOBt or related esters [5], into pentafluorophenyl esters [94], or into acyl halides [10] in the presence of amines (Figure 13.6). A highly reactive, *O*-acylisouronium salt is formed initially; this is more reactive than the *O*-acylisoureas formed during activation of carboxylic acids with carbodiimides. This intermediate cannot undergo intramolecular rearrangements and acts only as an acylating agent.

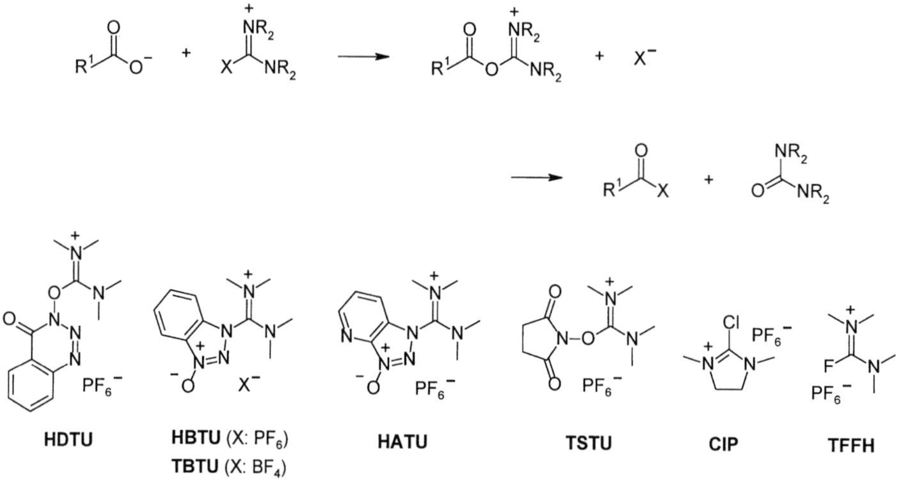

Fig. 13.6. Activation of carboxylates by uronium salts [5,10,17,18,83]. X: leaving group.

The reactivity of uronium- (and phosphonium-) based coupling reagents is mainly determined by the type of activated acid derivative formed during activation (see Figure 13.2). Unlike phosphonium salts, uronium salts can react with amines to yield guanidines [83]. This side reaction can interfere with amide formation if more uronium salt than carboxylic acid is used. Illustrative examples of the use of uronium salts are listed in Table 13.6.

Table 13.6. Preparation of amides by use of uronium salts as coupling agents.

Entry	Starting resin	Conditions	Product	Ref.
1		FmocHN—CO$_2$H (5 eq), TFFH (5 eq), DIPEA (10 eq), DMF, 0.5 h		[10]
2		Br—CO$_2$H HATU, DIPEA, DMF		[95]
3		Boc-Phe-OH (5 eq), HATU (5 eq), DIPEA (10 eq), DCM/NMP 1:1, 12 h		[96]
4		CIP (2.8 eq), HOAt (1.4 eq), Fmoc-Val-OH (2.8 eq), DIPEA (0.4 mol/L, 8.3 eq), DMF, 1.5 h		[97]
5		Fmoc-Ala-OH (3 eq), CIP (3 eq), DIPEA (6 eq), NMP, 20 °C, 16 h; repeat once		[98] see also [99,100]
6		Fmoc-amino acid (10 eq), HATU (10 eq), DIPEA (20 eq), DCM/DMF 9:1, 25 °C, 2 h; repeat once		[101]
7		(0.67 mol/L), HBTU (0.63 mol/L), DMF/ DIPEA 2:1, 45 min		[102]

Table 13.6. continued.

Entry	Starting resin	Conditions	Product	Ref.
8	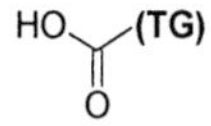	 $R\!\!\frown\!\!NH_2$ (0.07 mol/L), TBTU, HOBt, DIPEA (each 0.12 mol/L), DMF, 25 °C, 16 h	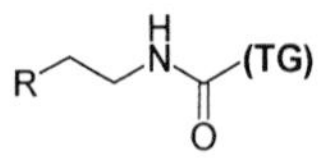	[103]
9	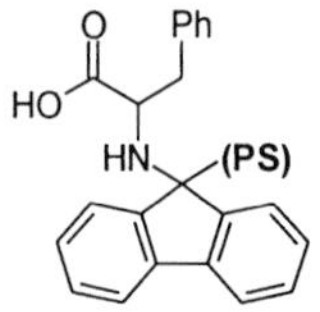	1. TBTU (5 eq), NMM (5 eq), DMF, 0.5 h, then wash resin 2. H-Phe-OMe, 4 h	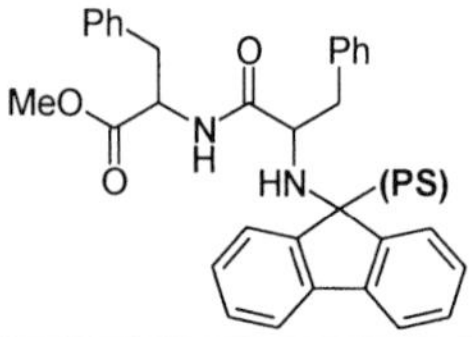	[104]

13.1.3 Miscellaneous Preparations of Amides

In addition to the reagents presented above, other acylating agents can be used to acylate support-bound amines (Table 13.7). Ketenes are highly reactive acylating agents, which can be generated in situ by Wolff-rearrangement of diazo ketones (Entry **1**, Table 13.7), by thermolysis of 2-acyl Meldrum's acids or substituted 1,3-dioxin-4-ones (Entry **2**, Table 13.7), or by photolysis of chromium(II) carbene complexes (Entry **3**, Table 13.7). Thermolysis of 1,3-dioxin-4-ones in the presence of amines is a convenient method for preparing 3-oxo amides, which are valuable intermediates for the synthesis of several types of heterocycle. 3-Oxo amides are difficult to prepare by conventional acylation with an acid, because most 3-oxo carboxylic acids are unstable and decarboxylate even at room temperature.

A further strategy used to prepare amides on insoluble supports is based on the Ugi reaction (Figure 13.7). Simple mixing of an amine, an aldehyde, an acid, and an isonitrile can lead to the formation of α-amino acid amides. The mechanism of this remarkable reaction is outlined in Figure 13.7. Sometimes the amine is first condensed with the aldehyde to form an imine, which is then combined with the acid and the isonitrile.

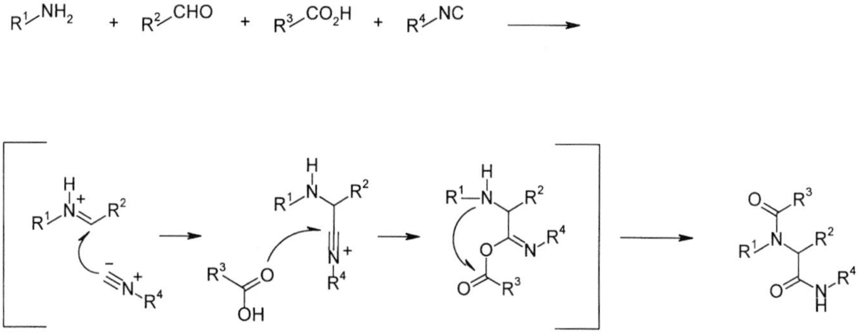

$$R^1\!\!\frown\!\!NH_2 \quad + \quad R^2\!\!\frown\!\!CHO \quad + \quad R^3\!\!\frown\!\!CO_2H \quad + \quad R^4\!\!\frown\!\!NC \quad \longrightarrow$$

Fig. 13.7. Mechanism of the Ugi reaction.

Although the Ugi reaction can in principle be performed with any of the four components linked to the support, in most of the examples reported support-bound amines and acids have been used (Entries **4–8**, Table 13.7).

Table 13.7. Miscellaneous preparations of amides.

Entry	Starting resin	Conditions	Product	Ref.
1		(4 eq), DMF/THF 1:1, PhCO₂Ag (0.35 eq), NEt₃, 0 °C, 4 h		[105]
2		(6 eq), PhMe, 65 °C, 6 h		[106] see also [107,108]
3		hν, CO, THF, 20 °C, 30 h	(89% de)	[109] see also [110]
4		(0.67 mol/L each), DCM, 8 h		[111]
5		(0.3 mol/L each), DCM/ MeOH 3:4, 1–8 h		[111]
6		1. BnNH₂, *i*PrCHO (5 eq of each), dioxane/ MeOH 4:1, MS 4 Å, 20 °C, 3 h 2. add BuNC (5 eq) and resin (1 eq), 55 °C, 48 h		[112] see also [113,114]
7		BnNH₂, EtCHO, DCM/ MeOH 1:1, 23 °C		[115]

Table 13.7. continued.

Entry	Starting resin	Conditions	Product	Ref.
8		PhCHO (0.5 mol/L), DCM, 25 °C, 1 h; then ArCO₂H, BuNC, 72 h		[116] see also [117-119]
9		histamine, ArCHO (each 0.17 mol/L, 10 eq), THF/ MeOH 1:1, 20 °C, 3 d		[37] see also [120]
10		KOH, EtOH, 80 °C		[121]

Support-bound C,H-acidic compounds, such as acetoacetamides, react with isocyanates under basic conditions to yield amides via C-carbamoylation [70]. This reaction is closely related to the C-thiocarbamoylation, which has been used for the solid-phase synthesis of thioamides (see Section 13.9).

13.1.4 Preparation of Amides by C-Alkylation of Other Amides

Support-bound amides with α-hydrogen can be deprotonated and C-alkylated with soft electrophiles. The base required will depend on the acidity of the amide. Amides from simple alkanoic acids can only be deprotonated stoichiometrically by strong bases such as LDA (Table 13.8). Glycine esters and amides have been C-alkylated on cross-linked polystyrene after conversion into the corresponding benzophenone imines. This strategy enables the solid-phase synthesis of unnatural α-amino acid derivatives.

Various chiral auxiliaries have been linked to insoluble supports and used to prepare enantiomerically enriched carboxylic acid derivatives (Entries **2** and **3**, Table 13.8). Although the recovery of support-bound chiral auxiliaries is certainly easier than in solution, there have been reports of variable results regarding the stereoselectivity of C-alkylations on insoluble supports [122]. It seems that the optimization of these diastereoselective alkylations is less straightforward on solid phase than in solution, and it remains to be determined if support-bound chiral auxiliaries of this type really represent a viable alternative to non-polymeric auxiliaries.

Table 13.8. Preparation of amides by C-alkylation of other amides.

Entry	Starting resin	Conditions	Product	Ref.
1		BnBr (2 eq), BEMP (2 eq), NMP, 20 °C, overnight		[123] see also [88,124–126]
2		LDA (2 eq), THF, 0 °C, 0.5 h, then allyl iodide (3 eq), 0 °C to 20 °C, 24 h	(87% de)	[127]
3		1. TiCl$_3$(OiPr) (5 eq), DIPEA (6 eq), DCM, 0 °C, 3 h; wash 2. acrylonitrile (10 eq), DCM, 0 °C, 24 h	(78% de)	[128] see also [122] [129]
4		LDA (15 eq), iPrI (50 eq), THF, 0–20 °C, 10 h Ar: 4-BrC$_6$H$_4$		[130]

13.1.5 Preparation of Amides by N-Alkylation of Other Amides

Acylated primary amines bound to insoluble supports can be deprotonated and N-alkylated. In solution this reaction generally only leads to acceptable results when using reactive, soft alkylating agents devoid of β-hydrogen, because of competing elimination and O-alkylation. β-Lactams and cyclic imides are often more readily N-alkylated than acyclic amides. Even under optimal conditions, however, the N-alkylation of amides in solution does not always proceed smoothly.

Elimination is a less severe problem when alkylating support-bound amides because a large excess of alkylating agents can be used. Examples of successful N-alkylations of amides are listed in Table 13.9. Strongly acidic amides, such as trifluoroacetanilides (and sulfonamides, see Section 8.4) can be N-alkylated with primary aliphatic alcohols under essentially neutral conditions by use of the Mitsunobu reaction (Entry **4**, Table 13.9).

Table 13.9. Preparation of amides by N-alkylation of other amides.

Entry	Starting resin	Conditions	Product	Ref.
1		1. LiOtBu (1 mol/L), THF 2. MeI, DMSO, 20 °C, 2 h		[131]
2		LiOtBu (10 eq), THF, 20 °C, 1.5 h, then drain and add (5 eq), DMSO, 5 h (repeat once)		[72]
3		PhN(Li)Ac, BuI, DMF, 20 °C, 4 h Ar: 2-(N_3)-5-BrC_6H_3		[132]
4		EtOH (0.2 mol/L), PPh₃ (0.4 mol/L), DEAD (0.2 mol/L), THF, −10 °C to 20 °C, 6 h		[133]
5		 [Bu₄N][HSO₄], BuLi, THF, −70 °C to 20 °C, overnight		[134]

13.2 Preparation of Hydroxamic Acids and Hydrazides

Hydroxylamines and hydrazines can be acylated on insoluble supports with the same type of acylating agent as is used for the acylation of amines [135–138]. Because of their higher nucleophilicity, acylations of hydroxylamines or hydrazines can proceed more readily than acylations of amines, and unreactive acylating agents, such as carboxylic esters, can sometimes be used with success (Table 13.10). Polystyrene-bound *O*-alkyl hydroxamic acids can be N-alkylated by treatment with reactive alkyl halides and bases such as DBU (Entry **5**, Table 13.10).

Table 13.10. Preparation of hydrazides and hydroxamic acid derivatives.

Entry	Starting resin	Conditions	Product	Ref.
1	MeO (benzoate)-PS	$N_2H_4 \cdot H_2O$/DMI 2:1, 90 °C, 6 h	H_2N-NH-(benzoyl)-PS	[139]
2	MeO (Leu derivative)-(PS)	N_2H_4/BuOH 2:8, 20 °C, 6.5 h	H_2N-NH (Leu derivative)-(PS)	[140]
3	EtO (benzoate)-PS	$NH_2OH \cdot HCl$, KOH, MeOH, EtOH, DCM, 6 h, then AcOH	HO-NH (benzoyl)-PS	[141]
4	HO_2C (AcHN derivative)-(PS)	H_2NOBn (8 eq), DIC (4 eq), HOBt (4 eq), DMF, overnight	Ph-O-NH (AcHN derivative)-(PS)	[142]
5	(aryl amide resin)-PS	$ArCH_2Br$ (0.4 mol/L, 6 eq), DBU (6 eq), PhMe, 20 °C, 3 d; Ar: 4-BrC_6H_4	(aryl amide with Ar resin)-PS	[135]

13.3 Preparation of Carboxylic Acids

Polymer-bound carboxylic acids have often been used as supports for alcohols or as intermediates for the preparation of other supports. Because supports with hydroxyl or amino groups are readily available, the most common strategy for preparing supports with free carboxyl groups is the acylation of polymeric alcohols or amines with succinic or glutaric anhydride. Carboxyl groups can, however, also be prepared by chemical transformation of several other functional groups on insoluble supports. These include the carboxylation of organometallic reagents, the oxidation of benzyl halides and alkenes, and the saponification of carboxylic acid derivatives (Figure 13.8).

One convenient means of access to carboxyl-functionalized, cross-linked polystyrene is metallation of the latter, followed by treatment with carbon dioxide [139,143–146]. Alternatively, better control of the loading and homogenicity of carboxylated polystyrene can be achieved by converting chloromethyl polystyrene into the corresponding polymeric benzoic acid. This can be accomplished by oxidation of Merrifield resin to the corresponding benzaldehyde by heating with DMSO, followed by exhaustive oxidation to the acid with MCPBA (Entry 1, Table 13.11). Polystyrene-bound

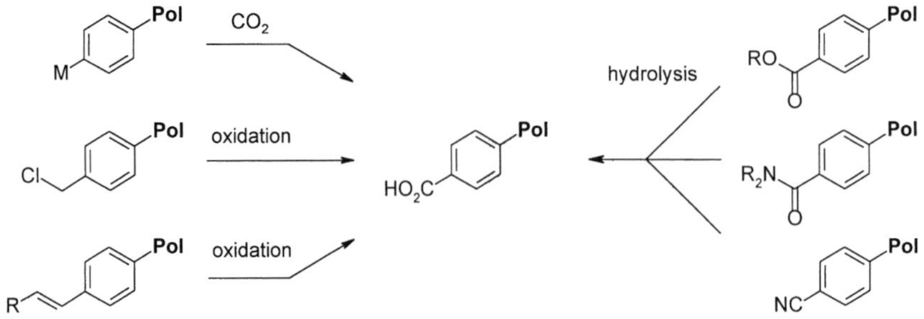

Fig. 13.8. Methods for the preparation of support-bound benzoic acids. M: metal.

alkenes can be oxidatively degraded to carboxylic acids by ozonolysis and oxidation of the intermediate ozonide with oxygen (Entry **2**, Table 13.11). Another means of access to carboxyl-functionalized polystyrene is the acid-mediated hydrolysis of nitriles, which are available from chloromethyl polystyrene by nucleophilic substitution with cyanide (Entry **3**, Table 13.11). Polyethylene has been partially carboxylated by treatment with $Cr_2O_3/H_2SO_4/H_2O$ [147].

Most of these procedures are not compatible with common linkers, and therefore not suitable for the transformation of support-bound substrates into carboxylic acids. A more versatile approach for this purpose is the saponification of carboxylic esters. Saponifications with KOH or NaOH usually proceed smoothly on hydrophilic supports, such as Tentagel [19] or polyacrylamides, but not on cross-linked polystyrene. Esters linked to hydrophobic supports are more conveniently saponified with LiOH [44] or $KOSiMe_3$ in THF or dioxane (Table 13.11). Alternatively, palladium(0)-mediated saponification of allyl esters [93] can be used to prepare acids on cross-linked polystyrene (Entries **9** and **10**, Table 13.11). Fmoc-protected amines are not deprotected under these conditions [148].

13.4 Preparation of Carboxylic Esters

Most studies of esterification reactions on insoluble supports had as goal the attachment of N-protected amino acids to polystyrene-derived alcohols. In recent years, however, the scope of esterifications has been expanded to other types of alcohol and acid. In the following section the most important strategies for the preparation of esters on solid phase will be discussed. These have been organized according to type of functionality initially linked to the support.

13.4.1 Preparation of Esters from Support-Bound Alcohols

Support-bound primary or secondary aliphatic alcohols can be acylated under conditions similar to those used in solution, as long as these conditions are compatible with the linker chosen. Acids can, for instance, be activated with a carbodiimide either

Table 13.11. Preparation of carboxylic acids.

Entry	Starting resin	Conditions	Product	Ref.
1		MCPBA, DME, 40 °C, 15 h		[149]
2		O$_3$, DCM/AcOH 7:1, −78 °C, 10 min, then O$_2$, 20 °C, overnight		[150]
3		H$_2$SO$_4$/AcOH/H$_2$O 1:1:1, 120 °C, 10 h		[151]
4		LiOH (5 eq), THF/H$_2$O 5:1, 20 °C, 1 h Ar: 2-naphthyl		[152]
5		NaOH, MeOH, THF, 60 °C, 24 h		[153]
6		KOH (0.6 mol/L), dioxane/H$_2$O 3:1, 20 °C, 12 h		[28]
7		1. Boc$_2$O, NEt$_3$, DMAP, DCM, 23 °C, 18 h 2. LiOH (0.8 mol/L), H$_2$O/H$_2$O$_2$/THF 19:1:5		[114] see also [154]
8		KOSiMe$_3$, THF, 20 °C		[74,155]
9		Pd(PPh$_3$)$_4$, Me$_3$SiN$_3$, DCE, 20 °C		[74,155] see also [148]
10		Pd(PPh$_3$)$_4$ (3 eq), CHCl$_3$/AcOH/NMM 37:2:1, 20 °C, 14 h		[156]

Table 13.11. continued.

Entry	Starting resin	Conditions	Product	Ref.
11		HCl, THF, 23 °C		[115]
12		LiOH, H_2O_2, THF, H_2O, 0 °C, 4 h		[157]

as symmetric anhydrides or as *O*-acylisoureas, which quickly react with alcohols in the presence of a catalyst, such as DMAP or another base, to yield esters (Table 13.12). Further acid derivatives suitable for esterification reactions on solid phase include acyl halides and imidazolides. HOBt esters react only slowly with alcohols, but enable the selective acylation of primary alcohols in the presence of secondary alcohols (Entry **5**, Table 13.12).

Esterification of resin-bound alcohols with 3-oxo carboxylic acids (which readily decarboxylate) or with malonic acid is best performed with the corresponding ketenes, which can be generated in situ by thermolysis of dioxinones or other precursors (Entries **6–9**, Table 13.12). PEG can also be acetoacetylated with acetylketene generated by thermolysis [158].

Transesterification under strongly basic reaction conditions has been used to acylate support-bound alcohols with alkyl esters (Entry **10**, Table 13.12). For sensitive acids the Mitsunobu reaction is a particularly mild method of esterification. This reaction gives high yields with support-bound primary aliphatic alcohols and proceeds under essentially neutral reaction conditions (Experimental Procedure 13.4). Mitsunobu esterification of PEG with *N*-Fmoc amino acids has also been reported [159].

Support-bound phenols [160–163] or other compounds with an acidic OH group (oximes [164–166], hydroxamic acids, *N*-hydroxy benzotriazoles, etc.) undergo rapid esterification with carbodiimides as coupling agents (Entries **13** and **14**, Table 13.12). The resulting products are susceptible to nucleophilic attack, and can be used as insoluble acylating agents.

Table 13.12. Acylation of support-bound alcohols.

Entry	Starting resin	Conditions	Product	Ref.
1		O$_2$N⌒CO$_2$H (0.67 mol/L, 4.5 eq), DCC (4.5 eq), THF, 0 °C, 3 h, then 20 °C, ultrasound, overnight		[167]
2		TsO$^-$ RCO$_2$H, DIPEA, DCM, 20 °C, 4 h R: ArOC(Me)$_2$CO$_2$H		[168]
3		EtO$_2$CCOCl (0.45 mol/L, 8 eq), NEt$_3$ (1 eq), DMAP (0.15 eq), DCM, 0–20 °C, overnight		[169]
4		phthalic anhydride, (0.44 mol/L, 6.7 eq), NEt$_3$ (6.7 eq), DMAP (1.3 eq), DMF, 20 °C, 18 h		[170]
5		BOP, HOBt (2 eq of each), DIPEA (3 eq), DCM, 20 °C, 24 h		[171]
6	PS–allyl alcohol copolymer	EtO$_2$C⌒CO$_2$Et (neat), microwaves (400 W), 5 min		[172] see also [173]
7		THF, 65 °C		[21] see also [76]
8		(10 eq), PhMe, 100 °C, 3 h		[174]
9		DMAP, PhMe, 110 °C, 18 h Ar: 2,4,5-F$_3$C$_6$H$_2$		[175]

Table 13.12. continued.

Entry	Starting resin	Conditions	Product	Ref.
10		LDA (1.1 eq), THF, −78 °C, 5 min, then 2 eq 20 °C, overnight		[176]
11		PPh$_3$, DEAD, THF, 24 h		[177] see also [178]
12		(0.6 mol/L, 0.9 eq), pyridine (1.6 eq), DCM, 20 °C, 72 h		[179]
13		(0.15 mol/L, 3 eq), EDC (3.3 eq), DMF/CHCl$_3$ 1:1, 20 °C, overnight		[180] see also [181]
14		R⁀CO$_2$H (0.03 mol/L, 1 eq), DCC (1 eq), DCM, 0 °C, 20 min		[182] see also [183-185]

Experimental Procedure 13.3: Esterification of Wang resin with acryloyl chloride [31]

Acryloyl chloride (143 mL, 1.76 mol) was added dropwise to a stirred mixture of Wang resin (1.0 kg, 0.88 mol), DCM (7.0 L), and triethylamine (0.36 L, 2.6 mol) within 20 min, keeping the temperature below 30 °C. After stirring for 2 h at room temperature the resin was filtered (inverse filtration) and washed three times with DCM. The acylation was repeated with the same amount of reagents and 4 L DCM. After 2 h the resin was removed by filtration, washed with DCM, methanol, and DMF, and capped by reaction with acetic anhydride (166 mL, 1.76 mol) and triethylamine (244 mL, 1.76 mol) in DMF (2 L) for 1 h at room temperature. Washing with DMF, methanol, DCM, and ether, followed by drying under reduced pressure, yielded 1.09 kg of resin-bound acrylate.

Experimental Procedure 13.4: Esterification of Wang resin under Mitsunobu conditions [186]

A solution of DEAD (0.27 mL, 1.73 mmol) in THF (2 mL) was added dropwise to a mixture of Wang resin (0.5 g, 0.35 mmol), THF (8 mL), bromoacetic acid (0.24 g, 1.73 mmol), and triphenylphosphine (0.45 g, 1.72 mmol). The resulting mixture was shaken at room temperature for 24 h, filtered, and the resin was washed with THF and DCM.

13.4.1.1 Esterification of Support-bound Alcohols with N-Protected α-Amino Acids

Standard solid-phase peptide synthesis requires the first (C-terminal) amino acid to be esterified with a polymeric alcohol. Partial racemization can occur during the esterification of N-protected amino acids with Wang resin or hydroxymethyl poly-styrene [187,188]. *N*-Fmoc amino acids are particularly problematic, because the bases required to catalyze the acylation of alcohols can also lead to deprotection. A comparative study of various esterification methods for the attachment of Fmoc amino acids to Wang resin [189] showed that the highest loadings with minimal racemization can be achieved under Mitsunobu conditions or by activation with 2,6-dichloroben-zoyl chloride (Experimental Procedure 13.5). *N*-Fmoc amino acid fluorides in the presence of DMAP also proved suitable for the racemization-free esterification of Wang resin (Entry **1**, Table 13.13). Most racemization was observed when DMF or THF were used as solvent, whereas little or no racemization occurred in toluene or DCM [190].

The esterification of *O*-glycosyl-*N*-Fmoc serine with Wang resin leads to extensive racemization of the amino acid if acylation is performed with HOBt, HOAt, or penta-fluorophenyl esters in the presence of DMAP [191]. Racemization is minimal, however, when MSNT is used as activating agent (in situ formation of a mixed carboxylic sulfonic anhydride) and 1-methylimidazole is used as base (Entry **2**, Table 13.13). 2,4,6-Trimethylpyridine (collidine) has been recommended as base for acylations with Fmoc-Ser(*t*Bu)-OH, which also undergoes extensive racemization when activated with, e.g., HATU/HOAt in the presence of NMM [34].

Tertiary alcohols can usually only be esterifed with highly reactive acylating agents, such as acyl chlorides, and few examples of such acylations have been reported (Entry **7**, Table 13.13).

Table 13.13. Acylation of support-bound aliphatic alcohols with α-amino acid derivatives.

Entry	Starting resin	Conditions	Product	Ref.
1	HO—⟨⟩—O–PS	NHFmoc / COF (0.15 mol/L, 3 eq), DMAP (2 eq), PhMe, 10 min	NHFmoc ... O–PS	[190]
2	HO—⟨⟩—O–PS	FmocHN, RO...CO$_2$H (3 eq), MSNT (0.1 mol/L, 3 eq), 1-methylimidazole (2.3 eq), DCM, 0.5 h (R: tetraacetyl-β-galactosyl)	FmocHN ... OAc OAc OAc OAc (+ 6% R-serine derivative)	[191] see also [188]
3	HO—⟨⟩—CH$_2$C(O)NH–PS	Boc-Val-OH (0.16 mol/L), CDI (0.14 eq), DCM, 20 °C, 8 h	NHBoc ... NH–PS	[62]
4	HO—CH(Ph)–⟨⟩–PS	Fmoc-Leu-OH (0.6 mol/L, 1.5 eq), PPh$_3$ (3 eq), DEAD (3 eq), THF, 20 °C, 0.5 h	FmocHN ... Ph PS	[192]
5	O$_2$N, OH –⟨⟩–PS	BocHN⌒CO$_2$H (3 eq), DCC (1.5 eq), pyridine, DCM, 20 °C, 24 h	BocHN ... O$_2$N ... PS	[193] see also [194]
6	HO...O–(PS)	BpocHN⌒CO$_2$H, DIC (3 eq), DMAP (0.1 eq), THF, 2 h	BpocHN ... O–(PS)	[195]
7	HO...(PS)	N-Fmoc pyrrolidine COCl (5 eq), pyridine/DCM 4:6, 25 °C, 10 h	N-Fmoc ... (PS)	[196] see also [197]
8	NO$_2$ HO—⟨⟩—CH$_2$–PS	BocHN...CO$_2$H (0.5 mol/L, 2 eq), DCC (2 eq), DMF, 0 °C, 1 h, 20 °C, 5 h	BocHN ... NO$_2$ ... PS	[198]
9	HO–NH–C(O)–⟨⟩–PS	Boc-Ala-OH (0.2 mol/L, 2 eq), DCC (1 eq), DCM/pyridine 7:1, 16 h	BocHN ... O–N–C(O)–⟨⟩–PS	[141]

Experimental Procedure 13.5: Esterification of Wang resin with Fmoc amino acids [189]

To a mixture of Wang resin (10 g, 7.4 mmol), Fmoc-Phe-OH (5.7 g, 14.7 mmol, 2 eq) and DMF (50 mL) were added pyridine (2 mL, 24.7 mmol, 3.3 eq) and 2,6-dichlorobenzoyl chloride (2.1 mL, 14.7 mmol, 2 eq). The suspension was shaken at room temperature for 15–20 h. After washing, the remaining hydroxyl groups were benzoylated by treatment of the resin with pyridine (3 mL) and benzoyl chloride (3 mL) in DCE (80 mL) for 2 h. The amount of epimeric amino acid ester remained < 1 % for the following *N*-Fmoc amino acids [189]: Asp(O*t*Bu), Arg(Mtr), Glu(O*t*Bu), Ile, Leu, Lys(Boc), Met, Phe, Ser(*t*Bu), Thr(*t*Bu), Tyr(*t*Bu).

13.4.2 Preparation of Esters from Support-Bound Alkylating Agents

As alternative to the acylation of support-bound alcohols, esters can be prepared by O-alkylation of carboxylates with resin-bound alkylating agents (Table 13.14). This strategy has often been used for linking *N*-Boc amino acids to Merrifield resin. Because the tendency of salts to diffuse into cross-linked polystyrene is a function of their lipophilicity and polarizability, the yield of O-alkylations of carboxylates depends strongly on the cation chosen. In a comparative study Gisin [199] found that highest substitution rates were observed when cesium salts of Boc amino acids were treated with chloromethyl polystyrene (1 % cross-linked), whereas rubidium, potassium, or sodium carboxylates [200,201] were less reactive. Surprisingly, alkylammonium and lithium carboxylates reacted about ten times slower with Merrifield resin than did the cesium salts. Benzyl halides with electron donating groups or benzyl bromides are more reactive than Merrifield resin, and can lead to O-alkylation of carboxylates under mild conditions (Experimental Procedure 13.6). Instead of benzyl halides, dialkyl(benzyl)sulfonium salts have also been examined as alkylating agents for carboxylates [202].

Polystyrene-bound benzhydryl- or trityl halides react much faster with carboxylates than chloromethyl polystyrene, and the base used to form the carboxylate no longer plays a decisive role in these reactions (see Experimental Procedure 13.7).

Support-bound phenyldiazomethanes have been used to prepare esters directly from carboxylic acids under mild reaction conditions. Unfortunately the diazomethanes required are not easy to prepare, and have not yet found widespread application.

Table 13.14. Preparation of esters from support-bound alkylating agents.

Entry	Starting resin	Conditions	Product	Ref.
1		BocHN$\overset{R}{\underset{}{}}CO_2$Cs (0.06 mol/L, 1 eq), DMF, 50 °C, overnight		[199] see also [203]
2		CO$_2$Cs, OH, DMF, 80 °C, 8 h		[204]
3		CO$_2$H (0.15 mol/L, 5 eq), NaH (5 eq), Bu$_4$NBr (0.2 eq), THF, 67 °C, 3 d		[205]
4		BocHN CO$_2$Cs (3 eq), DMF, 40 °C, overnight		[206]
5		FmocHN CO$_2$H DIPEA, DCE, 20 °C, 18–36 h		[207]
6		Fmoc-Ser-OH (1.2 eq), DCM, 20 °C, 5 min		[208]
7		Boc-Gly-OH, CHCl$_3$, 20 °C, < 1 h		[209] see also [210]

Experimental Procedure 13.6: Esterification by nucleophilic displacement [211]

p-Benzyloxybenzyl bromide resin (prepared from Wang resin (50.0 g, 36.5 mmol; for preparation see Experimental Procedure 6.1) was suspended in DMF and treated with 4-fluoro-3-nitrobenzoic acid (13.5 g, 72.9 mmol, 2 eq), cesium iodide

(19.0 g, 73.1 mmol, 2 eq), and DIPEA (9.43 g, 73.0 mmol, 2 eq) at room temperature overnight. Filtration, washing with water and then with DMF, DCM, *i*PrOH, DMF, DCM, and *i*PrOH (2 × 300 mL of each solvent), followed by drying under reduced pressure, yielded the polystyrene-bound ester.

Experimental Procedure 13.7: General procedure for the attachment of Fmoc amino acids to a 2-chlorotrityl linker [212]

A solution of DIPEA (0.44 mL, 2.5 mmol) in DCE (5 mL) was added dropwise to a mixture of 2-chlorotrityl chloride resin (1.0 g, 1.6 mmol), DCE (10 mL), and Fmoc amino acid (1.0 mmol). The resulting mixture was stirred for 20 min, and methanol (1 mL) was added. After stirring for a further 10 min the resin was removed by filtration, washed with DCE, DMF, *i*PrOH, methanol, and diethyl ether, and dried under reduced pressure.

13.4.3 Preparation of Esters from Support-Bound Carboxylic Acids

The esterification of support-bound carboxylic acids has not been investigated as thoroughly as the esterification of support-bound alcohols. Resin-bound activated acid derivatives which are well suited to the preparation of esters include *O*-acylisoureas (formed from acids and carbodiimides), acyl halides [23,213–215], and mixed anhydrides (Table 13.15). *N*-Acylurea formation does not seem to compete as efficiently with esterifications as it does with the formation of amides from support-bound acids. Esters can also be prepared on insoluble supports from carboxylic acids by acid-catalyzed esterification [141,216]. Alternatively, support-bound carboxylic acids can be esterified by O-alkylation either with primary or secondary aliphatic alcohols under Mitsunobu conditions or with reactive alkyl halides or sulfonates (Table 13.15).

Table 13.15. Preparation of esters from support-bound acids or acylating agents.

Entry	Starting resin	Conditions	Product	Ref.
1		CF$_3$CH$_2$OH, DIC, DMAP		[21] see also [205]
2		(5 eq), DIC (5 eq), pyridine, DMAP, DMF, 25 °C, 24 h		[217] see also [205]
3		DIC, DMAP, pyridine, DMF		[218]
4		NEt$_3$, DMAP, DCM		[20]
5		RR'CHCOH (0.2 eq), NEt$_3$ (2 eq), DMAP (0.1 eq), DCM, 25 °C, 8 h		[219]
6		Bu$_4$NI, DIPEA, PhMe, 80 °C		[153]
7		THF		[43]
8		FmocHN, PPh$_3$, DIAD		[220]
9		iPrOH, AcCl, 55 °C		[115]

13.4.4 Preparation of Esters by Chemical Modification of Other Esters

Esters linked to insoluble supports can be converted into ester enolates and then C-alkylated under conditions similar to those used in solution. Illustrative examples for such alkylations are listed in Table 13.16. Strong bases such as LDA, $KN(SiMe_3)_2$, or BEMP are required to achieve quantitative deprotonation. Benzophenone imines of glycine esterified with Wang resin have been C-alkylated with the aid of phosphazene bases (Entries **3** and **4**, Table 13.16). Alkylations of this type can be conducted enantioselectively (up to 89 % ee) by adding enantiomerically pure, quinine-derived quaternary ammonium salts to the mixture of base and alkylating agent [221]. An additional method for C-alkylating polystyrene-bound benzophenone imines of glycine consists in oxidation of the latter with $Pb(OAc)_4$ to an α-acetoxyimine, followed by alkylation with a borane under basic reaction conditions (Entry **5**, Table 13.16).

To avoid the formation of ketenes by elimination of alkoxide, ester enolates are often prepared at low temperatures. If unreactive alkyl halides are used, the addition of Bu_4NI to the reaction mixture can be beneficial [125]. Examples of the radical-mediated α-alkylation of support-bound α-haloesters are given in Table 5.4. Further methods for C-alkylating esters on insoluble supports include the Ireland–Claisen rearrangement of *O*-allyl keteneacetals (Entry **6**, Table 13.16). Malonic esters and similar, strongly C,H-acidic compounds have been C-alkylated with Merrifield resin [222,223].

13.5 Preparation of Thiol Esters

Thiol esters are stronger acylating agents than simple alkyl esters, and have been prepared on solid phase mainly as synthetic intermediates. The preparation of thiol esters as intermediates for the synthesis of support-bound thiols is discussed in Section 8.1. Further examples of the preparation of thiol esters on insoluble supports include the aldol addition of ketene thioacetals to polystyrene-bound aldehydes (Entry **1**, Table 13.17), the acylation of support-bound thiols (Entries **2** and **3**), and the acylation of thiols with support-bound carboxylic acids (Entry **4**, Table 13.17). Thiol esters are stable towards acids (e.g. 50 % TFA in DCM) but are readily cleaved by nucleophiles, or reduced to alcohols by treatment with $LiBH_4$ [157].

13.6 Preparation of Amidines and Imino Ethers

Amidines are strongly basic compounds which often show interesting biological activity. Few preparations of amidines on insoluble supports have, however, been reported (Table 13.18). Linkers for amidines are treated in Section 3.9.

Amidines have been prepared on insoluble supports from other amidines by nucleophilic displacement of one amine by another (Entry **1**, Table 13.18). Polystyrene-bound thioamides can, furthermore, be converted to amidines by treatment with an aliphatic or aromatic amine in the presence of EDC (Entry **2**, Table 13.18).

Table 13.16. Preparation of esters by C-alkylation of support-bound esters.

Entry	Starting resin	Conditions	Product	Ref.
1		KN(SiMe$_3$)$_2$, BnBr, THF, −78 °C to 0 °C		[49] see also [224,225]
2		KN(SiMe$_3$)$_2$, THF, −78 °C, allyl iodide		[226]
3		(0.7 mol/L, 10 eq), BEMP (10 eq), NMP, 20 °C, 24 h		[125] see also [88,124] [221]
4		(5 eq), BEMP (3 eq), NMP, 20 °C, 16 h		[126]
5		1. Pb(OAc)$_4$ (0.06 mol/L, 1.5 eq), DCM, 20 °C, overnight 2. 9-tBu-9-BBN (2 eq), ArOK (1.3 eq), THF, 20 °C, overnight		[227]
6		THF, 50 °C, 5 h		[228]
7		Me$_2$CuLi, Me$_3$SiCl (repeat once)		[226]

Table 13.17. Preparation of thiol esters.

Entry	Starting resin	Conditions	Product	Ref.
1	(aldehyde benzyl resin)	BF₃OEt₂, DCM, Et₂O, −78 °C to −5 °C, 1.5 h; tBuS; L: menthylmethyl	(thiol ester product) (88% ee)	[205]
2	HS~N(PS)	Boc-Gly-OH, DCC, DMAP (4 eq of each), DCM, 24 h	BocHN~S~N(PS)	[229]
3	HS~PS	BnO~Cl; NEt₃, DCM, 20 °C	Ph~O~S~PS	[230] see also [231,232]
4	HO~(PS) OTIPS	BnSH (30 eq), DCC, DMAP, THF, 20 °C, 18 h	Ph~S~(PS) OTIPS	[157]

Trichloroacetimidates are the only type of imino ethers which have found some application in solid phase synthesis. Trichloroacetimidates can be readily prepared from support-bound alcohols by treatment with trichloroacetonitrile and a base (Entry **3**, Table 13.18). Because trichloroacetimidates are good alkylating agents, this reaction is a convenient alternative for converting support-bound benzylic alcohols into electrophilic reagents (see Section 7.2.1). Trichloroacetimidates prepared from Wang resin or from hydroxymethyl polystyrene are quite stable and can be stored for several months without decomposition [233].

Table 13.18. Preparation of amidines and imino ethers.

Entry	Starting resin	Conditions	Product	Ref.
1	(amidine-O-PS)	NH (piperidone); PhMe, (NH₄)₂SO₄, 110 °C, 18 h	(product O-PS)	[234] see also [235]
2	S~NHPh; NC~O(PS)	ArCH₂NH₂ (0.6 mol/L), EDC (1 mol/L), DMF, 20 °C, 24 h	Ar~N~NHPh; NC~O(PS)	[236]
3	HO~PS	NaN(SiMe₃)₂ (0.9 eq), THF, 20 °C, 1 h, then Cl₃CCN (2 eq), 0–20 °C, 16 h, or (Wang resin) Cl₃CCN (1.5 mol/L), DBU/DCM 1:100, 0 °C, 40 min	Cl₃C~O~PS NH	[237] [238] see also [233]

13.7 Preparation of Nitriles and Isonitriles

Few preparations of nitriles have been performed on insoluble supports (Table 13.19). Aromatic and heteroaromatic nitriles have been prepared on solid phase from the corresponding iodoarenes by metallation and reaction with tosyl cyanide (Entry **1**, Table 13.19). The reaction of chloromethyl polystyrene with NaCN has, furthermore, been used to prepare support-bound benzyl cyanide (Entry **2**, Table 13.19). Cleavage with simultaneous formation of nitriles can be achieved by treating polystyrene-bound sulfonylhydrazones with KCN (Entry **3**, Table 13.19). Support-bound benzaldehydes have been converted into 3-aryl-2-propenenitriles by means of the Horner–Emmons reaction with $(EtO)_2P(O)CH_2CN$ [239].

Nitrile formation from unprotected asparagine or glutamine is an unwanted side reaction in solid-phase peptide synthesis [9,17]. This dehydration occurs during activation of asparagine or glutamine with carbodiimides (Figure 13.9), but can be prevented either by performing the coupling in the presence of HOBt, or by using asparagine or glutamine derivatives with a protected carboxamido group (e.g. *N*-trityl, *N*-4,4'-dimethoxybenzhydryl, *N*-2,4,6-trimethoxybenzyl, or *N*-9-xanthenyl amides [17]). These derivatives are also more soluble than non-amide-protected asparagine or glutamine.

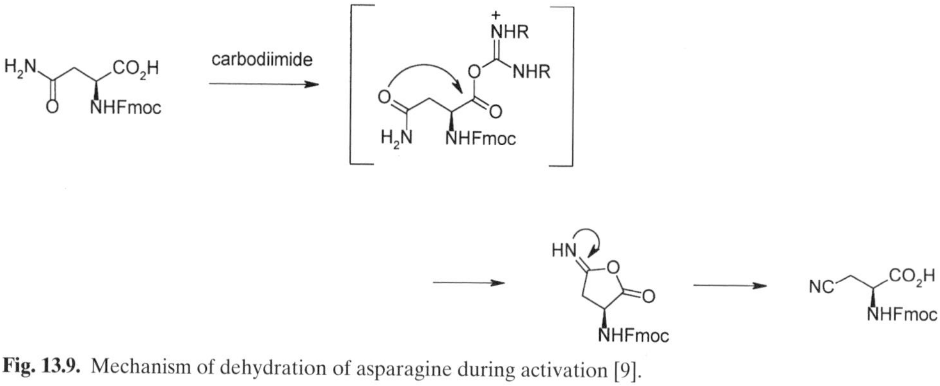

Fig. 13.9. Mechanism of dehydration of asparagine during activation [9].

Isonitriles are valuable synthetic intermediates, which can be used to prepare peptoids and various heterocycles. Because of their toxicity and strong, unpleasant smell, however, chemists tend to avoid working with isonitriles. Handling of isolated isonitriles can be elegantly circumvented by generating them on insoluble supports. Despite the obvious advantages of using immobilized isonitriles, few examples have been reported. In most of these the isonitrile functionality was formed by dehydration of formamides (Entries **4–6**, Table 13.19).

Table 13.19. Preparation of nitriles and isonitriles.

Entry	Starting resin	Conditions	Product	Ref.
1		*i*PrMgBr (0.19 mol/L, 7.3 eq), THF, −35 °C, 0.5 h, then TsCN		[240]
2		NaCN (1.13 mol/L), DMF/H$_2$O 10:2, 120 °C, 19.5 h		[151] see also [222]
3		KCN (0.2 mol/L, 3 eq), MeOH, 65 °C, 72 h		[241]
4		PPh$_3$, CCl$_4$, NEt$_3$, DCM		[242]
5		PPh$_3$ (4 eq), CCl$_4$ (4 eq), NEt$_3$ (10 eq), DCM, 20 °C, 16 h		[243]
6		PPh$_3$, CCl$_4$, NEt$_3$ (5 eq of each), DCM		[37] see also [244]

13.8 Preparation of Imides

The intermolecular N-acylation of amides generally requires treatment of the latter with strong acylating agents, such as ketenes, acyl halides, or anhydrides, in the presence of a base. Few examples of such N-acylations have been reported (Entries **1–3**, Table 13.20; see also saftey-catch linkers, Section 3.1.2.3).

During the solid-phase synthesis of peptides containing *n*-alkyl esters of aspartic acid, cyclic imides might form, either under acidic or under basic reaction conditions (Entry **5**, Table 13.20). Hydrolysis of the resulting succinimides can lead to the formation of peptides containing a β-amide bond (β-aspartyl peptides, Figure 13.10). This unwanted imide formation can be suppressed by protecting aspartic acid as *tert*-alkyl ester (Fmoc methodology) or as cyclohexyl ester (Boc methodology) [17,245]. Peptides containing glutamic acid esters do not undergo imide formation as readily as those containing aspartic acid esters.

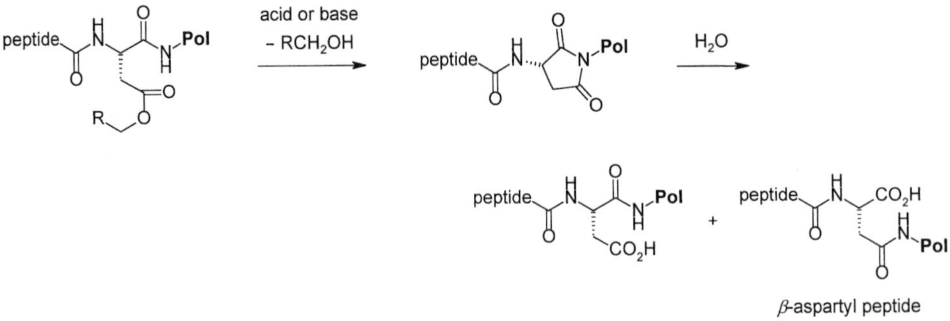

Fig. 13.10. Formation and hydrolysis of succinimides derived from aspartic acid.

Table 13.20. Preparation of imides and related compounds.

Entry	Starting resin	Conditions	Product	Ref.
1	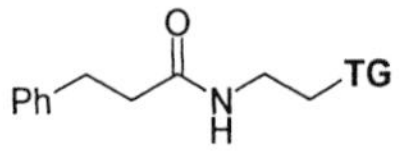 Ph—(O)—N(H)—TG	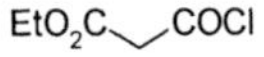 EtO$_2$C—COCl (3.9 mol/L), C$_6$H$_6$, 65 °C, 1.5 h	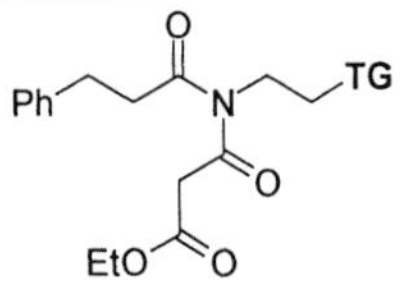 Ph—N—TG; EtO—O	[246-248]
2	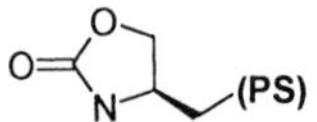 O=(oxazolidinone)—(PS)	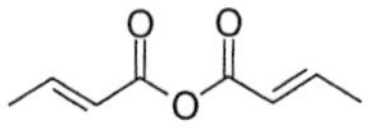 DMAP, NEt$_3$, THF, 70 °C, 72 h	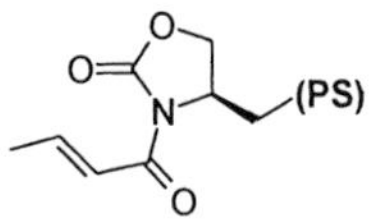 O=(oxazolidinone)—(PS)	[249] see also [129]
3	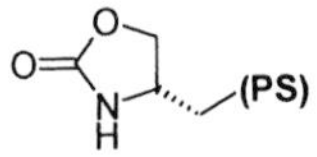 O=(oxazolidinone)—(PS)	LiN(SiMe$_3$)$_2$ (2 eq), THF, −20 °C, then 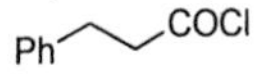 Ph—COCl	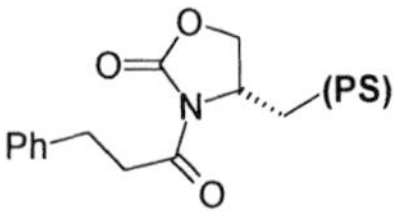 O=(oxazolidinone)—(PS); Ph	[250]
4	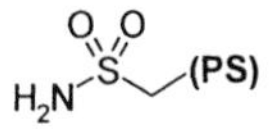 H$_2$N—S(O)$_2$—(PS)	Fmoc-Phe-OH (0.27 mol/L, 3 eq), DIPEA (5 eq), PyBOP (3 eq), CHCl$_3$, −20 °C, 8 h	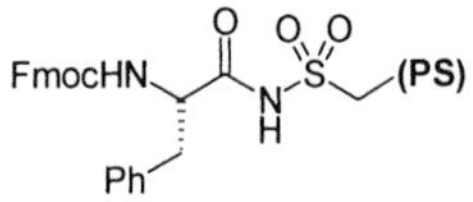 FmocHN—N(H)—S(O)$_2$—(PS); Ph	[82]
5	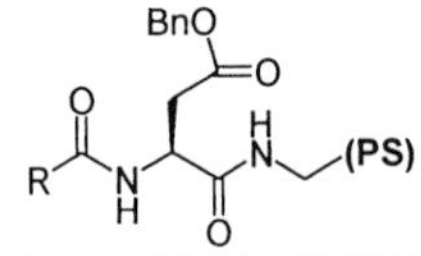 BnO—(O)—R—C(O)N(H)—N(H)—(PS)	NEt$_3$, DCM, 25 °C, 24 h	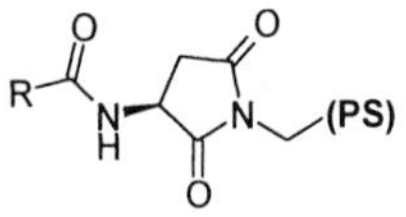 R—C(O)N(H)—(succinimide)—(PS)	[245]

13.9 Preparation of Thioamides

Thioamides are suitable intermediates for the preparation of amidines, thiazoles and thiophenes. Thioamides have mainly been prepared on insoluble supports by C-acylation of enamines or C,H-acidic compounds with isothiocyanates (Entries **1–3**, Table 13.21). Thioamides can, furthermore, be prepared on cross-linked polystyrene by addition of H_2S to nitriles (Entry **4**, Table 13.21).

Table 13.21. Preparation of thioamides.

Entry	Starting resin	Conditions	Product	Ref.
1		ArNCS, THF, 67 °C Ar: 4-(F$_3$CO)C$_6$H$_4$		[251]
2		PhNCS (0.7 mol/L), DBU/DMF 1:3.6, 20 °C, 18 h		[252] see also [236]
3		(1 mol/L), DBU/DMF 1:7, 20 °C, 15 h		[80]
4		EtO–P–OEt SH (1 mol/L), THF/H$_2$O 4:1, 70 °C, overnight		[253]

References for Chapter 13

[1] March, J. *Advanced Organic Chemistry*; John Wiley & Sons: New York, **1992**.
[2] Hudson, D. *J. Org. Chem.* **1988**, *53*, 617–624.
[3] Klose, J.; El-Faham, A.; Henklein, P.; Carpino, L. A.; Bienert, M. *Tetrahedron Lett.* **1999**, *40*, 2045–2048.
[4] Carpino, L. A.; El-Faham, A. *Tetrahedron* **1999**, *55*, 6813–6830.
[5] Knorr, R.; Trzeciak, A.; Bannwarth, W.; Gillessen, D. *Tetrahedron Lett.* **1989**, *30*, 1927–1930.
[6] Hudson, D. *Pept. Res.* **1990**, *3*, 51–55.
[7] Jones, J. *The Chemical Synthesis of Peptides*; Oxford University Press: Oxford, **1994**.
[8] Merrifield, B. In *Peptides; Synthesis, Structures, and Applications*; Gutte, B. Ed.; Academic Press: London, **1995**.
[9] Fields, G. B.; Noble, R. L. *Int. J. Pept. Prot. Res.* **1990**, *35*, 161–214.
[10] Carpino, L. A.; El-Faham, A. *J. Am. Chem. Soc.* **1995**, *117*, 5401–5402.
[11] Chan, T. Y.; Chen, A.; Allanson, N.; Chen, R.; Liu, D. S.; Sofia, M. J. *Tetrahedron Lett.* **1996**, *37*, 8097–8100.
[12] Carpino, L. A. *J. Am. Chem. Soc.* **1993**, *115*, 4397–4398.
[13] Falb, E.; Yechezkel, T.; Salitra, Y.; Gilon, C. *J. Pept. Res.* **1999**, *53*, 507–517.

[14] Wenschuh, H.; Beyermann, M.; Krause, E.; Brudel, M.; Winter, R.; Schumann, M.; Carpino, L. A.; Bienert, M. *J. Org. Chem.* **1994**, *59*, 3275–3280.

[15] Carpino, L. A.; Sadat-Aalaee, D.; Chao, H. G.; DeSelms, R. H. *J. Am. Chem. Soc.* **1990**, *112*, 9651–9652.

[16] Heimer, E. P.; Chang, C. D.; Lambros, T.; Meienhofer, J. *Int. J. Pept. Prot. Res.* **1981**, *18*, 237–241.

[17] Novabiochem Catalog and Peptide Synthesis Handbook, Läufelfingen, **1999**.

[18] Advanced Chemtech Handbook of Combinatorial and Solid-Phase Organic Chemistry, Louisville, **1998**.

[19] Gayo, L. M.; Suto, M. J. *Tetrahedron Lett.* **1997**, *38*, 211–214.

[20] Panek, J. S.; Zhu, B. *J. Am. Chem. Soc.* **1997**, *119*, 12022–12023.

[21] Hamper, B. C.; Kolodziej, S. A.; Scates, A. M. *Tetrahedron Lett.* **1998**, *39*, 2047–2050.

[22] Lorsbach, B. A.; Bagdanoff, J. T.; Miller, R. B.; Kurth, M. J. *J. Org. Chem.* **1998**, *63*, 2244–2250.

[23] Leznoff, C. C.; Dixit, D. M. *Can. J. Chem.* **1977**, *55*, 3351–3355.

[24] Yan, B.; Kumaravel, G.; Anjaria, H.; Wu, A. Y.; Petter, R. C.; Jewell, C. F.; Wareing, J. R. *J. Org. Chem.* **1995**, *60*, 5736–5738.

[25] Goldwasser, J. M.; Leznoff, C. C. *Can. J. Chem.* **1978**, *56*, 1562–1568.

[26] Leznoff, C. C.; Goldwasser, J. M. *Tetrahedron Lett.* **1977**, 1875–1878.

[27] Zhu, Z. M.; McKittrick, B. *Tetrahedron Lett.* **1998**, *39*, 7479–7482.

[28] Bilodeau, M. T.; Cunningham, A. M. *J. Org. Chem.* **1998**, *63*, 2800–2801.

[29] Murphy, M. M.; Schullek, J. R.; Gordon, E. M.; Gallop, M. A. *J. Am. Chem. Soc.* **1995**, *117*, 7029–7030.

[30] Bicknell, A. J.; Hird, N. W. *Bioorg. Med. Chem. Lett.* **1996**, *6*, 2441–2444.

[31] Hamper, B. C.; Kolodziej, S. A.; Scates, A. M.; Smith, R. G.; Cortez, E. *J. Org. Chem.* **1998**, *63*, 708–718.

[32] Plunkett, M. J.; Ellman, J. A. *J. Org. Chem.* **1997**, *62*, 2885–2893.

[33] Krchnák, V.; Flegelová, Z.; Weichsel, A. S.; Lebl, M. *Tetrahedron Lett.* **1995**, *36*, 6193–6196.

[34] Di Fenza, A.; Tancredi, M.; Galoppini, C.; Rovero, P. *Tetrahedron Lett.* **1998**, *39*, 8529–8532.

[35] Zaragoza, F.; Stephensen, H. *J. Org. Chem.* **1999**, *64*, 2555–2557.

[36] Tumelty, D.; Schwarz, M. K.; Needels, M. C. *Tetrahedron Lett.* **1998**, *39*, 7467–7470.

[37] Hulme, C.; Peng, J.; Morton, G.; Salvino, J. M.; Herpin, T.; Labaudiniere, R. *Tetrahedron Lett.* **1998**, *39*, 7227–7230.

[38] Chao, H. G.; Bernatowicz, M. S.; Reiss, P. D.; Klimas, C. E.; Matsueda, G. R. *J. Am. Chem. Soc.* **1994**, *116*, 1746–1752.

[39] Greenberg, M. M.; Gilmore, J. L. *J. Org. Chem.* **1994**, *59*, 746–753.

[40] Albericio, F.; Barany, G. *Int. J. Pept. Prot. Res.* **1985**, *26*, 92–97.

[41] Trautwein, A. W.; Jung, G. *Tetrahedron Lett.* **1998**, *39*, 8263–8266.

[42] Gordeev, M. F.; Luehr, G. W.; Hui, H. C.; Gordon, E. M.; Patel, D. V. *Tetrahedron* **1998**, *54*, 15879–15890.

[43] Léger, R.; Yen, R.; She, M. W.; Lee, V. J.; Hecker, S. J. *Tetrahedron Lett.* **1998**, *39*, 4171–4174.

[44] Linn, J. A.; Gerritz, S. W.; Handlon, A. L.; Hyman, C. E.; Heyer, D. *Tetrahedron Lett.* **1999**, *40*, 2227–2230.

[45] Gordeev, M. F.; Gordon, E. M.; Patel, D. V. *J. Org. Chem.* **1997**, *62*, 8177–8181.

[46] Li, W.; Yan, B. *J. Org. Chem.* **1998**, *63*, 4092–4097.

[47] Marti, R. E.; Yan, B.; Jarosinski, M. A. *J. Org. Chem.* **1997**, *62*, 5615–5618.

[48] Hanessian, S.; Huynh, H. K. *Synlett* **1999**, 102–104.

[49] Hanessian, S.; Xie, F. *Tetrahedron Lett.* **1998**, *39*, 737–740.

[50] Thompson, L. A.; Moore, F. L.; Moon, Y. C.; Ellman, J. A. *J. Org. Chem.* **1998**, *63*, 2066–2067.

[51] Tung, R. D.; Rich, D. H. *J. Am. Chem. Soc.* **1985**, *107*, 4342–4343.

[52] Van der Auwera, C.; Anteunis, M. J. O. *Int. J. Pept. Prot. Res.* **1987**, *29*, 574–588.

[53] Kemp, D. S.; Wrobel, S. J.; Wang, S.-W.; Bernstein, Z.; Rebek, J. *Tetrahedron* **1974**, *30*, 3969–3980.

[54] Sheehan, J. C.; Hess, G. P. *J. Am. Chem. Soc.* **1955**, *77*, 1067–1068.

[55] DeTar, D. F.; Silverstein, R.; Rogers, F. F. *J. Am. Chem. Soc.* **1966**, *88*, 1024–1030.

[56] Doleschall, G.; Lempert, K. *Tetrahedron Lett.* **1963**, 1195–1199.

[57] Smith, M.; Moffatt, J. G.; Khorana, H. G. *J. Am. Chem. Soc.* **1958**, *80*, 6204–6212.

[58] Kurzer, F.; Douraghi-Zadeh, K. *Chem. Rev.* **1967**, *67*, 107–152.

[59] Balcom, B. J.; Petersen, N. O. *J. Org. Chem.* **1989**, *54*, 1922–1927.

[60] DeTar, D. F.; Silverstein, R. *J. Am. Chem. Soc.* **1966**, *88*, 1013–1019.

[61] Sarin, V. K.; Kent, S. B. H.; Mitchell, A. R.; Merrifield, R. B. *J. Am. Chem. Soc.* **1984**, *106*, 7845–7850.

[62] Mitchell, A. R.; Erickson, B. W.; Ryabtsev, M. N.; Hodges, R. S.; Merrifield, R. B. *J. Am. Chem. Soc.* **1976**, *98*, 7357–7362.

[63] Matthews, J.; Rivero, R. A. *J. Org. Chem.* **1998**, *63*, 4808–4810.
[64] Weigelt, D.; Magnusson, G. *Tetrahedron Lett.* **1998**, *39*, 2839–2842.
[65] Simmonds, R. G. *Int. J. Pept. Prot. Res.* **1996**, *47*, 36–41.
[66] Quibell, M.; Packman, L. C.; Johnson, T. *J. Am. Chem. Soc.* **1995**, *117*, 11656–11668.
[67] Boojamra, C. G.; Burow, K. M.; Thompson, L. A.; Ellman, J. A. *J. Org. Chem.* **1997**, *62*, 1240–1256.
[68] Zuckermann, R. N.; Kerr, J. M.; Kent, S. B. H.; Moos, W. H. *J. Am. Chem. Soc.* **1992**, *114*, 10646–10647.
[69] Ouyang, X.; Tamayo, N.; Kiselyov, A. S. *Tetrahedron* **1999**, *55*, 2827–2834.
[70] Zaragoza, F. unpublished results.
[71] Zaragoza, F.; Petersen, S. V. *Tetrahedron* **1996**, *52*, 5999–6002.
[72] Heerding, D. A.; Takata, D. T.; Kwon, C.; Huffman, W. F.; Samanen, J. *Tetrahedron Lett.* **1998**, *39*, 6815–6818.
[73] Kim, S. W.; Hong, C. Y.; Koh, J. S.; Lee, E. J.; Lee, K. *Mol. Diversity* **1998**, *3*, 133–136.
[74] Hoekstra, W. J.; Greco, M. N.; Yabut, S. C.; Hulshizer, B. L.; Maryanoff, B. E. *Tetrahedron Lett.* **1997**, *38*, 2629–2632.
[75] Miller, P. C.; Owen, T. J.; Molyneaux, J. M.; Curtis, J. M.; Jones, C. R. *J. Comb. Chem.* **1999**, *1*, 223–234.
[76] Hamper, B. C.; Snyderman, D. M.; Owen, T. J.; Scates, A. M.; Owsley, D. C.; Kesselring, A. S.; Chott, R. C. *J. Comb. Chem.* **1999**, *1*, 140–150.
[77] Zhang, H. C.; Maryanoff, B. E. *J. Org. Chem.* **1997**, *62*, 1804–1809.
[78] Wess, G.; Bock, K.; Kleine, H.; Kurz, M.; Guba, W.; Hemmerle, H.; Lopez-Calle, E.; Baringhaus, K. H.; Glombik, H.; Enhsen, A.; Kramer, W. *Angew. Chem. Int. Ed. Engl.* **1996**, *35*, 2222–2224.
[79] Devraj, R.; Cushman, M. *J. Org. Chem.* **1996**, *61*, 9368–9373.
[80] Stephensen, H.; Zaragoza, F. *J. Org. Chem.* **1997**, *62*, 6096–6097.
[81] Coste, J.; Dufour, M.-N.; Pantaloni, A.; Castro, B. *Tetrahedron Lett.* **1990**, *31*, 669–672.
[82] Backes, B. J.; Ellman, J. A. *J. Org. Chem.* **1999**, *64*, 2322–2330.
[83] Albericio, F.; Bofill, J. M.; El-Faham, A.; Kates, S. A. *J. Org. Chem.* **1998**, *63*, 9678–9683.
[84] Castro, B.; Dormoy, J. R.; Evin, G.; Selve, C. *Tetrahedron Lett.* **1975**, 1219–1222.
[85] Coste, J.; Le-Nguyen, D.; Castro, B. *Tetrahedron Lett.* **1990**, *31*, 205–208.
[86] Kiso, Y.; Yajima, H. In *Peptides; Synthesis, Structures, and Applications*; Gutte, B. Ed.; Academic Press: London, **1995**.
[87] Ehrlich, A.; Heyne, H.-U.; Winter, R.; Beyermann, M.; Haber, H.; Carpino, L. A.; Bienert, M. *J. Org. Chem.* **1996**, *61*, 8831–8838.
[88] Griffith, D. L.; O'Donnell, M. J.; Pottorf, R. S.; Scott, W. L.; Porco, J. A. *Tetrahedron Lett.* **1997**, *38*, 8821–8824.
[89] Richter, L. S.; Zuckermann, R. N. *Bioorg. Med. Chem. Lett.* **1995**, *5*, 1159–1162.
[90] Zhu, T.; Boons, G. J. *Angew. Chem. Int. Ed. Engl.* **1998**, *37*, 1898–1900.
[91] Tan, D. S.; Foley, M. A.; Shair, M. D.; Schreiber, S. L. *J. Am. Chem. Soc.* **1998**, *120*, 8565–8566.
[92] Wang, F.; Hauske, J. R. *Tetrahedron Lett.* **1997**, *38*, 6529–6532.
[93] Roussel, P.; Bradley, M.; Matthews, I.; Kane, P. *Tetrahedron Lett.* **1997**, *38*, 4861–4864.
[94] Habermann, J.; Kunz, H. *Tetrahedron Lett.* **1998**, *39*, 265–268.
[95] Bhandari, A.; Jones, D. G.; Schullek, J. R.; Vo, K.; Schunk, C. A.; Tamanaha, L. L.; Chen, D.; Yuan, Z. Y.; Needels, M. C.; Gallop, M. A. *Bioorg. Med. Chem. Lett.* **1998**, *8*, 2303–2308.
[96] Swayze, E. E. *Tetrahedron Lett.* **1997**, *38*, 8465–8468.
[97] Akaji, K.; Hayashi, Y.; Kiso, Y.; Kuriyama, N. *J. Org. Chem.* **1999**, *64*, 405–411.
[98] van Loevezijn, A.; van Maarseveen, J. H.; Stegman, K.; Visser, G. M.; Koomen, G. J. *Tetrahedron Lett.* **1998**, *39*, 4737–4740.
[99] Fantauzzi, P. P.; Yager, K. M. *Tetrahedron Lett.* **1998**, *39*, 1291–1294.
[100] Mohan, R.; Chou, Y. L.; Morrissey, M. M. *Tetrahedron Lett.* **1996**, *37*, 3963–3966.
[101] del Fresno, M.; Alsina, J.; Royo, M.; Barany, G.; Albericio, F. *Tetrahedron Lett.* **1998**, *39*, 2639–2642.
[102] Baird, E. E.; Dervan, P. B. *J. Am. Chem. Soc.* **1996**, *118*, 6141–6146.
[103] Böhm, G.; Dowden, J.; Rice, D. C.; Burgess, I.; Pilard, J. F.; Guilbert, B.; Haxton, A.; Hunter, R. C.; Turner, N. J.; Flitsch, S. L. *Tetrahedron Lett.* **1998**, *39*, 3819–3822.
[104] Bleicher, K. H.; Wareing, J. R. *Tetrahedron Lett.* **1998**, *39*, 4591–4594.
[105] Marti, R. E.; Bleicher, K. H.; Bair, K. W. *Tetrahedron Lett.* **1997**, *38*, 6145–6148.
[106] Romoff, T. T.; Ma, L.; Wang, Y. W.; Campbell, D. A. *Synlett* **1998**, 1341–1342.
[107] Weber, L.; Iaiza, P.; Biringer, G.; Barbier, P. *Synlett* **1998**, 1156–1158.
[108] Trautwein, A. W.; Süssmuth, R. D.; Jung, G. *Bioorg. Med. Chem. Lett.* **1998**, *8*, 2381–2384.
[109] Zhu, J.; Hegedus, L. S. *J. Org. Chem.* **1995**, *60*, 5831–5837.
[110] Pulley, S. R.; Hegedus, L. S. *J. Am. Chem. Soc.* **1993**, *115*, 9037–9047.

[111] Szardenings, A. K.; Burkoth, T. S.; Lu, H. H.; Tien, D. W.; Campbell, D. A. *Tetrahedron* **1997**, *53*, 6573–6593.

[112] Obrecht, D.; Abrecht, C.; Grieder, A.; Villalgordo, J. M. *Helv. Chim. Acta* **1997**, *80*, 65–72.

[113] Tempest, P. A.; Brown, S. D.; Armstrong, R. W. *Angew. Chem. Int. Ed. Engl.* **1996**, *35*, 640–642.

[114] Mjalli, A. M. M.; Sarshar, S.; Baiga, T. J. *Tetrahedron Lett.* **1996**, *37*, 2943–2946.

[115] Strocker, A. M.; Keating, T. A.; Tempest, P. A.; Armstrong, R. W. *Tetrahedron Lett.* **1996**, *37*, 1149–1152.

[116] Cao, X. D.; Moran, E. J.; Siev, D.; Lio, A.; Ohashi, C.; Mjalli, A. M. M. *Bioorg. Med. Chem. Lett.* **1995**, *5*, 2953–2958.

[117] Paulvannan, K. *Tetrahedron Lett.* **1999**, *40*, 1851–1854.

[118] Hoel, A. M. L.; Nielsen, J. *Tetrahedron Lett.* **1999**, *40*, 3941–3944.

[119] Piscopio, A. D.; Miller, J. F.; Koch, K. *Tetrahedron* **1999**, *55*, 8189–8198.

[120] Short, K. M.; Ching, B. W.; Mjalli, A. M. M. *Tetrahedron* **1997**, *53*, 6653–6679.

[121] Srivastava, S. K.; Haq, W.; Chauhan, P. M. S. *Bioorg. Med. Chem. Lett.* **1999**, *9*, 965–966.

[122] Burgess, K.; Lim, D. *Chem. Commun.* **1997**, 785–786.

[123] O'Donnell, M. J.; Zhou, C. Y.; Scott, W. L. *J. Am. Chem. Soc.* **1996**, *118*, 6070–6071.

[124] Scott, W. L.; Zhou, C. Y.; Fang, Z. Q.; O'Donnell, M. J. *Tetrahedron Lett.* **1997**, *38*, 3695–3698.

[125] O'Donnell, M. J.; Lugar, C. W.; Pottorf, R. S.; Zhou, C. Y.; Scott, W. L.; Cwi, C. L. *Tetrahedron Lett.* **1997**, *38*, 7163–7166.

[126] Domínguez, E.; O'Donnell, M. J.; Scott, W. L. *Tetrahedron Lett.* **1998**, *39*, 2167–2170.

[127] Moon, H.; Schore, N. E.; Kurth, M. J. *Tetrahedron Lett.* **1994**, *35*, 8915–8918.

[128] Phoon, C. W.; Abell, C. *Tetrahedron Lett.* **1998**, *39*, 2655–2658.

[129] Allin, S. M.; Shuttleworth, S. J. *Tetrahedron Lett.* **1996**, *37*, 8023–8026.

[130] Backes, B. J.; Ellman, J. A. *J. Am. Chem. Soc.* **1994**, *116*, 11171–11172.

[131] Nefzi, A.; Ostresh, J. M.; Houghten, R. A. *Tetrahedron* **1999**, *55*, 335–344.

[132] Woolard, F. X.; Paetsch, J.; Ellman, J. A. *J. Org. Chem.* **1997**, *62*, 6102–6103.

[133] Norman, T. C.; Gray, N. S.; Koh, J. T.; Schultz, P. G. *J. Am. Chem. Soc.* **1996**, *118*, 7430–7431.

[134] Furman, B.; Thürmer, R.; Kaluza, Z.; Lysek, R.; Voelter, W.; Chmielewski, M. *Angew. Chem. Int. Ed. Engl.* **1999**, *38*, 1121–1123.

[135] Salvino, J. M.; Mervic, M.; Mason, H. J.; Kiesow, T.; Teager, D.; Airey, J.; Labaudiniere, R. *J. Org. Chem.* **1999**, *64*, 1823–1830.

[136] Mellor, S. L.; Chan, W. C. *Chem. Commun.* **1997**, 2005–2006.

[137] Chang, J. K.; Shimizu, M.; Wang, S. S. *J. Org. Chem.* **1976**, *41*, 3255–3258.

[138] Wang, S.; Merrifield, R. B. *J. Am. Chem. Soc.* **1969**, *91*, 6488–6491.

[139] Kobayashi, S.; Furuta, T.; Sugita, K.; Okitsu, O.; Oyamada, H. *Tetrahedron Lett.* **1999**, *40*, 1341–1344.

[140] Wilson, M. W.; Hernández, A. S.; Calvet, A. P.; Hodges, J. C. *Mol. Diversity* **1998**, *3*, 95–112.

[141] Sophiamma, P. N.; Sreekumar, K. *Indian J. Chem. Sect. B (Org. Chem.)* **1997**, *36*, 995–999.

[142] Chen, J. J.; Spatola, A. F. *Tetrahedron Lett.* **1997**, *38*, 1511–1514.

[143] Fyles, T. M.; Leznoff, C. C. *Can. J. Chem.* **1976**, *54*, 935–942.

[144] Itsuno, S.; Darling, G. D.; Stöver, H. D. H.; Fréchet, J. M. J. *J. Org. Chem.* **1987**, *52*, 4644–4645.

[145] O'Brien, R. A.; Chen, T.; Rieke, R. D. *J. Org. Chem.* **1992**, *57*, 2667–2677.

[146] Farrall, M. J.; Fréchet, J. M. J. *J. Org. Chem.* **1976**, *41*, 3877–3882.

[147] Luo, K. X.; Zhou, P.; Lodish, H. F. *Proc. Natl. Acad. Sci. USA* **1995**, *92*, 11761–11765.

[148] Annis, D. A.; Helluin, O.; Jacobsen, E. N. *Angew. Chem. Int. Ed. Engl.* **1998**, *37*, 1907–1909.

[149] Beebe, X.; Chiappari, C. L.; Kurth, M. J.; Schore, N. E. *J. Org. Chem.* **1993**, *58*, 7320–7321.

[150] Sylvain, C.; Wagner, A.; Mioskowski, C. *Tetrahedron Lett.* **1997**, *38*, 1043–1044.

[151] Kusama, T.; Hayatsu, H. *Chem. Pharm. Bull.* **1970**, *18*, 319–327.

[152] Kim, S. W.; Hong, C. Y.; Lee, K.; Lee, E. J.; Koh, J. S. *Bioorg. Med. Chem. Lett.* **1998**, *8*, 735–738.

[153] Yu, K. L.; Civiello, R.; Roberts, D. G. M.; Seiler, S. M.; Meanwell, N. A. *Bioorg. Med. Chem. Lett.* **1999**, *9*, 663–666.

[154] Flynn, D. L.; Zelle, R. E.; Grieco, P. A. *J. Org. Chem.* **1983**, *48*, 2424–2426.

[155] Hoekstra, W. J.; Maryanoff, B. E.; Andrade-Gordon, P.; Cohen, J. H.; Costanzo, M. J.; Damiano, B. P.; Haertlein, B. J.; Harris, B. D.; Kauffman, J. A.; Keane, P. M.; McComsey, D. F.; Villani, F. J.; Yabut, S. C. *Bioorg. Med. Chem. Lett.* **1996**, *6*, 2371–2376.

[156] Bourne, G. T.; Meutermans, W. D. F.; Alewood, P. F.; McGeary, R. P.; Scanlon, M.; Watson, A. A.; Smythe, M. L. *J. Org. Chem.* **1999**, *64*, 3095–3101.

[157] Reggelin, M.; Brenig, V.; Welcker, R. *Tetrahedron Lett.* **1998**, *39*, 4801–4804.

[158] Far, A. R.; Tidwell, T. T. *J. Org. Chem.* **1998**, *63*, 8636–8637.

[159] Blaskovich, M. A.; Kahn, M. *J. Org. Chem.* **1998**, *63*, 1119–1125.

[160] Parlow, J. J.; Normansell, J. E. *Mol. Diversity* **1996**, *1*, 266–269.

[161] Cohen, B. J.; Karoly-Hafeli, H.; Patchornik, A. *J. Org. Chem.* **1984**, *49*, 922–924.
[162] Marshall, D. L.; Liener, I. E. *J. Org. Chem.* **1970**, *35*, 867–868.
[163] Parlow, J. J.; Mischke, D. A.; Woodard, S. S. *J. Org. Chem.* **1997**, *62*, 5908–5919.
[164] Scarr, R. B.; Findeis, M. A. *Pept. Res.* **1990**, *3*, 238–241.
[165] DeGrado, W. F.; Kaiser, E. T. *J. Org. Chem.* **1982**, *47*, 3258–3261.
[166] Smith, R. A.; Bobko, M. A.; Lee, W. *Bioorg. Med. Chem. Lett.* **1998**, *8*, 2369–2374.
[167] Sylvain, C.; Wagner, A.; Mioskowski, C. *Tetrahedron Lett.* **1999**, *40*, 875–878.
[168] Brown, P. J.; Hurley, K. P.; Stuart, L. W.; Willson, T. M. *Synthesis* **1997**, 778–782.
[169] Cobb, J. M.; Fiorini, M. T.; Goddard, C. R.; Theoclitou, M. E.; Abell, C. *Tetrahedron Lett.* **1999**, *40*, 1045–1048.
[170] Barn, D. R.; Morphy, J. R. *J. Comb. Chem.* **1999**, *1*, 151–156.
[171] Whitehouse, D. L.; Savinov, S. N.; Austin, D. J. *Tetrahedron Lett.* **1997**, *38*, 7851–7852.
[172] Eynde, J. J. V.; Rutot, D. *Tetrahedron* **1999**, *55*, 2687–2694.
[173] Tietze, L. F.; Steinmetz, A. *Angew. Chem. Int. Ed. Engl.* **1996**, *35*, 651–652.
[174] Tietze, L. F.; Steinmetz, A. *Synlett* **1996**, 667–668.
[175] MacDonald, A. A.; DeWitt, S. H.; Hogan, E. M.; Ramage, R. *Tetrahedron Lett.* **1996**, *37*, 4815–4818.
[176] Karoyan, P.; Triolo, A.; Nannicini, R.; Giannotti, D.; Altamura, M.; Chassaing, G.; Perrotta, E. *Tetrahedron Lett.* **1999**, *40*, 71–74.
[177] Barbaste, M.; Rolland-Fulcrand, V.; Roumestant, M. L.; Viallefont, P.; Martinez, J. *Tetrahedron Lett.* **1998**, *39*, 6287–6290.
[178] Fancelli, D.; Fagnola, M. C.; Severino, D.; Bedeschi, A. *Tetrahedron Lett.* **1997**, *38*, 2311–2314.
[179] Hahn, H. G.; Chang, K. H.; Nam, K. D.; Bae, S. Y.; Mah, H. *Heterocycles* **1998**, *48*, 2253–2261.
[180] Adamczyk, M.; Fishpaugh, J. R.; Mattingly, P. G. *Tetrahedron Lett.* **1999**, *40*, 463–466.
[181] Adamczyk, M.; Fishpaugh, J. R.; Mattingly, P. G. *Bioorg. Med. Chem. Lett.* **1999**, *9*, 217–220.
[182] Huang, W.; Kalivretenos, A. G. *Tetrahedron Lett.* **1995**, *36*, 9113–9116.
[183] Kalir, R.; Warshawsky, A.; Fridkin, M.; Patchornik, A. *Eur. J. Biochem.* **1975**, *59*, 55–61.
[184] Dendrinos, K. G.; Kalivretenos, A. G. *Tetrahedron Lett.* **1998**, *39*, 1321–1324.
[185] Pop, I. E.; Déprez, B. P.; Tartar, A. L. *J. Org. Chem.* **1997**, *62*, 2594–2603.
[186] Nouvet, A.; Lamaty, F.; Lazaro, R. *Tetrahedron Lett.* **1998**, *39*, 3469–3470.
[187] Mergler, M.; Tanner, R.; Gosteli, J.; Grogg, P. *Tetrahedron Lett.* **1988**, *29*, 4005–4008.
[188] Blankemeyer-Menge, B.; Nimtz, M.; Frank, R. *Tetrahedron Lett.* **1990**, *31*, 1701–1704.
[189] Sieber, P. *Tetrahedron Lett.* **1987**, *28*, 6147–6150.
[190] Granitza, D.; Beyermann, M.; Wenschuh, H.; Haber, H.; Carpino, L. A.; Truran, G. A.; Bienert, M. *J. Chem. Soc. Chem. Commun.* **1995**, 2223–2224.
[191] Harth-Fritschy, E.; Cantacuzène, D. *J. Pept. Res.* **1997**, *50*, 415–420.
[192] Barlos, K.; Gatos, D.; Kallitsis, J.; Papaioannou, D.; Sotiriu, P.; Schäfer, W. *Liebigs Ann. Chem.* **1987**, 1031–1035.
[193] Ajayaghosh, A.; Pillai, V. N. R. *J. Org. Chem.* **1987**, *52*, 5714–5717.
[194] Routledge, A.; Stock, H. T.; Flitsch, S. L.; Turner, N. J. *Tetrahedron Lett.* **1997**, *38*, 8287–8290.
[195] Kuisle, O.; Quiñoá, E.; Riguera, R. *Tetrahedron Lett.* **1999**, *40*, 1203–1206.
[196] Akaji, K.; Kiso, Y.; Carpino, L. A. *J. Chem. Soc. Chem. Commun.* **1990**, 584–586.
[197] Blackburn, C.; Pingali, A.; Kehoe, T.; Herman, L. W.; Wang, H. Q.; Kates, S. A. *Bioorg. Med. Chem. Lett.* **1997**, *7*, 823–826.
[198] Kalir, R.; Fridkin, M.; Patchornik, A. *Eur. J. Biochem.* **1974**, *42*, 151–156.
[199] Gisin, B. F. *Helv. Chim. Acta* **1973**, *56*, 1476–1482.
[200] Yoo, S. E.; Seo, J. S.; Yi, K. Y.; Gong, Y. D. *Tetrahedron Lett.* **1997**, *38*, 1203–1206.
[201] Kurth, M. L.; Randall, L. A. A.; Chen, C.; Melander, C.; Miller, R. B.; McAlister, K.; Reitz, G.; Kang, R.; Nakatsu, T.; Green, C. *J. Org. Chem.* **1994**, *59*, 5862–5864.
[202] Dorman, L. C.; Love, J. *J. Org. Chem.* **1969**, *34*, 158–165.
[203] Frenette, R.; Friesen, R. W. *Tetrahedron Lett.* **1994**, *35*, 9177–9180.
[204] Hanessian, S.; Yang, R. Y. *Tetrahedron Lett.* **1996**, *37*, 5835–5838.
[205] Gennari, C.; Ceccarelli, S.; Piarulli, U.; Aboutayab, K.; Donghi, M.; Paterson, I. *Tetrahedron* **1998**, *54*, 14999–15016.
[206] Nicolás, E.; Clemente, J.; Ferrer, T.; Albericio, F.; Giralt, E. *Tetrahedron* **1997**, *53*, 3179–3194.
[207] Garigipati, R. S. *Tetrahedron Lett.* **1997**, *38*, 6807–6810.
[208] Bhalay, G.; Dunstan, A. R. *Tetrahedron Lett.* **1998**, *39*, 7803–7806.
[209] Chapman, P. H.; Walker, D. *J. Chem. Soc. Chem. Commun.* **1975**, 690–691.
[210] Mergler, M.; Dick, F.; Gosteli, J.; Nyfeler, R. *Tetrahedron Lett.* **1999**, *40*, 4663–4664.
[211] Morales, G. A.; Corbett, J. W.; DeGrado, W. F. *J. Org. Chem.* **1998**, *63*, 1172–1177.
[212] Barlos, K.; Chatzi, O.; Gatos, D.; Stavropoulos, G. *Int. J. Pept. Prot. Res.* **1991**, *37*, 513–520.

[213] Wong, J. Y.; Leznoff, C. C. *Can. J. Chem.* **1973**, *51*, 2452–2456.
[214] Kantorowski, E. J.; Kurth, M. J. *J. Org. Chem.* **1997**, *62*, 6797–6803.
[215] Nizi, E.; Botta, M.; Corelli, F.; Manetti, F.; Messina, F.; Maga, G. *Tetrahedron Lett.* **1998**, *39*, 3307–3310.
[216] Fréchet, J. M.; Schuerch, C. *J. Am. Chem. Soc.* **1971**, *93*, 492–496.
[217] Meyers, H. V.; Dilley, G. J.; Durgin, T. L.; Powers, T. S.; Winssinger, N. A.; Zhu, H.; Pavia, M. R. *Mol. Diversity* **1995**, *1*, 13–20.
[218] Barber, A. M.; Hardcastle, I. R.; Rowlands, M. G.; Nutley, B. P.; Marriott, J. H.; Jarman, M. *Bioorg. Med. Chem. Lett.* **1999**, *9*, 623–626.
[219] Nicolaou, K. C.; Winssinger, N.; Vourloumis, D.; Ohshima, T.; Kim, S.; Pfefferkorn, J.; Xu, J. Y.; Li, T. *J. Am. Chem. Soc.* **1998**, *120*, 10814–10826.
[220] Krchnák, V.; Weichsel, A. S.; Lebl, M.; Felder, S. *Bioorg. Med. Chem. Lett.* **1997**, *7*, 1013–1016.
[221] O'Donnell, M. J.; Delgado, F.; Pottorf, R. S. *Tetrahedron* **1999**, *55*, 6347–6362.
[222] Fréchet, J. M. J.; de Smet, M. D.; Farrall, M. J. *J. Org. Chem.* **1979**, *44*, 1774–1779.
[223] Wolters, E. T. M.; Tesser, G. I.; Nivard, R. J. F. *J. Org. Chem.* **1974**, *39*, 3388–3392.
[224] Camps, F.; Castells, J.; Ferrando, M. J.; Font, J. *Tetrahedron Lett.* **1971**, 1713–1714.
[225] Kraus, M. A.; Patchornik, A. *Israel J. Chem.* **1971**, *9*, 269–271.
[226] Hanessian, S.; Ma, J.; Wang, W. *Tetrahedron Lett.* **1999**, *40*, 4631–4634.
[227] O'Donnell, M. J.; Delgado, F.; Drew, M. D.; Pottorf, R. S.; Zhou, C. Y.; Scott, W. L. *Tetrahedron Lett.* **1999**, *40*, 5831–5835.
[228] Hu, Y.; Porco, J. A. *Tetrahedron Lett.* **1999**, *40*, 3289–3292.
[229] Vlattas, I.; Dellureficio, J.; Dunn, R.; Sytwu, I. I.; Stanton, J. *Tetrahedron Lett.* **1997**, *38*, 7321–7324.
[230] Kobayashi, S.; Wakabayashi, T.; Yasuda, M. *J. Org. Chem.* **1998**, *63*, 4868–4869.
[231] Kobayashi, S.; Hachiya, I.; Suzuki, S.; Moriwaki, M. *Tetrahedron Lett.* **1996**, *37*, 2809–2812.
[232] Kobayashi, S.; Hachiya, I.; Yasuda, M. *Tetrahedron Lett.* **1996**, *37*, 5569–5572.
[233] Phoon, C. W.; Oliver, S. F.; Abell, C. *Tetrahedron Lett.* **1998**, *39*, 7959–7962.
[234] Furth, P. S.; Reitman, M. S.; Gentles, R.; Cook, A. F. *Tetrahedron Lett.* **1997**, *38*, 6643–6646.
[235] Furth, P. S.; Reitman, M. S.; Cook, A. F. *Tetrahedron Lett.* **1997**, *38*, 5403–5406.
[236] Zaragoza, F. *Tetrahedron Lett.* **1997**, *38*, 7291–7294.
[237] Craig, D.; Robson, M. J.; Shaw, S. J. *Synlett* **1998**, 1381–1383.
[238] Hanessian, S.; Xie, F. *Tetrahedron Lett.* **1998**, *39*, 733–736.
[239] Lyngsø, L. O.; Nielsen, J. *Tetrahedron Lett.* **1998**, *39*, 5845–5848.
[240] Boymond, L.; Rottländer, M.; Cahiez, G.; Knochel, P. *Angew. Chem. Int. Ed. Engl.* **1998**, *37*, 1701–1703.
[241] Kamogawa, H.; Kanzawa, A.; Kadoya, M.; Naito, T.; Nanasawa, M. *Bull. Chem. Soc. Jpn.* **1983**, *56*, 762–765.
[242] Blackburn, C. *Tetrahedron Lett.* **1998**, *39*, 5469–5472.
[243] Kulkarni, B. A.; Ganesan, A. *Tetrahedron Lett.* **1999**, *40*, 5633–5636.
[244] Zhang, C. Z.; Moran, E. J.; Woiwode, T. F.; Short, K. M.; Mjalli, A. M. M. *Tetrahedron Lett.* **1996**, *37*, 751–754.
[245] Tam, J. P.; Riemen, M. W.; Merrifield, R. B. *Pept. Res.* **1988**, *1*, 6–18.
[246] Gowravaram, M. R.; Gallop, M. A. *Tetrahedron Lett.* **1997**, *38*, 6973–6976.
[247] Whitehouse, D. L.; Nelson, K. H.; Savinov, S. N.; Austin, D. J. *Tetrahedron Lett.* **1997**, *38*, 7139–7142.
[248] Whitehouse, D. L.; Nelson, K. H.; Savinov, S. N.; Löwe, R. S.; Austin, D. J. *Bioorg. Med. Chem.* **1998**, *6*, 1273–1282.
[249] Winkler, J. D.; McCoull, W. *Tetrahedron Lett.* **1998**, *39*, 4935–4936.
[250] Purandare, A. V.; Natarajan, S. *Tetrahedron Lett.* **1997**, *38*, 8777–8780.
[251] Albert, R.; Knecht, H.; Andersen, E.; Hungerford, V.; Schreier, M. H.; Papageorgiou, C. *Bioorg. Med. Chem. Lett.* **1998**, *8*, 2203–2208.
[252] Zaragoza, F. *Tetrahedron Lett.* **1996**, *37*, 6213–6216.
[253] Goff, D.; Fernandez, J. *Tetrahedron Lett.* **1999**, *40*, 423–426.

14 Preparation of Carbonic Acid Derivatives

14.1 Preparation of Carbodiimides

Carbodiimides have been prepared on insoluble supports mainly as polymeric dehydrating agents or as intermediates for the synthesis of guanidines and various heterocycles. Synthetic strategies which enable the preparation of carbodiimides on insoluble supports include the dehydration of ureas (Entry 1, Table 14.1), and the condensation of isocyanates or isothiocyanates with iminophosphoranes RN=PPh$_3$ (Entries 2–4, Table 14.1). The latter are obtained either by treating azides with phosphines or by treatment of primary amines with phosphines in the presence of an oxidant, such as DEAD.

Carbodiimides are generally less reactive than other heterocumulenes, and can be stored for long periods in closed containers at room temperature.

Table 14.1. Preparation of carbodiimides.

Entry	Starting resin	Conditions	Product	Ref.
1		TsCl (0.15 mol/L), NEt$_3$ (0.57 mol/L), DCM, 40 °C, 50 h		[1]
2		1. PPh$_3$ (0.5 mol/L, 5 eq), THF, 20 °C, 6 h 2. PhNCO, THF, 20 °C, 8 h		[2]
3		1. PPh$_3$ (0.4 mol/L, 5 eq), DEAD (5 eq), THF, 23 °C, 36 h 2. PhNCO (0.4 mol/L, 5 eq), PhMe, 23 °C, 4 h		[3]
4		BnNCS (10 eq), PPh$_3$ (35 eq), THF, 25 °C, 4 h		[4] see also [5]

14.2 Preparation of Isocyanates and Isothiocyanates

Isocyanates and isothiocyanates are highly reactive heterocumulenes, which are usually only prepared on solid phase as intermediates for the synthesis of ureas, carbamates, thioureas, etc., using methods similar to those used in solution (Table 14.2).

Support-bound primary amines can be converted into isocyantes by treatment with phosgene or with synthetic equivalents thereof (e.g. bis(trichloromethyl) carbonate 'triphosgene' or trichloromethyl chloroformate 'diphosgene'). Isocyanates can also be prepared on insoluble supports by oxidative degradation of amides (Hofmann degradation), thermolysis of acyl azides (Curtius degradation), and by cycloreversion of certain heterocycles ([6,7]; see Section 15.5). The Curtius degradation (Entry **3**, Table 14.2) enables the conversion of support-bound dicarboxylic acids into derivatives of unnatural amino acids, offering thereby interesting possibilities for the design of peptide mimetics (see also Entries **4–6**, Table 14.8). Isocyanates can, furthermore, be prepared by treating polystyrene-bound Fmoc amines with methyltrichlorosilane and triethylamine (Entry **4**, Table 14.2). The mechanism of this reaction is probably amine-induced cleavage of the Fmoc group to yield a carbamic acid, which is trapped and dehydrated by the chlorosilane.

Isothiocyanates can be prepared from support-bound primary amines by treatment with thiophosgene [8] or synthetic analogs thereof (Entry **5**, Table 14.2). In an alternative two-step procedure the amine is first treated with CS_2 and a tertiary amine, to yield an ammonium dithiocarbamate, which is finally desulfurized with TsCl or a chloroformate (Entry **6**, Table 14.2; Experimental Procedure 14.1). Highly reactive acyl isothiocyanates have been prepared from support-bound acyl chlorides and tetrabutylammonium thiocyanate (Entry **7**, Table 14.2). These acyl isothiocyanates react with amines to the corresponding *N*-acylthioureas, which can be used to prepare guanidines on insoluble supports [9] (Entry **6**, Table 14.3).

Experimental Procedure 14.1: Conversion of polystyrene-bound primary aliphatic amines into isothiocyanates [19]

Wang-resin-bound 1,3-diamino-2,2-dimethylpropane (0.60 g, approx. 0.6 mmol; for preparation see Experimental Procedure 14.2) was swollen for 1 min in DCE (7.0 mL). The solvent was filtered off, and DCE (5.2 mL), carbon disulfide (0.8 mL), and DIPEA (0.52 mL) were added. After shaking for 45 min a solution of tosyl chloride (1.32 g, 6.91 mmol, 12 eq) in DCE (1.5 mL) was added and shaking was continued for 15 h. The mixture was filtered and the resin was washed with DCM (5 × 8.0 mL). An IR spectrum of the dried support (KBr pellet) showed a strong absorption at 2091 cm^{-1}.

Table 14.2. Preparation of isocyanates and isothiocyanates.

Entry	Starting resin	Conditions	Product	Ref.
1	H₂N—...—O—(PS), Ph	COCl₂ (0.36 mol/L, 5 eq), pyridine (5 eq), DCM, reflux, 1 h	OCN—...—O—(PS), Ph	[10] see also [11,12]
2	H₂N—...—N(H)—(PS), Ar	triphosgene (3.3 eq), DIPEA (10 eq), DCM, 0.5 h Ar: 4-(BnO)C₆H₄	OCN—...—N(H)—(PS), Ar	[13]
3	(structure with O—TG and OH)	DPPA (4 eq), NEt₃, PhMe, 20 °C, 2 h, then PhMe, 90 °C, 4 h	(structure with O—TG and NCO)	[14] see also [15]
4	FmocHN—...—O—(PS), Ph	MeSiCl₃ (0.45 mol/L, 10 eq), NEt₃ (20 eq), DCM, 20 °C, 10 h	OCN—...—O—(PS), Ph	[16]
5	H₂N—...—N(H)—(PS)	(pyridyl carbonate structure) (0.14 mol/L), DMF, 20 °C, 1 h	SCN—...—N(H)—(PS)	[17]
6	H₂N—...—PS	CS₂, NEt₃, DMF, 0–20 °C, 4 h, then ClCO₂Et, 20 °C, overnight	SCN—...—PS	[18]
7	HO—...—PS	1. (COCl)₂ (2.5 eq), DCE/DMF 95:5, 8 h (repeat once; 14 h) 2. Bu₄NNCS (0.4 mol/L), DCE/THF 1:1, 4 h (repeat once)	SCN—...—PS	[9]

14.3 Preparation of Guanidines

Most solid-phase strategies for the preparation and protection of guanidines were developed to enable the use of arginine in solid-phase peptide synthesis. In recent years, however, the biological activity of many guanidines has spurred the search for more versatile solid-phase syntheses of this class of compound.

Several reagents can be used for the conversion of support-bound amines into guanidines (Figure 14.1, Table 14.3). Aliphatic amines react with N,N'-di(alkoxycarbonyl)thioureas in the presence of carbodiimides as condensing agents to yield protected guanidines. Instead of the in situ activation of thioureas with carbodiimides, isolated S-alkyl- or S-arylisothioureas can sometimes also be used to effect this transformation. If faster conversion is desired, or if sterically demanding amines are to be transformed

Table 14.3. Preparation of protected and unprotected guanidines.

Entry	Starting resin	Conditions	Product	Ref.
1	H₂N~~~(PS)	SEt / Boc-N,N-Boc (2 eq), DIPEA (5 eq), DMF, 20 °C, 15 h	BocHN~NH~(PS), NBoc	[21] see also [27,28]
2	(piperidine ester resin)	Boc-N,N-Boc thiourea (1.5 eq), DIC (1.5 eq), DCE, 4 d	BocHN, BocN~(PS)	[29]
3	H₂N~benzoate~(PS)	Ph-C(O)-NH-C(S)-NH~, FmocHN (3 eq), EDC (3 eq), DMF, 25 °C, 48 h	FmocHN~, Ph-guanidine~(PS)	[4]
4	Boc-NH-C(S)-NH-C(O)O~(PS)	BnNHMe (3 eq), NMP	Ph-CH₂-N(Me), Boc-N-guanidine~(PS)	[25]
5	Boc-NH-C(S)-NH-C(O)O~(PS)	2-chloropyridinium I⁻, PhNH₂, NEt₃, NMP	Ph-NH, Boc-N-guanidine~(PS)	[25]
6	cyclohexyl-NH-C(S)-NH-C(O)~benzoate~PS	tetrahydroisoquinoline NH, EDC, DIPEA, DMF	cyclohexyl guanidine~PS	[9]
7	HN=C(SMe)-NH-aryl~(PS)	morpholine (2 mol/L), DMSO, 80 °C, 12 h	HN=C, morpholine guanidine~(PS)	[26]
8	Ph-C(O)-N=C(SMe)-NH-C(O)-N~(PS)	BuNH₂ (0.56 mol/L, 10 eq), HgCl₂ (2 eq), NEt₃ (15 eq), DMF/DCM 5:1, 3 d	Ph-C(O)-N, BuNH guanidine~(PS)	[30] see also [25]
9	Ph substituted diamine~(PS)	thiocarbonyldiimidazole, DCM, 16 h	Ph, cyclic guanidinium~(PS)	[31]
10	Ph-N=C(N=)-N~benzamide~(PS)	PrNH₂ (10 eq), DMSO, 25 °C, 18 h	Pr, Ph guanidine~(PS)	[4] see also [3,5]

into guanidines, more reactive guanylating agents are required. These include *S*-nitroaryl- and *S*-(1-methyl-2-pyridinium)isothioureas, and various pyrazole derivatives, of which the 4-nitropyrazole derivatives are among the most reactive. The latter reagents can even be used to convert aniline or diisopropylamine into guanidines (in solution; 25 °C, DMF [20]).

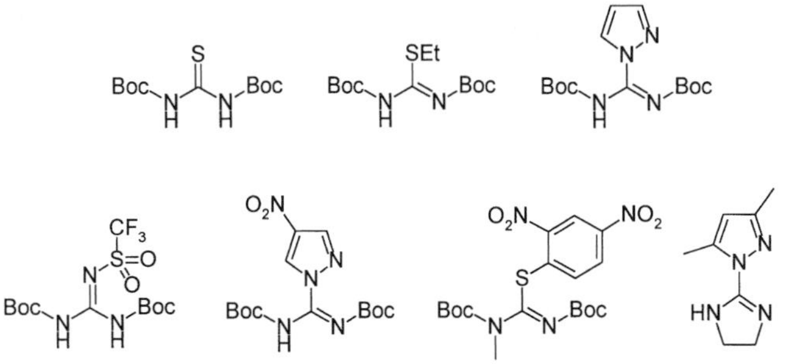

Fig. 14.1. Reagents for the conversion of support-bound amines into protected or unprotected guanidines [4,20–24].

As alternative to the guanylation of resin-bound amines, support-bound activated urea derivatives can be prepared, which upon reaction with amines yield guanidines. *N,N'*-Di(alkoxycarbonyl)thioureas can be converted into guanidines by treatment with aliphatic amines, without the need for another reagent (Entry **2**, Table 14.3). The addition of condensing agents, however, extends the scope of this reaction to include anilines (fair yields only [25]). Activation of support-bound thioureas can be achieved by S-alkylation [26] or S-arylation [4]. Carbodiimides have also been used as starting material for the preparation of guanidines on cross-linked polystyrene (Entry **10**, Table 14.3).

14.4 Preparation of Ureas

The most common strategies for preparing ureas on insoluble supports are outlined in Figure 14.2. The reaction conditions are similar to thoes used in solution, because most of the reagents required are compatible with common polymeric supports.

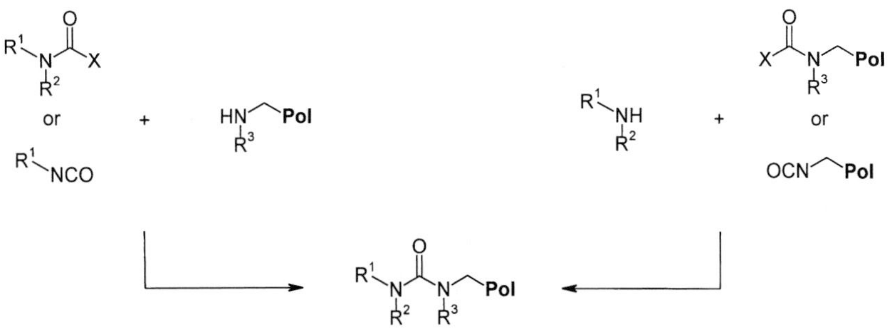

Fig. 14.2. Strategies for the preparation of ureas on solid phase. X: leaving group.

Isocyanates are the most efficient reagents for transforming support-bound amines into ureas. A broad range of amines, including sterically hindered aliphatic or aromatic amines, undergo clean aminocarbonylation with aliphatic and aromatic isocyanates (Table 14.4). The only disadvantage of isocyanates as reagents for solid-phase synthesis is their limited stability. Less reactive agents for aminocarbonylation include carbamoyl chlorides (which can be prepared from secondary amines and phosgene), *N*-(aminocarbonyl)benzotriazoles (Entry **2**, Table 14.5), and aryl carbamates (Entry **4**, Table 14.4).

Table 14.4. Preparation of ureas from support-bound amines and aminocarbonylating reagents.

Entry	Starting resin	Conditions	Product	Ref.
1	Ph-CH(Ph)-NH-CO-CH-O-CH₂-iPr-PS	PhNCO (10 eq), DCE, 50 °C, 3 d	Ph-HN-CO-N(...)-Ph structure	[32] see also [33-39]
2	tetrahydro-β-carboline with NH, O-(PS) ester	OCN-CH₂CH₂-Cl (0.3 mol/L, 3 eq), DCM, 2 h	N-carbamoyl product, HN-CH₂CH₂-Cl	[40]
3	Ph-O-N(Ph)-CO-CH-O-CH₂-PS	PhNCO, DCE, 83 °C, 24 h	Ph-O-N(Ph)-...-CO-NH-Ph	[41]
4	N₃-CH(CH₃)-NH-CO-OAr aminomethyl resin, MeO, OMe, PS	(5 eq, 0.07 mol/L), DIPEA (7 eq), DCM, 20 °C, 4 h; Ar: 4-(O₂N)C₆H₄	N₃-CH(CH₃)-NH-CO-NH-... product, MeO, OMe, PS	[42]
5	H₂N-aryl(CO₂Me)-O-CH₂CH₂-(TG)	PhNCO (0.58 mol/L, 20 eq), DCM, 18 h	Ph-NH-CO-NH-aryl(CO₂Me)-O-(TG)	[43] see also [44]

Ureas can also be prepared from support-bound aminocarbonylating reagents and amines. Suitable aminocarbonylating reagents, which can also be prepared on insoluble supports, are isocyanates (Section 14.2) and aryl carbamates, which can be obtained from support-bound amines and phosgene or aryl chloroformates, respectively (Table 14.5). Support-bound 4-nitrophenyl carbamates react with most aliphatic amines at room temperature or upon warming, but only slowly [12,45] or not at all with anilines (see, however, Entry **4**, Table 14.5). Aromatic amines can usually be converted efficiently into *N*-arylureas only by use of support-bound isocyanates or carbamoyl chlorides (Entry **8**, Table 14.5).

Table 14.5. Preparation of ureas from support-bound aminocarbonylating reagents.

Entry	Starting resin	Conditions	Product	Ref.
1	(structure)	Ph–CH₂–CH₂–NH₂ (0.5 mol/L), THF, 20 °C, overnight	(structure)	[46] see also [47]
2	(structure)	H-Ser(tBu)-OMe (0.13 mol/L, 3 eq), DIPEA (6 eq), DCM/DMF 1:1, 20 °C, overnight	(structure)	[48] see also [49]
3	(structure)	O₂N-C₆H₄-O-CO-Cl, DIPEA (0.5 mol/L each), THF/DCE 1:1, 20 °C, then HexNH₂ (0.35 mol/L), DIPEA, DMF, 20 °C, 12 h	(structure)	[45] see also [50]
4	(structure)	3-aminopyridine (0.5 mol/L), DCM, 12 h	(structure)	[43] see also [51]
5	(structure)	MeS-C(=NH)-NH₂ (0.56 mol/L, 10 eq), NEt₃ (10 eq), DMF, 24 h	(structure)	[30]
6	(structure)	1. COCl₂ (10 eq), DIPEA, DCM, 0–20 °C, 2 h 2. piperidin-4-ol (10 eq), pyridine (10 eq), DCM, 2 h	(structure)	[52] see also [53]
7	(structure)	1. Me₃SiCl (10 eq), NEt₃ (20 eq), DCM, 20 °C, 24 h, then drain 2. piperidine/DCM 1:9, 0.5 h	(structure)	[16]
8	(structure)	anthranilic acid, pyridine/DMF 1:9, 20 °C	(structure)	[12]

Table 14.5. continued.

Entry	Starting resin	Conditions	Product	Ref.
9		(5 eq), DCM, 1 h Ar: 4-(BnO)C$_6$H$_4$		[13] see also [10]
10		DMA, 4 h		[14]

The reaction of immobilized aminocarbonylating reagents with amines to yield non-resin-bound ureas is treated in Section 3.3.3.

14.5 Preparation of Thioureas and Isothioureas

Thioureas can be prepared on solid phase in a manner similar as ureas but using, instead, the corresponding thiocarbonyl derivatives. Hence, thioureas have been prepared on insoluble supports either by treating support-bound amines with isothiocyanates or by reaction of amines with support-bound aminothiocarbonylating reagents. Isothiocyanates are less reactive than isocyanates and even their reaction with primary aliphatic amines can take several days (Table 14.6). A high concentration of isothiocyanate and high reaction temperatures can sometimes be helpful in such instances. More reactive than alkyl- or arylisothiocyanates are N-acyl- and N-alkoxy-carbonylisothiocyanates (Entries **3**, **4**, and **6**; Table 14.6). Fmoc and Alloc isothiocyanate have been used as synthetic equivalents of H–NCS to prepare monosubstituted thioureas from support-bound amines (Entries **3** and **4**, Table 14.6). These thioureas can be readily converted into thiazoles by treatment with α-haloketones. Support-bound isothiocyanates (Section 14.2) can be converted into thioureas by treatment with amines. Few examples of this reaction have, however, been reported (Entries **5** and **6**, Table 14.6).

Isothioureas can be prepared on insoluble supports by S-alkylation of S-arylation of thioureas (Entry **7**, Table 14.6). Further methods for the preparation of isothioureas on insoluble supports include the N-alkylation of polystyrene-bound, N,N'-di(alkoxy-carbonyl)isothioureas with aliphatic alcohols by Mitsunobu reaction (Entry **7**, Table 14.6) and the addition of thiols to resin-bound carbodiimides [2].

The conversion of support-bound α-amino acids into thioureas can be accompanied by the release of thiohydantoins into solution (see Section 15.9). The rate of this cyclization depends, however, on the type of linker used and on the nucleophilicity of the intermediate thiourea.

Table 14.6. Preparation of thioureas.

Entry	Starting resin	Conditions	Product	Ref.
1	H₂N–CH₂–C₆H₄–(PS)	BocHN–cyclohexyl–NCS (10 eq), DCM, 45 °C, 2–4 d	NHBoc cyclohexyl, HN–C(S)–NH–CH₂–C₆H₄–(PS)	[54]
2	H₂N–CH₂–C₆H₄–(PS)	BnNCS (10 eq), DCM, 25 °C, 48 h	Ph, HN–C(S)–NH–CH₂–C₆H₄–(PS)	[4]
3	cyclopentyl–NH–CH₂–(TG)	FmocNCS (0.2 mol/L), DCM, 20 min, then piperidine/DMF 2:8, 3 × 2.5 min	H₂N–C(S)–N(cyclopentyl)–CH₂–(TG)	[55] see also [26]
4	H₂N–CH(Ph)–C(O)–O–CH₂CH₂–(PS)	allyl–O–C(O)–NCS, PhMe, 1 h; then Pd(PPh₃)₄, Me₂NSiMe₃, CF₃CO₂SiMe₃, DCM, 6 h	H₂N–C(S)–NH–CH(Ph)–C(O)–O–CH₂CH₂–(PS)	[56]
5	SCN–CH₂CH₂–(PS)	–C(O)–NH–NH₂ (0.3 mol/L), DMF, overnight	–C(O)–NH–NH–C(S)–NH–CH₂CH₂–(PS)	[17]
6	SCN–C(O)–C₆H₄–PS	2,6-dichloroaniline, DMF, 20 °C	Cl₂C₆H₃–NH–C(S)–NH–C(O)–C₆H₄–PS	[9]
7	Cl–CH₂–C₆H₄–PS	1. (H₂N)₂CS (0.5 mol/L, 5 eq), DMF, 75 °C, 16 h 2. Boc₂O (0.26 mol/L, 6 eq), DIPEA (10 eq), DCM, 40 h 3. PhO(CH₂)₂OH (0.3 mol/L, 5 eq), PPh₃ (5 eq), DIAD (5 eq), THF, 14 h	PhO–CH₂CH₂–N(Boc)–C(NBoc)–S–CH₂–C₆H₄–PS	[57] see also [58]

14.6 Preparation of Carbamates

Carbamates have mainly been used in solid-phase synthesis as linkers and protective groups for amines (see Sections 3.6.2 and 10.1.10.1). Carbamates are generally prepared by treating amines with aryl carbonates or chloroformates, which can be prepared from alcohols and phosgene or synthetic equivalents thereof. The alternative route, in which carbamates, isocyanates, or carbamoyl chlorides are reacted with alcohols, is less common, but does also lead to satisfactory results on insoluble supports (Tables 14.7 and 14.8).

Support-bound aliphatic alcohols react smoothly with isocyanates in the presence of catalytic amounts of a base to yield carbamates (Table 14.7). In an interesting variant of this reaction, isocyanates were generated in situ by Curtius degradation of acyl azides (Entry **2**, Table 14.7).

Wang resin bound 4-nitrophenyl carbonate is a convenient intermediate for the attachment of amines to polystyrene as carbamates (see Experimental Procedure 14.2). Aliphatic amines [59–64] and ammonia [65] undergo exothermic reaction with this support, whereas anilines generally require catalysis and/or long reaction times (Entry **3**, Table 14.7). For the immobilization of anilines as carbamates, Wang resin derived chloroformate [66–68] generally leads to better results than resin-bound 4-nitrophenyl carbonates. Amidines also react with polystyrene-bound 4-nitrophenyl carbonates to yield *N*-alkoxycarbonyl amidines ([69–71], Section 3.9). Support-bound alkoxycarbonyl hydrazines can be prepared by treating polystyrene-bound phenyl carbamate with hydrazine [72–74].

Experimental Procedure 14.2: Attachment of piperazine to Wang resin as carbamate [75]

A solution of 4-nitrophenyl chloroformate (43.0 g, 231 mmol, 5.5 eq) in DCM (200 mL) was dropwise added within 0.5 h to a stirred suspension of Wang resin (45.0 g, 42.3 mmol) in DCM (600 mL) and pyridine (52.0 mL, 644 mmol, 15 eq). When addition was complete, stirring at room temperature was continued for 3 h, and the mixture was filtered. The resin was washed with DCM (5 × 300 mL) and added in portions to a stirred solution of piperazine (38.2 g, 444 mmol, 11 eq) in DMF (600 mL), whereby the temperature of the mixture rised and its color changed to yellow–orange. The resulting mixture was shaken at room temperature for 13 h, filtered, and the resin was washed extensively with DMF, DCM, methanol, and finally with DCM. After drying in air about 45 g resin-bound piperazine was obtained.

As an alternative to 4-nitrophenyl chloroformate, carbonyl diimidazole [76–78] or di-*N*-succinimidyl carbonate [79,80] can be used to convert polymeric alcohols into alkoxycarbonylating reagents suitable for the preparation of support-bound carbamates. Polystyrene-bound alkoxycarbonyl imidazole is less reactive than the corresponding 4-nitrophenyl carbonate, and sometimes requires heating to undergo reac-

Table 14.7. Preparation of carbamates from support-bound alcohols or alkoxycarbonylating reagents.

Entry	Starting resin	Conditions	Product	Ref.
1		MeO, OMe, NCO (0.83 mol/L, 6 eq), NEt₃, DCM, 6.5 h		[81] see also [82]
2		CO₂H (pyridine) (0.36 mol/L, 3 eq), DPPA (5 eq), NEt₃ (10 eq), PhMe, 100 °C, 16 h		[83]
3		NO₂, Cl, NH₂, BSA, DMAP, DMF, 24 h		[84] see also [66,85]
4		Ph, LiO₂C, NH₂ (5 eq), DMF, 20 °C		[86] see also [24,87]
5		BnNH₂ (2 eq), DCM or DMF, 20 °C, Ar: 4-(O₂N)C₆H₄		[88]
6		R-N, NH₃Cl, Ph (0.16 mol/L, 5 eq), NMM, THF/NMP 1:1, 60 °C, 4 h		[76]
7		MeOTf (0.11 mol/L, 1.7 eq), DCE, NEt₃ (5 eq), 10–20 °C, 15 min, then H-Leu-OMe (6 eq), 20 °C, 5.5 h		[89]
8		1. COCl₂, DCM, pyridine, 20 °C, 2 h 2. H₂NNHCO₂Me, NEt₃, DCM, 20 °C, 2 h	(can be oxidized to the azo compound: NBS (1.1 eq), pyridine (1 eq), DCM, 20 °C, 1 h)	[90]
9		triphosgene (0.13 mol/L, 4 eq), 20 °C, overnight, then 2-aminopyridine (0.19 mol/L, 3.1 eq), DCM, 20 °C, overnight		[91] see also [67,92]

tion with amines. Additional activation of these imidazolides can be achieved by N-methylation (Entry **7**, Table 14.7).

Carbamates can also be prepared by treating support-bound amines with alkoxycar-bonylating reagents, such as chloroformates or aryl carbonates. Chloroformates or dicarbonates (e.g. Boc_2O) should not be used in large excess for the alkoxycarbonyla-tion of primary amines, because double derivatization can occur [93]. Less reactive reagents include 4-nitrophenyl carbonates and *N*-succinimidyl carbonates (Entries **2** and **3**, Table 14.8).

Support-bound isocyanates can be conveniently prepared from carboxylic acids by Curtius degradation. Because the reaction of the intermediate acyl azides with alco-hols to yield esters is slow, Curtius degradation can be conducted in the presence of alcohols to yield carbamates directly (Entries **4–6**, Table 14.8).

Some carbamates can be cleanly N-alkylated on insoluble supports, by treatment either with strong bases and alkylating agents, or with aliphatic alcohols under Mitsu-nobu conditions. Alkylation under Mitsunobu conditions proceeds smoothly only with acidic carbamates, such as aryl carbamates substituted with electron-withdrawing groups, but not with *N*-alkyl carbamates.

Polystyrene-bound allylsilanes react upon Lewis acid catalysis with *N*-alkoxycarbo-nylimines to yield *N*-homoallylcarbamates (Entry **10**, Table 14.8). Similarly, Wang resin bound carbamates have been successfully N-alkylated with allylsilanes and alde-hydes in a Mannich-type reaction (Entry **11**, Table 14.8).

14.7 Preparation of Carbonates and Miscellaneous Carbonic Acid Derivatives

The preparation of carbonates is mechanistically closely related to the synthesis of carbamates, and similar reagents can be used for this purpose (Table 14.9). Resin-bound alcohols can be converted into carbonates directly by reaction with a chlorofor-mate (see Experimental Procedure 14.2), or in two steps by treatment with phosgene or a synthetic equivalent thereof and then with an alcohol in the presence of a base. Thiocarbonates can be prepared from Wang resin and *S,S'*-di-2-pyridyl dithiocarbo-nate (Entry **4**, Table 14.9), and have been used as alkylating agents for the immobiliza-tion of aliphatic alcohols as benzyl ethers [98].

Table 14.8. Preparation of carbamates from support-bound amines or isocyanates, or by N-alkylation of carbamates.

Entry	Starting resin	Conditions	Product	Ref.
1		(1.1 eq), DIPEA (1.1 eq), DCM, 20 °C, 12 h		[94]
2		(5 eq), HOBt (10 eq), DIPEA (11 eq), THF, 50 °C, 5 h, repeat once Ar: 4-(O$_2$N)C$_6$H$_4$		[95] see also [96]
3		DIPEA, DCM		[53]
4		1. DPPA (0.83 mol/L, 10 eq), NEt$_3$ (15 eq), NMP, 20 °C, 1.5 h 2. 9-fluorenylmethanol (0.83 mol/L), *m*-xylene, 90 °C, 16 h		[15]
5		same conditions as Entry **4**		[15]
6		same conditions as Entry **4**		[15]
7		BnBr (0.5 mol/L, 5 eq), DBU (5 eq), PhMe, 20 °C, 3 d		[94] see also [83]
8		(1.4 mol/L, 10 eq), LiI (1 eq), LiN(SiMe$_3$)$_2$ (1.1 eq), THF/NMP, 20 °C, 24 h		[81]

Table 14.8. continued.

Entry	Starting resin	Conditions	Product	Ref.
9		EtOH, PBu$_3$, TMAD, THF, 60 °C, 15 h		[83]
10		BocN (23 eq), BF$_3$OEt$_2$ (4 eq), DCM, 3 h		[97]
11		PhCHO, MeCN, BF$_3$OEt$_2$, −5 °C, 2 h		[65]

Table 14.9. Preparation of carbonates and thiocarbonates.

Entry	Starting resin	Conditions	Product	Ref.
1		pyridine, 0 °C		[86] see also [87]
2		PhOCOCl (0.9 mol/L, 9 eq), pyridine (10 eq), DCM, 0 °C, overnight		[72]
3		(3 eq), DBU (3 eq), DCM, 20 °C, 15 min		[99] see also [100]
4		(0.12 mol/L, 3 eq), NEt$_3$ (3 eq), DCM, 24 h		[98]

References for Chapter 14

[1] Weinshenker, N. M.; Shen, C. M.; Wong, J. Y. *Org. Synth.* **1988**, *Coll. Vol. VI*, 951–954.
[2] Villalgordo, J. M.; Obrecht, D.; Chucholowski, A. *Synlett* **1998**, 1405–1407.
[3] Wang, F.; Hauske, J. R. *Tetrahedron Lett.* **1997**, *38*, 8651–8654.
[4] Schneider, S. E.; Bishop, P. A.; Salazar, M. A.; Bishop, O. A.; Anslyn, E. V. *Tetrahedron* **1998**, *54*, 15063–15086.
[5] Drewry, D. H.; Gerritz, S. W.; Linn, J. A. *Tetrahedron Lett.* **1997**, *38*, 3377–3380.
[6] Whitehouse, D. L.; Nelson, K. H.; Savinov, S. N.; Austin, D. J. *Tetrahedron Lett.* **1997**, *38*, 7139–7142.
[7] Gowravaram, M. R.; Gallop, M. A. *Tetrahedron Lett.* **1997**, *38*, 6973–6976.
[8] Zaragoza, F. unpublished results.
[9] Wilson, L. J.; Klopfenstein, S. R.; Li, M. *Tetrahedron Lett.* **1999**, *40*, 3999–4002.
[10] Matthews, J.; Rivero, R. A. *J. Org. Chem.* **1997**, *62*, 6090–6092.
[11] Annis, D. A.; Helluin, O.; Jacobsen, E. N. *Angew. Chem. Int. Ed. Engl.* **1998**, *37*, 1907–1909.
[12] Gordeev, M. F. *Biotechnology and Bioengineering* **1998**, *61*, 13–16.
[13] Limal, D.; Semetey, V.; Dalbon, P.; Jolivet, M.; Briand, J. P. *Tetrahedron Lett.* **1999**, *40*, 2749–2752.
[14] Shao, H.; Colucci, M.; Tong, S.; Zhang, H.; Castelhano, A. L. *Tetrahedron Lett.* **1998**, *39*, 7235–7238.
[15] Richter, L. S.; Andersen, S. *Tetrahedron Lett.* **1998**, *39*, 8747–8750.
[16] Chong, P. Y.; Petillo, P. A. *Tetrahedron Lett.* **1999**, *40*, 4501–4504.
[17] Wilson, M. W.; Hernández, A. S.; Calvet, A. P.; Hodges, J. C. *Mol. Diversity* **1998**, *3*, 95–112.
[18] Dowling, L. M.; Stark, G. R. *Biochemistry* **1969**, *8*, 4728–4734.
[19] Stephensen, H.; Zaragoza, F. *J. Org. Chem.* **1997**, *62*, 6096–6097.
[20] Yong, Y. F.; Kowalski, J. A.; Thoen, J. C.; Lipton, M. A. *Tetrahedron Lett.* **1999**, *40*, 53–56.
[21] Corbett, J. W.; Graciani, N. R.; Mousa, S. A.; DeGrado, W. F. *Bioorg. Med. Chem. Lett.* **1997**, *7*, 1371–1376.
[22] Shey, J. Y.; Sun, C. M. *Synlett* **1998**, 1423–1425.
[23] Ho, K. C.; Sun, C. M. *Bioorg. Med. Chem. Lett.* **1999**, *9*, 1517–1520.
[24] Lee, Y.; Silverman, R. B. *Synthesis* **1999**, 1495–1499.
[25] Josey, J. A.; Tarlton, C. A.; Payne, C. E. *Tetrahedron Lett.* **1998**, *39*, 5899–5902.
[26] Kearney, P. C.; Fernandez, M.; Flygare, J. A. *Tetrahedron Lett.* **1998**, *39*, 2663–2666.
[27] Rockwell, A. L.; Rafalski, M.; Pitts, W. J.; Batt, D. G.; Petraitis, J. J.; DeGrado, W. F.; Mousa, S.; Jadhav, P. K. *Bioorg. Med. Chem. Lett.* **1999**, *9*, 937–942.
[28] Kowalski, J.; Lipton, M. A. *Tetrahedron Lett.* **1996**, *37*, 5839–5840.
[29] Robinson, S.; Roskamp, E. J. *Tetrahedron* **1997**, *53*, 6697–6705.
[30] Lin, P.; Ganesan, A. *Tetrahedron Lett.* **1998**, *39*, 9789–9792.
[31] Ostresh, J. M.; Schoner, C. C.; Hamashin, V. T.; Nefzi, A.; Meyer, J. P.; Houghten, R. A. *J. Org. Chem.* **1998**, *63*, 8622–8623.
[32] Lee, S. H.; Chung, S. H.; Lee, Y. S. *Tetrahedron Lett.* **1998**, *39*, 9469–9472.
[33] Boeijen, A.; Kruijtzer, J. A. W.; Liskamp, R. M. J. *Bioorg. Med. Chem. Lett.* **1998**, *8*, 2375–2380.
[34] Park, K. H.; Olmstead, M. M.; Kurth, M. J. *J. Org. Chem.* **1998**, *63*, 6579–6585.
[35] Swayze, E. E. *Tetrahedron Lett.* **1997**, *38*, 8465–8468.
[36] Fivush, A. M.; Willson, T. M. *Tetrahedron Lett.* **1997**, *38*, 7151–7154.
[37] Kim, S. W.; Ahn, S. Y.; Koh, J. S.; Lee, J. H.; Ro, S.; Cho, H. Y. *Tetrahedron Lett.* **1997**, *38*, 4603–4606.
[38] Kolodziej, S. A.; Hamper, B. C. *Tetrahedron Lett.* **1996**, *37*, 5277–5280.
[39] Burgess, K.; Ibarzo, J.; Linthicum, D. S.; Russell, D. H.; Shin, H.; Shitangkoon, A.; Totani, R.; Zhang, A. J. *J. Am. Chem. Soc.* **1997**, *119*, 1556–1564.
[40] Mohan, R.; Chou, Y. L.; Morrissey, M. M. *Tetrahedron Lett.* **1996**, *37*, 3963–3966.
[41] Hanessian, S.; Yang, R. Y. *Tetrahedron Lett.* **1996**, *37*, 5835–5838.
[42] Kim, J. M.; Bi, Y. Z.; Paikoff, S. J.; Schultz, P. G. *Tetrahedron Lett.* **1996**, *37*, 5305–5308.
[43] Buckman, B. O.; Mohan, R. *Tetrahedron Lett.* **1996**, *37*, 4439–4442.
[44] Wijkmans, J. C. H. M.; Culshaw, A. J.; Baxter, A. D. *Mol. Diversity* **1998**, *3*, 117–120.
[45] Wilson, L. J.; Li, M.; Portlock, D. E. *Tetrahedron Lett.* **1998**, *39*, 5135–5138.
[46] Xiao, X. Y.; Ngu, K.; Chao, C.; Patel, D. V. *J. Org. Chem.* **1997**, *62*, 6968–6973.
[47] Hutchins, S. M.; Chapman, K. T. *Tetrahedron Lett.* **1994**, *35*, 4055–4058.
[48] Nieuwenhuijzen, J. W.; Conti, P. G. M.; Ottenheijm, H. C. J.; Linders, J. T. M. *Tetrahedron Lett.* **1998**, *39*, 7811–7814.
[49] Bauser, M.; Winter, M.; Valenti, C. A.; Wiesmüller, K. H.; Jung, G. *Mol. Diversity* **1998**, *3*, 257–260.

[50] Hutchins, S. M.; Chapman, K. T. *Tetrahedron Lett.* **1995**, *36*, 2583–2586.
[51] Gordeev, M. F.; Hui, H. C.; Gordon, E. M.; Patel, D. V. *Tetrahedron Lett.* **1997**, *38*, 1729–1732.
[52] Wang, G. T.; Chen, Y. W.; Wang, S. D.; Sciotti, R.; Sowin, T. *Tetrahedron Lett.* **1997**, *38*, 1895–1898.
[53] Kick, E. K.; Ellman, J. A. *J. Med. Chem.* **1995**, *38*, 1427–1430.
[54] Smith, J.; Liras, J. L.; Schneider, S. E.; Anslyn, E. V. *J. Org. Chem.* **1996**, *61*, 8811–8818.
[55] Kearney, P. C.; Fernandez, M.; Flygare, J. A. *J. Org. Chem.* **1998**, *63*, 196–200.
[56] Stadlwieser, J.; Ellmerer-Müller, E. P.; Takó, A.; Maslouh, N.; Bannwarth, W. *Angew. Chem. Int. Ed. Engl.* **1998**, *37*, 1402–1404.
[57] Dodd, D. S.; Wallace, O. B. *Tetrahedron Lett.* **1998**, *39*, 5701–5704.
[58] Lee, J.; Gauthier, D.; Rivero, R. A. *Tetrahedron Lett.* **1998**, *39*, 201–204.
[59] Dixit, D. M.; Leznoff, C. C. *J. Chem. Soc. Chem. Commun.* **1977**, 798–799.
[60] Marsh, I. R.; Smith, H.; Bradley, M. *Chem. Commun.* **1996**, 941–942.
[61] Ho, C. Y.; Kukla, M. J. *Tetrahedron Lett.* **1997**, *38*, 2799–2802.
[62] Kim, S. W.; Hong, C. Y.; Lee, K.; Lee, E. J.; Koh, J. S. *Bioorg. Med. Chem. Lett.* **1998**, *8*, 735–738.
[63] Brady, S. F.; Stauffer, K. J.; Lumma, W. C.; Smith, G. M.; Ramjit, H. G.; Lewis, S. D.; Lucas, B. J.; Gardell, S. J.; Lyle, E. A.; Appleby, S. D.; Cook, J. J.; Holahan, M. A.; Stranieri, M. T.; Lynch, J. J.; Lin, J. H.; Chen, I. W.; Vastag, K.; Naylor-Olsen, A. M.; Vacca, J. P. *J. Med. Chem.* **1998**, *41*, 401–406.
[64] Tomasi, S.; Le Roch, M.; Renault, J.; Corbel, J. C.; Uriac, P.; Carboni, B.; Moncoq, D.; Martin, B.; Delcros, J. G. *Bioorg. Med. Chem. Lett.* **1998**, *8*, 635–640.
[65] Meester, W. J. N.; Rutjes, F. P. J. T.; Hermkens, P. H. H.; Hiemstra, H. *Tetrahedron Lett.* **1999**, *40*, 1601–1604.
[66] Raju, B.; Kogan, T. P. *Tetrahedron Lett.* **1997**, *38*, 3373–3376.
[67] Burdick, D. J.; Struble, M. E.; Burnier, J. P. *Tetrahedron Lett.* **1993**, *34*, 2589–2592.
[68] Smith, A. L.; Thomson, C. G.; Leeson, P. D. *Bioorg. Med. Chem. Lett.* **1996**, *6*, 1483–1486.
[69] Mohan, R.; Yun, W. Y.; Buckman, B. O.; Liang, A.; Trinh, L.; Morrissey, M. M. *Bioorg. Med. Chem. Lett.* **1998**, *8*, 1877–1882.
[70] Kim, S. W.; Hong, C. Y.; Koh, J. S.; Lee, E. J.; Lee, K. *Mol. Diversity* **1998**, *3*, 133–136.
[71] Roussel, P.; Bradley, M.; Matthews, I.; Kane, P. *Tetrahedron Lett.* **1997**, *38*, 4861–4864.
[72] Wang, S. *J. Am. Chem. Soc.* **1973**, *95*, 1328–1333.
[73] Wang, S. *J. Org. Chem.* **1975**, *40*, 1235–1239.
[74] Wang, S.; Merrifield, R. B. *J. Am. Chem. Soc.* **1969**, *91*, 6488–6491.
[75] Zaragoza, F.; Petersen, S. V. *Tetrahedron* **1996**, *52*, 10823–10826.
[76] Hauske, J. R.; Dorff, P. *Tetrahedron Lett.* **1995**, *36*, 1589–1592.
[77] Rotella, D. P. *J. Am. Chem. Soc.* **1996**, *118*, 12246–12247.
[78] Munson, M. C.; Cook, A. W.; Josey, J. A.; Rao, C. *Tetrahedron Lett.* **1998**, *39*, 7223–7226.
[79] Alsina, J.; Rabanal, F.; Chiva, C.; Giralt, E.; Albericio, F. *Tetrahedron* **1998**, *54*, 10125–10152.
[80] Alsina, J.; Chiva, C.; Ortiz, M.; Rabanal, F.; Giralt, E.; Albericio, F. *Tetrahedron Lett.* **1997**, *38*, 883–886.
[81] Buchstaller, H. P. *Tetrahedron* **1998**, *54*, 3465–3470.
[82] Fitzpatrick, L. J.; Rivero, R. A. *Tetrahedron Lett.* **1997**, *38*, 7479–7482.
[83] Sunami, S.; Sagara, T.; Ohkubo, M.; Morishima, H. *Tetrahedron Lett.* **1999**, *40*, 1721–1724.
[84] Huang, W.; Scarborough, R. M. *Tetrahedron Lett.* **1999**, *40*, 2665–2668.
[85] Gouilleux, L.; Fehrentz, J. A.; Winternitz, F.; Martinez, J. *Tetrahedron Lett.* **1996**, *37*, 7031–7034.
[86] Léger, R.; Yen, R.; She, M. W.; Lee, V. J.; Hecker, S. J. *Tetrahedron Lett.* **1998**, *39*, 4171–4174.
[87] Singh, R.; Nuss, J. M. *Tetrahedron Lett.* **1999**, *40*, 1249–1252.
[88] Dressman, B. A.; Singh, U.; Kaldor, S. W. *Tetrahedron Lett.* **1998**, *39*, 3631–3634.
[89] Hernández, A. S.; Hodges, J. C. *J. Org. Chem.* **1997**, *62*, 3153–3157.
[90] Arnold, L. D.; Assil, H. I.; Vederas, J. C. *J. Am. Chem. Soc.* **1989**, *111*, 3973–3976.
[91] Scialdone, M. A.; Shuey, S. W.; Soper, P.; Hamuro, Y.; Burns, D. M. *J. Org. Chem.* **1998**, *63*, 4802–4807.
[92] Scialdone, M. A. *Tetrahedron Lett.* **1996**, *37*, 8141–8144.
[93] Hulme, C.; Peng, J.; Morton, G.; Salvino, J. M.; Herpin, T.; Labaudiniere, R. *Tetrahedron Lett.* **1998**, *39*, 7227–7230.
[94] Salvino, J. M.; Mervic, M.; Mason, H. J.; Kiesow, T.; Teager, D.; Airey, J.; Labaudiniere, R. *J. Org. Chem.* **1999**, *64*, 1823–1830.
[95] Paikoff, S. J.; Wilson, T. E.; Cho, C. Y.; Schultz, P. G. *Tetrahedron Lett.* **1996**, *37*, 5653–5656.
[96] Cho, C. Y.; Youngquist, R. S.; Paikoff, S. J.; Beresini, M. H.; Hébert, A. R.; Berleau, L. T.; Liu, C. W.; Wemmer, D. E.; Keough, T.; Schultz, P. G. *J. Am. Chem. Soc.* **1998**, *120*, 7706–7718.
[97] Brown, R. C. D.; Fisher, M. *Chem. Commun.* **1999**, 1547–1548.
[98] Hanessian, S.; Huynh, H. K. *Tetrahedron Lett.* **1999**, *40*, 671–674.
[99] Routledge, A.; Stock, H. T.; Flitsch, S. L.; Turner, N. J. *Tetrahedron Lett.* **1997**, *38*, 8287–8290.
[100] Routledge, A.; Abell, C.; Balasubramanian, S. *Tetrahedron Lett.* **1997**, *38*, 1227–1230.

15 Preparation of Heterocycles

For more than two decades most developments in solid-phase chemistry focussed on the preparation of biopolymers. Only in recent years has interest in other synthetic targets, including heterocyclic compounds, begun to grow. Today the field of solid-phase heterocyclic chemistry is rapidly expanding, and numerous preparations have been reported.

Interest in solid-phase heterocyclic chemistry originated mainly in the pharmaceutical industry. Heterocycles not only enable the spatial fixation of a set of structural elements relevant to reversible binding to proteins, but can also have a strong influence on the solubility and on other physicochemical properties of a compound. Because substituted heterocycles are often more easy to prepare than the corresponding carbocycles, heterocyclic chemistry played, and still plays, an important role in the development of new drugs (anti-inflammatory agents, antibiotics, antifungals, analgesics, etc.). Syntheses which enable the quick production of arrays of heterocycles, useful for the identification of new lead structures, are of critical importance to the pharmaceutical industry.

Several review articles have appeared covering the synthesis of heterocycles on insoluble supports for the production of compound libraries for drug discovery [1–3]. In this chapter these syntheses have been organized according to the type of heterocycle prepared.

15.1 Preparation of Epoxides and Aziridines

Epoxides are reactive electrophiles, which enable the facile preparation of substituted alcohols by reaction with a broad range of nucleophiles. Epoxides can be prepared on insoluble supports either by epoxidation of alkenes or from aldehydes (Table 15.1).

Olefins bound to cross-linked polystyrene can be epoxidized under conditions similar to those used in solution. The most common reagent is m-chloroperbenzoic acid in DCM, but other reagents have also been used (Table 15.1). Because excess oxidant is usually required to furnish clean products, care must be taken with linkers or other functional groups prone to oxidation (ketones, amines, benzyl ethers, etc.).

The reaction of support-bound aldehydes with trimethylsulfonium halides in the presence of a strong base affords non-oxidative access to oxiranes (Entry **5**, Table 15.1). The same reaction has been performed with polystyrene-bound trialkyl-sulfonium salts, which react with aldehydes under basic reaction conditions to yield (non-support-bound) epoxides [4].

Aziridines have been prepared on insoluble supports by addition of primary amines to α-bromo acrylates and acrylamides (Entry **6**, Table 15.1). These aziridines are sufficiently stable to tolerate treatment with TFA [5].

Table 15.1. Preparation of epoxides and aziridines.

Entry	Starting resin	Conditions	Product	Ref.
1		MCPBA, NaHCO₃, DCM, 40 °C, 20 h	(82:18, main isomer shown)	[6] see also [7,8]
2		H₂O₂, K₂HPO₄, DCM, Cl₃CCN, H₂O, 40 °C, 3 h		[9]
3		O–O (0.1 mol/L, 3 eq), Me₂CO, 20 °C, 1 h		[10]
4		O–O (0.33 mol/L, 25 eq), DCM/Me₂CO, 0 °C, 1 h; repeat once		[11]
5		Me₃SCl, DCM, BnMe₃NCl, NaOH, H₂O, 20 °C, 3 h; or Me₃SI, KN(SiMe₃)₂, DMF, 20 °C, 10 h		[8,12]
6		BnNH₂, NMM, THF		[5]

15.2 Preparation of Azetidines and Thiazetidines

β-Lactams can be prepared from support-bound imines by reaction with ketenes (Table 15.2). The required imines are readily accessible by condensation of primary aliphatic or aromatic amines with carbonyl compounds, optionally in the presence of a dehydrating agent or catalytic amounts of an acid (see also Section 10.1.4). Ketenes can be generated by dehydrohalogenation with tertiary amines of acyl halides with an α-hydrogen. Alternatively, titanium ester enolates from 2-pyridinethiol esters also undergo [2 + 2] cycloaddition to imines, and can be used to convert support-bound imines into β-lactams (Entry **3**, Table 15.2). Because ketenes are highly reactive intermediates, prone to numerous side reactions, support-bound ketenes are less suitable intermediates for the preparation of β-lactams.

Sulfonyl chlorides with α-hydrogen eliminate hydrogen chloride when treated with bases to yield sulfenes. Like ketenes, sulfenes can undergo [2 + 2] cycloaddition to support-bound imines to yield β-sultams (Entry **5**, Table 15.2).

Table 15.2. Preparation of azetidines and thiazetidines.

Entry	Starting resin	Conditions	Product	Ref.
1	Ph–N=...O–(PS)	PhO–CH₂–COCl (0.8 mol/L), NEt₃ (1.1 mol/L), DCM, 0–25 °C, 16 h	(PS)	[13] see also [14-16]
2	Ph–N=... (PEG)	Ph–O–CH₂–COCl (20 eq), NEt₃ (20 eq), DCM, 23 °C, 15 h	(PEG)	[17] see also [18]
3	...N= (PEG)	(7 eq), N(octyl)₃, TiCl₄, DCM, −78 °C to 20 °C	(PEG)	[19]
4	Ph–...–S–PS, NH	Hg(O₂CCF₃)₂, DCM, Me₂CO	MeO–	[20]
5	Ph–N=...O–(PS)	MeO₂C–CH₂–SO₂Cl (0.36 mol/L, 14 eq), pyridine (17 eq), THF, −78 °C, 3 h, then 20 °C, 24 h	(PS)	[21]

15.3 Preparation of Pyrroles and Pyrrolidines

Various approaches have been used to prepare pyrroles on insoluble supports (Figure 15.1). These include the condensation of α-haloketones or nitroalkenes with enamines (Hantzsch pyrrole synthesis) and the decarboxylative condensation of *N*-acyl α-amino acids with alkynes (Table 15.3). The enamines required for the Hanztsch pyrrole synthesis are obtained by treating support-bound acetoacetamides with primary aliphatic amines. Unfortunately, 3-keto amides other than acetoacetamides are not readily accessible; this imposes some limitations on the scope of substituents in the products.

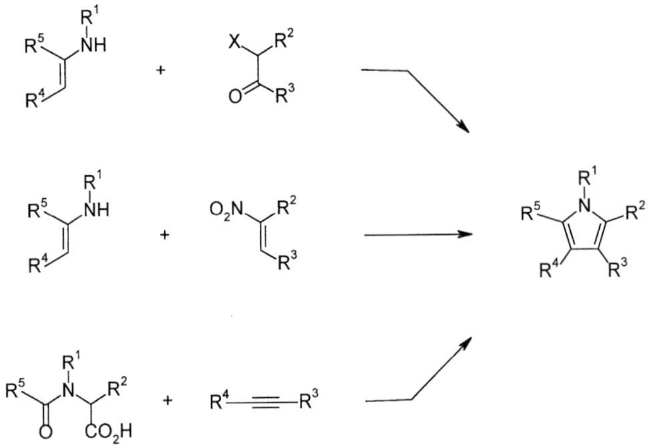

Fig. 15.1. Strategies used for the preparation of pyrroles on insoluble supports.

Table 15.3. Preparation of pyrroles and 2,5-dihydropyrroles.

Entry	Starting resin	Conditions	Product	Ref.
1	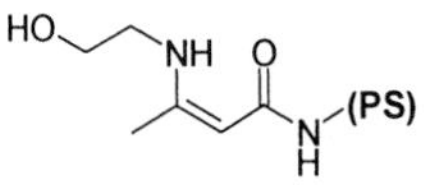	ArCOCH$_2$Br, DTBP, DMF, 20 °C, 17 h Ar: 4-(Et$_2$N)C$_6$H$_4$	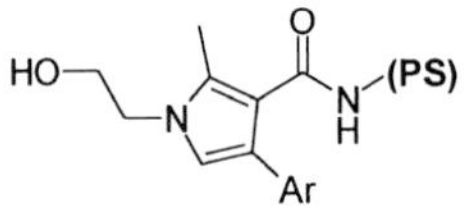	[24]
2	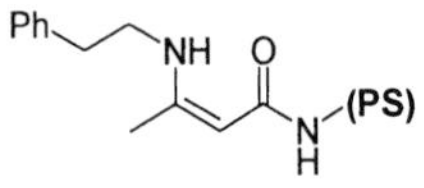	ArCHO, EtNO$_2$, HC(OMe)$_3$, piperidine, DMF/EtOH 1:1, 70 °C, 5 h Ar: 4-(F$_3$C)C$_6$H$_4$	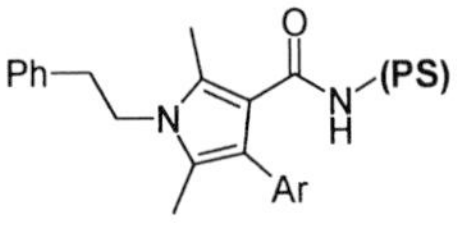	[25]
3	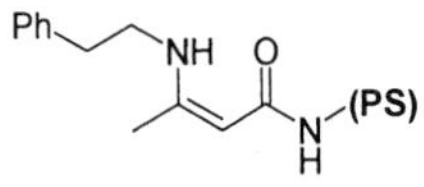	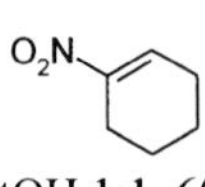 DMF/EtOH 1:1, 60 °C, 2 h	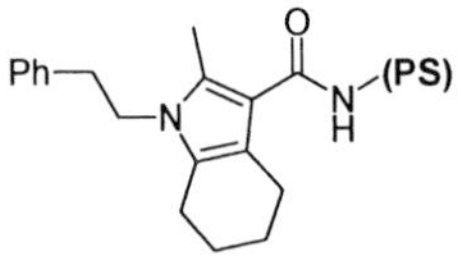	[25]
4	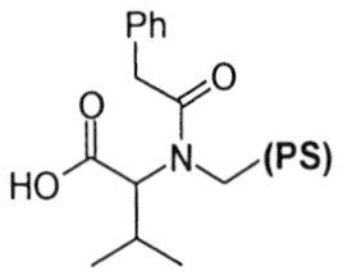	MeO$_2$C≡CO$_2$Me Ac$_2$O, 100 °C, 1–2 d	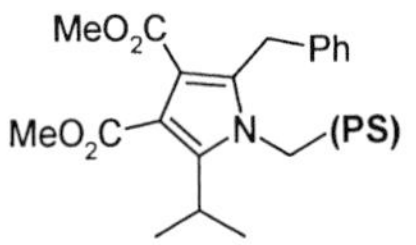	[26] see also [27]
5	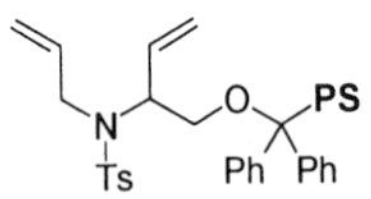	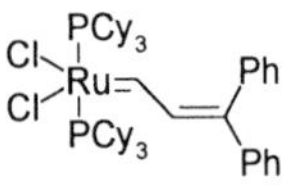 (0.15 eq), C$_6$H$_6$, 30 °C, 5 d	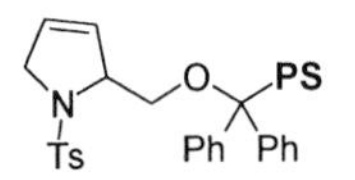	[28]
6	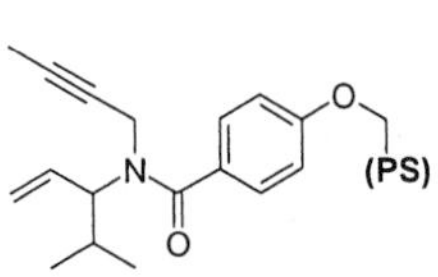	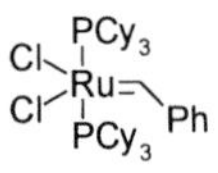 (0.05 eq), C$_6$H$_6$, 75 °C, 18 h	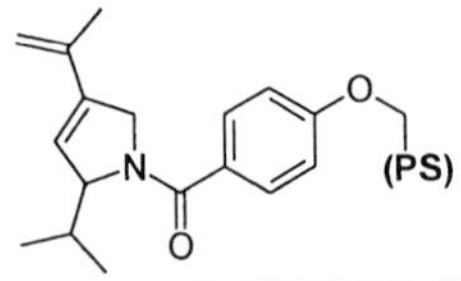	[29]

Table 15.4. Preparation of pyrrolidines and pyrrolidinones.

Entry	Starting resin	Conditions	Product	Ref.
1		1. LDA (3 eq), THF, −78 °C to 0 °C 2. ZnBr$_2$ (5 eq) 3. H$_2$O, citric acid		[30]
2		SiMe$_3$ (10 eq), BF$_3$OEt$_2$ (3 eq), DCM, 20 °C, 17 h		[31]
3		Pd(acac)$_2$ (0.1 eq), dppe (0.3 eq), THF, heat 59–95% yield		[32]
4		CO$_2$H, H$_2$N Ph (0.3 mol/L, 5 eq of each), CHCl$_3$/MeOH 2:1, 20 °C, 48 h		[33]
5		CHO, Ph NC, EtO$_2$C CO$_2$H (0.2 mol/L, 10 eq of each), MeOH/DCM 2:1, 36 h		[34] see also [35]
6		KOH (1 eq), DCM/MeOH 1:1, 0.5 h; or DIPEA/dioxane 3:7, 80 °C, 16 h		[36,37] see also [38]
7		NaOEt (0.1 mol/L), 85 °C, 24 h (Wang resin)		[39]
8		NaOEt (0.1 mol/L, 2 eq), 85 °C, 24 h (Wang resin)		[39]

Table 15.4. continued.

Entry	Starting resin	Conditions	Product	Ref.
9		ZnCl$_2$ (2 eq), LiN(SiMe$_3$)$_2$ (2 eq), THF, −78 °C to 0 °C, 2 h Ar: 4-(MeO)C$_6$H$_4$		[40]
10		AcOH/PhMe 25:75, 110 °C, 24 h		[41]
11		AcOH/DMF 1:20, 130 °C, 18 h		[41]

2,5-Dihydropyrroles have recently become readily available by ring-closing metathesis. For this purpose N-acylated or N-sulfonylated bis(allyl)amines are treated with catalytic amounts of a ruthenium carbene complex, whereby cyclization to the dihydropyrrole occurs (Entries **5** and **6**, Table 15.3 [22,23]). Catalysis by carbene complexes is most efficient in aprotic, non-nucleophilic solvents, and can be also conducted on hydrophobic supports, such as cross-linked polystyrene. Free amines or other soft nucleophiles might, however, compete with the alkene for electrophilic attack by the catalyst, and should therefore be avoided.

Pyrrolidines and pyrrolidinones can be synthesized on insoluble supports by intramolecular carbometallation of alkenes (Entry **1**, Table 15.4) and by Ugi reaction of support-bound isonitriles with amines and 4-oxo acids (Entry **4**, Table 15.4). In Entry **5** (Table 15.4) the formation of the pyrrolidinone ring proceeds via intramolecular Diels–Alder reaction of a furan with a fumaric acid derivative.

Numerous examples of the preparation of tetramic acids from N-acylated amino acid esters via a Dieckmann-type cyclocondensation have been reported (Entries **6–8**, Table 15.4). Deprotonated 1,3-dicarbonyl compounds and unactivated amide enolates can be used as carbon nucleophiles. In most of these examples the ester which acts as electrophile also links the substrate to the support, so that cyclization and cleavage from the support occur simultaneously.

Five-membered cyclic imides are accessible by a similar approach, namely by intramolecular nucleophilic cleavage of support-bound phthalic or succinic acid amides (Entries **10** and **11**, Table 15.4). The formation of cyclic imides (aspartimides) during the solid-phase synthesis of peptides containing aspartic acid esters is discussed in Section 13.8.

A powerful means of access to pyrrolidines, which is also suitable for the solid-phase synthesis of this important class of heterocycles, is the cycloaddition of 2-aza-allyl anions to olefins (Figure 15.2). Various modifications of this reaction have been reported, the most common being the addition of deprotonated, amino acid derived imines to acceptor-substituted alkenes (Entries **1–5**, Table 15.5).

Table 15.5. Preparation of pyrrolidines from 2-azaallyl anions.

Entry	Starting resin	Conditions	Product	Ref.
1		AgNO₃, NEt₃ (1 mol/L of each), MeCN, 8 h		[15] see also [43]
2		NEt₃ (0.05 mol/L, 3 eq of each), AcOH, DMF, 100 °C, 18 h		[44]
3	Ar: 4-(MeO)C₆H₄	(0.16 mol/L, 10 eq), PhMe, 110 °C, 24 h; Ar: 4-(MeO)C₆H₄	(racemic)	[45] see also [43]
4	Ar: 4-(MeO)C₆H₄	MeO₂C–N=–Ph, LiBr, DBU, THF, 3 d; Ar: 4-(MeO)C₆H₄		[46]
5		AgOAc (0.2 mol/L, 3 eq), DIPEA (3 eq), MeCN, 20 °C, 48 h		[47]
6		(5 eq), NEt₃ (5 eq), THF, 20 °C, 1 h		[48]
7		(0.11 mol/L, 1.7 eq), NEt₃ (1.7 eq), PhMe, 110 °C, 18 h		[49]
8		1. BuLi, THF, PhSeCH=CH₂, −40 °C 2. MeI 3. Bu₃SnH, AIBN, PhH, 80 °C		[42]

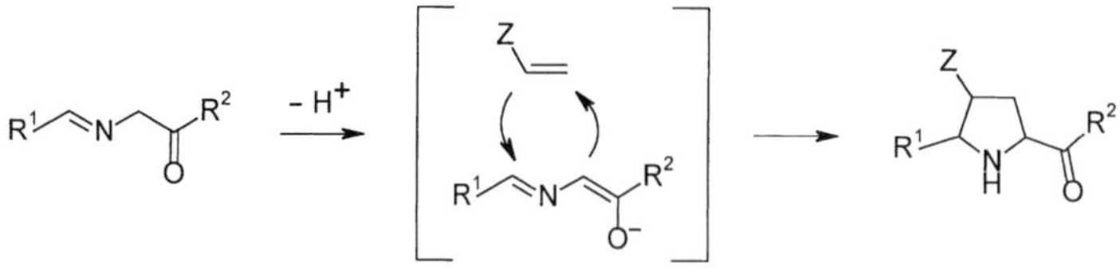

Fig. 15.2. Formation of pyrrolidines by cycloaddition of 2-azaallyl anions to alkenes.

This cycloaddition can be performed with either the amino acid, the aldehyde, or the electron-poor alkene linked to the support. Intramolecular azaallyl anion cycloadditions have been used to prepare polycyclic systems on solid phase (Entries **5** and **7**, Table 15.5).

Less stabilized (and more reactive) azaallyl anions are formed upon transmetallation of α-stannylamine-derived imines with butyllithium. The resulting intermediates react even at low temperature with non-activated alkenes, such as stilbenes or vinyl sulfides, to yield N-metallated pyrrolidines [42].

15.4 Preparation of Indoles and Indolines

Linkers suitable for the attachment of indoles to insoluble supports are discussed in Section 3.10.

The most important synthetic approaches to indoles which have been realized on insoluble supports are outlined in Figure 15.3. These include the Fischer indole synthesis, which has been performed successfully with polystyrene- or Tentagel-bound ketones and aldehydes (Entries **1** and **2**, Table 15.6). Indoles and indolines have, furthermore, been prepared by palladium-mediated reaction of support-bound 2-iodo- or 2-bromoaniline derivatives with alkynes or 1,3-dienes (Table 15.6). These reactions, and the closely related cyclization of 2-alkynylanilines, probably proceed via an intermediate palladacycle, which undergoes reductive elimination to yield the observed indoles (Figure 15.3).

Additional reactions suitable for synthesizing indoles on insoluble supports include intramolecular Heck and Wittig reactions. The latter have been performed with phosphonium salts derived from polystyrene-bound triphenylphosphine, whereby indole formation and cleavage of the product from the support take place simultaneously (Entry **10**, Table 15.6).

Support-bound indoles can be modified in several ways. N-Alkylation of polystyrene-bound indoles has been achieved by treatment with reactive alkylating agents (MeI, BnBr, BrCH$_2$CO$_2$R) and with NaH or KOtBu as base in DMF at room temperature (Entries **1** and **2**, Table 15.7). Aminomethylation of indole at C-3 proceeds smoothly on cross-linked polystyrene. The resulting (aminomethyl)indoles are thermally unstable and undergo substitution reactions with various carbon nucleophiles (e.g. cyanide or nitroacetates) at higher temperatures (Entry **4**, Table 15.7). C-Alkylation of indoles can also be performed via palladium-catalyzed coupling reactions, such as Suzuki coupling (Entry **5**, Table 15.7). 2-Iodoindoles have been prepared on polystyrene by oxidation of 2-silylindoles with NIS (Entry **6**, Table 15.7).

Table 15.6. Preparation of indoles and indolines.

Entry	Starting resin	Conditions	Product	Ref.
1	Ph, (PS)	ZnCl$_2$ (each 0.5 mol/L), AcOH, 70 °C, 20 h	H N 2-Ph (PS)	[50] see also [51]
2	(TG)	O$_2$N– (0.06 mol/L, 10 eq), TFA/DCM/PhOMe 5:15:1, 40 °C, 17 h	O$_2$N (TG)	[52]
3	I, HN, O (PS)	HO SiMe$_3$; Pd(PPh$_3$)$_2$Cl$_2$ (0.2 eq), TMG (10 eq), DMF, 110 °C, 5 h (repeat once, 16 h)	HO SiMe$_3$ (PS)	[53]
4	H N-Ac, I, O TG	PhS; Pd(PPh$_3$)$_2$Cl$_2$, CuI, TMG, dioxane, 90 °C, 18 h	PhS H N O TG	[54] see also [55,56]
5	I, HN, Tol–S=O (PS)	AcO (8 eq), Pd(OAc)$_2$ (0.1 eq), LiCl (2 eq), DIPEA (8 eq), DMF, 100 °C, 2 d	AcO NH (PS) Tol–S=O	[57]
6	Ph, F$_3$C, (PS)	CO$_2$Me, OTf; K$_2$CO$_3$, DMF, Pd(PPh$_3$)$_4$, 20 °C, 24 h	CO$_2$Me, Ph (PS)	[58]
7	Ph, N, (PS), I	Pd(PPh$_3$)$_2$Cl$_2$ (1 mmol/L, 0.15 eq), Bu$_4$NCl (1.5 eq), NEt$_3$ (8 eq), DMF/H$_2$O 9:1, 80 °C, 24 h	Ph (PS)	[59]
8	N, (TG), Br	Pd(PPh$_3$)$_4$ (0.5 eq), PPh$_3$ (2 eq), NEt$_3$ (13 eq), DMA, 90 °C, 5 h (repeat once)	(TG)	[60]

Table 15.6. continued.

Entry	Starting resin	Conditions	Product	Ref.
9	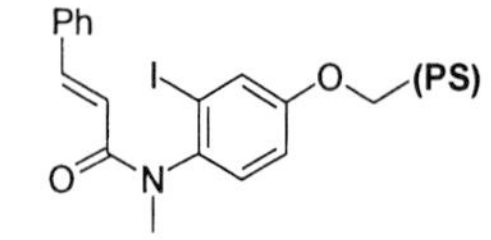	Pd(OAc)$_2$ (7 mmol/L, 0.3 eq), Ag$_2$CO$_3$ (2 eq), PPh$_3$ (0.6 eq), DMF, 100 °C, 16 h	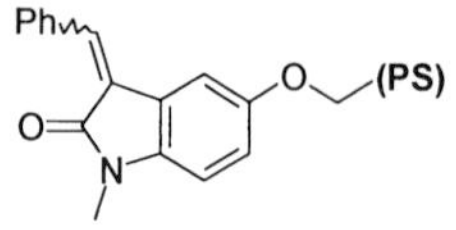	[61]
10	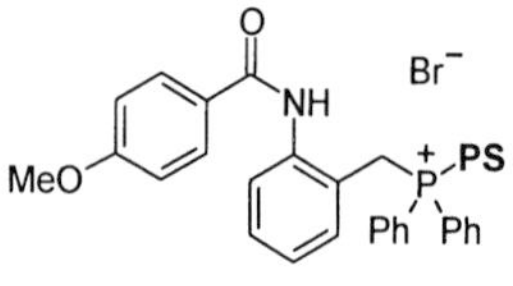	KOtBu, PhMe, DMF, 110 °C, 45 min	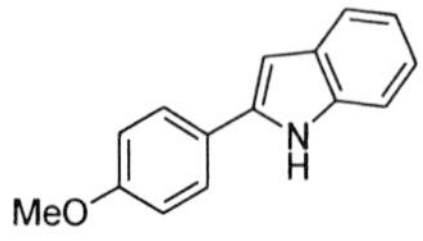	[62]
11	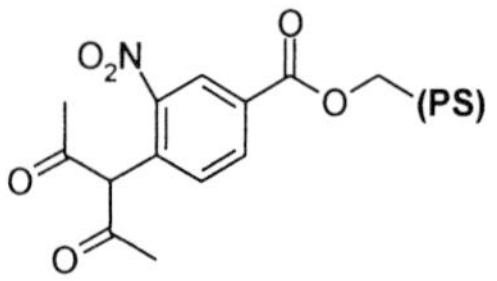	SnCl$_2$•H$_2$O (2.4 mol/L, 53 eq), NMP, 20 °C, 10 h	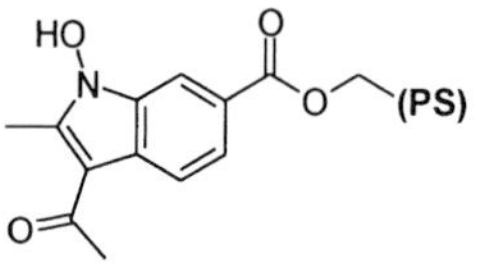	[63]
12	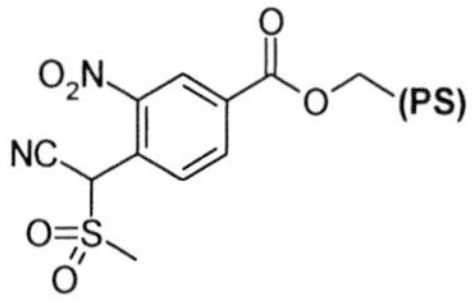	SnCl$_2$•H$_2$O (2.4 mol/L, 53 eq), NMP, 20 °C, 10 h	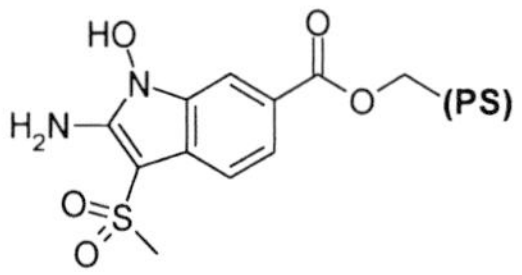	[63]

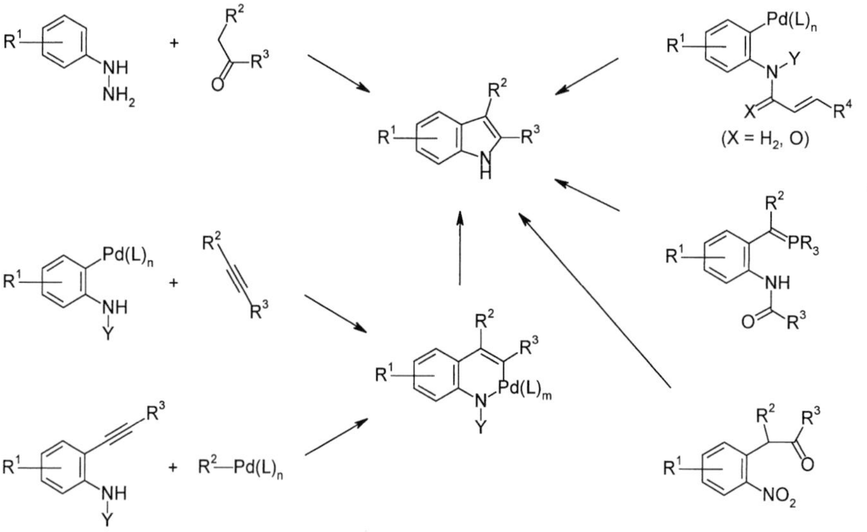

Fig. 15.3. Strategies for the preparation of indoles on insoluble supports. Y: Protective group.

Table 15.7. Derivatization of indoles.

Entry	Starting resin	Conditions	Product	Ref.
1		NaH, DMF, 0.5 h, then BrCH₂CO₂Et, 20 °C, 4 h		[58]
2		BnBr (3 eq), KOtBu (3 eq), DMF, 3 h		[56]
3		BnNH₂ (8 eq), HCHO (8 eq), AcOH/dioxane 1:4, 23 °C, 1.5 h		[56]
4		O₂N CO₂Et (5 eq), PhMe, 100 °C, 9 h		[56]
5		B(OH)₂ / SO₂Ph (9 eq), Na₂CO₃ (9 eq), Pd(PPh₃)₄ (0.04 eq), H₂O/EtOH/DME 3:2:8, microwaves (45 W), 3.8 min		[64]
6	Me₃Si—	NIS, DCM		[55]

15.5 Preparation of Furans and Tetrahydrofurans

Few solid phase syntheses of furans have been reported. Most of these are limited to furans with a specific substitution pattern and lack general applicability.

Furans can be obtained by thermolysis of support-bound 7-oxa-2-azanorbornenes, which are prepared by 1,3-dipolar cycloaddition of isomünchnones and electron-poor alkynes (Figure 15.4).

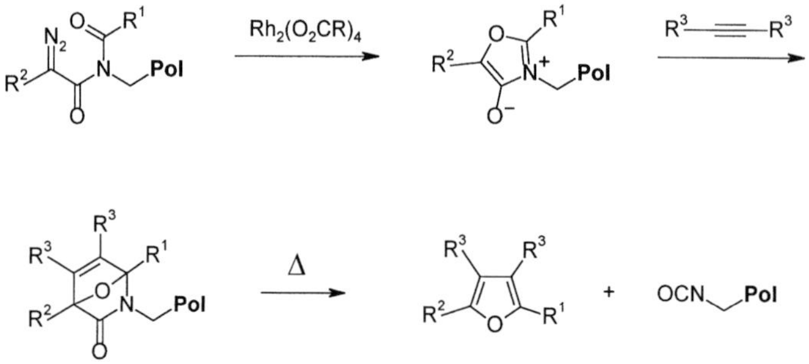

Fig. 15.4. Generation of furans by thermolysis of support-bound 7-oxa-2-azanorbornenes [65–67].

Metallated furans can be generated on cross-linked polystyrene either by direct lithiation with butyllithium [68] or by halogen–metal exchange with Grignard reagents [69]. The resulting furyllithium or -magnesium compounds are strongly nucleophilic and react smoothly with a number of electrophiles (Entries **3** and **4**, Table 15.8).

Vinylations and arylations of polystyrene-bound 2-bromofurans have been accomplished by treatment of the latter with stannanes in the presence of palladium complexes (Stille coupling [70]). Alternatively, 2-furylstannanes can be coupled with support-bound aryl iodides or bromides in the presence of palladium or copper complexes (Entries **5–7**, Table 15.8).

Tetrahydrofurans have been prepared on insoluble supports by treatment of 5-(3-butenyl)-4,5-dihydroisoxazole with electrophiles, whereby a remarkable C–C bond scission takes place to yield 2-(cyanomethyl)tetrahydrofurans (Figure 15.5; Entries **8** and **9**, Table 15.8).

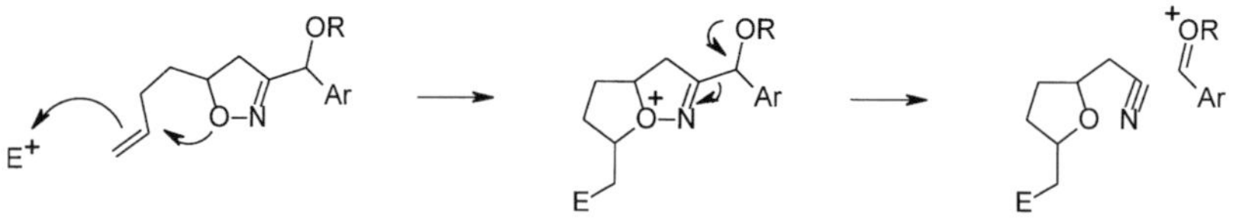

Fig. 15.5. Mechanism of the formation of 2-cyanomethyltetrahydrofurans from butenylisoxazolines.

Tetrahydrofurans can also be prepared from support-bound haloalkenes by radical cyclization (Entries **10** and **11**, Table 15.8). Halogen abstraction by tin radicals leads to the formation of carbon-centered radicals, which undergo fast, intramolecular addition to alkenes and alkynes.

2,5-Dihydrofurans have been prepared from 2-butene-1,4-diols etherified with Wang resin (Entry **12**, Table 15.8). TFA-mediated cleavage leads to simultaneous for-

Table 15.8. Preparation of furans and tetrahydrofurans.

Entry	Starting resin	Conditions	Product	Ref.
1		C_6H_6, 70 °C, 0.5 h		[65]
2		C_6H_6, 80 °C, 1.5 h		[66]
3		iPrMgBr (0.19 mol/L, 7.3 eq), THF, −35 °C, 0.5 h, then add TsCN		[69]
4		BuLi (5 eq), THF, −30 °C, 4 h, then add DMA (10 eq), 20 °C, 2 h		[68]
5		NMP, CuI, NaCl, 100 °C, 24 h		[71]
6		(3 eq), Pd(PPh$_3$)$_4$ (0.05 eq), DMF, 60 °C, 24 h		[70]
7		(6 eq), Pd(PPh$_3$)$_4$ (0.2 eq), dioxane, 100 °C, 24 h		[10]
8		ICl (0.17 mol/L), DCM, −78 °C, 15 min		[72]
9		IBr (0.3 mol/L), DCM, −78 °C, 2.5 h	(mixture of diastereomers)	[73]

Table 15.8. continued.

Entry	Starting resin	Conditions	Product	Ref.
10		Bu₃SnH (25 eq), AIBN (0.05 eq), PhMe, 70–80 °C, 2 h		[74] see also [75]
11		Bu₃SnH (0.1 mol/L, 5 eq), AIBN (0.5 eq), C₆H₆, 80 °C, 18 h		[76]
12		TFA/DCM 1:9, 20 °C, 10 min (Wang resin)		[77]

mation of the cyclic ethers. 1,3-Dihydroisobenzofurans can also be prepared using this strategy (Entry **11**, Table 15.9). The required alcohols were prepared by metallating support-bound vinyl iodides with *i*PrMgBr followed by reaction of the resulting vinyl Grignard compound with aldehydes (see Entry **4**, Table 7.5).

Linkers which enable the preparation of γ-lactones by cleavage of hydroxy esters from insoluble supports are discussed in Section 3.5.2.

15.6 Preparation of Benzofurans and Dihydrobenzofurans

Benzofurans and dihydrobenzofurans have been prepared on polymeric supports by the palladium-mediated reaction of 2-iodophenols with dienes or alkynes (Entries **1** and **2**, Table 15.9). This reaction is closely related to the synthesis of indoles from 2-iodoanilines, and probably proceeds via an intermediate palladacycle (Figure 15.3). Benzofuran and isobenzofuran derivatives have, furthermore, been prepared on cross-linked polystyrene by intramolecular addition of aryl radicals to C–C double bonds and by intramolecular Heck reaction.

2-(Allyloxy)iodoarenes react with SmI₂ to yield radical anions which undergo thermal fragmentation. The resulting aryl radicals can cyclize to dihydrobenzofurans (Entries **8–10**, Table 15.9). After cyclization, the resulting radical can be reduced and treated with a proton source, such as water or alcohols, to yield alkanes, or with carbonyl compounds to yield alcohols (Entry **10**, Table 15.9).

Table 15.9. Preparation of benzofurans.

Entry	Starting resin	Conditions	Product	Ref.
1		Pd(OAc)$_2$, LiCl, DIPEA, DMF, 100 °C		[57]
2		BocHN⟋ Pd(PPh$_3$)$_2$Cl$_2$, CuI, DMF, 50 °C, 16 h		[78]
3		Bu$_3$SnH (3 eq), AIBN (0.6 eq), C$_6$H$_6$, 80 °C, 46 h		[79]
4		Bu$_3$SnH (3 eq), AIBN (0.6 eq), C$_6$H$_6$, 80 °C, 46 h		[79] see also [75]
5		Bu$_3$SnH (3 eq), AIBN (0.6 eq), C$_6$H$_6$, 80 °C, 48 h		[80]
6		Pd(PPh$_3$)$_2$Cl$_2$ (1 mmol/L, 0.15 eq), Bu$_4$NCl (1.5 eq), NEt$_3$ (8 eq), DMF/H$_2$O 9:1, 80 °C, 24 h		[59]
7		Pd(OAc)$_2$ (0.3 eq), PPh$_3$ (0.6 eq), Bu$_4$NCl (2 eq), K$_2$CO$_3$ (4 eq), DMA, 100 °C, 27 h		[80]
8		HMPA (40 eq), SmI$_2$ (10 eq), THF, 20 °C, 1 h		[81]
9		HMPA (40 eq), SmI$_2$ (10 eq), THF, 20 °C, 1 h		[81]

Table 15.9. continued.

Entry	Starting resin	Conditions	Product	Ref.
10	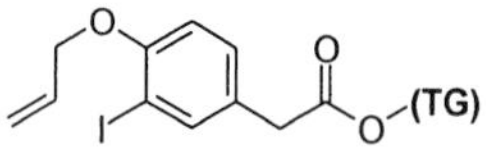	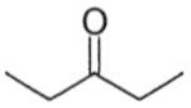 (20 eq), HMPA (40 eq), SmI$_2$ (10 eq), THF, 20 °C, 2 h	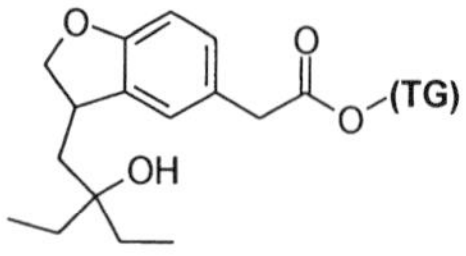	[82]
11	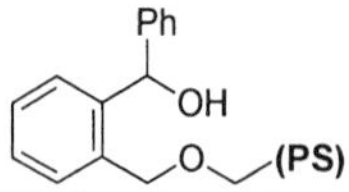	TFA/DCM 1:9, 20 °C, 10 min (Wang resin)	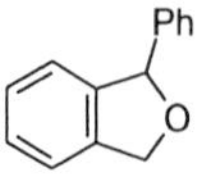	[77]

15.7 Preparation of Thiophenes

The few thiophene syntheses reported in which the formation of the heterocycle is realized on an insoluble support (Entries **1** and **2**, Table 15.10) are based on the intra-molecular addition of C,H-acidic compounds to nitriles (Thorpe–Ziegler reaction). The mechanism of this cyclization is outlined in Figure 15.6. In thiophene preparations performed on solid phase the required α-(cyano)thiocarbonyl compounds were thio-amides, which can be readily prepared from nitriles and isothiocyanates. Either of these components can be linked to the support.

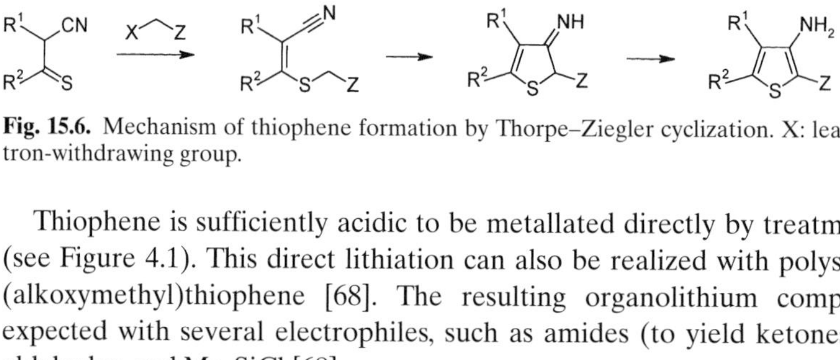

Fig. 15.6. Mechanism of thiophene formation by Thorpe–Ziegler cyclization. X: leaving group; Z: electron-withdrawing group.

Thiophene is sufficiently acidic to be metallated directly by treatment with *n*-BuLi (see Figure 4.1). This direct lithiation can also be realized with polystyrene-bound 3-(alkoxymethyl)thiophene [68]. The resulting organolithium compounds react as expected with several electrophiles, such as amides (to yield ketones), alkyl halides, aldehydes, and Me$_3$SiCl [68].

Polystyrene-bound 2-bromothiophene can be metallated by treatment with Grignard reagents. The resulting thienylmagnesium compounds can be directly trea-ted with carbon electrophiles to yield the corresponding derivatized thiophenes. For some types of electrophile, transmetallation with CuCN might be required to obtain clean products (Entry **5**, Table 15.10).

Thiophenes have been prepared on insoluble supports mainly by arylation or vinylation of halothiophenes and thienylstannanes (Table 15.10). Heck, Suzuki, and Stille couplings usually proceed with thiophenes as smoothly as with substituted ben-zenes, and arylations or vinylations of thiophenes have often been used as examples to illustrate new conditions for the realization of these coupling reactions on solid phase.

Table 15.10. Preparation of thiophenes.

Entry	Starting resin	Conditions	Product	Ref.
1		1. ArCOCH₂Br (1 mol/L, 10 eq), DMF/AcOH 20:1, 20 °C, 15 h 2. DMF/DBU 9:2, 20 °C, 15 h Ar: 4-PhC₆H₄		[83]
2		1. PhCOCH₂Cl (1 mol/L, 10 eq), DMF/AcOH 10:1, 20 °C, 20 h 2. DMF/DBU 2:1, 20 °C, 20 h		[84]
3		BuLi, THF, −30 °C, 4 h, then add allyl bromide (10 eq), −30 °C to 20 °C, 2 h		[68]
4		iPrMgBr (0.19 mol/L, 7.3 eq), THF, −35 °C, 0.5 h, then add TsCN		[69]
5		iPrMgBr (0.19 mol/L, 7.3 eq), THF, −35 °C, 15 min, then add CuCN·2 LiCl (10 eq), 15 min, then add EtO₂C—⟋⟍—Br		[69]
6		MeO—⟨⟩—ZnBr Pd(dba)₂ (0.2 eq), P(2-furyl)₃ (0.4 eq), THF, 25 °C, 20 h		[85]
7		2-bromothiophene, Pd₂(dba)₃, NEt₃, P(o-Tol)₃, DMF, 100 °C, 20 h		[86]

Table 15.10. continued.

Entry	Starting resin	Conditions	Product	Ref.
8		ZnBr PdCl₂(dppf), THF, 20 °C, 18 h		[87]
9		 (0.3 mol/L, 1.7 eq), Pd(PPh₃)₄ (0.03 eq), Na₂CO₃ (1.7 eq), DME/H₂O 6:1, 85 °C, overnight		[88]
10		B(OH)₂ Pd₂(dba)₃ (0.1 eq), K₂CO₃ (2 eq), DMF, 20 °C, 20 h		[89]
11		SnBu₃ (3 eq), Pd(PPh₃)₄ (0.05 eq), DMF, 60 °C, 24 h		[70]
12		SnBu₃ (3 eq), Pd(PPh₃)₄ (0.05 eq), DMF, 60 °C, 24 h		[70]
13		BF₄⁻ ⁺Ph (0.05 mol/L, 1.5 eq), Pd(PPh₃)₄ (0.05 eq), Na₂CO₃ (3 eq), DMF, 20 °C, 20 h	 (+ 20% biphenyl derivative)	[90] see also [89]
14		(HO)₂B (3 eq), Pd(PPh₃)₄ (0.05 eq), Na₂CO₃ (8 eq), DME, 85 °C, 48 h		[91]
15		H₁₇C₈ SnMe₃ Pd(PPh₃)₂Cl₂, DMF, 80 °C		[92]

15.8 Preparation of Imidazoles

Several types of cyclocondensation have been used to prepare imidazoles on insoluble supports (Figure 15.7). These include the condensation of 1,2-dicarbonyl compounds with aldehydes and amines, the 1,3-dipolar cycloaddition of *N*-sulfonylimines to münchnones, and the condensation of N-acylated α-aminoketones with ammonia (Table 15.11). Imidazoles have, moreover, been alkylated and arylated on cross-linked polystyrene at nitrogen and at C-2. Illustrative examples of these transformations are listed in Table 15.11.

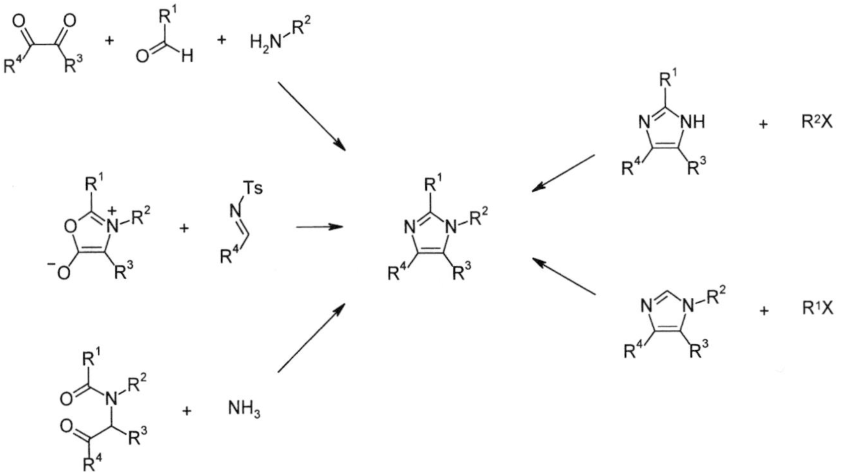

Fig. 15.7. Strategies for the preparation of imidazoles on insoluble supports.

Support-bound 1,2-diamines can be readily converted into imidazolidinones by treatment with carbonyl diimidazole [93,94]. The required diamines have been prepared on cross-linked polystyrene by reduction of peptides with diborane (MBHA resin). Similarly, bicyclic imidazolines have been prepared from triamines and thiocarbonyl diimidazole (Entry **9**, Table 14.3).

Table 15.11. Preparation of imidazoles and imidazolidinones.

Entry	Starting resin	Conditions	Product	Ref.
1		 (0.41 mol/L, 10 eq), EDC (10 eq), DCM, 20 °C, 12 h		[95]
2		NH₄OAc (60 eq), AcOH, 100 °C, 20 h		[96]
3		 (1.2 mol/L, 20 eq), NH₃ (20 eq), NH₄OAc (1.4 eq), AcOH, 100 °C, 4 h		[97]
4		 (1.2 mol/L each, 20 eq), NH₄OAc (40 eq), AcOH, 100 °C, 4 h Ar: 2-(HO)-4-BrC₆H₃		[97]
5		BuLi (3 eq), THF, −50 °C, 20 min, then add PhCHO (10 eq), −60 °C, 1 h, 20 °C, 2 h		[98]
6		Tol-B(OH)₂ (3 eq), Cu(OAc)₂ (5 eq), pyridine/NMP 1:1, microwaves, 30 s (repeat 5 times)		[99]
7		Boc-His-OH (0.23 mol/L, 5 eq), DIPEA (5 eq), DMF, 2 d		[100]
8		 FmocHN DIPEA, DCM, 20 °C, 2 h		[101]

Table 15.11. continued.

Entry	Starting resin	Conditions	Product	Ref.
9		FmocHN (0.2 mol/L, 2 eq), DIPEA (2 eq), DCM, 1 h		[102]
10		imidazole (8 eq), dioxane, 80 °C, 68 h		[10]
11		N-SiMe₃ (2.1 mol/L, 20 eq), AgOTf (1.3 eq), DMF, 85 °C, 16 h (repeat once)		[103]
12		imidazole (10 eq), MeCN, 100 °C, 20 h		[104]
13		PhNCO (1 mol/L, 60 eq), NEt₃ (60 eq), DMF, 20–55 °C, overnight		[105]

15.9 Preparation of Hydantoins (2,4-Imidazolidinediones) and Thiohydantoins

One of the first examples of intramolecular nucleophilic cleavage of compounds from insoluble supports was the preparation of hydantoins from α-amino acids. The latter, esterified with hydroxymethyl polystyrene or a similar support, were converted into ureas by treatment with an isocyanate, and cyclization to hydantoins and simultaneous cleavage from the support occurred upon heating or treatment with a base (Figure 15.8; Entries 1–4, Table 15.12). Neat diisopropylamine seems to be particularly well suited for promoting this formation of hydantoins [106]. Thioureas can be cyclized in the same way to yield thiohydantoins. The formation of thiohydantoins by this method usually proceeds more readily than the formation of hydantoins [107].

This strategy of preparing hydantoins is hampered by the fact that the nucleophilicity of the intermediate urea depends on the type of isocyanate used (R^2 in the first equation in Figure 15.8) and different yields might be obtained for each member of a hydantoin library. During the preparation of thiohydantoins by this method, thiourea formation and cyclization usually occur simultaneously, and if excess isothiocyanate is used the crude thiohydantoins will be contaminated with isothiocyanate.

Alternatively, hydantoins can be prepared on solid phase without simultaneous cleavage from the support. Two possible strategies are sketched in Figure 15.8 (Entries **11**–**14**, Table 15.12).

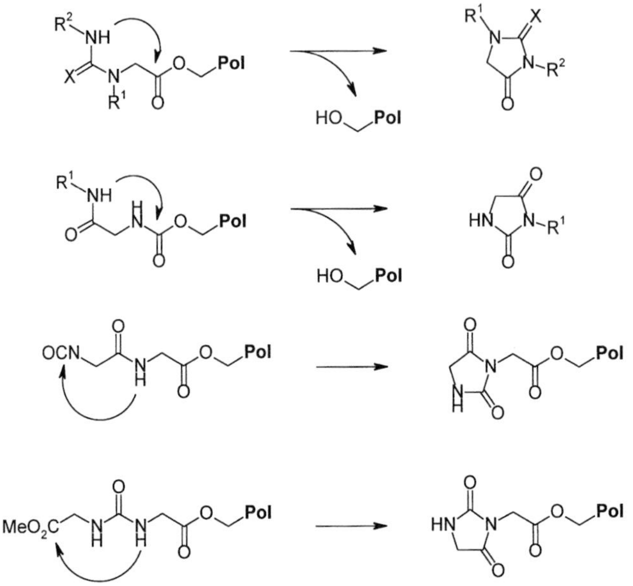

Fig. 15.8. Strategies for the preparation of hydantoins and thiohydantoins on insoluble supports. X: O, S.

Further methods for preparing hydantoins includ N-alkylation of hydantoins with reactive alkyl halides in the presence of strong bases [108]. Hydantoinimines have been synthesized from polystyrene-bound isonitriles by an Ugi-type multicomponent condensation (Entry **15**, Table 15.12).

Table 15.12. Preparation of hydantoins and related heterocycles.

Entry	Starting resin	Conditions	Product	Ref.
1		*i*PrNH$_2$ (neat), 20 °C, 6 h		[109] see also [106] [110]
2		NEt$_3$ (0.82 mol/L), CHCl$_3$, 61 °C, 72 h (Wang resin)		[111] see also [106] [112] [113]
3		1. triphosgene (3 eq), pyridine (10 eq), DCM, 3 h 2. pyridine (10 eq), (0.44 mol/L, 10 eq), DCM, 17 h (Wang resin)		[114]
4		NEt$_3$/MeOH 1:9, 20 °C, 3 h		[115]
5		(0.1 eq), DCE, 80 °C, 8–24 h		[116] see also [117]
6		DIPEA (0.5 mol/L), DMF, 80 °C, 24 h		[118]
7		1. PhNCO (3 eq), DCM, 20 °C, 24 h 2. DIPEA (1 mol/L, 10 eq), DMF, 100 °C, 36 h		[47]
8		PhNCS (0.05 mol/L, 0.7 eq), MeCN/CHCl$_3$ 1:1, reflux, 16 h (Wang resin)		[111]

Table 15.12. continued.

Entry	Starting resin	Conditions	Product	Ref.
9		KO*t*Bu, EtOH, 20 °C Ar: 3-ClC$_6$H$_4$		[119]
10		NEt$_3$ (14 eq), MeOH, 90 °C, 48 h		[120]
11		MeSiCl$_3$ (0.6 mol/L, 20 eq), NEt$_3$ (40 eq), CHCl$_3$, 70 °C, 24 h		[121]
12		CDI (15 eq) or triphosgene (5 eq), DIPEA (15 eq), DCM		[122] see also [123]
13		NaOMe (10 eq), THF/MeOH 1:1, 3 h (complete racemization)		[108]
14		*i*Pr$_2$NH (neat), 20 °C, 1 h		[110]
15		 (0.35 mol/L, 5 eq of each), KOCN (10 eq), pyridine•HCl (10 eq), CHCl$_3$/MeOH/H$_2$O 5:5:1, 20 °C, 24 h		[33]
16		allyl bromide (10 eq), BTPP (10 eq), THF, 12 h		[108]

15.10 Preparation of Benzimidazoles

Several procedures have been reported for the preparation of benzimidazoles on insoluble supports (Table 15.13). The main synthetic strategies are the cyclocondensation of support-bound 1,2-diaminobenzenes with carboxylic acid derivatives and the oxidation of dihydrobenzimidazoles, which are readily formed from aldehydes and 1,2-diaminobenzenes (Figure 15.9). The approach based on carboxylic acid derivatives generally requires heating and/or strong acids (e.g. TFA; Entry **4**, Table 15.13), which limits the choice of supports and linkers. Preparations based on aldehydes, however, often proceed under significantly milder conditions. Oxidation of the cyclic aminals can often be achieved with weak oxidants, and sometimes occurs even in the absence of additional oxidants if excess aldehyde is used [12,124]. When support-bound 1-amino-2-alkylaminobenzenes are treated with excess aldehyde, benzimidazolium salts (Figure 15.9) are occasionally formed as byproducts [12].

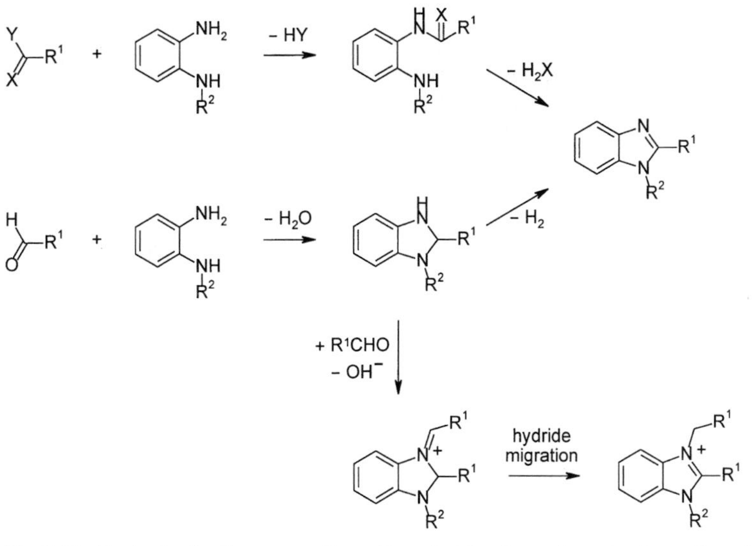

Fig. 15.9. Strategies for the preparation of benzimidazoles and mechanism of benzimidazolium salt formation [12]. X: O, S; Y: leaving group.

The 1,2-diaminobenzenes required are generally prepared on solid phase from 2-fluoronitrobenzenes. Aromatic nucleophilic substitution with an aliphatic or aromatic primary amine, followed by reduction of the nitro group (e.g. with $SnCl_2$), yields the desired intermediates. Because formic acid readily converts 1,2-diaminobenzenes into the corresponding benzimidazoles, the reduction of support-bound 2-aminonitrobenzenes should not be conducted in DMF but in NMP [125].

Support-bound benzimidazoles can be N-arylated by treatment with arylboronic acids in the presence of $Cu(OAc)_2$ (Entry **12**, Table 15.13). Benzimidazolones, which can be prepared from resin-bound 1,2-diaminobenzenes and di-*N*-succinimidyl carbonate (Entry **6**, Table 15.13), can also be N-alkylated on insoluble supports (Entry **13**,

Table 15.13. Preparation of benzimidazoles and related heterocycles.

Entry	Starting resin	Conditions	Product	Ref.
1		Ph~~CHO (4 eq), DDQ (2 eq), DMF, 20 °C, 5 h		[126]
2		(0.15 mol/L, 10 eq), NMP, 20 °C, 5 h, then 50 °C, 8 h		[124]
3		(0.25 mol/L, 30 eq), DMF/BuOH 5:6, 90 °C, 24 h		[127]
4		1. (BrCH$_2$CO)$_2$O (10 eq), DMF, 1 h 2. pyrrolidine (2 mol/L), DMSO, 20 °C, 4 h 3. TFA (neat), 16 h		[125]
5		HC(OMe)$_3$/TFA/ DCM 1:1:2, 3 h		[128]
6		(2 eq), THF/DCM/ DMF 8:2:1, 24 h		[129]
7		(10 eq), THF, 20 °C, overnight		[130] see also [131]
8		(0.6 mol/L, 10 eq), PhNO$_2$, 180 °C, 3 d		[132] see also [133]

Table 15.13. continued.

Entry	Starting resin	Conditions	Product	Ref.
9	(PEG) aldehyde resin	(0.6 mol/L, 10 eq), PhNO$_2$, 180 °C, 3 d	(PEG)	[132]
10	(PS), Tol, NH	isopentyl nitrite, AcOH	(PS), Tol	[134]
11	HN, (PS), F	PPh$_3$, DEAD, THF/DCM 1:1, 20 °C, 48 h	(PS), F	[135]
12	(TG)	Tol-B(OH)$_2$ (3 eq), Cu(OAc)$_2$ (5 eq), pyridine/NMP 1:1, microwaves, 30 s (repeat 5 times)	(TG) (two isomers, 1:1)	[99]
13	Ph, O=, (PS)	NaH (20 eq), DMF, 50 min; then iBuBr (40 eq), DMF, 24 h	Ph, O=, (PS)	[129]
14	Ph, S=, (PS)	BnBr, DIPEA, DMF, 20 °C, 16 h	Ph, S, Ph, (PS)	[130] see also [131]

Table 15.13). Alkylation of benzimidazole-2-thiones, on the other hand, generally leads to clean S-alkylation (Entry **14**, Table 15.13).

15.11 Preparation of Isoxazoles

Isoxazoles and their partially or fully saturated analogs have mainly been prepared, both in solution and on insoluble supports, by 1,3-dipolar cycloaddition of nitrile oxides and nitrones to alkenes or alkynes (Figure 15.10). Nitrile oxides can be generated in situ on insoluble supports by dehydration of nitroalkanes with isocyanates, or by conversion of aldehyde-derived oximes to α-chloroximes and dehydrohalogenation of the latter. Nitrile oxides react smoothly with a broad variety of alkenes and alkynes to yield the corresponding isoxazoles. A less convergent approach to isoxazoles is the cyclocondensation of hydroxylamine with 1,3-dicarbonyl compounds or α,β-unsaturated ketones.

Dihydroisoxazoles with a substituent at nitrogen are most conveniently prepared by 1,3-dipolar cycloaddition of nitrones to alkenes or alkynes. Nitrones are usually prepared in situ from carbonyl compounds and *N*-(alkyl)hydroxylamines (Figure 15.10).

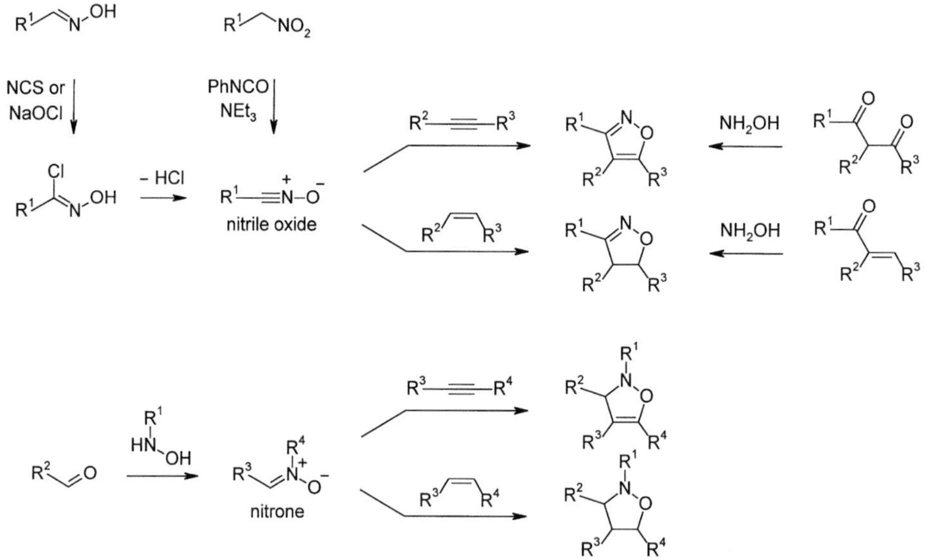

Fig. 15.10. Synthetic strategies for the preparation of isoxazoles, isoxazolines, and isoxazolidines.

Most approaches sketched in Figure 15.10 have been realized successfully on insoluble supports, either with alkenes or alkynes linked to the support, or with support-bound 1,3-dipoles (Table 15.14). Intramolecular 1,3-dipolar cycloadditions of nitrile oxides and nitrones to alkenes have been used to prepare polycyclic isoxazolidines on solid phase (Entries **7** and **9**, Table 15.14).

Strategies which lead to the formation of isoxazoles during cleavage from an insoluble support include the oxidative cleavage of *N*-(4-alkoxybenzyl)isoxazolidines with DDQ to yield isoxazolines (Entry **14**, Table 15.14), the nucleophilic cleavage of 2-acylenamines with hydroxylamine (Entry **15**, Table 15.14), and the acidolysis of 2-cyanophenols etherified with an oxime resin (Entry **17**, Table 15.14). The required oxime ethers for the latter synthesis were prepared by reaction of the corresponding 2-fluorobenzonitriles with Kaiser oxime resin (oxime from 4-nitrobenzoylated cross-linked polystyrene [136]).

Table 15.14. Preparation of isoxazoles, isoxazolines, and isoxazolidines.

Entry	Starting resin	Conditions	Product	Ref.
1		BnNO$_2$ (0.05 mol/L), PhNCO (0.1 mol/L), NEt$_3$, THF, 60 °C, 20 h		[112] see also [137-139]
2		(5 eq), NaOCl (bleach; 0.4 mol/L, 10 eq), THF, 4 h		[140]
3		THPO—NO$_2$ (5 eq), PhNCO (11 eq), NEt$_3$, C$_6$H$_6$, 80 °C, 30 h		[141] see also [139]
4		(5 eq), NEt$_3$ (5 eq), THF, 20 °C, 2 h		[48]
5		NaOCl (bleach; 0.4 mol/L, 10 eq), allyl alcohol (5 eq), THF, 4 h		[140]
6		NCS (4 eq), DCM, 2 h; then 1-benzyl-2,5-dihydropyrrole (10 eq), NEt$_3$, DCM, overnight		[142]
7		PhNCO (1.2 mol/L), NEt$_3$ (0.12 mol/L), C$_6$H$_6$, 20 °C, 3 d		[73]
8		NH$_2$OH (2.5 mol/L), DMA, 80 °C, 24 h		[143]
9		HATU, DIPEA, DMAP, DCM, 20 °C		[144]

Table 15.14. continued.

Entry	Starting resin	Conditions	Product	Ref.
10		Ar²CHO (0.33 mol/L, 10 eq), HONHMe•HCl (10 eq), DIPEA (10 eq), PhMe, 80 °C, 5 h	(mixture of diastereomers)	[145]
11		HONHMe•HCl (0.87 mol/L, 10 eq), DIPEA (10 eq), PhMe, 80 °C, 0.5 h, then ArCH=CH₂ (10 eq), 80 °C, 5 h Ar: 1-naphthyl		[145]
12		1. ArCHO (0.45 mol/L, 10 eq), PhMe, 80 °C, 0.5 h 2. PhSO₂CH=CH₂ (0.45 mol/L, 10 eq), PhMe, 80 °C, 5 h Ar: 4-(O₂N)C₆H₄		[145]
13		Yb(OTf)₃ (0.2 eq), PhMe, 20 °C, 20 h		[146]
14		DDQ (3 eq), DCM/H₂O 10:1, 12 h		[146]
15		HONH₃Cl, NaHCO₃, EtOH/H₂O, 50 °C		[147]
16		SnCl₂•H₂O (1 mol/L, 31 eq), NMP, 20 °C, 18 h Ar: 4-(AcHN)C₆H₄		[63]
17		TFA/HCl (5 mol/L in H₂O) 4:1, 55 °C, 2 h		[136]

15.12 Preparation of Oxazoles and Oxazolidines

Few solid phase syntheses for oxazoles have been reported (Table 15.15). Polystyrene-bound 2-(acylamino)phenols have been cyclized to benzoxazoles by treatment with PPh₃/DEAD (Entry **1**, Table 15.15). Oxazolidin-2-ones have been prepared by intramolecular, nucleophilic cleavage of carbamates from insoluble supports (Table 15.15).

Table 15.15. Preparation of oxazoles and oxazolidines.

Entry	Starting resin	Conditions	Product	Ref.
1	(structure) (PS)	PPh₃ (0.38 mol/L, 5 eq), DEAD (5 eq), THF, 20 °C, 17 h	(structure) (PS)	[148]
2	(structure) PS	DBN (3 eq), DCM, 20 °C, overnight	(structure)	[149]
3	(structure) (PS)	pyrrolidine (0.5 mol/L, 5 eq), LiClO₄ (5 eq), THF, 20 °C	(structure)	[150]

15.13 Preparation of Thiazoles and Thiazolidines

Most solid-phase thiazole syntheses reported are based on the cyclocondensation of thioamides or thioureas with α-haloketones (Table 15.16). These reactions generally proceed smoothly under basic or acidic conditions, and are compatible with most common linkers.

Thiazolidin-4-ones have been prepared by condensation of support-bound imines with α-mercapto carboxylic acids. These thioaminals are quite stable and tolerate, e.g., treatment with trifluoroacetic acid [151,152]. Also unacylated thiazolidines, which can be prepared from resin-bound cysteine and aldehydes, are remarkably stable towards acid-promoted hydrolysis. Libraries of thiazolidinones have been used to identify new cyclooxygenase-1 inhibitors [153].

Table 15.16. Preparation of thiazoles and thiazolidines.

Entry	Starting resin	Conditions	Product	Ref.
1		(0.5 mol/L), DMF, 70 °C, 4 h		[154]
2		(0.2 mol/L, 5 eq), dioxane, 3 × 1 h		[155]
3		dioxane, 60 °C, 16 h		[156]
4		PhCOCH₂Cl (1 mol/L, 10 eq), DMF/AcOH 10:1, 20 °C, 20 h		[84]
5		thiourea (4 eq), MeOH, 65 °C, 12 h		[157]
6		HS⌒CO₂H (0.7 mol/L), THF, 70 °C, 3 h		[15] see also [151-153]
7		ArCHO, PhMe/MeCN/AcOH 45:45:10		[158]

15.14 Preparation of Pyrazoles

Most solid-phase syntheses of pyrazoles are based on the cyclocondensation of hydrazines with suitable 1,3-dielectrophiles. The examples reported include the reaction of hydrazines with support-bound α,β-unsaturated ketones, 1,3-diketones, 3-keto esters, α-(cyano)carbonyl compounds, and α,β-unsaturated nitriles (Table 15.17). Pyrazoles have also been prepared from polystyrene-bound 3-(hydrazino)esters, which result from the addition of ester enolates to hydrazones (Entry 7, Table 15.17; see also Section 10.3). Benzopyrazoles can also be prepared from support-bound

Table 15.17. Preparation of pyrazoles and pyrazolones.

Entry	Starting resin	Conditions	Product	Ref.
1		(0.5 mol/L, 10 eq), DMSO, 100 °C, 16 h		[160]
2		N$_2$H$_4$ (2.5 mol/L), DMA, 80 °C, 24 h		[143]
3		PhMe, 100 °C, 5 h		[161]
4		BnN$_2$H$_3$, AcOH, EtOH, 70 °C		[162]
5		BnN$_2$H$_3$•HCl, AcOH/EtOH 1:9		[163]
6		PhN$_2$H$_3$ (0.14 mol/L, 10 eq), NaOEt (100 eq), EtOH, 78 °C, 24 h		[164]
7		NaOMe (0.42 mol/L), MeOH, 60 °C, 8 h		[165]
8		pyrazole (0.35 mol/L, 15 eq), K$_2$CO$_3$ (6 eq), MeCN, 60 °C, 2 d		[166]
9		Tol-B(OH)$_2$ (3 eq), Cu(OAc)$_2$ (5 eq), pyridine/NMP 1:1, microwaves, 30 s (repeat 5 times)		[99]

hydrazones, using the reaction sequence outlined in Figure 15.11. Oxidation of a poly-styrene-bound benzophenone hydrazone yields an α-(acyloxy)azo compound. Upon treatment with a Lewis acid this intermediate is converted into a 1,2-diazaallyl cation, which undergoes intramolecular aromatic electrophilic amination to yield the ob-served benzopyrazole.

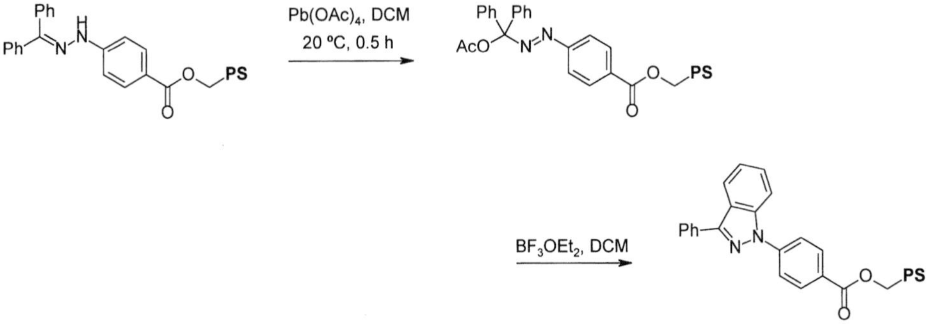

Fig. 15.11. Synthesis of benzopyrazoles by intramolecular, electrophilic amination [159].

Examples of N-alkylation and N-arylation of pyrazoles on insoluble supports have also been reported. Unsubstituted pyrazole undergoes Michael addition to resin-bound α-acetamido acrylates to yield 1-pyrazolylalanine derivatives (Entry **8**, Table 15.17). N-Ary-lation of support-bound pyrazole has been accomplished with arylboronic acids in the presence of Cu(OAc)$_2$ upon microwave irradiation (Entry **9**, Table 15.17).

15.15 Preparation of Triazoles, Tetrazoles, Oxadiazoles, and Thiadiazoles

1,2,3-Triazoles have been prepared on insoluble supports by 1,3-dipolar cycloaddi-tion of alkylazides to support-bound alkynes, and by diazo group transfer to 2-(acyl) enamines (Entries **1** and **2**, Table 15.18). The second approach avoids handling of hazardous reagents, and yields triazoles with complete control of regioselectivity.

1-Hydroxybenzotriazoles can be prepared on cross-linked polystyrene from 2-chloro-nitroarenes, and their reduction to the corresponding benzotriazoles can be achieved by treatment with PCl$_3$ (Entries **3** and **4**, Table 15.18). Benzotriazoles have been N-arylated on insoluble supports by treatment with arylboronic acids in the presence of catalytic amounts of copper salts. 1,2,4-Triazoles undergo Michael addition to polystyrene-bound α-acetamido acrylates to yield triazole-derived α-amino acids (Entry **5**, Table 15.18).

The multicomponent, Ugi-type condensation of aldehydes, secondary amines, azide and polystyrene-bound isonitriles has been reported to yield tetrazoles in acceptable yields (Entry **7**, Table 15.18). The reaction of Tentagel-bound carboxylic esters with amidoximes has been used to prepare oxadiazoles (Entry **8**, Table 15.18). Thiadiazoles have been prepared from support-bound *N*-sulfonylhydrazones by treatment with thionyl chloride. Thiadiazole formation and cleavage from the support occurred simultaneously (Entry **9**, Table 15.18).

Table 15.18. Preparation of triazoles, tetrazoles, and thiadiazoles.

Entry	Starting resin	Conditions	Product	Ref.
1		RCH_2N_3 (0.07 mol/L, 2 eq), PhMe, 110 °C, 12 h	 (+ regioisomer)	[167] see also [168]
2		TsN_3 (0.5 mol/L, 6 eq), DMF/DIPEA 3:1, 20 °C, 24 h		[169]
3		1. N_2H_4 (0.1 mol/L), $EtOCH_2CH_2OH$, 114 °C, 20 h 2. dioxane/conc aq HCl 1:1, 100 °C, 20 h		[170]
4		PCl_3 (1.2 mol/L), $CHCl_3$, 61 °C, 20 h		[170]
5		1,2,4-triazole (0.14 mol/L, 6 eq), K_2CO_3 (3 eq), MeCN, 25 °C, 2 d		[166]
6		Tol-B(OH)$_2$ (3 eq), Cu(OAc)$_2$ (5 eq), pyridine/NMP 1:1, microwaves, 30 s (repeat 5 times)	 (two isomers, 1:1)	[99]
7		PhCHO, piperidine (0.39 mol/L, 5 eq of each), NaN_3 (10 eq), pyridine·HCl (10 eq), $CHCl_3$/MeOH/H_2O 5:5:2, 20 °C, 4 d		[33]
8		 NaOEt (10 eq of each), EtOH/DCM 1:1, 20 °C, 3 d		[171]
9		$SOCl_2$ (4.2 mol/L, 20 eq), DCE, 60 °C, 5 h		[172]

15.16 Preparation of Pyridines and Dihydropyridines

Various approaches can be used to synthesize pyridines and partially saturated pyridines on insoluble supports. Dihydropyridines can be readily prepared by cyclocondensation of amines with ketones and aldehydes (Hantzsch pyridine synthesis, Figure 15.12). This synthesis proceeds particularly smoothly with β-keto carboxylic acid derivatives as ketone component. On insoluble supports this cyclocondensation has been realized either with the amine or with the β-keto carboxylic acid linked to the insoluble support (Entries **1** and **2**, Table 15.19). If ammonia is used as amine, the initially formed dihydropyridines can be oxidized to the corresponding pyridines (Entries **3** and **4**, Table 15.19).

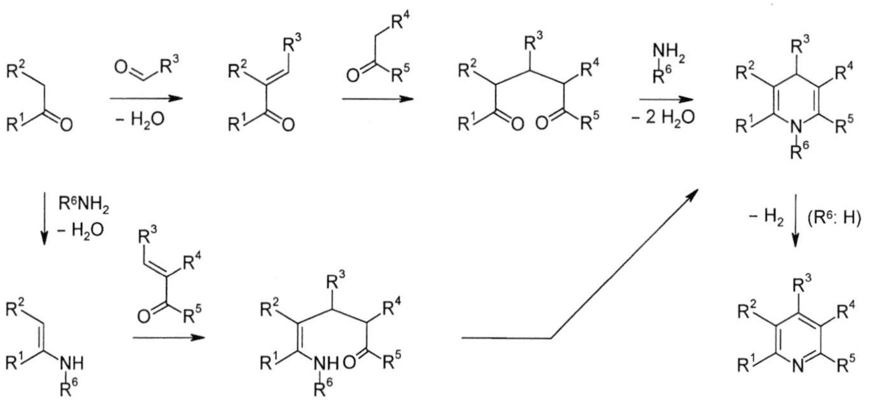

Fig. 15.12. Two possible mechanisms of formation of dihydropyridines and pyridines by cyclocondensation.

Closely related to the Hantzsch pyridine synthesis is the cyclocondensation of acetylketene (formed by thermolysis of Meldrum's acid) with PEG-bound enamines, whereby 1,4-dihydro-4-pyridinones are obtained (Entry **5**, Table 15.19).

Dihydropyridines can be prepared on cross-linked polystyrene by addition of organometallic reagents to pyridinium salts (Entry **6**, Table 15.19). These reactions do not always give high yields because of several competing processes (e.g. cleavage of the linker, deacylation of the N-acylpyridinium salt).

Support-bound alkylating agents have been used to N-alkylate pyridines and dihydropyridines (Entries **7** and **8**, Table 15.19). Polystyrene-bound 1-[(alkoxycarbonyl)-methyl]pyridinium salts can be prepared by N-alkylating pyridine with immobilized haloacetates (Entry **8**, Table 15.19). These pyridinium salts react with acceptor-substituted olefins to yield cyclopropanes (Section 5.1.3.6).

A convenient method for preparing vinylated or arylated pyridines on insoluble supports is palladium-mediated cross-coupling. The Heck, Suzuki, and Stille reactions have been successfully used for this purpose (Table 15.20). The conditions are essentially the same as in the related coupling of arenes (see Section 5.2.4).

Table 15.19. Preparation of pyridines and dihydropyridines.

Entry	Starting resin	Conditions	Product	Ref.
1		DMF, MS 4 Å, 80 °C, 10 h		[173] see also [174]
2		PhCHO, pentane-2,4-dione, pyridine, MS 4 Å, 45 °C, 24 h, then TFA/DCM (3:97), 45 min (Rink amide resin)		[175] see also [176]
3		(0.23 mol/L, 4 eq), CsF (1 eq), DMSO, 70 °C, 3 h; wash; NH$_4$OAc, AcOH/ DMF 1:20, 100 °C, 18 h Ar: 3,4-F$_2$C$_6$H$_3$		[177]
4		CAN, DMA, 20 °C, 15 min		[173] see also [174]
5		PhMe, 110 °C		[178]
6		PhMgCl (4 eq), THF, 2 h		[179] see also [180]
7		(2 eq), THF, HMPA, 20 °C, 12 h		[181]
8		pyridine (4.3 eq), DCM, 20 °C, 24 h		[48] see also [182] [183]
9		TFA/DCM 2:1, O$_2$, 48 h		[180]

Table 15.20. Preparation of pyridines by palladium-mediated vinylation and arylation of other pyridines.

Entry	Starting resin	Conditions	Product	Ref.
1		3-bromopyridine, Pd$_2$(dba)$_3$, NEt$_3$, DMF, P(o-Tol)$_3$, 100 °C, 20 h		[86]
2		Pd$_2$(dba)$_3$ (0.1 eq), K$_2$CO$_3$ (2 eq), DMF, 20 °C, 20 h		[89]
3		Pd$_2$(dba)$_3$, AsPh$_3$, dioxane, 50 °C, 24 h		[184]
4		(0.19 mol/L, 10 eq), Pd(PPh$_3$)$_4$ (0.05 eq), Na$_2$CO$_3$ (21 eq), PhMe/EtOH 9:1, 90 °C, 24 h		[182]
5		Pd(dba)$_2$ (0.2 eq), P(2-furyl)$_3$ (0.4 eq), THF, 25 °C, 20 h		[85]
6		PhSnBu$_3$, Pd(PPh$_3$)$_4$, DMF, 60 °C, 24 h		[70]

15.17 Preparation of Tetrahydropyridines and Piperidines

Ring-closing metathesis is well suited for the preparation of five- or six-membered heterocycles, and has also been used with success to prepare tetrahydropyridines on insoluble supports (Entries **1** and **2**, Table 15.21). Because metathesis catalysts (ruthenium or molybdenum carbene complexes) are electrophilic, reactions should be conducted with acylated amines to avoid poisoning of the catalyst.

One approach to tetrahydropyridinones is the Lewis acid mediated hetero-Diels–Alder reaction of Danishefsky's diene with polystyrene-bound imines (Entry **3**, Table 15.21). The Ugi reaction of 5-oxo carboxylic acids and primary amines with support-bound isonitriles has been used to prepare piperidinones on insoluble supports

(Entry **4**, Table 15.21). Entry **5** in Table 15.21 is an example of the preparation of a 4-piperidinone by amine-induced β-elimination of a resin-bound sulfinate, followed by Michael addition of the amine to the newly generated divinylketone. The intramolecular Pauson–Khand reaction of propargyl(3-butenyl)amines, which yields cyclopenta[*c*]pyridin-6-ones, is sketched in Table 12.4.

Table 15.21. Preparation of pyridines, partially saturated pyridines, and piperidines.

Entry	Starting resin	Conditions	Product	Ref.
1		(0.13 eq), DCM, 40 °C, 12 h		[28] see also [9]
2		(0.05 eq), styrene (1 eq), PhMe, 50 °C, 18 h Ar: 4-(MeO)C₆H₄		[185] see also [186]
3		(0.9 mol/L, 5 eq), Yb(OTf)₃ (0.1 eq), THF, 60 °C, 3 h		[187]
4		(0.3 mol/L, 5 eq), BnNH₂ (5 eq), CHCl₃/MeOH 2:1, 20 °C, 48 h		[33]
5		BnNH₂, THF, 20 °C, 3 d		[188]
6		ClMg CuI, BF₃OEt₂, THF, 4 h		[189]
7		(1 mol/L, 10 eq), THF, 65 °C, overnight		[190]

Support-bound pyridines and partially saturated pyridines can be valuable synthetic intermediates, enabling various types of chemical transformation. Piperidinones can be prepared on cross-linked polystyrene by addition of organometallic reagents to tetrahydropyridinones (Entry **6**, Table 15.21). 1,2-Dihydropyridines are electron-rich dienes which can undergo Diels–Alder reaction with electron-poor dienophiles. Diels–Alder cycloaddition of support-bound 1,2-dihydropyridines has been used to prepare nitrogen-containing polycyclic systems (Entry **7**, Table 15.21).

15.18 Preparation of Fused Pyridines

Tetrahydroisoquinolines can be readily prepared in solution and on insoluble supports by condensation of phenethylamines with aldehydes (Figure 15.13; Pictet–Spengler synthesis). The precise reaction conditions depend on the substitution pattern of the arene, but smooth cyclization generally occurs only with electron-rich arenes. Most of the examples reported were performed under acidic reaction conditions, but phenols or imidazoles can also be alkylated intramolecularly by imines under basic or neutral conditions. Illustrative examples of Pictet–Spengler reactions on solid phase are listed in Table 15.22.

Closely related to the Pictet–Spengler reaction is the Bischler–Napieralski synthesis of isoquinolines (Figure 15.13). In this reaction an acylated phenethylamine is cyclized by treatment with a strong dehydrating agent. The resulting dihydroisoquinoline can either be reduced to the tetrahydroisoquinoline (e.g. with NaBH$_4$) or oxidized to yield an isoquinoline.

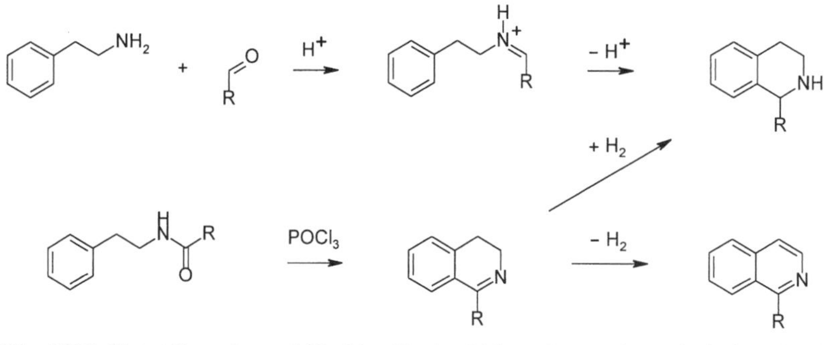

Fig. 15.13. Pictet–Spengler and Bischler–Napieralski syntheses of tetrahydroisoquinolines and quinolines from 2-arylethylamines.

Histamine, tryptamine, and the corresponding amino acids histidine and tryptophan, can also be cyclized by treatment with aldehydes to yield annulated piperidines (Table 15.22). Numerous examples of the preparation of carbolines from support-bound tryptophan derivatives have been reported [191–194].

Isoquinolines have been prepared on insoluble supports by radical-mediated cyclizations and by intramolecular Heck reaction (Table 15.23). Entry **1** in Table 15.23 is a rare example of the formation of biaryls by intramolecular addition of an aryl radical to an arene. Oxidative aromatization was achieved by using a large excess of AIBN.

Table 15.22. Pictet–Spengler and Bischler–Napieralski reactions on insoluble supports.

Entry	Starting resin	Conditions	Product	Ref.
1		(0.5 mol/L), pyridine, 100 °C, 14 h		[195]
2		PhCH$_2$CHO (10 eq), TFA/DCM 1:9, 20 °C, overnight		[196]
3		(4 eq), DCM/TFA 99:1, 20 °C, 48–72 h (Wang resin)		[197]
4		OHC (0.5 mol/L), pyridine, 100 °C, 14 h		[195]
5		1. POCl$_3$ (20 eq), PhMe, 90 °C, 18 h 2. NaBH$_4$ (20 eq), DCM, MeOH, 0–22 °C, 4 h		[198] see also [199]

Polystyrene-bound *o*-quinodimethanes, which are formed upon thermolysis of benzocyclobutanes, can be converted into 1,2,3,4-tetrahydroisoquinoline derivatives by reaction with *N*-sulfonylimines. Reaction of *o*-quinodimethanes with electron-poor nitriles leads to the formation of 1,4-dihydroisoquinolines, which undergo elimination with simultaneous release of isoquinolines into solution (Entry **7**, Table 15.23).

Reissert compounds (1-acyl-1,2-dihydro-2-quinolinecarbonitriles) have been prepared on cross-linked polystyrene and C-alkylated in the presence of strong bases (Entry **9**, Table 15.23). Treatment of polystyrene-bound C-alkylated Reissert compounds with KOH leads to the release of isoquinolines into solution. The reaction of support-bound quinoline- and isoquinoline *N*-oxides with acylating agents followed by treatment with electron-rich heteroarenes and enamines has been used to prepare alkylated and arylated derivatives of these heterocycles (Entry **10**, Table 15.23; see also Table 15.24).

Various different types of cyclization have been used to prepare quinolines on insoluble supports. One reaction which enables the rapid preparation of substituted tetra-

Table 15.23. Preparation of isoquinolines.

Entry	Starting resin	Conditions	Product	Ref.
1		Bu₃SnH (16.2 eq), AIBN (13.5 eq), C₆H₆, 80 °C		[80]
2		aq HCl (3 mol/L)/ dioxane 1:1, 80 °C, 48 h		[91]
3		Pd(OAc)₂ (0.2 eq), PPh₃ (0.4 eq), Bu₄NCl (2 eq), K₂CO₃ (4 eq), DMA, 100 °C, 24 h		[80]
4		Pd(OAc)₂ (0.2 eq), PPh₃ (0.4 eq), Bu₄NCl (2 eq), K₂CO₃ (4 eq), DMA, 100 °C, 24 h		[80]
5		Pd(PPh₃)₄, NaOAc, PPh₃, DMA, 85 °C, 5 h		[200]
6		Ar⁓NTs (2 eq), EtCN, 108 °C, 14 h Ar: 4-(O₂N)C₆H₄		[201]
7		Cl₃CCN (2 eq), PhMe, 105 °C, 14 h		[201]
8		LDA, EtI, THF, −78 °C to 20 °C, 48 h		[202]
9		KOH (1 mol/L)/THF 1:1, 67 °C, 12 h		[202]

Table 15.23. continued.

Entry	Starting resin	Conditions	Product	Ref.
10	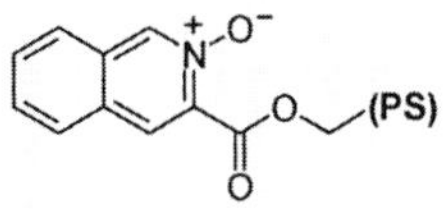	PhCOCl (0.11 mol/L, 3 eq), DCM, 0 °C, 15 min, then add indole, 0–20 °C, 2 h	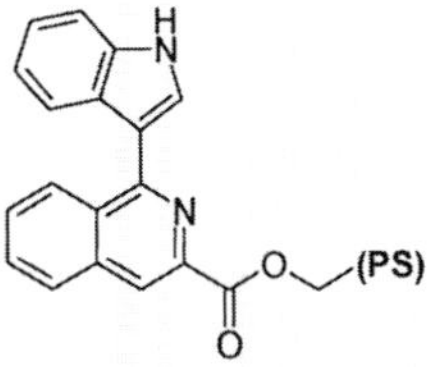	[203]
11	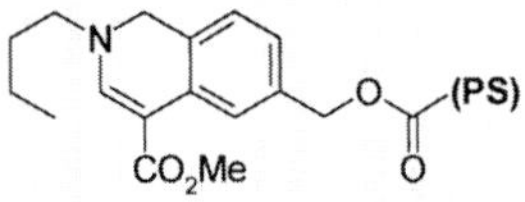	O₂, DCE, 20 °C, 24 h	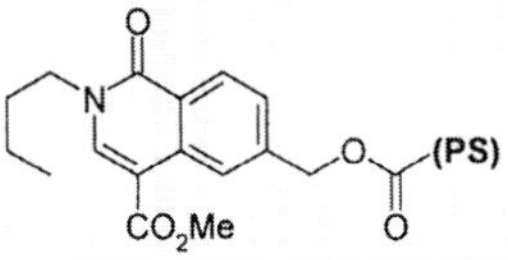	[80]

hydroquinolines is the acid-catalyzed condensation of anilines with aldehydes and alkenes (Figure 15.14). Each of the required components can be linked to the support, whereby products with a large number of different substitution patterns become accessible (Entries **1–3**, Table 15.24).

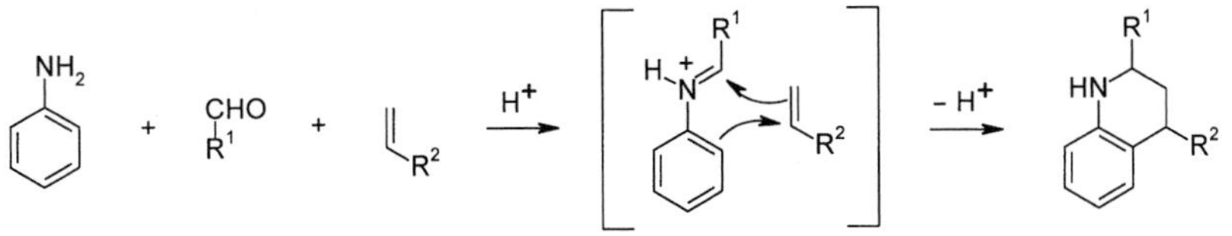

Fig. 15.14. Preparation of tetrahydroquinolines from anilines, aldehydes, and alkenes.

Quinolines have also been prepared on insoluble supports by cyclocondensation reactions and by intramolecular aromatic nucleophilic substitution (Table 15.24). Entry **10** in Table 15.24 is an example of a remarkable palladium-mediated cycloaddition of support-bound 2-iodoanilines to 1,4-dienes.

The reduction of the nitro group of polystyrene-bound 2-nitro-1-(3-oxoalkyl)benzenes with SnCl₂ (Entry **11**, Table 15.24) leads to the formation of quinoline-*N*-oxides. These intermediates can be reduced to the quinolines, on solid phase, by treatment with TiCl₃.

Table 15.24. Preparation of quinolines.

Entry	Starting resin	Conditions	Product	Ref.
1		Yb(OTf)$_3$, MeCN/DCM 2:1, 24 h		[204] see also [205]
2		PhNH$_2$ (3.3 eq), PhCHO (3.3 eq), MeCN/TFA 99:1, 12 h		[206]
3		PhNH$_2$ (0.15 eq), cyclopentadiene (0.15 eq), MeCN/TFA 99:1, 24 h		[206]
4		TMG, DCM, 55 °C, 18 h		[207]
5		PhMe, 80 °C, 24 h (Wang resin)		[208]
6		EtOCH=C(CO$_2$Et)$_2$, then 260 °C		[209]
7		SnCl$_2$ (2 mol/L), DMF, 20 °C		[16]
8		(0.11 mol/L, 10 eq of each), C$_6$H$_6$, 80 °C, 8 h Ar: 4-(O$_2$N)C$_6$H$_4$		[210]

Table 15.24. continued.

Entry	Starting resin	Conditions	Product	Ref.
9	HO (structure) (PS)	(10 eq), HOBt (10 eq), EDC (10 eq), DMF, 16 h, repeat once, then add pyridine, piperidine, 16 h	(structure) (PS)	[211]
10	(structure) (PS)	Pd(OAc)$_2$, LiCl, DIPEA, DMF, 100 °C	(structure) (PS)	[57]
11	Ph (structure) (PS)	1. SnCl$_2$•2 H$_2$O, EtOH, 78 °C, 4 h (formation of quinoline-*N*-oxide) 2. TiCl$_3$, DCM, PhMe, 22 °C, overnight	Ph (structure) (PS)	[212]
12	(structure) (PS)	PhCOCl (0.11 mol/L, 3 eq), DCM, 0 °C, 15 min, then *N*-methylpyrrole, 0–20 °C, 2 h	(structure) (PS)	[203]

15.19 Preparation of Pyridazines (1,2-Diazines)

Pyridazines and their partially saturated analogs have been prepared on insoluble supports by Diels–Alder reaction of electron-rich alkenes or alkynes with 1,2,4,5-tetrazines (Entries **1–3**, Table 15.25). The mechanism of this reaction is sketched in Figure 15.15. An additional approach, also based on the Diels–Alder reaction, is the cycloaddition of azo compounds to 1,3-dienes (Entries **4** and **5**, Table 15.25). The resulting tetrahydropyridazines (Entry **4**) have been used as constrained β-strand mimetics for the discovery of new protease inhibitors [213]. An example of the N-alkylation of hexahydropyridazines on solid phase is given in Section 10.3.

4-Halocinnolines can be prepared by treatment of polystyrene-bound (2-alkynyl-phenyl)triazenes with hydrogen chloride or hydrogen bromide (Entry **6**, Table 15.25). This reaction proceeds via initial release into solution of 2-alkynylphenyldiazonium salts, which then cyclize to the halocinnolines. The corresponding 4-hydroxycinnolines were obtained as byproducts in variable amounts. Hydrogen fluoride or hydrogen iodide did not lead to the formation of cinnolines [214].

Table 15.25. Preparation of pyridazines and tetrahydropyridazines.

Entry	Starting resin	Conditions	Product	Ref.
1		(0.2 mol/L, 10 eq), 20 °C or 100 °C, 24 h		[217]
2		(0.2 mol/L, 10 eq), 100 °C, 16 h		[217]
3		(0.2 mol/L, 10 eq), 100 °C, 16 h		[217]
4		(4 eq), PhI(OAc)$_2$ (4 × 0.5 eq), dioxane, 4 h (repeat once)		[218] see also [213]
5		DEAD (1 mol/L, 10 eq), THF, 65 °C, overnight		[190]
6		HCl, H$_2$O, Me$_2$CO, 20 °C, 1 h		[214]
7		H$_2$N–N OH, DCM, 20 °C		[215]
8		1. N$_2$H$_4$•H$_2$O (0.28 mol/L, 5 eq), DCM/MeOH 3:2, 20 °C, 4 h 2. TsOH (5 eq), MeCN, 81 °C, 24 h (Wang resin)		[219]

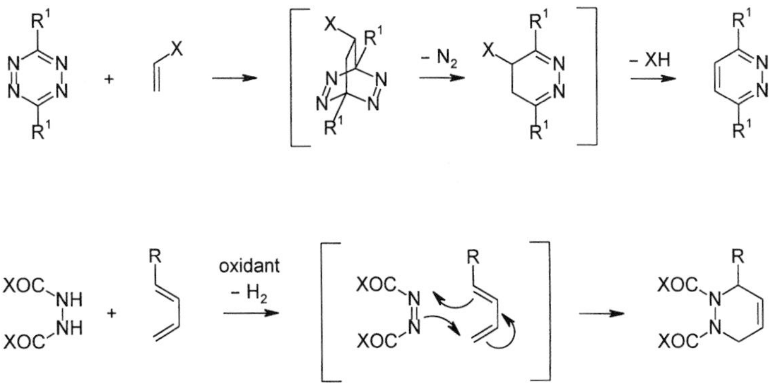

Fig. 15.15. Preparation of 1,2-diazines by hetero-Diels–Alder reaction. X: OR, NR$_2$.

Phthalazine-1,4-diones have been prepared by nucleophilic cleavage of polystyrene-bound phthalimides (Entry **7**, Table 15.25). The reaction rate was highly dependent on the solvent chosen: it was low in EtOH but high in DCM. With DMF as solvent no phthalazinediones could be isolated [215].

Hexahydropyridazines have been prepared on cross-linked polystyrene in low yields by treatment of support-bound 1-acyloxy-1,3-dienes with diimide [216].

15.20 Preparation of Pyrimidines (1,3-Diazines)

Ureas can be cyclized to hexahydro- and tetrahydro-2-pyrimidinones on insoluble supports either intramolecularly or by treatment with α,β-unsaturated carbonyl compounds (Entries **1–3**, **5**; Table 15.26). Because of the low nucleophilicity of ureas, harsh reaction conditions are generally required for these cyclizations. Similarly, polystyrene-bound guanidines can be condensed with 3-oxo esters to yield 2-amino-4-pyrimidinones (Entry **4**, Table 15.26). If support-bound monoalkylguanidines are the starting material used, the reaction with 3-oxo esters does not always proceed with high regioselectivity, and mixtures of regioisomers (2-alkylamino-4-pyrimidinones and 2-amino-3-alkyl-4-pyrimidinones) can result [220]. The reversed strategy, in which amidines or guanidines undergo oxidative cyclocondensation with support-bound enones to yield pyrimidines, has also been reported (Entries **6–8**, Table 15.26).

Polystyrene-bound isothiuronium chloride, which is readily prepared from chloromethyl polystyrene and thiourea, has been used as starting material for the preparation of various substituted pyrimidines (Entries **9–11**, Table 15.26). After oxidation to the corresponding sulfones, nucleophilic cleavage with amines proceeds smoothly to yield substituted 2-aminopyrimidine derivatives (see Section 3.8). Additional examples of the cleavage of pyrimidines from supports are given in Tables 3.43 and 3.46.

Palladium-mediated (Sonogashira) coupling [221] of 5-iodouracil with alkynes can be performed under mild reaction conditions, and is compatible with the support, linkers, and protective groups used in solid-phase oligonucleotide synthesis (Entries **12**

Table 15.26. Preparation of pyrimidines.

Entry	Starting resin	Conditions	Product	Ref.
1		PhMe saturated with HCl, sealed tube, 95 °C, 4 h (Wang resin)		[222]
2		PhMe saturated with HCl, sealed tube, 95 °C, 4 h (Wang resin)		[222]
3		(0.13 mol/L, 4 eq of each), THF, HCl, 55 °C, 36 h		[223]
4		NaOMe (10 eq), MeOH, 65 °C, 16 h		[220]
5		(0.5 mol/L, 10 eq), NaOEt (10 eq), DMA, 20 °C, 16 h		[160]
6		(0.5 mol/L, 10 eq), DMA, air, 100 °C, overnight		[160]
7		guanidine (0.5 mol/L, 10 eq), DMA, air, 100 °C, overnight		[160]
8		(10 eq), K$_2$CO$_3$, DMA, 70 °C, 8 h, then CAN (0.2 mol/L), DMA, 20 °C, 2 h		[224]
9		tBuO (0.4 mol/L, 1.2 eq), DIPEA (1.5 eq), DMF, 20 °C, 48 h		[225]

Table 15.26. continued.

Entry	Starting resin	Conditions	Product	Ref.
10	Cl⁻ H₂N⁺─C(NH₂)─S─PS	MeS SMe / NC CN (0.17 mol/L, 2 eq), DIPEA (5 eq), DMF, 0–20 °C, 96 h	MeS─N═C─S─PS ... NC, NH₂	[226]
11	Cl⁻ H₂N⁺─C(NH₂)─S─PS	OEt / NC CN (0.19 mol/L, 1.8 eq), DIPEA (5 eq), DMF, 0–20 °C, 96 h	N─S─PS ... NC, NH₂	[226]
12	(propargyl amide) O─N─O─(PS) Ph Ph(2-Cl)	5-iodouridine, CuI, Pd(PPh₃)₄, NEt₃, DMF, 25 °C	HN─uridine alkyne─N─O─(PS) Ph Ph(2-Cl)	[227]
13	I─uridine DMTO─(sugar)─O─P(O)─O─(CPG) NC	CF₃─C(O)─NH─propargyl / Pd(PPh₃)₄, CuI, DMF/NEt₃ 7:3, 3 h	CF₃─C(O)─NH─alkyne─uridine DMTO─(sugar)─O─P(O)─O─(CPG) NC	[228]

and **13**, Table 15.26). This coupling reaction has been used to prepare modified oligo-nucleotides on polystyrene and on CPG.

15.21 Preparation of Quinazolines

Numerous solid-phase preparations of quinazolinones have been reported. The main synthetic strategies used are summarized in Figure 15.16. Quinazolin-2,4-diones can be prepared from anthranilic acid derived ureas or from *N*-(alkoxycarbonyl) anthranilamides. On insoluble supports these reactions have been performed either in such a way that the cyclized product remains linked to the support or with simultaneous ring formation and cleavage from the support. Quinazolin-4-ones can be prepared by cyclocondensation of anthranilamides with aldehydes or carboxylic acid derivatives.

The selection of examples listed in Table 15.27 illustrates the scope of substitution patterns accessible with these cyclizations.

Table 15.27. Preparation of quinazolines.

Entry	Starting resin	Conditions	Product	Ref.
1		DMF, 125 °C, 16 h		[229]
2		NEt₃ (10 eq), MeOH, 60 °C, 24 h		[230]
3		TMG/NMP 5:95, 60 °C		[231] see also [232]
4		KOH (1 mol/L), EtOH, 1 h		[233]
5		K₂CO₃, MeCN, 60 °C, 24 h		[234] see also [235]
6		(10 eq), AcOH/DMA 5:95, 100 °C, 24 h Ar: 2-(MeO)C₆H₄		[236]
7		(1.8 mol/L, 3.5 eq), CSA (1 eq), dioxane, 100 °C, 72 h		[237]
8		piperidine (0.4 mol/L, 5 eq), m-xylene, 23 °C, 2 h, 80 °C, 4 h		[238]

Table 15.27. continued.

Entry	Starting resin	Conditions	Product	Ref.
9		HCO₂H, 20 °C		[239]
10		Bu₄NI, TMG, DMSO, 20 °C, overnight		[231]
11		ArCH₂OH, DIAD, PPh₃, THF, 20 °C, overnight Ar: 4-(MeO)C₆H₄		[231]
12		(16 eq), THF, 20 °C, 1.5 h, then EtI (40 eq), DMF, 18 h		[233]
13		KMnO₄ (10 eq), Me₂CO, overnight		[236]

Support-bound quinazolin-2,4-diones can be N-alkylated either with alkyl halides under basic conditions or with aliphatic alcohols by means of the Mitsunobu reaction (Entries **10–12**, Table 15.27).

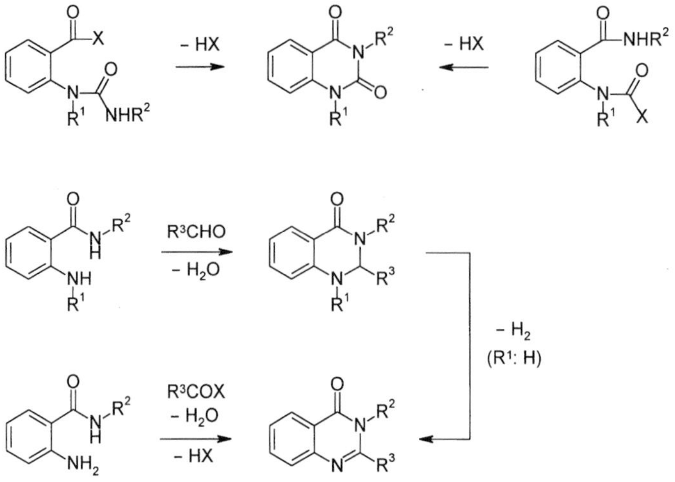

Fig. 15.16. Strategies for the preparation of quinazolinones.

15.22 Preparation of Pyrazines and Piperazines (1,4-Diazines), and Fused Derivatives thereof

15.22.1 Preparation of Diketopiperazines

An unwanted side reaction in solid-phase peptide synthesis is the formation of diketopiperazines during deprotection of the second amino acid (Figure 15.17). This reaction occurs particularly readily with dipeptides containing *N*-methylamino acids, proline [240], glycine [241], and with dipeptides consisting of a D- and an L-amino acid, and is one of the preferred strategies for preparing diketopiperazines. A second strategy consists in linking an α-amino acid ester via the amino group to a backbone amide linker (see Section 3.3.1), and acylating the resulting secondary amine with an N-protected α-amino acid. Deprotection will usually lead to spontaneous diketopiperazine formation, although without simultaneous cleavage from the support (Figure 15.17).

Diketopiperazine formation by intramolecular nucleophilic cleavage proceeds particularly smoothly if a linker prone to facile nucleophilic cleavage is chosen. In most of the examples reported hydroxymethyl polystyrene or polystyrene with the PAM linker were used. These linkers tolerate treatment with trifluoroacetic acid, and are therefore compatible with the Boc methodology. Cleavage and simultaneous diketopiperazine formation can be performed under either basic or acidic conditions (Table 15.28). Because cleavage from the support only occurs during cyclization, most byproducts formed during solid-phase synthesis of the dipeptide remain attached to the support, and the diketopiperazines obtained are usually very pure.

Table 15.28. Preparation of diketopiperazines (piperazine-2,5-diones).

Entry	Starting resin	Conditions	Product	Ref.
1		DIPEA (2.2 eq), AcOH (5 eq), DCM, 16 h		[244] see also [245]
2		AcOH (1.25 mol/L), PhMe, 90 °C		[246]
3		AcOH (0.1 mol/L), DCM, 25 °C ($t_{1/2}$: 8.1 min)		[240]
4		AcOH/PhMe 1:99 or NEt$_3$/PhMe 4:96, 20 °C, 12 h; (PAM resin)		[247]
5		isobutylamine (2 mol/L, 40 eq), DMSO, 20 °C, 21 h (Wang resin)		[248]
6		isobutylamine (2 mol/L, 20 eq), DMSO, 70 °C, 24 h, wash; TFA/H$_2$O 95:5, 1 h (Wang resin)		[248]
7		1. TFA, 3 h, concentrate 2. PhMe, 110 °C, 5 h (Wang resin; Ar: 4-(MeO)C$_6$H$_4$)		[249]
8		piperidine/THF 5:95, 20 °C, 16 h		[250]

Table 15.28. continued.

Entry	Starting resin	Conditions	Product	Ref.
9	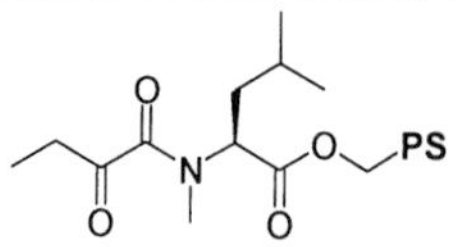	NH$_4$OAc, AcOH, PhMe, heat	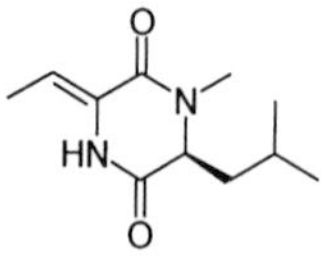	[251]
10	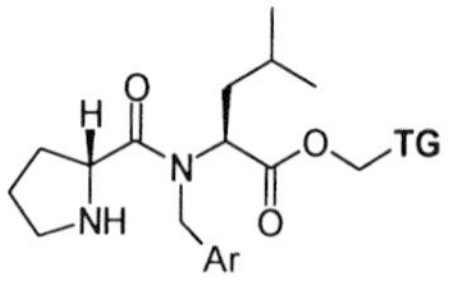	NEt$_3$/PhMe 4:96, 20 °C, 12 h Ar: 4-(BnO)C$_6$H$_4$	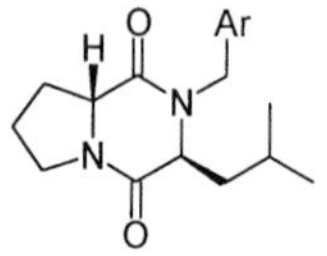	[247]
11	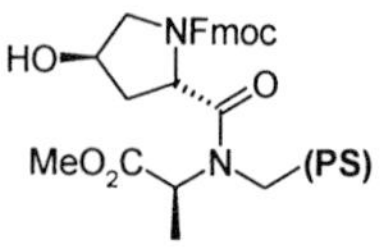	piperidine/DMF 1:4, 25 °C, 3 × 1 min, 3 × 5 min	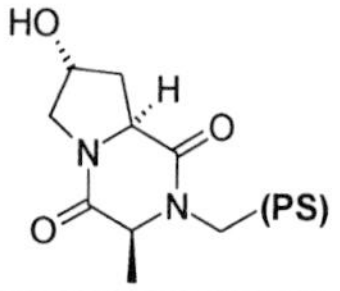	[242] see also [102]

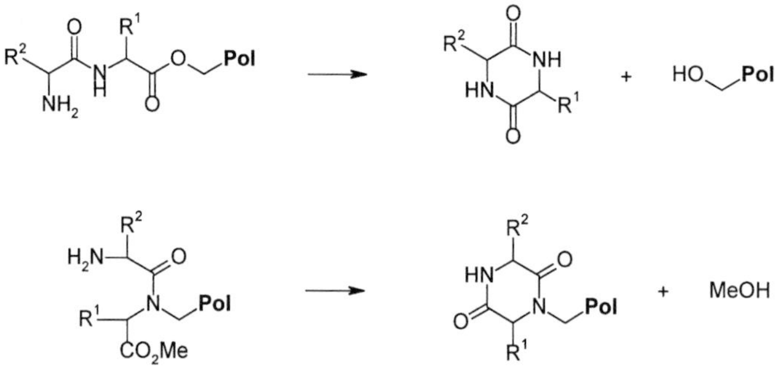

Fig. 15.17. Strategies for the preparation of diketopiperazines from support-bound α-amino acid derivatives.

The Wang linker, although being slightly more resistant towards nucleophilic cleavage than Merrifield or PAM resin, can also be used for the preparation of diketopiperazines. If spontaneous cyclization does not occur during TFA-mediated cleavage of a dipeptide from the support, heating of the crude product in toluene will usually bring about ring closure (Entry **7**, Table 15.28).

Diketopiperazines can also be prepared by cyclizing *N*-alkylamino acid derivatives, e.g. those available from α-haloacids and primary amines (Entries **5** and **6**, Table 15.28). In this case the linker chosen should be sufficiently stable towards aminolysis to avoid premature cleavage of the intermediates from the support. Otherwise a volatile amine should be chosen for the last substitution, to facilitate removal of excess amine from the product (Entry **5**, Table 15.28).

Polystyrene-bound diketopiperazines, such as those prepared with the aid of a backbone amide linker, can be *N*-alkylated by treatment with alkyl halides in the presence of strong bases [242]. The conditions required for N-alkylation are similar to those for the N-alkylation of hydantoins.

As illustrated by the examples in Table 15.28, diketopiperazines enable many structural variations. These heterocycles are therefore well suited for the preparation of libraries for lead discovery, and have, for instance, been used for the identification of new enzyme inhibitors [243].

15.22.2 Preparation of Miscellaneous 1,4-Diazines and Quinoxalines

1,4-Diazines other than diketopiperazines can also be prepared on insoluble supports (Table 15.29). Several examples of the solid-phase preparation of quinoxalinones have been reported. Most were prepared from support-bound 2-fluoronitrobenzenes according to the strategies outlined in Figure 15.18. 1,4-Diazines have, moreover, been chemically modified by N- or C-alkylation on insoluble supports (Entries **6** and **7**, Table 15.29).

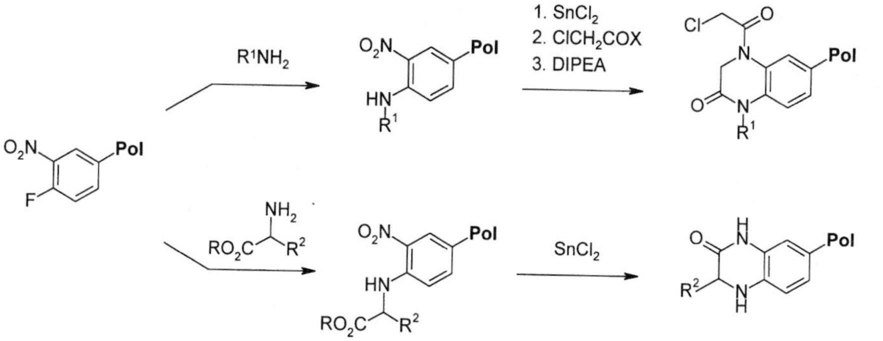

Fig. 15.18. Strategies for the preparation of 1,2,3,4-tetrahydroquinoxalinones on insoluble supports.

Table 15.29. Preparation of 2-piperazinones and quinoxalinones.

Entry	Starting resin	Conditions	Product	Ref.
1		DIPEA/NMP 1:9, 20 °C, 5 h		[252]
2		SnCl₂ (2 mol/L, 20 eq), DMF, 80 °C, overnight	(no racemization)	[253] see also [254]
3		H₂N— (2 mol/L), DMSO, 20 °C, 2 h		[255] see also [256]
4		PPh₃ (4 eq), DEAD (4 eq), MeCN/DMF		[257]
5		K₂CO₃, DMF, 60 °C Ar: 2-(O₂N)C₆H₄		[258]
6		BnBr (1 mol/L, 25 eq), K₂CO₃ (25 eq), Me₂CO, 55 °C, 24 h		[254]
7		BnBr (5 eq), LiN(SiMe₃)₂ (5 eq), THF, 20 °C, overnight		[259]

15.23 Preparation of Triazines

Few examples of the solid-phase synthesis of triazines have been reported (Table 15.30). These include the cyclocondensation of polystyrene-bound isothiuronium salts with *N*-(cyano)iminodithiocarbonates and the cyclization with simultaneous cleavage from the support of derivatives of α-amino acid hydrazides (Table 15.30).

(Benzylthio)triazines such as those listed in Table 15.30 (Entries **1** and **2**) were cleaved from the support by oxidation to sulfoxides or sulfones with *N*-benzenesulfonyl-3-phenyloxaziridine, followed by nucleophilic cleavage with primary or secondary aliphatic amines or with electron-rich anilines (see Section 3.8).

Table 15.30. Preparation of triazines.

Entry	Starting resin	Conditions	Product	Ref.
1		CN, N SMe, SMe (0.2 mol/L, 1.2 eq), DIPEA (1.5 eq), DMA, 80 °C, 72 h		[260]
2		NH (0.37 mol/L, 2.5 eq), DIPEA (5 eq), DMA, 45 °C, 17 h		[260]
3		DIPEA (0.5 mol/L, 10 eq), DMF, 80 °C, 24 h		[118]
4		DIPEA (0.5 mol/L, 10 eq), DMF, 80 °C, 24 h		[118]

15.24 Preparation of Pyrans and Benzopyrans

The synthetic strategies used for the preparation of pyrans on insoluble supports have mainly been hetero-Diels–Alder reactions of enones with enol ethers and ring-closing olefin metathesis (Table 15.31). Benzopyrans have been prepared by hetero-Diels–Alder reaction of polystyrene-bound *o*-quinodimethanes with aldehydes. The required quinodimethanes were formed upon thermolysis of benzocyclobutanes, which were prepared in solution [201]. Other solid-phase procedures for the preparation of benzopyrans are the palladium-mediated reaction of support-bound 2-iodophenols with 1,4-dienes (Entry **4**, Table 15.31) and the intramolecular Knoevenagel condensation of malonic esters of salicylaldehyde (Entry **5**, Table 15.31).

Table 15.31. Preparation of pyrans and benzopyrans.

Entry	Starting resin	Conditions	Product	Ref.
1	*(structure: O=C–C(=CH)–C(=O)O–CH₂–(PS))*	EtO⎯⎯ (10 eq), DCM, 60 °C, 3 d	*(structure: EtO-substituted dihydropyran with (PS))*	[261]
2	*(allyl–CH(Ph)–O–CH₂–CH=CH–PS)*	Cl–Ru(=CHPh)(PCy₃)₂Cl (0.05 eq), DCM, 20 °C, 16 h	*(3,6-dihydro-2H-pyran with Ph, dihydropyran)*	[186]
3	*(benzocyclobutene–O–CH₂–PS)*	ArCHO (2 mol/L, 2 eq), PhMe, 108 °C, 14 h Ar: 4-(O₂N)C₆H₄	*(isochroman with O–CH₂–PS and Ar)*	[201]
4	*(iodo-hydroxybenzamide: I, HO on benzene, C(=O)N(H)(PS))*	Pd(OAc)₂, LiCl, DIPEA, DMF, 100 °C	*(chroman-fused amide with N(H)(PS))*	[57]
5	EtO–C(=O)–CH₂–C(=O)–O–CH₂–(PS)	salicylaldehyde (20 eq), piperidine, pyridine, 20 °C, 16 h (repeat once); then TFA/DCM 1:2, 1 h (Wang resin)	*(coumarin-3-carboxylic acid, C(=O)OH)*	[262]

15.25 Preparation of Oxazines and Thiazines

Examples of the preparation of oxazines and thiazines on insoluble supports are listed in Table 15.32. 3,1-Benzoxazin-4-ones can be prepared by intramolecular O-acylation of *N*-aminocarbonyl anthranilic acids (Entry **1**, Table 15.32). The resulting benzoxazinones are sufficiently stable towards acids to enable TFA-mediated cleavage from a Wang linker [263]. 1,3-Oxazines have also been obtained by acidolytic cleavage of functionalized 3-amino-1-propanols from Wang resin (Table 3.29).

Morpholine-2,5-diones have been prepared by intramolecular, nucleophilic cleavage of polystyrene-bound amino acids acylated with α-hydroxy acids (Entry **2**, Table 15.32). This preparation is analogous to the formation of diketopiperazines from dipeptide esters. In a similar approach, morpholine-2,5-diones were obtained by acidolytic cleavage of *N*-(bromoacetyl)amino acids from Wang resin (Entry **3**, Table 15.32). In a remarkable cycloaddition, 1,2-oxazines have been prepared from polystyrene-bound acrylates by reaction under high-pressure with enol ethers and nitroalkenes (Entry **4**, Table 15.32).

1,3-Thiazin-4-ones are rather stable towards acids, and can be cleaved from Wang resin by treatment with TFA/DCM 1:1 without decomposition [151]. The

Table 15.32. Preparation of oxazines and thiazines.

Entry	Starting resin	Conditions	Product	Ref.
1		DIC, THF; or TsCl, pyridine; or Ac₂O, THF, 20 °C, overnight		[263]
2		NEt₃/DCM 5:95, 3 h		[247]
3		TFA/H₂O 95:5, 1 h (Wang resin)		[248]
4		(2 eq of each), DCM, 15 kbar, 20 °C, 48 h		[264]
5		(0.5 mol/L), PhCHO (0.25 mol/L), MS 3 Å, THF, 70 °C, 2 h		[151]
6		HCO₂H, 20 °C		[239]
7		1. TFA/(iBu)₃SiH/ DCM 5:5:90 2. NMM, DMF		[265] see also [266]
8		HATU (5 eq), DIPEA (5 eq), DMF, overnight		[265]

preparation of 1,3-thiazin-4-ones can be performed either by one-pot condensation of 3-mercaptopropionic acids with primary amines and aldehydes (Entry **5**, Table 15.32), or by condensation of 3-mercaptopropionamides with aldehydes (Entry **6**, Table 15.32).

3-Thiomorpholinones have been prepared on cross-linked polystyrene by intramolecular thioether formation and by lactamization of suitable amino acids (Entries **7** and **8**, Table 15.32).

15.26 Preparation of Azepines and Larger Heterocycles with one Nitrogen Atom

Whereas five- and six-membered systems are usually easy to prepare and often form spontaneously from suitable intermediates, seven-membered heterocycles can be quite difficult to synthesize. Only if the precursor has little conformational flexibility and a conformation suitable for ring closure can readily be attained, will the cyclization proceed smoothly (as, e.g., in the formation of benzodiazepines from 2-(aminoacetylamino)benzophenones). Otherwise, ring closure must be forced by chemical activation or high reaction temperatures, and a low loading might be required to supress polymerization.

Tetrahydroazepines can be prepared on insoluble supports by ring-closing olefin metathesis (Entries **1**–**3**, Table 15.33), either with or without simultaneous cleavage from the support. The yields of this cyclization are not as high as for the formation of five- or six-membered rings, but under optimal conditions pure products can be obtained. Larger nitrogen-containing heterocycles are also available by ring-closing metathesis on solid phase (Entry **4**, Table 15.33).

If one of the two alkenes which undergo metathesis also acts as linker, ring-closing metathesis and cleavage from the support will occur simultaneously. For this type of cyclization yields are usually higher if an additional alkene (e.g. styrene, see Entry **2**, Table 15.33) is added to the reaction mixture. This alkene enables efficient regeneration of the catalyst by reaction with the support-bound carbene complex, and enables fast ring-closing metathesis to occur with only small amounts of catalyst (see Figure 3.37).

Benzazepinones can be prepared on cross-linked polystyrene by intramolecular Heck reaction (Entry **5**, Table 15.33). In the presence of sodium formate the intramolecular Heck reaction of iodoarenes with alkynes yields methylene benzazepinones (Entry **6**, Table 15.33). Surprisingly, when this reaction was performed in solution the main product (65 % yield) was a dehalogenated, non-cyclized benzamide. In the synthesis on cross-linked polystyrene, however, this product was not observed [267].

Caprolactams can be prepared by intramolecular nucleophilic cleavage of support-bound 6-aminohexanoates (Entry **8**, Table 15.33). Yields of such reactions are usually not high, even if reactive esters are used as attachment to the support. High loading should be avoided to prevent the formation of oligomers by intermolecular nucleophilic cleavage.

Table 15.33. Preparation of azepines and azocines.

Entry	Starting resin	Conditions	Product	Ref.
1	BocHN, O, MeO$_2$C, N, Ph, PS	Cl-Ru=Ph with PCy$_3$ (0.05 eq), DCE, 80 °C, 16 h	BocHN, O, MeO$_2$C, N, Ph	[186] see also [268,269]
2	(PS), MeO$_2$C, N, O=S=O, Ar	Cl-Ru=Ph with PCy$_3$ (0.05 eq), styrene (1 eq), PhMe, 50 °C, 18 h	MeO$_2$C, N, O=S=O, Ar	[185]
3	Ph, N, F$_3$C, O, Ph Ph, PS	Cl-Ru=Ph with PCy$_3$ (0.11 eq), DCM, 40 °C, overnight	Ph, N, F$_3$C, O, Ph Ph, PS	[9]
4	Ph, N, F$_3$C, O, Ph Ph, PS	Cl-Ru=Ph with PCy$_3$ (0.08 eq), DCM, 40 °C, overnight	Ph, N, F$_3$C, O, Ph Ph, PS	[9]
5	O, N, O, (PS)	Pd(OAc)$_2$ (0.2 eq), PPh$_3$, Bu$_4$NCl, KOAc, DMF, 70 °C, 5 h	O, N, O, (PS)	[267]
6	Ph, O, N, O, (PS)	Pd(OAc)$_2$ (0.2 eq), PPh$_3$, Bu$_4$NCl, HCO$_2$Na, DMF, 70 °C	Ph, O, N, O, (PS)	[267]
7	H$_2$N-(PS) (Rink amide resin)	(0.27 mol/L, 5.4 eq), DCM/MeOH 1:1, 2 d	HO, N-(PS), O, OH	[270]
8	O, O-N, N=N, PS, NHBoc	TFA, DCM, then 5% NEt$_3$ in DCM, 25 °C, 24 h	O, NH	[271]

15.27 Preparation of Diazepines, Thiazepines, and larger Heterocycles with more than one Heteroatom

Diazepanones have been prepared on insoluble supports by intramolecular nucleophilic cleavage, by intramolecular Mitsunobu reaction of sulfonamides with alcohols, and by intramolecular acylations (Table 15.34). As for azepines these reactions do not always proceed smoothly, and care must be taken to prevent potential side reactions from occurring. For instance, intramolecular acylations in peptides containing aspartic acid (Entry **2**, Table 15.34) will generally lead to the formation of succinimides (see Table 13.20), unless *N*-alkylamino acids are used.

Table 15.34. Preparation of diazepanes and related heterocycles.

Entry	Starting resin	Conditions	Product	Ref.
1		DIPEA (2.2 eq), AcOH (5 eq), DCM, 16 h		[244]
2		DPPA, DIPEA, overnight; or HATU (3 eq), DIPEA (3 eq), DMF		[272,273]
3		PPh₃, DEAD, THF, 0 °C, 2 h, 20 °C, 36 h		[274]
4		NEt₃, THF, 67 °C, 5 d		[275]

Benzodiazepines, on the other hand, are easier to prepare than non-fused diazepines, because the precursors are less flexible and well-preoriented to facilitate ring closure. Most benzodiazepines do not form spontaneously, however, and additional activation is usually required to promote the cyclization (Table 15.35). Even under forcing conditions, cyclizations can sometimes still be difficult. For instance, the cyclization shown in Entry **2** (Table 15.35) only proceeded when pure DIC was used as coupling agent, whereby the desired benzodiazepinone together with 20–50 % DIC-derived *N*-acylurea were obtained. Additives such as HOBt or DMAP, or other coupling reagents (EDC, HATU, DECP) did not lead to cyclization in this case [276].

Table 15.35. Preparation of benzodiazepines and related heterocycles.

Entry	Starting resin	Conditions	Product	Ref.
1		HOBt (0.38 mol/L, 4 eq), DIC (4 eq), DMF, 20 °C, overnight		[277] see also [278]
2		DIC, DCM/C$_6$H$_6$ 1:1, 20 °C, 6 h		[276]
3		HBTU (3 eq), DIPEA (3 eq), DMF, overnight		[279]
4		Ph–N=C(OLi) (0.22 mol/L, 20 eq), DMF/THF 1:1, 20 °C, 30 h, then allyl bromide (40 eq), 6 h		[280]
5		*p*-xylene, 130 °C, 7 h		[281]
6		AcOH/DMF 5:95, 60 °C, 12 h Ar: 4-(MeO)C$_6$H$_4$		[282]
7		(0.24 mol/L, 1.4 eq), pyridine, 115 °C, 48 h		[283]
8		NaO*t*Bu, THF, 60 °C, 24 h		[284]

Table 15.35. continued.

Entry	Starting resin	Conditions	Product	Ref.
9		NEt₃/DCM 1:1, 27 °C, 4 h		[191]
10		NaOMe (0.4 mol/L, 2 eq), THF/MeOH 4:1, 20 °C, 10 h (Wang resin)		[285]
11		1. SnCl₂ (0.07 mol/L, 3 eq), NEt₃ (8 eq), PhSH (11 eq), C₆H₆, 20 °C, 1 h 2. DMF/TFA/H₂O 7:2:1, 20 °C, 14 h		[286]
12		DBU/DMF 5:95, 20 °C, 24 h		[287] see also [288]
13		hexyl iodide (0.25 mol/L), THF/ DMF 1:1, 20 °C, 1 h Ar: 2-MeC₆H₄		[278]
14		BnBr (0.9 mol/L, 20 eq), K₂CO₃ (20 eq), Me₂CO, 55 °C, overnight		[277]

Examples of cyclizations yielding benzodiazepines with simultaneous cleavage from the support have also been reported (Entries 7–8, 10; Table 15.35). Unfortunately, in most of these examples the desired products are contaminated with non-volatile byproducts, and purification of the products will probably be required for most applications.

Support-bound benzodiazepines can be chemically modified by N-alkylation (Entries **13** and **14**, Table 15.35). Strong bases and a large excess of reactive alkylating agents are often required to achieve complete conversion of the support-bound substrates.

References for Chapter 15

[1] Corbett, J. W. *Org. Prep. Proc. Int.* **1998**, *30*, 489–550.
[2] Nefzi, A.; Ostresh, J. M.; Houghten, R. A. *Chem. Rev.* **1997**, *97*, 449–472.
[3] Nuss, J. M.; Desai, M. C.; Zuckermann, R. N.; Singh, R.; Renhowe, P. A.; Goff, D. A.; Chinn, J. P.; Wang, L.; Dorr, H.; Brown, E. G.; Subramanian, S. *Pure Appl. Chem.* **1997**, *69*, 447–452.
[4] Farrall, M. J.; Durst, T.; Fréchet, J. M. *Tetrahedron Lett.* **1979**, 203–206.
[5] Filigheddu, S. N.; Masala, S.; Taddei, M. *Tetrahedron Lett.* **1999**, *40*, 6503–6506.
[6] Rotella, D. P. *J. Am. Chem. Soc.* **1996**, *118*, 12246–12247.
[7] Le Hetet, C.; David, M.; Carreaux, F.; Carboni, B.; Sauleau, A. *Tetrahedron Lett.* **1997**, *38*, 5153–5156.
[8] Fréchet, J. M. J.; Eichler, E. *Polym. Bull.* **1982**, *7*, 345–351.
[9] Pernerstorfer, J.; Schuster, M.; Blechert, S. *Synthesis* **1999**, 138–144.
[10] Wendeborn, S.; De Mesmaeker, A.; Brill, W. K. D. *Synlett* **1998**, 865–868.
[11] Savin, K. A.; Woo, J. C. G.; Danishefsky, S. J. *J. Org. Chem.* **1999**, *64*, 4183–4186.
[12] Zaragoza, F. unpublished results.
[13] Ruhland, B.; Bhandari, A.; Gordon, E. M.; Gallop, M. A. *J. Am. Chem. Soc.* **1996**, *118*, 253–254.
[14] Ruhland, B.; Bombrun, A.; Gallop, M. A. *J. Org. Chem.* **1997**, *62*, 7820–7826.
[15] Ni, Z. J.; Maclean, D.; Holmes, C. P.; Murphy, M. M.; Ruhland, B.; Jacobs, J. W.; Gordon, E. M.; Gallop, M. A. *J. Med. Chem.* **1996**, *39*, 1601–1608.
[16] Pei, Y. Z.; Houghten, R. A.; Kiely, J. S. *Tetrahedron Lett.* **1997**, *38*, 3349–3352.
[17] Molteni, V.; Annunziata, R.; Cinquini, M.; Cozzi, F.; Benaglia, M. *Tetrahedron Lett.* **1998**, *39*, 1257–1260.
[18] Singh, R.; Nuss, J. M. *Tetrahedron Lett.* **1999**, *40*, 1249–1252.
[19] Benaglia, M.; Cinquini, M.; Cozzi, F. *Tetrahedron Lett.* **1999**, *40*, 2019–2020.
[20] Kobayashi, S.; Moriwaki, M.; Akiyama, R.; Suzuki, S.; Hachiya, I. *Tetrahedron Lett.* **1996**, *37*, 7783–7786.
[21] Gordeev, M. F.; Gordon, E. M.; Patel, D. V. *J. Org. Chem.* **1997**, *62*, 8177–8181.
[22] Zaragoza, F. *Metal Carbenes in Organic Synthesis*; Wiley-VCH: Weinheim, New York, **1999**.
[23] Ivin, K. J.; Mol, J. C. *Olefin Metathesis and Metathesis Polymerization*; Academic Press: London, **1997**.
[24] Trautwein, A. W.; Süssmuth, R. D.; Jung, G. *Bioorg. Med. Chem. Lett.* **1998**, *8*, 2381–2384.
[25] Trautwein, A. W.; Jung, G. *Tetrahedron Lett.* **1998**, *39*, 8263–8266.
[26] Mjalli, A. M. M.; Sarshar, S.; Baiga, T. J. *Tetrahedron Lett.* **1996**, *37*, 2943–2946.
[27] Strocker, A. M.; Keating, T. A.; Tempest, P. A.; Armstrong, R. W. *Tetrahedron Lett.* **1996**, *37*, 1149–1152.
[28] Schuster, M.; Pernerstorfer, J.; Blechert, S. *Angew. Chem. Int. Ed. Engl.* **1996**, *35*, 1979–1980.
[29] Heerding, D. A.; Takata, D. T.; Kwon, C.; Huffman, W. F.; Samanen, J. *Tetrahedron Lett.* **1998**, *39*, 6815–6818.
[30] Karoyan, P.; Triolo, A.; Nannicini, R.; Giannotti, D.; Altamura, M.; Chassaing, G.; Perrotta, E. *Tetrahedron Lett.* **1999**, *40*, 71–74.
[31] Veerman, J. J. N.; Rutjes, F. P. J. T.; van Maarseveen, J. H.; Hiemstra, H. *Tetrahedron Lett.* **1999**, *40*, 6079–6082.
[32] Brown, R. C. D.; Fisher, M. *Chem. Commun.* **1999**, 1547–1548.
[33] Short, K. M.; Ching, B. W.; Mjalli, A. M. M. *Tetrahedron* **1997**, *53*, 6653–6679.
[34] Paulvannan, K. *Tetrahedron Lett.* **1999**, *40*, 1851–1854.
[35] Sun, S.; Murray, W. V. *J. Org. Chem.* **1999**, *64*, 5941–5945.
[36] Romoff, T. T.; Ma, L.; Wang, Y. W.; Campbell, D. A. *Synlett* **1998**, 1341–1342.
[37] Weber, L.; Iaiza, P.; Biringer, G.; Barbier, P. *Synlett* **1998**, 1156–1158.
[38] Kulkarni, B. A.; Ganesan, A. *Tetrahedron Lett.* **1998**, *39*, 4369–4372.
[39] Matthews, J.; Rivero, R. A. *J. Org. Chem.* **1998**, *63*, 4808–4810.
[40] Miller, P. C.; Owen, T. J.; Molyneaux, J. M.; Curtis, J. M.; Jones, C. R. *J. Comb. Chem.* **1999**, *1*, 223–234.

[41] Barn, D. R.; Morphy, J. R. *J. Comb. Chem.* **1999**, *1*, 151–156.
[42] Pearson, W. H.; Clark, R. B. *Tetrahedron Lett.* **1997**, *38*, 7669–7672.
[43] Murphy, M. M.; Schullek, J. R.; Gordon, E. M.; Gallop, M. A. *J. Am. Chem. Soc.* **1995**, *117*, 7029–7030.
[44] Hamper, B. C.; Dukesherer, D. R.; South, M. S. *Tetrahedron Lett.* **1996**, *37*, 3671–3674.
[45] Bicknell, A. J.; Hird, N. W. *Bioorg. Med. Chem. Lett.* **1996**, *6*, 2441–2444.
[46] Hollinshead, S. P. *Tetrahedron Lett.* **1996**, *37*, 9157–9160.
[47] Gong, Y. D.; Najdi, S.; Olmstead, M. M.; Kurth, M. J. *J. Org. Chem.* **1998**, *63*, 3081–3086.
[48] Bicknell, A. J.; Hird, N. W.; Readshaw, S. A. *Tetrahedron Lett.* **1998**, *39*, 5869–5872.
[49] Marx, M. A.; Grillot, A. L.; Louer, C. T.; Beaver, K. A.; Bartlett, P. A. *J. Am. Chem. Soc.* **1997**, *119*, 6153–6167.
[50] Hutchins, S. M.; Chapman, K. T. *Tetrahedron Lett.* **1996**, *37*, 4869–4872.
[51] Kim, R. M.; Manna, M.; Hutchins, S. M.; Griffin, P. R.; Yates, N. A.; Bernick, A. M.; Chapman, K. T. *Proc. Natl. Acad. Sci. USA* **1996**, *93*, 10012–10017.
[52] Cheng, Y.; Chapman, K. T. *Tetrahedron Lett.* **1997**, *38*, 1497–1500.
[53] Smith, A. L.; Stevenson, G. I.; Swain, C. J.; Castro, J. L. *Tetrahedron Lett.* **1998**, *39*, 8317–8320.
[54] Fagnola, M. C.; Candiani, I.; Visentin, G.; Cabri, W.; Zarini, F.; Mongelli, N.; Bedeschi, A. *Tetrahedron Lett.* **1997**, *38*, 2307–2310.
[55] Zhang, H. C.; Brumfield, K. K.; Maryanoff, B. E. *Tetrahedron Lett.* **1997**, *38*, 2439–2442.
[56] Zhang, H. C.; Brumfield, K. K.; Jaroskova, L.; Maryanoff, B. E. *Tetrahedron Lett.* **1998**, *39*, 4449–4452.
[57] Wang, Y.; Huang, T. N. *Tetrahedron Lett.* **1998**, *39*, 9605–9608.
[58] Collini, M. D.; Ellingboe, J. W. *Tetrahedron Lett.* **1997**, *38*, 7963–7966.
[59] Zhang, H. C.; Maryanoff, B. E. *J. Org. Chem.* **1997**, *62*, 1804–1809.
[60] Yun, W. Y.; Mohan, R. *Tetrahedron Lett.* **1996**, *37*, 7189–7192.
[61] Arumugam, V.; Routledge, A.; Abell, C.; Balasubramanian, S. *Tetrahedron Lett.* **1997**, *38*, 6473–6476.
[62] Hughes, I. *Tetrahedron Lett.* **1996**, *37*, 7595–7598.
[63] Stephensen, H.; Zaragoza, F. *Tetrahedron Lett.* **1999**, *40*, 5799–5802.
[64] Larhed, M.; Lindeberg, G.; Hallberg, A. *Tetrahedron Lett.* **1996**, *37*, 8219–8222.
[65] Whitehouse, D. L.; Nelson, K. H.; Savinov, S. N.; Löwe, R. S.; Austin, D. J. *Bioorg. Med. Chem.* **1998**, *6*, 1273–1282.
[66] Gowravaram, M. R.; Gallop, M. A. *Tetrahedron Lett.* **1997**, *38*, 6973–6976.
[67] Whitehouse, D. L.; Nelson, K. H.; Savinov, S. N.; Austin, D. J. *Tetrahedron Lett.* **1997**, *38*, 7139–7142.
[68] Li, Z.; Ganesan, A. *Synlett* **1998**, 405–406.
[69] Boymond, L.; Rottländer, M.; Cahiez, G.; Knochel, P. *Angew. Chem. Int. Ed. Engl.* **1998**, *37*, 1701–1703.
[70] Chamoin, S.; Houldsworth, S.; Snieckus, V. *Tetrahedron Lett.* **1998**, *39*, 4175–4178.
[71] Kang, S. K.; Kim, J. S.; Yoon, S. K.; Lim, K. H.; Yoon, S. S. *Tetrahedron Lett.* **1998**, *39*, 3011–3012.
[72] Beebe, X.; Schore, N. E.; Kurth, M. J. *J. Org. Chem.* **1995**, *60*, 4196–4203.
[73] Beebe, X.; Chiappari, C. L.; Olmstead, M. M.; Kurth, M. J.; Schore, N. E. *J. Org. Chem.* **1995**, *60*, 4204–4212.
[74] Routledge, A.; Abell, C.; Balasubramanian, S. *Synlett* **1997**, 61–62.
[75] Berteina, S.; De Mesmaeker, A.; Wendeborn, S. *Synlett* **1999**, 1121–1123.
[76] Watanabe, Y.; Ishikawa, S.; Takao, G.; Toru, T. *Tetrahedron Lett.* **1999**, *40*, 3411–3414.
[77] Rottländer, M.; Knochel, P. *J. Comb. Chem.* **1999**, *1*, 181–183.
[78] Fancelli, D.; Fagnola, M. C.; Severino, D.; Bedeschi, A. *Tetrahedron Lett.* **1997**, *38*, 2311–2314.
[79] Berteina, S.; De Mesmaeker, A. *Tetrahedron Lett.* **1998**, *39*, 5759–5762.
[80] Berteina, S.; Wendeborn, S.; De Mesmaeker, A. *Synlett* **1998**, 1231–1233.
[81] Du, X.; Armstrong, R. W. *J. Org. Chem.* **1997**, *62*, 5678–5679.
[82] Du, X.; Armstrong, R. W. *Tetrahedron Lett.* **1998**, *39*, 2281–2284.
[83] Stephensen, H.; Zaragoza, F. *J. Org. Chem.* **1997**, *62*, 6096–6097.
[84] Zaragoza, F. *Tetrahedron Lett.* **1996**, *37*, 6213–6216.
[85] Rottländer, M.; Knochel, P. *Synlett* **1997**, 1084–1086.
[86] Yu, K. L.; Deshpande, M. S.; Vyas, D. M. *Tetrahedron Lett.* **1994**, *35*, 8919–8922.
[87] Marquais, S.; Arlt, M. *Tetrahedron Lett.* **1996**, *37*, 5491–5494.
[88] Han, Y.; Giroux, A.; Lépine, C.; Laliberté, F.; Huang, Z.; Perrier, H.; Bayly, C. I.; Young, R. N. *Tetrahedron* **1999**, *55*, 11669–11685.
[89] Guiles, J. W.; Johnson, S. G.; Murray, W. V. *J. Org. Chem.* **1996**, *61*, 5169–5171.
[90] Kang, S. K.; Yoon, S. K.; Lim, K. H.; Son, H. J.; Baik, T. G. *Synthetic Commun.* **1998**, *28*, 3645–3655.

[91] Chamoin, S.; Houldsworth, S.; Kruse, C. G.; Bakker, W. I.; Snieckus, V. *Tetrahedron Lett.* **1998**, *39*, 4179–4182.
[92] Malenfant, P. R. L.; Fréchet, J. M. J. *Chem. Commun.* **1998**, 2657–2658.
[93] Nefzi, A.; Ostresh, J. M.; Meyer, J. P.; Houghten, R. A. *Tetrahedron Lett.* **1997**, *38*, 931–934.
[94] Nefzi, A.; Ostresh, J. M.; Giulianotti, M.; Houghten, R. A. *J. Comb. Chem.* **1999**, *1*, 195–198.
[95] Bilodeau, M. T.; Cunningham, A. M. *J. Org. Chem.* **1998**, *63*, 2800–2801.
[96] Zhang, C. Z.; Moran, E. J.; Woiwode, T. F.; Short, K. M.; Mjalli, A. M. M. *Tetrahedron Lett.* **1996**, *37*, 751–754.
[97] Sarshar, S.; Siev, D.; Mjalli, A. M. M. *Tetrahedron Lett.* **1996**, *37*, 835–838.
[98] Havez, S.; Begtrup, M.; Vedsø, P.; Andersen, K.; Ruhland, T. *J. Org. Chem.* **1998**, *63*, 7418–7420.
[99] Combs, A. P.; Saubern, S.; Rafalski, M.; Lam, P. Y. S. *Tetrahedron Lett.* **1999**, *40*, 1623–1626.
[100] Stahl, G. L.; Walter, R.; Smith, C. W. *J. Am. Chem. Soc.* **1979**, *101*, 5383–5394.
[101] Eleftheriou, S.; Gatos, D.; Panagopoulos, A.; Stathopoulos, S.; Barlos, K. *Tetrahedron Lett.* **1999**, *40*, 2825–2828.
[102] Sabatino, G.; Chelli, M.; Mazzucco, S.; Ginanneschi, M.; Papini, A. M. *Tetrahedron Lett.* **1999**, *40*, 809–812.
[103] Tortolani, D. R.; Biller, S. A. *Tetrahedron Lett.* **1996**, *37*, 5687–5690.
[104] Rueter, J. K.; Nortey, S. O.; Baxter, E. W.; Leo, G. C.; Reitz, A. B. *Tetrahedron Lett.* **1998**, *39*, 975–978.
[105] Goff, D. *Tetrahedron Lett.* **1998**, *39*, 1477–1480.
[106] Kim, S. W.; Koh, J. S.; Lee, E. J.; Ro, S. *Mol. Diversity* **1998**, *3*, 129–132.
[107] Sim, M. M.; Ganesan, A. *J. Org. Chem.* **1997**, *62*, 3230–3235.
[108] Bauser, M.; Winter, M.; Valenti, C. A.; Wiesmüller, K. H.; Jung, G. *Mol. Diversity* **1998**, *3*, 257–260.
[109] Lee, S. H.; Chung, S. H.; Lee, Y. S. *Tetrahedron Lett.* **1998**, *39*, 9469–9472.
[110] Kim, S. W.; Ahn, S. Y.; Koh, J. S.; Lee, J. H.; Ro, S.; Cho, H. Y. *Tetrahedron Lett.* **1997**, *38*, 4603–4606.
[111] Matthews, J.; Rivero, R. A. *J. Org. Chem.* **1997**, *62*, 6090–6092.
[112] Park, K. H.; Abbate, E.; Najdi, S.; Olmstead, M. M.; Kurth, M. J. *Chem. Commun.* **1998**, 1679–1680.
[113] Boeijen, A.; Liskamp, R. M. J. *Eur. J. Org. Chem.* **1999**, 2127–2135.
[114] Scicinski, J. J.; Barker, R. D.; Murray, P. J.; Jarvie, E. M. *Bioorg. Med. Chem. Lett.* **1998**, *8*, 3609–3614.
[115] Boeijen, A.; Kruijtzer, J. A. W.; Liskamp, R. M. J. *Bioorg. Med. Chem. Lett.* **1998**, *8*, 2375–2380.
[116] Wilson, L. J.; Li, M.; Portlock, D. E. *Tetrahedron Lett.* **1998**, *39*, 5135–5138.
[117] Yoon, J.; Cho, C. W.; Han, H.; Janda, K. D. *Chem. Commun.* **1998**, 2703–2704.
[118] Hamuro, Y.; Marshall, W. J.; Scialdone, M. A. *J. Comb. Chem.* **1999**, *1*, 163–172.
[119] Hanessian, S.; Yang, R. Y. *Tetrahedron Lett.* **1996**, *37*, 5835–5838.
[120] Dressman, B. A.; Spangle, L. A.; Kaldor, S. W. *Tetrahedron Lett.* **1996**, *37*, 937–940.
[121] Chong, P. Y.; Petillo, P. A. *Tetrahedron Lett.* **1999**, *40*, 2493–2496.
[122] Nefzi, A.; Ostresh, J. M.; Giulianotti, M.; Houghten, R. A. *Tetrahedron Lett.* **1998**, *39*, 8199–8202.
[123] Xiao, X. Y.; Ngu, K.; Chao, C.; Patel, D. V. *J. Org. Chem.* **1997**, *62*, 6968–6973.
[124] Tumelty, D.; Schwarz, M. K.; Cao, K.; Needels, M. C. *Tetrahedron Lett.* **1999**, *40*, 6185–6188.
[125] Tumelty, D.; Schwarz, M. K.; Needels, M. C. *Tetrahedron Lett.* **1998**, *39*, 7467–7470.
[126] Mayer, J. P.; Lewis, G. S.; McGee, C.; Bankaitis-Davis, D. *Tetrahedron Lett.* **1998**, *39*, 6655–6658.
[127] Phillips, G. B.; Wei, G. P. *Tetrahedron Lett.* **1996**, *37*, 4887–4890.
[128] Huang, W.; Scarborough, R. M. *Tetrahedron Lett.* **1999**, *40*, 2665–2668.
[129] Wei, G. P.; Phillips, G. B. *Tetrahedron Lett.* **1998**, *39*, 179–182.
[130] Lee, J.; Gauthier, D.; Rivero, R. A. *Tetrahedron Lett.* **1998**, *39*, 201–204.
[131] Yeh, C. M.; Sun, C. M. *Synlett* **1999**, 810–812.
[132] Blettner, C. G.; König, W. A.; Rühter, G.; Stenzel, W.; Schotten, T. *Synlett* **1999**, 307–310.
[133] Sun, Q.; Yan, B. *Bioorg. Med. Chem. Lett.* **1998**, *8*, 361–364.
[134] Heizmann, G.; Eberle, A. N. *Mol. Diversity* **1997**, *2*, 171–174.
[135] Gray, N. S.; Kwon, S.; Schultz, P. G. *Tetrahedron Lett.* **1997**, *38*, 1161–1164.
[136] Lepore, S. D.; Wiley, M. R. *J. Org. Chem.* **1999**, *64*, 4547–4550.
[137] Park, K. H.; Olmstead, M. M.; Kurth, M. J. *J. Org. Chem.* **1998**, *63*, 6579–6585.
[138] Kurth, M. J.; Randall, L. A. A.; Takenouchi, K. *J. Org. Chem.* **1996**, *61*, 8755–8761.
[139] Pei, Y. H.; Moos, W. H. *Tetrahedron Lett.* **1994**, *35*, 5825–5828.
[140] Cheng, J. F.; Mjalli, A. M. M. *Tetrahedron Lett.* **1998**, *39*, 939–942.
[141] Kantorowski, E. J.; Kurth, M. J. *J. Org. Chem.* **1997**, *62*, 6797–6803.
[142] Shankar, B. B.; Yang, D. Y.; Girton, S.; Ganguly, A. K. *Tetrahedron Lett.* **1998**, *39*, 2447–2448.

[143] Marzinzik, A. L.; Felder, E. R. *Tetrahedron Lett.* **1996**, *37*, 1003–1006.
[144] Tan, D. S.; Foley, M. A.; Shair, M. D.; Schreiber, S. L. *J. Am. Chem. Soc.* **1998**, *120*, 8565–8566.
[145] Haap, W. J.; Kaiser, D.; Walk, T. B.; Jung, G. *Tetrahedron* **1998**, *54*, 3705–3724.
[146] Kobayashi, S.; Akiyama, R. *Tetrahedron Lett.* **1998**, *39*, 9211–9214.
[147] Albert, R.; Knecht, H.; Andersen, E.; Hungerford, V.; Schreier, M. H.; Papageorgiou, C. *Bioorg. Med. Chem. Lett.* **1998**, *8*, 2203–2208.
[148] Wang, F.; Hauske, J. R. *Tetrahedron Lett.* **1997**, *38*, 6529–6532.
[149] ten Holte, P.; Thijs, L.; Zwanenburg, B. *Tetrahedron Lett.* **1998**, *39*, 7407–7410.
[150] Buchstaller, H. P. *Tetrahedron* **1998**, *54*, 3465–3470.
[151] Holmes, C. P.; Chinn, J. P.; Look, G. C.; Gordon, E. M.; Gallop, M. A. *J. Org. Chem.* **1995**, *60*, 7328–7333.
[152] Munson, M. C.; Cook, A. W.; Josey, J. A.; Rao, C. *Tetrahedron Lett.* **1998**, *39*, 7223–7226.
[153] Look, G. C.; Schullek, J. R.; Holmes, C. P.; Chinn, J. P.; Gordon, E. M.; Gallop, M. A. *Bioorg. Med. Chem. Lett.* **1996**, *6*, 707–712.
[154] Goff, D.; Fernandez, J. *Tetrahedron Lett.* **1999**, *40*, 423–426.
[155] Kearney, P. C.; Fernandez, M.; Flygare, J. A. *J. Org. Chem.* **1998**, *63*, 196–200.
[156] Stadlwieser, J.; Ellmerer-Müller, E. P.; Takó, A.; Maslouh, N.; Bannwarth, W. *Angew. Chem. Int. Ed. Engl.* **1998**, *37*, 1402–1404.
[157] Ball, C. P.; Barrett, A. G. M.; Commercon, A.; Compère, D.; Kuhn, C.; Roberts, R. S.; Smith, M. L.; Venier, O. *Chem. Commun.* **1998**, 2019–2020.
[158] Pátek, M.; Drake, B.; Lebl, M. *Tetrahedron Lett.* **1995**, *36*, 2227–2230.
[159] Yan, B.; Gstach, H. *Tetrahedron Lett.* **1996**, *37*, 8325–8328.
[160] Marzinzik, A. L.; Felder, E. R. *J. Org. Chem.* **1998**, *63*, 723–727.
[161] Tietze, L. F.; Steinmetz, A. *Synlett* **1996**, 667–668.
[162] Wilson, R. D.; Watson, S. P.; Richards, S. A. *Tetrahedron Lett.* **1998**, *39*, 2827–2830.
[163] Watson, S. P.; Wilson, R. D.; Judd, D. B.; Richards, S. A. *Tetrahedron Lett.* **1997**, *38*, 9065–9068.
[164] Lyngsø, L. O.; Nielsen, J. *Tetrahedron Lett.* **1998**, *39*, 5845–5848.
[165] Kobayashi, S.; Furuta, T.; Sugita, K.; Okitsu, O.; Oyamada, H. *Tetrahedron Lett.* **1999**, *40*, 1341–1344.
[166] Barbaste, M.; Rolland-Fulcrand, V.; Roumestant, M. L.; Viallefont, P.; Martinez, J. *Tetrahedron Lett.* **1998**, *39*, 6287–6290.
[167] Moore, M.; Norris, P. *Tetrahedron Lett.* **1998**, *39*, 7027–7030.
[168] Freeze, S.; Norris, P. *Heterocycles* **1999**, *51*, 1807–1817.
[169] Zaragoza, F.; Petersen, S. V. *Tetrahedron* **1996**, *52*, 10823–10826.
[170] Schiemann, K.; Showalter, H. D. H. *J. Org. Chem.* **1999**, *64*, 4972–4975.
[171] Liang, G.-B.; Qian, X. *Bioorg. Med. Chem. Lett.* **1999**, *9*, 2101–2104.
[172] Hu, Y. H.; Baudart, S.; Porco, J. A. *J. Org. Chem.* **1999**, *64*, 1049–1051.
[173] Gordeev, M. F.; Patel, D. V.; Wu, J.; Gordon, E. M. *Tetrahedron Lett.* **1996**, *37*, 4643–4646.
[174] Tadesse, S.; Bhandari, A.; Gallop, M. A. *J. Comb. Chem.* **1999**, *1*, 184–187.
[175] Gordeev, M. F.; Patel, D. V.; England, B. P.; Jonnalagadda, S.; Combs, J. D.; Gordon, E. M. *Bioorg. Med. Chem.* **1998**, *6*, 883–889.
[176] Gordeev, M. F.; Patel, D. V.; Gordon, E. M. *J. Org. Chem.* **1996**, *61*, 924–928.
[177] Chiu, C.; Tang, Z.; Ellingboe, J. W. *J. Comb. Chem.* **1999**, *1*, 73–77.
[178] Far, A. R.; Tidwell, T. T. *J. Org. Chem.* **1998**, *63*, 8636–8637.
[179] Chen, C.; Munoz, B. *Tetrahedron Lett.* **1998**, *39*, 6781–6784.
[180] Chen, C.; Munoz, B. *Tetrahedron Lett.* **1998**, *39*, 3401–3404.
[181] Obika, S.; Nishiyama, T.; Tatematsu, S.; Nishimoto, M.; Miyashita, K.; Imanishi, T. *Heterocycles* **1998**, *49*, P261–267, 261–267.
[182] Lago, M. A.; Nguyen, T. T.; Bhatnagar, P. *Tetrahedron Lett.* **1998**, *39*, 3885–3888.
[183] Vo, N. H.; Eyermann, C. J.; Hodge, C. N. *Tetrahedron Lett.* **1997**, *38*, 7951–7954.
[184] Wendeborn, S.; Berteina, S.; Brill, W. K. D.; De Mesmaeker, A. *Synlett* **1998**, 671–675.
[185] Veerman, J. J. N.; van Maarseveen, J. H.; Visser, G. M.; Kruse, C. G.; Schoemaker, H. E.; Hiemstra, H.; Rutjes, F. P. J. T. *Eur. J. Org. Chem.* **1998**, 2583–2589.
[186] Piscopio, A. D.; Miller, J. F.; Koch, K. *Tetrahedron Lett.* **1997**, *38*, 7143–7146.
[187] Wang, Y. H.; Wilson, S. R. *Tetrahedron Lett.* **1997**, *38*, 4021–4024.
[188] Barco, A.; Benetti, S.; De Risi, C.; Marchetti, P.; Pollini, G. P.; Zanirato, V. *Tetrahedron Lett.* **1998**, *39*, 7591–7594.
[189] Chen, C.; McDonald, I. A.; Munoz, B. *Tetrahedron Lett.* **1998**, *39*, 217–220.
[190] Chen, C.; Munoz, B. *Tetrahedron Lett.* **1999**, *40*, 3491–3494.
[191] Fantauzzi, P. P.; Yager, K. M. *Tetrahedron Lett.* **1998**, *39*, 1291–1294.
[192] Chou, Y. L.; Morrissey, M. M.; Mohan, R. *Tetrahedron Lett.* **1998**, *39*, 757–760.

[193] Mohan, R.; Chou, Y. L.; Morrissey, M. M. *Tetrahedron Lett.* **1996**, *37*, 3963–3966.

[194] Kaljuste, K.; Undén, A. *Tetrahedron Lett.* **1995**, *36*, 9211–9214.

[195] Hutchins, S. M.; Chapman, K. T. *Tetrahedron Lett.* **1996**, *37*, 4865–4868.

[196] Yang, L. H.; Guo, L. Q. *Tetrahedron Lett.* **1996**, *37*, 5041–5044.

[197] Mayer, J. P.; Bankaitis-Davis, D.; Zhang, J. W.; Beaton, G.; Bjergarde, K.; Andersen, C. M.; Goodman, B. A.; Herrera, C. J. *Tetrahedron Lett.* **1996**, *37*, 5633–5636.

[198] Rölfing, K.; Thiel, M.; Künzer, H. *Synlett* **1996**, 1036–1038.

[199] Meutermans, W. D. F.; Alewood, P. F. *Tetrahedron Lett.* **1995**, *36*, 7709–7712.

[200] Goff, D. A.; Zuckermann, R. N. *J. Org. Chem.* **1995**, *60*, 5748–5749.

[201] Craig, D.; Robson, M. J.; Shaw, S. J. *Synlett* **1998**, 1381–1383.

[202] Lorsbach, B. A.; Bagdanoff, J. T.; Miller, R. B.; Kurth, M. J. *J. Org. Chem.* **1998**, *63*, 2244–2250.

[203] Hoemann, M. Z.; Melikian-Badalian, A.; Kumaravel, G.; Hauske, J. R. *Tetrahedron Lett.* **1998**, *39*, 4749–4752.

[204] Kiselyov, A. S.; Smith, L.; Armstrong, R. W. *Tetrahedron* **1998**, *54*, 5089–5096.

[205] Kiselyov, A. S.; Armstrong, R. W. *Tetrahedron Lett.* **1997**, *38*, 6163–6166.

[206] Kiselyov, A. S.; Smith, L.; Virgilio, A.; Armstrong, R. W. *Tetrahedron* **1998**, *54*, 7987–7996.

[207] MacDonald, A. A.; DeWitt, S. H.; Hogan, E. M.; Ramage, R. *Tetrahedron Lett.* **1996**, *37*, 4815–4818.

[208] Sim, M. M.; Lee, C. L.; Ganesan, A. *Tetrahedron Lett.* **1998**, *39*, 6399–6402.

[209] Srivastava, S. K.; Haq, W.; Murthy, P. K.; Chauhan, P. M. S. *Bioorg. Med. Chem. Lett.* **1999**, *9*, 1885–1888.

[210] Gopalsamy, A.; Pallai, P. V. *Tetrahedron Lett.* **1997**, *38*, 907–910.

[211] Watson, B. T.; Christiansen, G. E. *Tetrahedron Lett.* **1998**, *39*, 9839–9840.

[212] Ruhland, T.; Künzer, H. *Tetrahedron Lett.* **1996**, *37*, 2757–2760.

[213] Ogbu, C. O.; Qabar, M. N.; Boatman, P. D.; Urban, J.; Meara, J. P.; Ferguson, M. D.; Tulinsky, J.; Lum, C.; Babu, S.; Blaskovich, M. A.; Nakanishi, H.; Ruan, F. Q.; Cao, B. L.; Minarik, R.; Little, T.; Nelson, S.; Nguyen, M.; Gall, A.; Kahn, M. *Bioorg. Med. Chem. Lett.* **1998**, *8*, 2321–2326.

[214] Bräse, S.; Dahmen, S.; Heuts, J. *Tetrahedron Lett.* **1999**, *40*, 6201–6203.

[215] Nielsen, J.; Rasmussen, P. H. *Tetrahedron Lett.* **1996**, *37*, 3351–3354.

[216] Gaviña, F.; Gil, P.; Palazón, B. *Tetrahedron Lett.* **1979**, 1333–1336.

[217] Panek, J. S.; Zhu, B. *Tetrahedron Lett.* **1996**, *37*, 8151–8154.

[218] Boldi, A. M.; Johnson, C. R.; Eissa, H. O. *Tetrahedron Lett.* **1999**, *40*, 619–622.

[219] Steger, M.; Young, D. W. *Tetrahedron* **1999**, *55*, 7935–7956.

[220] Nizi, E.; Botta, M.; Corelli, F.; Manetti, F.; Messina, F.; Maga, G. *Tetrahedron Lett.* **1998**, *39*, 3307–3310.

[221] Sonogashira, K.; Tohda, Y.; Hagihara, N. *Tetrahedron Lett.* **1975**, 4467–4470.

[222] Kolodziej, S. A.; Hamper, B. C. *Tetrahedron Lett.* **1996**, *37*, 5277–5280.

[223] Wipf, P.; Cunningham, A. *Tetrahedron Lett.* **1995**, *36*, 7819–7822.

[224] Hamper, B. C.; Gan, K. Z.; Owen, T. J. *Tetrahedron Lett.* **1999**, *40*, 4973–4976.

[225] Obrecht, D.; Abrecht, C.; Grieder, A.; Villalgordo, J. M. *Helv. Chim. Acta* **1997**, *80*, 65–72.

[226] Masquelin, T.; Sprenger, D.; Baer, R.; Gerber, F.; Mercadal, Y. *Helv. Chim. Acta* **1998**, *81*, 646–660.

[227] Khan, S. I.; Grinstaff, M. W. *Tetrahedron Lett.* **1998**, *39*, 8031–8034.

[228] Khan, S. I.; Grinstaff, M. W. *J. Am. Chem. Soc.* **1999**, *121*, 4704–4705.

[229] Smith, A. L.; Thomson, C. G.; Leeson, P. D. *Bioorg. Med. Chem. Lett.* **1996**, *6*, 1483–1486.

[230] Gouilleux, L.; Fehrentz, J. A.; Winternitz, F.; Martinez, J. *Tetrahedron Lett.* **1996**, *37*, 7031–7034.

[231] Gordeev, M. F.; Hui, H. C.; Gordon, E. M.; Patel, D. V. *Tetrahedron Lett.* **1997**, *38*, 1729–1732.

[232] Gordeev, M. F.; Luehr, G. W.; Hui, H. C.; Gordon, E. M.; Patel, D. V. *Tetrahedron* **1998**, *54*, 15879–15890.

[233] Buckman, B. O.; Mohan, R. *Tetrahedron Lett.* **1996**, *37*, 4439–4442.

[234] Shao, H.; Colucci, M.; Tong, S.; Zhang, H.; Castelhano, A. L. *Tetrahedron Lett.* **1998**, *39*, 7235–7238.

[235] Villalgordo, J. M.; Obrecht, D.; Chucholowski, A. *Synlett* **1998**, 1405–1407.

[236] Mayer, J. P.; Lewis, G. S.; Curtis, M. J.; Zhang, J. W. *Tetrahedron Lett.* **1997**, *38*, 8445–8448.

[237] Cobb, J. M.; Fiorini, M. T.; Goddard, C. R.; Theoclitou, M. E.; Abell, C. *Tetrahedron Lett.* **1999**, *40*, 1045–1048.

[238] Wang, F.; Hauske, J. R. *Tetrahedron Lett.* **1997**, *38*, 8651–8654.

[239] Vojkovsky, T.; Weichsel, A.; Pátek, M. *J. Org. Chem.* **1998**, *63*, 3162–3163.

[240] Gisin, B. F.; Merrifield, R. B. *J. Am. Chem. Soc.* **1972**, *94*, 3102–3106.

[241] Fields, G. B.; Noble, R. L. *Int. J. Pept. Prot. Res.* **1990**, *35*, 161–214.

[242] del Fresno, M.; Alsina, J.; Royo, M.; Barany, G.; Albericio, F. *Tetrahedron Lett.* **1998**, *39*, 2639–2642.

[243] Szardenings, A. K.; Antonenko, V.; Campbell, D. A.; DeFrancisco, N.; Ida, S.; Shi, L.; Sharkov, N.; Tien, D.; Wang, Y.; Navre, M. *J. Med. Chem.* **1999**, *42*, 1348–1357.

[244] Smith, R. A.; Bobko, M. A.; Lee, W. *Bioorg. Med. Chem. Lett.* **1998**, *8*, 2369–2374.

[245] Flanigan, E.; Marshall, G. R. *Tetrahedron Lett.* **1970**, 2403–2406.

[246] Kowalski, J.; Lipton, M. A. *Tetrahedron Lett.* **1996**, *37*, 5839–5840.

[247] Szardenings, A. K.; Burkoth, T. S.; Lu, H. H.; Tien, D. W.; Campbell, D. A. *Tetrahedron* **1997**, *53*, 6573–6593.

[248] Scott, B. O.; Siegmund, A. C.; Marlowe, C. K.; Pei, Y.; Spear, K. L. *Mol. Diversity* **1995**, *1*, 125–134.

[249] Steele, J.; Gordon, D. W. *Bioorg. Med. Chem. Lett.* **1995**, *5*, 47–50.

[250] van Loevezijn, A.; van Maarseveen, J. H.; Stegman, K.; Visser, G. M.; Koomen, G. J. *Tetrahedron Lett.* **1998**, *39*, 4737–4740.

[251] Li, W. R.; Peng, S. Z. *Tetrahedron Lett.* **1998**, *39*, 7373–7376.

[252] Zaragoza, F.; Stephensen, H. *J. Org. Chem.* **1999**, *64*, 2555–2557.

[253] Morales, G. A.; Corbett, J. W.; DeGrado, W. F. *J. Org. Chem.* **1998**, *63*, 1172–1177.

[254] Lee, J.; Murray, W. V.; Rivero, R. A. *J. Org. Chem.* **1997**, *62*, 3874–3879.

[255] Goff, D. A. *Tetrahedron Lett.* **1998**, *39*, 1473–1476.

[256] Goff, D. A.; Zuckermann, R. N. *Tetrahedron Lett.* **1996**, *37*, 6247–6250.

[257] Kung, P. P.; Swayze, E. *Tetrahedron Lett.* **1999**, *40*, 5651–5654.

[258] Mohamed, N.; Bhatt, U.; Just, G. *Tetrahedron Lett.* **1998**, *39*, 8213–8216.

[259] Zhu, Z. M.; McKittrick, B. *Tetrahedron Lett.* **1998**, *39*, 7479–7482.

[260] Masquelin, T.; Meunier, N.; Gerber, F.; Rossé, G. *Heterocycles* **1998**, *48*, 2489–2505.

[261] Tietze, L. F.; Hippe, T.; Steinmetz, A. *Synlett* **1996**, 1043–1044.

[262] Watson, B. T.; Christiansen, G. E. *Tetrahedron Lett.* **1998**, *39*, 6087–6090.

[263] Gordeev, M. F. *Biotechnology and Bioengineering* **1998**, *61*, 13–16.

[264] Kuster, G. J.; Scheeren, H. W. *Tetrahedron Lett.* **1998**, *39*, 3613–3616.

[265] Nefzi, A.; Giulianotti, M.; Houghten, R. A. *Tetrahedron Lett.* **1998**, *39*, 3671–3674.

[266] Mortezaei, R.; Ida, S.; Campbell, D. A. *Mol. Diversity* **1999**, *4*, 143–148.

[267] Bolton, G. L.; Hodges, J. C. *J. Comb. Chem.* **1999**, *1*, 130–133.

[268] Piscopio, A. D.; Miller, J. F.; Koch, K. *Tetrahedron Lett.* **1998**, *39*, 2667–2670.

[269] van Maarseveen, J. H.; den Hartog, J. A. J.; Engelen, V.; Finner, E.; Visser, G.; Kruse, C. G. *Tetrahedron Lett.* **1996**, *37*, 8249–8252.

[270] Gauzy, L.; Le Merrer, Y.; Depezay, J. C.; Clerc, F.; Mignani, S. *Tetrahedron Lett.* **1999**, *40*, 6005–6008.

[271] Huang, W.; Kalivretenos, A. G. *Tetrahedron Lett.* **1995**, *36*, 9113–9116.

[272] Krchnák, V.; Weichsel, A. S. *Tetrahedron Lett.* **1997**, *38*, 7299–7302.

[273] Nefzi, A.; Ostresh, J. M.; Houghten, R. A. *Tetrahedron Lett.* **1997**, *38*, 4943–4946.

[274] Nouvet, A.; Binard, M.; Lamaty, F.; Martinez, J.; Lazaro, R. *Tetrahedron* **1999**, *55*, 4685–4698.

[275] de Bont, D. B. A.; Moree, W. J.; Liskamp, R. M. J. *Bioorg. Med. Chem.* **1996**, *4*, 667–672.

[276] Schwarz, M. K.; Tumelty, D.; Gallop, M. A. *J. Org. Chem.* **1999**, *64*, 2219–2231.

[277] Lee, J.; Gauthier, D.; Rivero, R. A. *J. Org. Chem.* **1999**, *64*, 3060–3065.

[278] Schwarz, M. K.; Tumelty, D.; Gallop, M. A. *Tetrahedron Lett.* **1998**, *39*, 8397–8400.

[279] Nefzi, A.; Ong, N. A.; Giulianotti, M. A.; Ostresh, J. M.; Houghten, R. A. *Tetrahedron Lett.* **1999**, *40*, 4939–4942.

[280] Boojamra, C. G.; Burow, K. M.; Thompson, L. A.; Ellman, J. A. *J. Org. Chem.* **1997**, *62*, 1240–1256.

[281] Goff, D. A.; Zuckermann, R. N. *J. Org. Chem.* **1995**, *60*, 5744–5745.

[282] Plunkett, M. J.; Ellman, J. A. *J. Org. Chem.* **1997**, *62*, 2885–2893.

[283] Camps, F.; Cartells, J.; Pi, J. *Anales de Química* **1974**, *70*, 848–849.

[284] Mayer, J. P.; Zhang, J. W.; Bjergarde, K.; Lenz, D. M.; Gaudino, J. J. *Tetrahedron Lett.* **1996**, *37*, 8081–8084.

[285] Bhalay, G.; Blaney, P.; Palmer, V. H.; Baxter, A. D. *Tetrahedron Lett.* **1997**, *38*, 8375–8378.

[286] Woolard, F. X.; Paetsch, J.; Ellman, J. A. *J. Org. Chem.* **1997**, *62*, 6102–6103.

[287] Ouyang, X.; Kiselyov, A. S. *Tetrahedron* **1999**, *55*, 8295–8302.

[288] Ouyang, X.; Kiselyov, A. S. *Tetrahedron Lett.* **1999**, *40*, 5827–5830.

16 Preparation of Oligomeric Compounds

In this chapter the term 'oligomeric compound' refers to products prepared by repeatedly linking one or several types of monomer to a growing chain of monomers. Included are biopolymers, such as peptides or oligonucleotides, and unnatural products, such as peptide nucleic acids, peptoids, or other synthetic polyamides.

Biopolymers play a central role in all living systems. The functions which proteins and peptides can perform are unparalleled by any other class of substance, and generally far exceed the efficiency and specificity of any non-peptide-mediated process. These functions include catalysis (enzymes), molecular recognition (enzymes, receptors, hormones), signal transduction (receptors), conversion of chemical into mechanical energy (muscle tissue), transport and storage of other compounds, etc. Many of these functions depend on the precise arrangement and the correct dynamic behavior of a set of functional groups within the protein (e.g. mobility of these groups, the ability to 'collapse' on to a ligand or substrate, the ability to bind a substrate but quickly release a product). These dynamic parameters, which are crucial for the function of proteins, are a result of the rigidity of the peptide backbone and the three-dimensional (tertiary) structure of the protein.

Although attempts to produce non-peptidic oligomers with properties similar to those of peptides or proteins have so far met with only limited success at most, the continuous development of more efficient strategies for parallel solid-phase synthesis and screening of compound libraries has made it possible to prepare and screen millions of compounds rather quickly [1–4]. With the aid of these technologies it might be possible in the near future to identify synthetic oligomeric compounds with the capability of performing functions similar to those of proteins. Several potential applications of such synthetic oligomers can be conceived, such as their use as synthetic enzymes or as synthetic receptors for small molecules [5–11]. The development of new synthetic sequences which enable the solid-phase synthesis of non-natural, functionalized oligomeric compounds will, therefore, remain an active and challenging area of research in years to come.

16.1 Peptides

16.1.1 Merrifield's Peptide Synthesis

The first successful solid-phase synthesis of a peptide (H-Leu-Ala-Gly-Val-OH) was reported by Merrifield in 1963 [12]. The strategy used is outlined in Figure 16.1. The support was chloromethylated, partially nitrated (or brominated), 2 % cross-linked polystyrene. Nitration was necessary to suppress acidolytic cleavage of the benzyl ester attachment during Z-group removal. Acylations of deprotected, support-bound amino acids were performed with DCC, and at the end of the synthesis the peptide was cleaved from the support by saponification with sodium hydroxide.

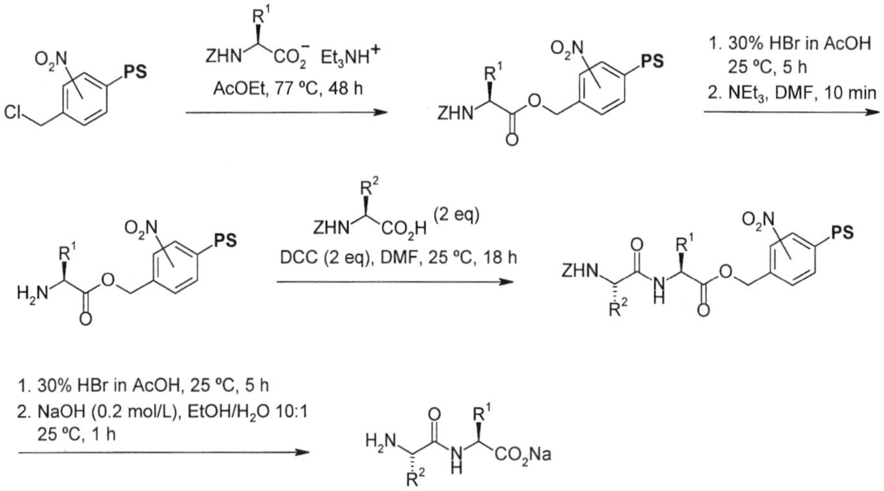

Fig. 16.1. The first solid-phase peptide synthesis, developed by Merrifield [12].

This peptide synthesis, although hampered by some disadvantages, already featured most of the key elements of solid-phase peptide synthesis which ultimately led to the development of the powerful synthetic protocols used today.

One critical issue is the direction in which the peptide is prepared. In Merrifield's strategy the peptide is linked to the support via the carboxyl group, and coupling of monomers is effected by acylation of a support-bound amine. This strategy has two main advantages, which were decisive for its success. Acylations of support-bound amines readily reach yields > 99.95 % when excess of a sufficiently activated acylating agent is used. The opposite variant, i.e. the reaction of a support-bound acylating agent with an amine, cannot usually be performed in such high yields, because strongly activated acylating agents will also undergo hydrolysis to some extent, and less activated acylating agents will react only slowly and not to completion. Acylating agents can, furthermore, be susceptible to intramolecular reactions leading to the formation of byproducts (e.g. N-acylureas), which would accumulate on the support and reduce the yield and purity of the desired peptide. This small difference between yields might not be relevant for the synthesis of short peptides or small non-peptides, but for the preparation of large peptides quantitative coupling yields are absolutely essential to keep the amount of deletion peptides (peptides missing one or more amino acids) as low as possible.

A second advantage of preparing peptides by sequential acylation of support-bound amines arises from the fact that activated *N*-acyl amino acids readily form oxazolones, which quickly racemize under basic conditions, e.g. in the presence of excess amine. Hence, carboxyl group activation of support-bound peptides in the presence of an amine will readily lead to its racemization (Figure 16.2).

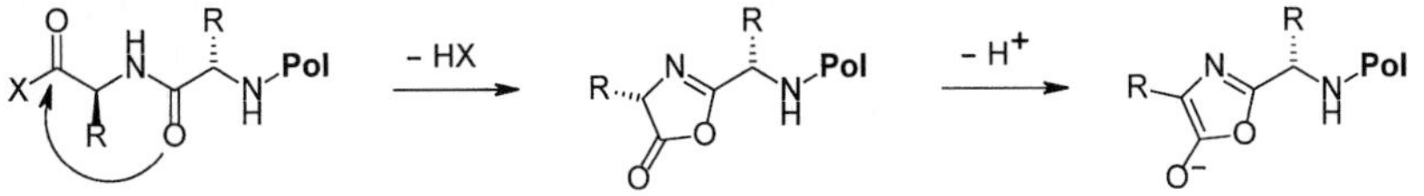

Fig. 16.2. Racemization of activated *N*-acyl amino acids as a result of oxazolone formation. X: leaving group.

Activated *N*-alkoxycarbonyl amino acid derivatives, on the other hand, do not cyclize as readily as *N*-acyl amino acids, and therefore racemize more slowly. Accordingly, solid-phase peptide synthesis is generally performed by acylation of support-bound amines with activated, *N*-alkoxycarbonyl amino acids. Examples of the preparation of peptides by the 'inverse' strategy (first amino acid linked via its amino group as carbamate to the support, activation of support-bound *N*-acylamino acids) have nevertheless been reported [13,14].

The first solid-phase peptide synthesis reported by Merrifield had some disadvantages, which were corrected in later versions of this synthesis. The main improvements were the replacement of benzyloxycarbonyl protective groups by the TFA-labile Boc-protection, and the use of TFA-resistant linkers cleavable by HF.

16.1.2 The Boc Strategy

Figure 16.3 shows a representative example of the synthesis of peptides by use of the Boc strategy. Hydroxymethyl polystyrene can be used as support or, if large peptides are to be prepared, polystyrene with the more acid-resistant PAM linker. Polyacrylamides [15] with a benzyl alcohol linker are also compatible with the Boc strategy.

Attachment of the first amino acid is usually performed with a symmetric anhydride in the presence of DMAP, but mixed anhydrides or acid fluorides can also be used (see Section 13.4.1). Alternatively, chloromethyl polystyrene can be treated with the cesium salt of the first amino acid to yield the corresponding benzyl ester (Section 13.4.2). All proteinogenic Boc-protected α-amino acids linked to hydroxymethyl polystyrene or PAM resin are commercially available [16,17] and can be stored for long periods without deterioration.

Boc-group removal usually requires treatment of the resin-bound carbamate with 50 % TFA in DCM for 30 min (Section 10.1.10.1). Under these conditions neither cleavage of the peptide from the support nor hydrolysis of side-chain protective groups should occur. Loss of peptide by premature cleavage from the support can be minimized by use of the PAM resin, which is approximately 100 times more stable towards acidolysis than hydroxymethyl polystyrene [18]. N-Alkylated peptides are generally less stable towards acids than normal peptides, and hydrolysis of N-alkyl peptides during Boc-group removal has been reported [19].

After deprotection, the amino group can be neutralized by washing with a tertiary amine. The main reason for this neutralization step is the removal of TFA, which could otherwise lead to the formation of trifluoroacetamides when a coupling reagent is added. It has, however, been reported that the neutralization can be omitted if the coupling is performed in the presence of a base (e.g. RCO_2H (4 eq), TBTU (4 eq), DIPEA (2 eq), DMF, 20 min [20]). Surprisingly, the formation of trifluoroacetamides does not seem to occur under these conditions, despite the presence of TFA in the reaction mixture.

During neutralization of dipeptides linked to alcohol resins as esters, diketopiperazines can be formed by intramolecular, nucleophilic cleavage (Section 15.22.1). Diketopiperazine formation during peptide synthesis can be avoided by acylating the first, support-bound amino acid with a preformed, N-protected dipeptide, or by using the 2-chlorotrityl linker, which cannot be readily cleaved by nucleophiles.

Standard coupling is generally conducted with active esters formed in situ, but for difficult couplings other acylating agents, such as symmetric anhydrides or acid fluorides, longer reaction times, and twofold couplings might be required (Section 13.1).

Final cleavage of the peptide from the support requires HF or other acids with high ionizing power (TfOH, HBr/AcOH [16]). HF is a good solvent for amino acids and proteins, and does usually not lead to the cleavage of peptide bonds [21]. Because of its toxicity and capacity to dissolve glass, HF must be handled with great care in special, HF-resistant containers.

Before cleavage with HF, those side-chain protective groups which will not be cleaved by HF must be removed (e.g. N-formyl groups from tryptophan, N-dinitrophenyl groups from histidine). The N-terminal Boc group is usually also removed

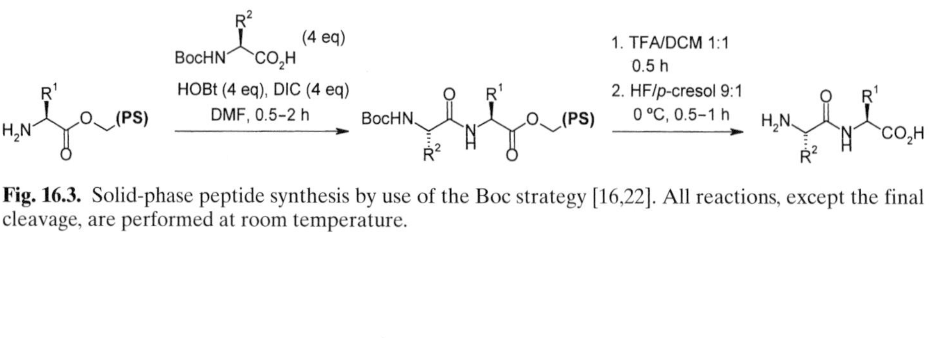

Fig. 16.3. Solid-phase peptide synthesis by use of the Boc strategy [16,22]. All reactions, except the final cleavage, are performed at room temperature.

before treatment with HF, to minimize the risk of alkylation of electron-rich functionalities in the peptide by carbocations. Some of the side-chain protective groups will, however, be removed from the peptide during cleavage from the support. These include, for instance, benzyl esters of aspartic and glutamic acid, benzyl ethers of serine, threonine, and cysteine, and benzyloxycarbonyl groups from lysine, tyrosine, or histidine [16]. Because the resulting carbocations can readily alkylate the peptide irreversibly at susceptible positions (e.g. at tyrosine or tryptophan) the use of scavengers is required to trap carbocations during cleavage from the support. Anisol, dimethyl sulfide, EDT, thiocresol, triethylsilane, or other compounds which react faster with alkylating agents than the peptide, can be used as scavengers.

The Boc strategy has been used with success for the preparation of peptides containing up to about 140 amino acids. Peptides of this size or larger peptides are, however, more conveniently prepared by condensation of protected fragments in solution [23]. One particular advantage of the Boc strategy, compared with the Fmoc strategy (see below), is that treatment with TFA after each coupling leads to the destruction of β-sheets, which can readily form during solid-phase peptide synthesis, and which often cause coupling difficulties [20]. One of the main disadvantages of the Boc strategy is that coupling efficiencies cannot be as readily monitored as in peptide synthesis with Fmoc protection. The repeated treatment of the peptide with TFA also limits the choice of special amino acids which can be incorporated into a peptide. For instance, glycopeptides containing O-glycosylated serine or threonine will be difficult to prepare with the Boc strategy, because of the acid-lability of glycosidic bonds. Most importantly, the Boc strategy requires the handling of HF, which implies the availability of trained staff and special equipment. Because such equipment is not always readily available, many chemists choose the Fmoc strategy rather than the Boc strategy.

16.1.3 The Fmoc Strategy

In the 1970s milder methods for the solid-phase synthesis of peptides were developed. To avoid repetitive treatment of the growing peptide with strong acids, which could lead to premature release of the peptide from the support and to cleavage of sensitive amide bonds (e.g. in N-alkyl peptides), the use of base-labile N-protection was investigated. The Fmoc group, developed by Carpino in 1972 [24], proved particularly suitable because of the mild conditions required for its removal and because of the formation of a readily detectable byproduct (9-(1-piperidinylmethyl)fluorene) which enables the automated monitoring of each deprotection step (for the mechanism of Fmoc deprotection, see Figure 10.7). A convenient solid-phase protocol for peptide synthesis based on Fmoc protection was developed by Atherton and Sheppard [25], and has become the most common strategy for synthesizing peptides [26] (Figure 16.4).

Because no treatment with acid is required during peptide assembly, peptide synthesis with Fmoc amino acids can be conducted on acid-sensitive supports (e.g. Tentagel) and with acid-labile linkers. Wang resin is suitable for most purposes, but other supports, such as Sasrin or 2-chlorotrityl resin, can also be used. CPG, macroporous polystyrene, and polyacrylamides are also compatible with Fmoc peptide synthesis [27]. Premature cleavage of the peptide from Wang resin during Fmoc removal (nucleophilic cleavage with piperidine) is negligible.

As for Boc amino acids, all proteinogenic and several unnatural Fmoc amino acids esterified with Wang or similar resins are commercially available. Deprotection with 20 % piperidine in DMF is usually complete within a few minutes, and the release of the fluorene derivative is easily monitored at 300–320 nm (see, e.g., [28]). The peak area can be used to determine the amount of chromophore released, which is proportional to the efficiency of the preceding coupling reaction. At the end of an automated peptide synthesis inspection of the chromatogram of all Fmoc releases enables rapid assessment of the quality of the resin-bound peptide and quick location of positions in the peptide where coupling was unsuccessful or difficult.

During acylations with Fmoc-protected amino acids addition of bases should be avoided, because these could lead to partial deprotection and thence to multiple incorporation of the amino acid. Small amounts of DIPEA or pyridine, however, do not usually cause major problems (see Experimental Procedure 13.5).

The conditions used for final cleavage of the peptide from the support depend on the type of support and linker chosen. Because side-chain protective groups must also be removed, the use of scavengers is required during cleavage to avoid alkylation of the peptide.

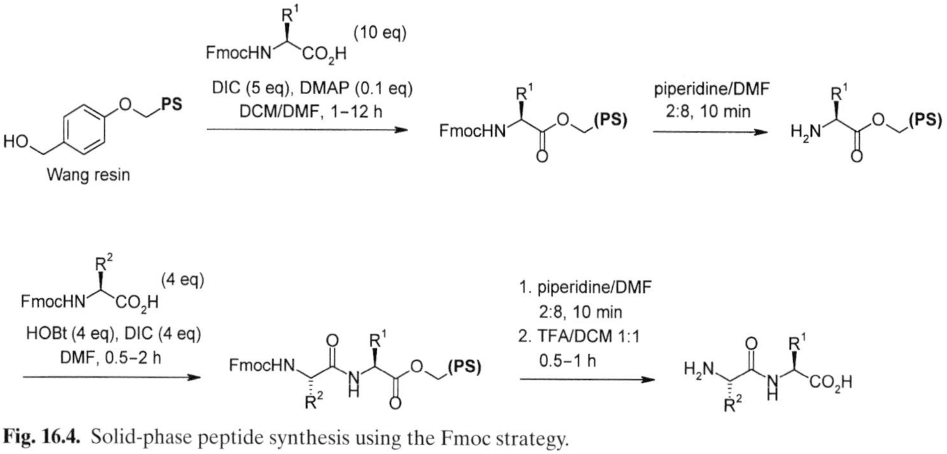

Fig. 16.4. Solid-phase peptide synthesis using the Fmoc strategy.

Peptides with up to 166 amino acids (aglycon of erythropoietin [29]) have been prepared using the Fmoc strategy on cross-linked polystyrene. The preparation of peptides containing several *N*-methylamino acids (e.g. cyclosporin [30]) has, furthermore, been successfully performed with Fmoc protection. However, such difficult peptides require careful planning of the synthesis and often the use of special coupling techniques.

The release of a readily detectable compound during each deprotection and the absence of corrosive reagents during peptide assembly makes the Fmoc strategy particularly well suited for automation. A typical programmable, fully automated peptide synthesizer can perform one complete coupling and deprotection cycle in approximately 1 h (20–40 min coupling, 10 min Fmoc removal, 10 min washing). Cleavage of the peptide from the support is generally not performed on the synthesizer but manually in a hood.

The mild reaction conditions during peptide synthesis with the Fmoc strategy also enable the use of sensitive amino acid derivatives, such as glycosylated serine, threonine, and asparagine (Figure 16.5) for the solid-phase synthesis of glycopeptides [31–33]. Although glycopeptides can be prepared by means of the Boc strategy [32], the harsh conditions used for Boc group removal can lead to hydrolysis of glycosidic bonds, and the Fmoc strategy will usually be more applicable for glycopeptide synthesis.

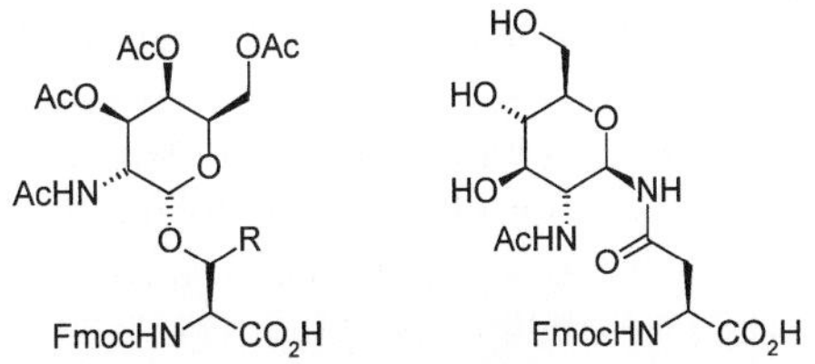

Fig. 16.5. Typical protected, glycosylated amino acid derivatives, suitable for the solid-phase synthesis of glycopeptides [34–39]. R: H, Me.

16.1.4 Side-Chain Protection

Some amino acids contain functional groups which need to be masked during peptide synthesis. Protective groups suitable for the Boc strategy must be resistant towards TFA, whereas for the Fmoc strategy stability towards piperidine must be ensured.

Choosing the right side-chain protection for solid-phase peptide synthesis can be a difficult task, in particular if large peptides are to be prepared. Many factors play a

Table 16.1. Recommended side-chain protective groups for solid-phase peptide synthesis [16].

Amino Acid	Boc strategy	Fmoc strategy
Arg	Ts, Mts	Mtr, Pmc, Pbf
Asn	-	Tr
Asp	Bn, Cy	t-Bu
Cys	4-MeBn, Acm	Tr, Acm
Gln	-	Tr
Glu	Bn	t-Bu
His	Bom, Dnp	Tr
Lys	2-Cl-Z	Boc
Ser	Bn	t-Bu
Thr	Bn	t-Bu
Trp	formyl	Boc
Tyr	2-Br-Z	t-Bu

role, and careful planning of the synthesis and deprotection is usually required. Details about the compatibility of various protective groups and strategies for their removal can be found in more specialized literature [16,22,26,40]. Table 16.1 lists some of the most common side-chain protective groups for solid-phase peptide synthesis, which will be suitable for most applications.

16.1.5 Backbone Protection

It has often been observed that during solid-phase synthesis the growing peptide forms aggregates with the support, with other peptides, or with itself. These aggregates can, for instance, be β-sheets or other secondary structures, stabilized by hydrogen bonds between carboxamido groups RCONHR [16,41]. The formation of such aggregates occurs particularly readily in peptides containing an uninterrupted sequence of amino acids with short, lipophilic side chains (Ala, Val, Leu, etc.), and usually significantly reduces the rate of couplings and deprotections.

The formation of such aggregates can be partly prevented by use of dipolar aprotic solvents instead of DCM during N-acylation, by addition of salts (LiBr, LiCl, NaClO$_4$, KSCN) or protic solvents (e.g. hexafluoroisopropanol), by use of a polar supports (e.g. polyacrylamide), or by performing the synthesis with Boc strategy [16,20]. Most elegantly, hydrogen-bond formation will be precluded by introducing an *N*-alkylamino acid at certain positions of the peptide. This will disrupt secondary structures and keep the growing peptide well solvated and accessible to reagents. One type of backbone protection for peptides, suitable for peptide synthesis with Fmoc strategy, is Hmb protection [41–44] (2-*hydroxy-4-methoxybenzyl*; Figure 16.6). The N-alkylated amino acids required are available commercially [16], and the incorporation at every sixth to eighth position is usually sufficient to prevent aggregation of the peptide [45]. Alternatively, support-bound amino acids can be Hmb-protected by reductive alkylation [46].

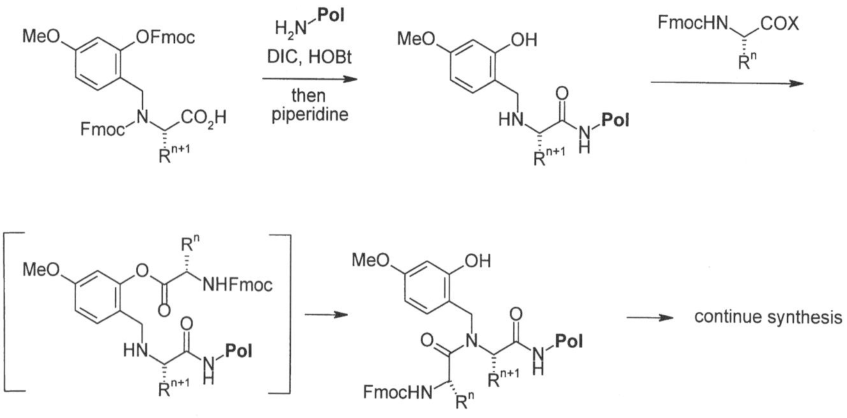

Fig. 16.6. Incorporation and acylation of *N*-(Hmb) amino acids during solid-phase peptide synthesis.

The acylation of support-bound *N*-benzylamino acids is difficult, and requires symmetric anhydrides or similarly activated amino acids [44]. (2-Hydroxybenzyl)amino acids are probably acylated via intramolecular *N,O*-acyl migration (Figure 16.6). After introduction of Hmb protection the free phenolic hydroxyl group is not acylated unless highly reactive acylating agents are used. The resulting esters are not stable, however, and will be cleaved during Fmoc-group removal. Deprotection of the peptide is achieved during cleavage from the support by treatment with TFA (and a scavenger).

16.1.6 Cyclic Peptides

Interest in cyclic peptides as turn mimetics or conformationally constrained peptide analogs is increasing. The cyclization of peptides can either be performed after cleavage from the support in solution (see, e.g., [47]), or on the support. In addition to cyclizations of the peptide backbone, peptides can be cyclized via the side chains or by means of an unnatural spacer. To cyclize the backbone of peptides on insoluble supports the peptide must be attached to the support either by a backbone amide linker [48] or via the side chain of an amino acid [49] (Figure 16.7).

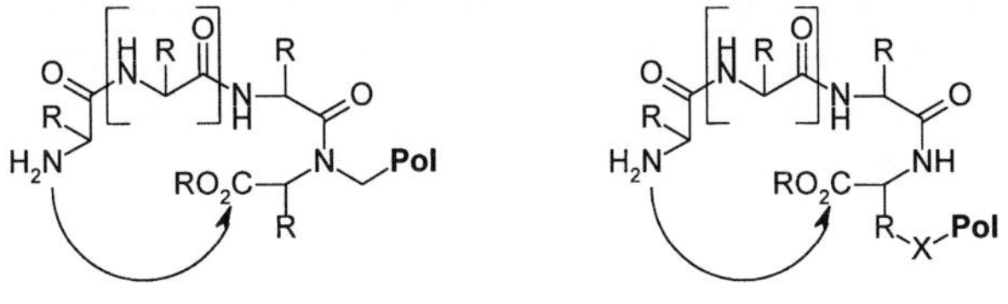

Fig. 16.7. Strategies for the backbone cyclization of peptides on solid phase without simultaneous cleavage from the support.

Peptides with more than seven amino acids can usually be cyclized without problems, but smaller cyclopeptides can be difficult to prepare. Few examples of the preparation of cyclotetrapeptides have been reported [50,51]. The preparation of cyclopentapeptides proceeds satisfactorily in solution with HOAt-derived coupling reagents [52,53].

As shown in Figure 16.7, when using backbone attachment, the first amino acid is usually protected as an ester, which needs to be saponified before cyclization. The use of allyl esters enables the palladium-mediated hydrolysis under mild reaction conditions [48]. During the cyclization, basic reaction conditions should be avoided to prevent racemization of the activated *N*-acylamino acid. Because such racemization can readily occur (see Figure 16.2), it is advisable to use coupling reagents with a low tendency to induce racemization. These include HOAt-derived reagents [52–55], DPPA, and PyBOP/HOBt [56].

Peptides can also be cyclized via side-chain functionalities (e.g. using lysine, cysteine, glutamic acid, or aspartic acid [57–60]), or with the aid of synthetic spacers ([61]; see also Table 8.5). Examples of the preparation of cyclic peptides using a safety-catch linker ((4-alkylmercapto)phenol [51]), 2-nitrophenyl ester attachment

[50], or oxime ester attachment [62–65], in which cyclization and release of the cyclo-peptide by intramolecular nucleophilic cleavage occurred spontaneously after removal of the terminal amine protection, have also been reported.

Amino acids suitable for side-chain attachment to supports, which have been used to prepare cyclic peptides on insoluble supports include, for instance, tyrosine (etherified with Wang resin [49]), histidine (N-alkylated with trityl resin [66]), aspartic acid [55,56,67], asparagine [67], glutamic acid [67,68], lysine [54], and serine [54].

The main side reaction during the cyclization of peptides on insoluble supports is their oligomerization. This competing process can be prevented only by choosing a support with a lower loading and thereby reducing the concentration of the peptide. The use of *N*-alkylamino acids (proline, Hmb-protected amino acids [52]) will, furthermore, increase the flexibility of the peptide, by reducing the energy barrier to amide bond rotation, and thereby facilitate cyclization.

16.2 Oligonucleotides

The synthesis of oligonucleotides and analogs thereof on insoluble supports has become an extensive area of research, and detailed treatment would far exceed the scope of this book. In the following sections only the most important features of oligonucleotide synthesis will be discussed. For further details more specialized sources should be consulted, e.g. [69]. Review articles have appeared, which discuss supports for the synthesis of oligonucleotides [70], and the uses of synthetic oligonucleotides in gene technology [71] and in drug discovery [72,73].

16.2.1 Historical Overview

Unlike solid-phase peptide synthesis, the synthesis of oligonucleotides on supports was complicated by numerous severe problems, and the development of useful synthetic protocols took approximately 15 years.

The first attempts to prepare oligodeoxynucleotides on insoluble supports were reported by Letsinger in 1965 [74] (Figure 16.8). He chose cross-linked polystyrene as support, and immobilized the first nucleoside via the heterocyclic base. Phosphorylation was achieved by coupling cyanoethyl phosphate to the 5′-hydroxyl group of the resin-bound nucleoside with the aid of DCC. The resulting phosphodiester was then activated by treatment with mesitylenesulfonyl chloride and then coupled with a second nucleoside. Cleavage from the support was realized by saponification, whereby simultaneous removal of the cyanoethyl protective group occurred (via β-elimination).

Several groups began to work on improving Letsinger's approach, not only by modifying the protective group strategy, but also by choosing other attachment sites and by testing different supports.

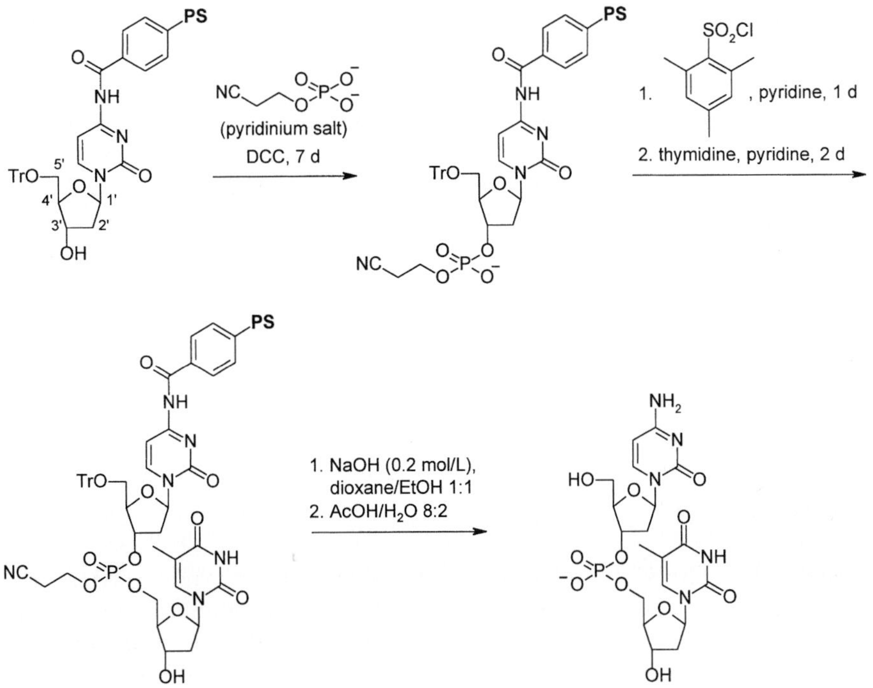

Fig. 16.8. The first solid-phase synthesis of a dinucleotide [74].

In 1966 two groups presented a synthesis [75–77], in which the first nucleoside was attached to non-cross-linked (soluble) polystyrene with a methoxytrityl linker via its 5'-hydroxyl group. The resulting, support-bound nucleoside was then coupled with 3'-O-acetylthymidine-5'-phosphate in pyridine with mesitylenesulfonyl chloride as coupling agent, and, after saponification of the 3'-acetoxy group, the 3'-hydroxyl group became available for a new coupling (Figure 16.9).

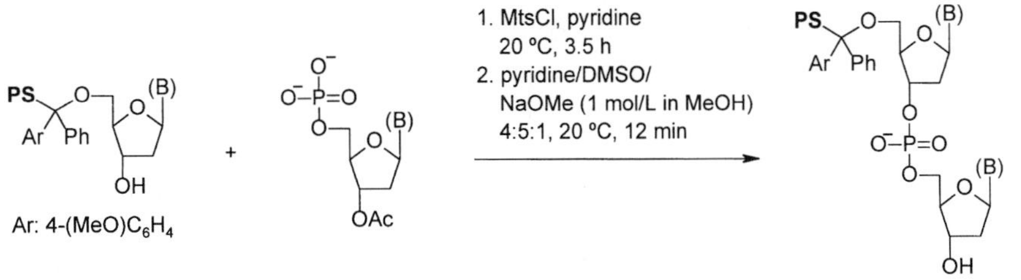

Fig. 16.9. Synthesis of oligonucleotides on non-cross-linked polystyrene using nucleoside 5'-phosphates [77].

This so-called phosphodiester method [78] was later also performed on cross-linked polystyrene [79], macroporous, 25 % cross-linked polystyrene [80], silica [81], poly(vinyl alcohol) [82,83], proteins [84], PEG [85], and polyacrylamide [86]. This extensive evaluation of different supports was because of the observation that cross-linked polystyrene or other, gel-type supports were not ideal for the preparation of phosphoric esters [79]. The highly polar phosphate backbone of DNA seemed not to be compatible with hydrophobic supports. Diffusion of activated phosphates into gel-type supports seemed, furthermore, to be too slow to enable the efficient and high-yielding coupling of nucleotides. Significantly higher reaction rates could be attained on macroporous supports and on silica, in which the growing oligonucleotide is only located on the surface of the support material. Such non-swellable supports are compatible with a broader choice of solvents than gel-type supports, and are also better suited to continuous-flow synthesis because the volume of the support material remains constant throughout the synthesis.

16.2.2 The Phosphotriester Method

In the 1970s a significant modification of previous oligonucleotide syntheses was implemented by several groups [87–89]; this later became known as the phosphotriester method (Figure 16.10; for a review see [90]). In this new strategy not a simple nucleoside phosphate but a phosphodiester was coupled with the 5'-hydroxyl group of the support-bound nucleoside, to yield a non-charged phosphotriester (as in Letsinger's approach, Figure 16.8). The lower polarity of the phosphotriester, compared with a phosphodiester, made the triester method more appropriate for hydrophobic supports, such as polystyrene [78], and enabled the preparation of larger oligonucleotides.

Alcohols suitable for phosphate protection are, for instance, chlorophenols [88,89,91] or 2,2,2-trichloroethanol [87], which enable selective, nucleophilic deprotection by treatment with NaOH, without destruction of the oligonucleotide. Thiophenol has also been used as an (oxidant-labile) protective group of the phosphate backbone [92].

Supports used for the phosphotriester approach include cross-linked polystyrene [91], kieselguhr–polyamide [93], cellulose [93], polystyrene grafted on PTFE [94], and LCAA-CPG [95]. In most examples of the triester method the first nucleoside was no longer attached to the support via the 5'-hydroxyl group but via its 3'-hydroxyl group as hemisuccinate (Figure 16.10).

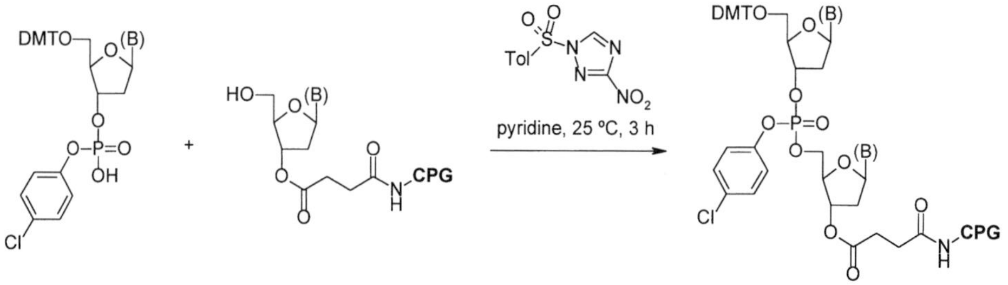

Fig. 16.10. Phosphoric ester formation in the triester method [95].

16.2.3 The Phosphoramidite and *H*-Phosphonate Methods

In 1975 a further significant improvement of the synthesis of oligonucleotides was reported by Letsinger [96]. He found that the formation of phosphites from phosphinic chlorides (RO)PCl$_2$ or (RO)$_2$PCl proceeds much faster than the formation of phosphates from chlorophosphates, and that the oxidation of phosphites to phosphates by iodine also proceeds quickly and under mild reaction conditions which are compatible with protected oligonucleotides. This new strategy, i.e. phosphite formation followed by oxidation, was later used with success on silica for the preparation of oligoribonucleotides [97] and oligodeoxyribonucleotides (Figure 16.11). The phosphinic chlorides required were prepared shortly before use by treatment of 5′-DMT-protected nucleosides with methyldichloro phosphite. These activated nucleosides react quickly under mild conditions with the support-bound nucleoside to yield a phosphite, which must be oxidized to the more stable trialkyl phosphate after each coupling. After capping with phenylisocyanate (to derivatize all non-phosphorylated 5′-hydroxyl groups) the newly introduced nucleotide is detritylated by brief treatment with an acid. At the end of a synthesis the methyl phosphates are demethylated by treatment with thiophenolate, and the oligonucleotide is finally cleaved from the support by treatment with ammonia.

Later it was found that dialkyl(dialkylamino) phosphites (phosphoramidites, (RO)$_2$PNR$_2$ [100]), which are stable towards air and water and can be stored for longer, can readily be converted into the phosphinic chlorides (RO)$_2$PCl by treatment with dimethylaniline hydrochloride, or into the corresponding tetrazolides by treatment with tetrazole. Tetrazolides (RO)$_2$P–(1-tetrazolyl) had proven excellent reagents for the phosphorylation of nucleosides [99], and the treatment of phosphoramidites with alcohols in the presence of tetrazole was found to be a satisfactory method for the rapid preparation of trialkyl phosphites [101,102].

A typical solid-phase synthesis of oligodeoxyribonucleotides from phosphoramidites, as usually performed on a modern synthesizer, is sketched in Figure 16.12 [69,103]. The synthesis is conducted under continuous flow on non-swellable support materials, such as CPG, macroporous polystyrene [104], polymethacrylate [105], or polystyrene grafted on PTFE [106,107]. Larger amounts of oligonucleotide can be prepared on PEG–polystyrene graft supports [108–111] or on soluble supports [112–114].

The most remarkable aspect of the phosphoramidite approach is its speed: taking into account the time required for washing (30 s with acetonitrile after each step) it takes only approximately 5 min to couple one nucleotide to the growing oligonucleotide.

Cyanoethyl groups are usually used as protection of the phosphate backbone; these groups are removed with concentrated aqueous ammonia during or after cleavage of the oligonucleotide from the support. Some of the heterocyclic bases (adenine, guanine, cytosine) contain primary amino groups; these are generally also protected to avoid problems during solid-phase assembly of oligonucleotides. For this purpose the amino groups in the monomers are acylated with benzoic or isobutyric acid. The corresponding amides are cleaved by ammonolysis during final cleavage of the product from the support. It was recently found [115] that base-protection of nucleosides is not required if imidazolium triflate is used as promoter for phosphite formation.

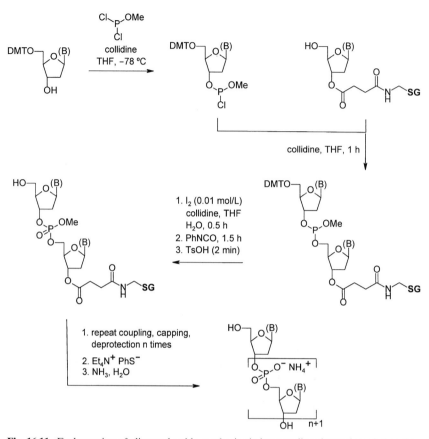

Fig. 16.11. Early version of oligonucleotide synthesis via intermediate formation of phosphites [98,99].

The synthesis sketched in Figure 16.12 can be used to prepare oligodeoxynucleotides of up to about 150 bases, and examples for the preparation of cyclic oligodeoxynucleotides [108] and oligoribonucleotides [116] with this technique have also been reported.

In a closely related strategy nucleoside-3'-*H*-phosphonates RO–P(OH)(H)=O, which are tautomers of monoalkyl phosphites RO–P(OH)$_2$, are used as the monomers and are activated with coupling agents such as HOBt-derived phosphonium salts [117] or with pivaloyl or 1-adamantoyl chloride [111] (Figure 16.13; for the preparation of *H*-phosphonates see Figure 16.28). The resulting *H*-phosphonate diesters are more stable than trialkyl phosphites, and the oxidation with iodine to the phosphates only needs to be performed once (at the end of the synthesis). Because *H*-phosphonates are readily converted into phosphoric acid derivatives other than phosphates, *H*-phosphonates are valuable intermediates for the preparation of modified DNA-analogs, such as phosphorothioates [69], phosphoramidates [118], or boranophosphates [119] (see Section 16.2.4). The phosphonate method is, furthermore, cheaper than the amidite method [111], and should be considered if large amounts of oligonucleotides must be prepared. Coupling yields with activated *H*-phosphonates are, however, usually lower (approx. 95 %) than with phosphoramidites [69].

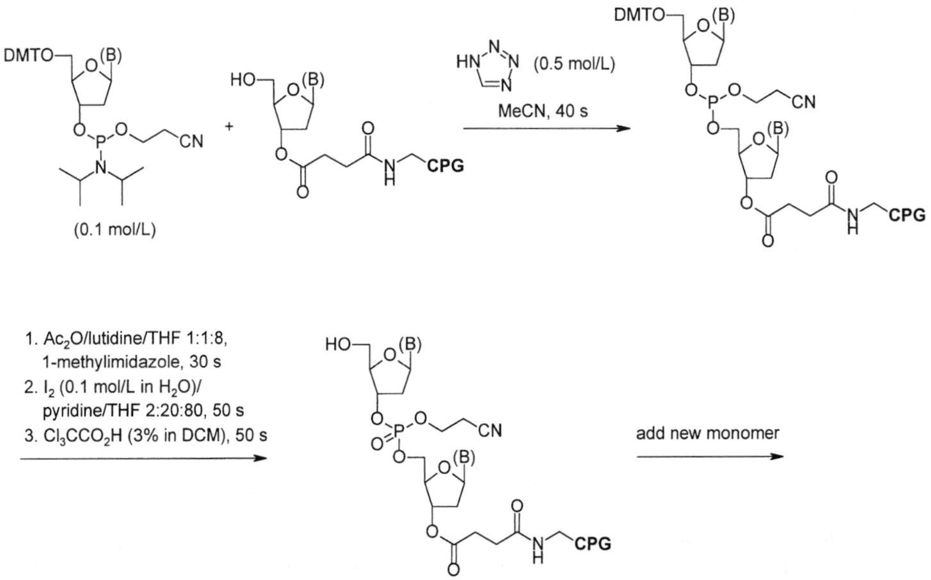

Fig. 16.12. Solid-phase oligonucleotide synthesis using phosphoramidites [69].

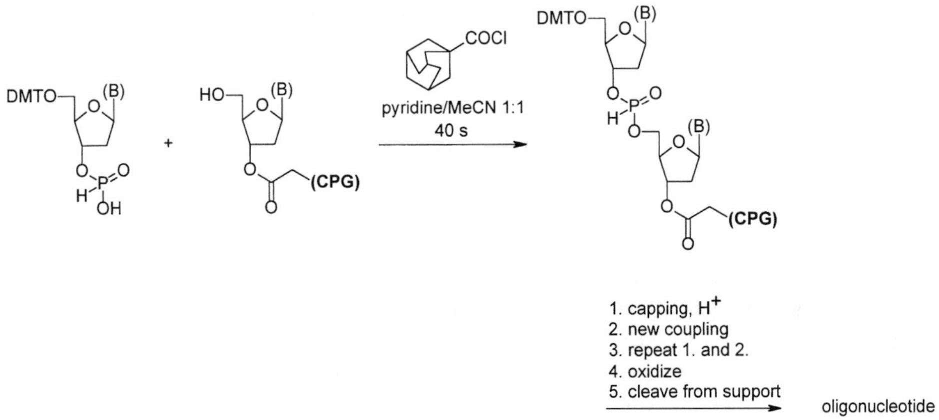

Fig. 16.13. Preparation of oligonucleotides by the *H*-phosphonate method [69].

Oligoribonucleotides are more difficult to handle than oligodeoxyribonucleotides, mainly because of their sensitivity towards nucleases. For synthesis of oligoribonucleotides on insoluble supports protection of the 2′-hydroxyl group is required which must be stable towards oxidation with iodine and the acidic conditions of DMT-deprotection, but which should be easy to remove under mild conditions during or after cleavage of the oligonucleotide from the support. Of the many protective groups evaluated the TBS group seems to fulfil these requirements, and oligoribonucleotides with up to 30–40 bases can be prepared on supports by use of TBS-protected ribonucleotides and the phosphoramidite method [69,120]. Photosensitive protective groups (e.g. 2-nitrobenzyloxymethyl [121]) have also been evaluated.

16.2.4 Oligonucleotide Analogs

Oligonucleotides and analogs thereof can interact with proteins, DNA, and RNA in a highly specific manner, and might therefore be useful for the treatment of various human diseases [69,72,73,122,123]. Several different approaches have been described for modification of oligonucleotides with the aim of modulating their biological activity and pharmacokinetic properties.

One important modification of oligonucleotides is the replacement of phosphates by phosphorodithioates $(RO)_2P(S)S^-$ [69,124] and phosphorothioates $(RO)_2P(O)S^-$ [125–129] (Figure 16.14), producing analogs with increased stability towards enzymatic and chemical degradation. Other examples of backbone modification of oligonucleotides, compatible with solid-phase synthesis, are the replacement of phosphates by phosphonates [69,130], phosphoramidates [118,131], boranophosphates [119,132], guanidines [133], and hydroxylamines [134] (Figure 16.14).

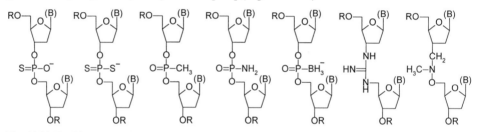

Fig. 16.14. Backbone-modified DNA analogs which can be prepared on insoluble supports.

Other known oligonucleotide analogs include oligomers in which the heterocyclic bases have been modified [135] or replaced by other types of compound [136]. The solid-phase synthesis of hybrids of DNA with peptides [137–140], with carbohydrates [141], and with PNA (peptide nucleic acids, see Section 16.4.1.2 [142,143]) has also been reported and several strategies have been developed which enable the preparation of oligonucleotides with a modified 5′ or 3′-terminus [144–149]. Various techniques for the parallel solid-phase synthesis of oligonucleotides have been developed for the preparation of compound libraries [150–154].

16.3 Oligosaccharides

The development of strategies for the solid-phase synthesis of oligosaccharides received little attention in the 1960s and 1970s, compared with the synthesis of peptides and oligonucleotides. One reason for this might have been the inherent difficulty of oligosaccharide synthesis, because, unlike peptide bond formation or phosphodiester formation, glycosylation can lead to mixtures of diastereomers. In recent years, however, powerful methods have been developed which enable the realization of highly stereoselective glycosidation on supports [32,155,156].

Two different strategies can be envisioned for preparation of glycosides on insoluble supports (Figure 16.15). Support-bound, partially protected carbohydrates can be glycosylated by addition of a suitable glycosyl donor and a promoter. Alternatively, a glycosyl donor can be generated on solid phase and coupled with a partially protected carbohydrate. It seems that the first strategy, based on the glycosidation of support-bound alcohols, is better suited to the preparation of oligosaccharides, because most examples reported are based on this scheme. The second strategy has mainly been realized by Danishefsky and coworkers, using glycals as latent, support-bound glycosyl donors (see below).

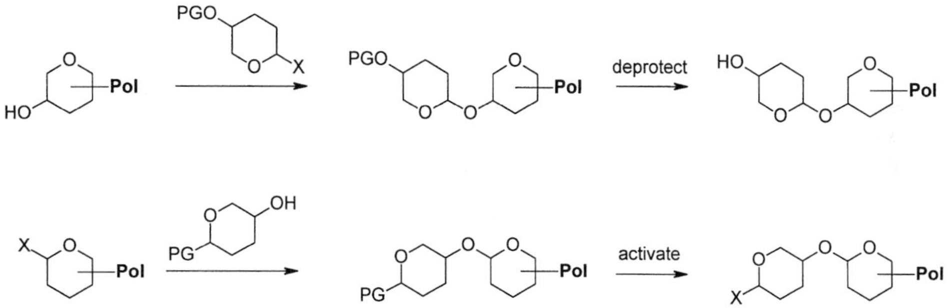

Fig. 16.15. Strategies for the assembly of oligosaccharides on insoluble supports. PG: protective group; X: leaving group.

The synthesis of oligosaccharides has been performed successfully on various types of support, including gel-type, hydrophobic or hydrophilic supports, and non-swellable hydrophobic or hydrophilic supports.

16.3.1 Glycosylation with Glycosyl Bromides

In early attempts at solid-phase oligosaccharide synthesis the formation of glycosidic bonds was achieved by treating pyranosyl bromides with partially protected carbohydrates (Figure 16.16). These syntheses could be performed either with the pyranosyl bromide (glycosyl donor) [157,158] or the alcohol (glycosyl acceptor) [159–161] linked to the support.

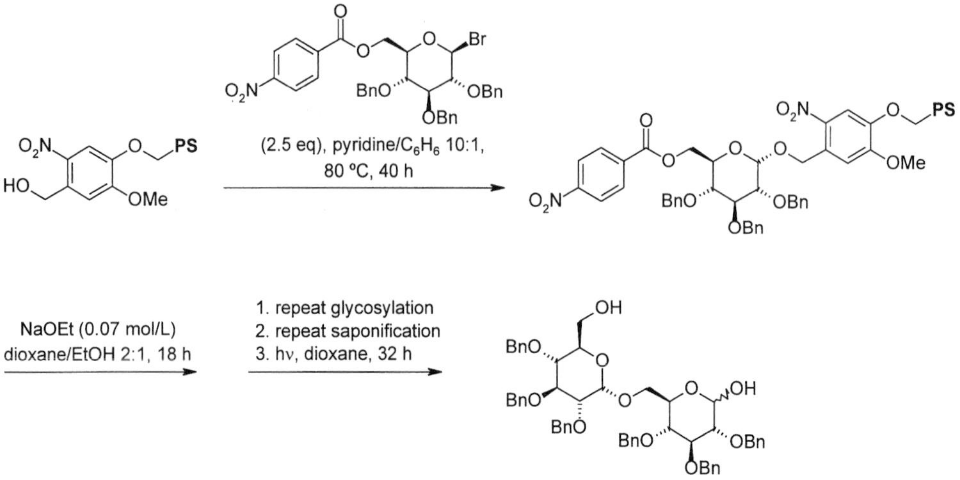

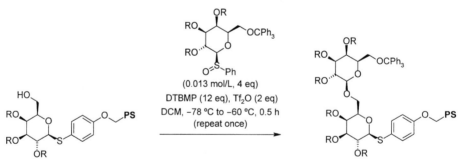

Fig. 16.16. Early solid-phase synthesis of oligosaccharides using glycosyl bromides [160].

16.3.2 Glycosylation with Glycosyl Sulfoxides

More efficient methods for the preparation of glycosides were developed later and used with success on insoluble supports. One of these is based on the use of glycosyl sulfoxides, which react with alcohols upon treatment with Tf$_2$O at low temperatures to yield the corresponding glycosides with high stereoselectivity [162–164]. With this strategy di- and trisaccharides have been prepared on cross-linked polystyrene using a thioacetal-based linkage (Figure 16.17). Cleavage from the support was accomplished by treatment with Hg(O$_2$CCF$_3$)$_2$/DCM/H$_2$O [162]. This strategy has also been applied to the solid-phase synthesis of oligosaccharide libraries [165].

Fig. 16.17. Glycosylation of polystyrene-bound alcohols with pyranosyl sulfoxides [162,166]. R: pivaloyl.

16.3.3 Glycosylation with Glycosyl Thioethers

Glycosyl thioethers can be converted into highly reactive glycosylating agents by S-methylation with MeOTf or [Me$_2$SSMe$^+$][OTf$^-$] [167,168]. Glycosylations with pyranosyl thioethers have been performed both with the glycosyl donor [169,170] or with the alcohol bound to the support [167]. Illustrative examples of this type of glycosylation are sketched in Figures 16.18 and 16.20. The glycosylation of support-bound alcohols with glycosyl thioethers is one of the most efficient and flexible strategies for the preparation of oligosaccharides, and even branched heptasaccharides can be synthesized on polystyrene by use of this methodology [167].

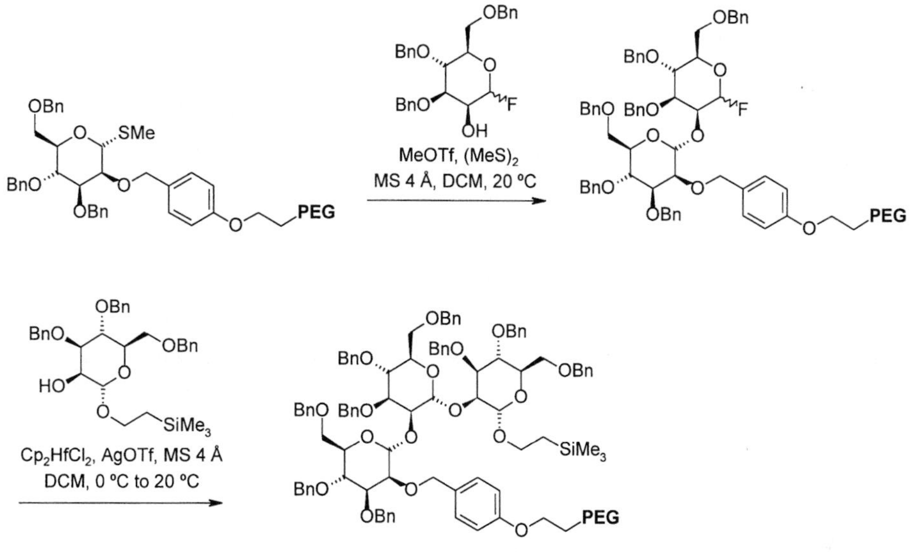

Fig. 16.18. Glycosidations with support-bound glycosyl thioethers and glycosyl fluorides [171].

16.3.4 Glycosylation with Miscellaneous Glycosyl Donors

Other glycosyl derivatives suitable for the preparation of glycosides on insoluble supports include fluorides (Figure 16.18), pent-4-en-1-yl glycosides [172] (activation by NIS), and trichloroacetimidates (Figure 16.19). Trichloroacetimidates, in particular, seem to be convenient reagents for glycosidation, and numerous examples have been reported. Glycosyl trichloroacetimidates also react with support-bound thiols, upon Lewis acid catalysis, to yield glycosyl thioethers. Such thioethers can be used as linkers for carbohydrates. Oxidative cleavage with NBS in the presence of water leads to the release of the carbohydrate [173,174]. Cross-linked polystyrene [168,174,175], PEGA [176], CPG [141,173,177], polystyrene–PEG graft polymers [178], and PEG [179,180] have been used with success as supports for glycosylation with trichloroacetimidates .

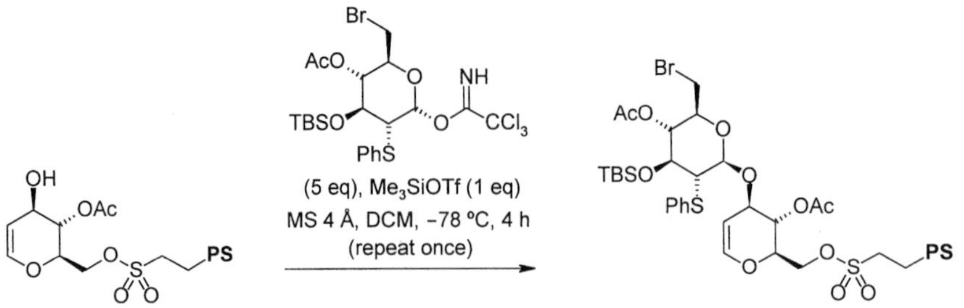

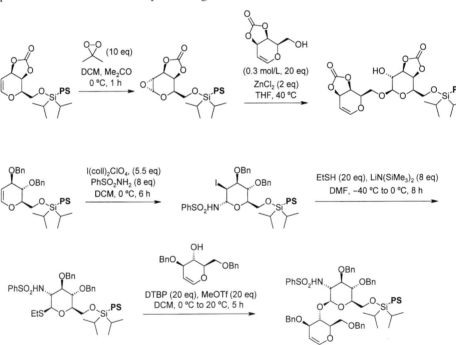

Fig. 16.19. Glycosidation of support-bound alcohols with trichloroacetimidates [181].

The use of glycals as monomers for the solid-phase synthesis of oligosaccharides has been explored by Danishefsky (Figure 16.20; [156,182]). Glycals can be converted into glycosyl donors by epoxidation or haloamination, which usually proceed with high diastereoselectivity. The ensuing, Lewis acid mediated glycosylations generally proceed in a stereochemically unambiguous manner.

Fig. 16.20. Solid-phase synthesis of oligosaccharides from glycals [183,184].

Some types of support, such as CPG or PEGA, enable the use of enzymes to promote various chemical transformations. Enzymes are also well suited to the regio- and stereoselective preparation of glycosides from unprotected carbohydrates, and several examples of enzyme-mediated preparations of glycosides on insoluble supports have been reported [185–187].

16.4 Miscellaneous Oligomeric Compounds

Several other types of compound have been prepared on insoluble supports by sequential addition of monomers to a growing chain [188,189]. Some of the most important oligomers of this kind will be discussed in the next sections.

16.4.1 Oligoamides

Because amides can be readily prepared on solid phase, the synthesis of oligomeric amides other than peptides has received much attention in recent years, in particular with regard to the production of compound libraries. Several protected unnatural amino acids have been prepared and used for the synthesis of oligomers by use of the standard procedures of solid-phase peptide synthesis. A selection of such artificial amino acids is given in Table 16.2.

Table 16.2. Synthetic amino acids or precursors thereof, suitable for the solid-phase preparation of oligoamides.

Entry	Monomer	Application	Reference
1	(Boc)FmocHN ... CO_2H	hybrids with peptides as sequence-specific DNA ligands	[190-193]
2	BocHN ... CO_2H	hybrids with peptides as sequence-specific DNA ligands	[192]
3	FmocHN CO_2H R^1CO-N R^2	peptide mimetics	[194,195]
4	N_3 ... CO_2H AcO OAc		[196]
5	R FmocN CO_2H	peptide mimetics	[28]
6	OH HO CO_2H FmocHN O OBn	oligosaccharide mimetics	[197]
7	Boc N CO_2H OH	oligosaccharide mimetics	[198]

16.4.1.1 Oligoglycines

One type of oligoamide which can be readily prepared on supports without the need for any partially protected monomers (which are often tedious and expensive to synthesize) are N-substituted oligoglycines (Figure 16.21). These compounds are prepared by acylating a support-bound amine with bromoacetic acid and then displacing bromide with a primary aliphatic or aromatic amine, followed by a new acylation with bromoacetic acid. Because primary amines are cheap and available in large number, this approach enables the cost-efficient production of large, diverse compound libraries. Alternatively, protected N-substituted glycines can also be prepared in solution and then assembled on insoluble supports (Entry **5**, Table 16.2).

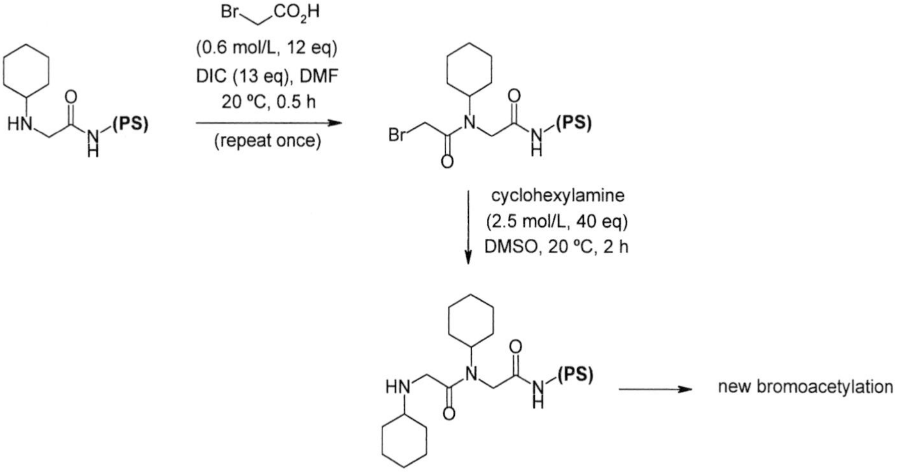

Fig. 16.21. Preparation of N-substituted oligoglycines on cross-linked polystyrene [199].

The scope of the chemistry outlined in Figure 16.21 can be further expanded by including other acids susceptible to irreversible reaction with a primary amine. These can be other halocarboxylic acids, such as (chloromethyl)benzoic acids, or acrylic acid, which reacts with amines to yield N-substituted β-alanines. N-Substituted oligo(β-alanines) have been successfully prepared by sequentially acylating support-bound amines with acrylic acid and then performing a Michael addition with primary amines [200].

16.4.1.2 Peptide Nucleic Acids (PNA)

In 1991 [201] Nielsen and coworkers introduced a new type of DNA-mimetic, called peptide nucleic acid (PNA), in which the backbone of DNA had been replaced by oligo-[*N*-(2-aminoethyl)glycine] (Figure 16.22). Surprisingly, PNA forms very stable Watson–Crick duplexes with DNA. Because PNA is stable towards nucleases, PNA and hybrids of PNA and DNA are convenient DNA-mimetics with a number of potentially interesting applications, such as diagnostic tools or antisense therapeutics.

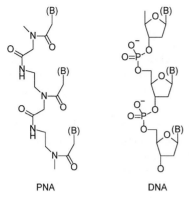

PNA   DNA

Fig. 16.22. Structure of peptide nucleic acids and deoxyribonucleic acids.

PNA can be prepared on insoluble supports either from protected monomers [143,202–204] (e.g. first equation, Figure 16.23), or by assembly on solid phase of the required *N*-(2-aminoethyl)glycines [205] as outlined in Figure 16.23.

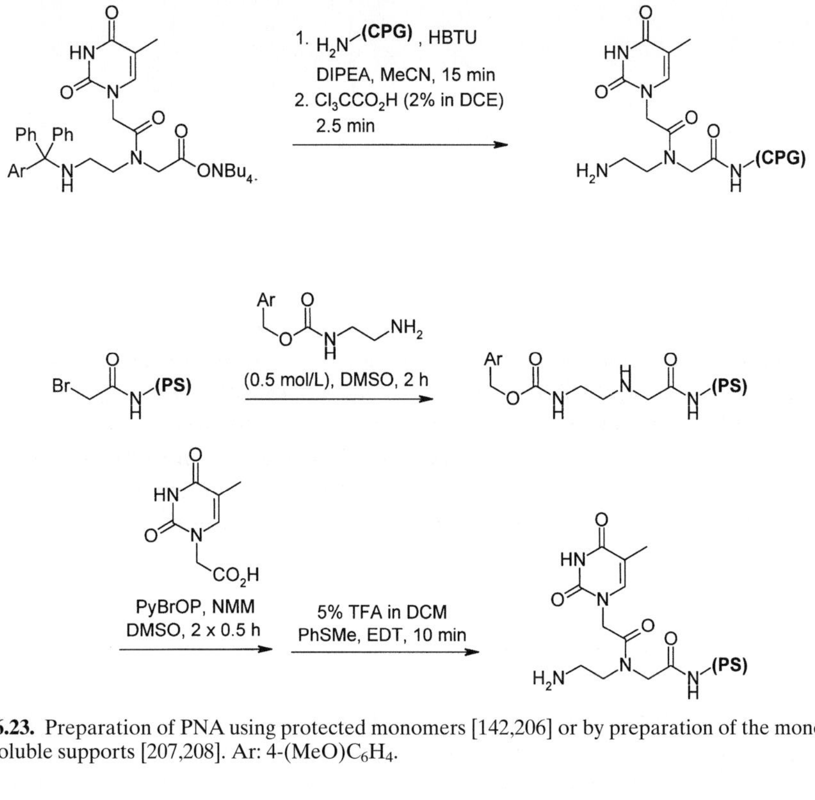

Fig. 16.23. Preparation of PNA using protected monomers [142,206] or by preparation of the monomers on insoluble supports [207,208]. Ar: 4-(MeO)C₆H₄.

16.4.2 Oligoesters

Although oligomers of α-hydroxy acids (depsides) and mixed oligomers of α-hydroxy and α-amino acids (depsipeptides) are found in nature (antibiotics, ion-transporters [209]), little solid-phase methodology has been developed for their preparation. One recent strategy is based on sequential acylation with THP-protected α-hydroxy acids (DIC, DMAP, THF, 2 h; see Entry **6**, Table 13.13), followed by deprotection with TsOH/MeOH [209]. Under these conditions no racemization could be observed. A similar approach to the solid-phase synthesis of depsipeptides is outlined in Figure 16.24.

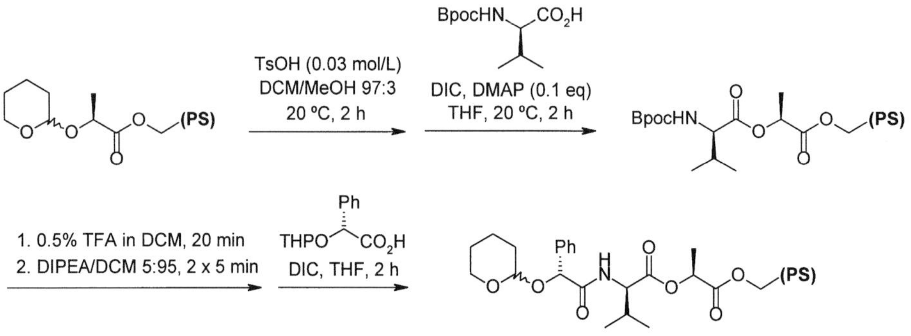

Fig. 16.24. Solid-phase synthesis of depsipeptides [209].

16.4.3 Oligoureas and Oligothioureas

α-Amino acids can be converted into monomers suitable for the solid-phase synthesis of oligoureas, which are interesting peptide mimetics with greater resistance to enzymatic degradation. Two approaches for realizing this concept are outlined in Figure 16.25.

In the first approach an amino acid derived 4-nitrophenyl carbamate is used as the carbamoylating reagent. 4-Nitrophenyl carbamates are generally not very reactive, and carbamoylations with these reagents proceed smoothly only with sufficiently nucleophilic, aliphatic primary or secondary amines.

The second approach uses the more reactive isocyanates, which can be prepared from the corresponding primary amines by treatment with phosgene [210]. Oligomers of hydrazine-derived ureas have been prepared as peptide mimetics on PEG in solution [211]. For this purpose protected pentafluorophenyl carbazates have been used to convert support-bound hydrazines into ureas (Figure 16.25).

Oligothioureas have been prepared on cross-linked polystyrene from Boc-protected ω-aminoisothiocyanates, which were synthesized in solution from symmetric diamines [214]. The formation of thioureas from support-bound amines and isothiocyanates is a rather sluggish reaction (e.g. 3 d, 45 °C, mean coupling yield: 90 % [214]), and only short oligomers could be prepared with this strategy.

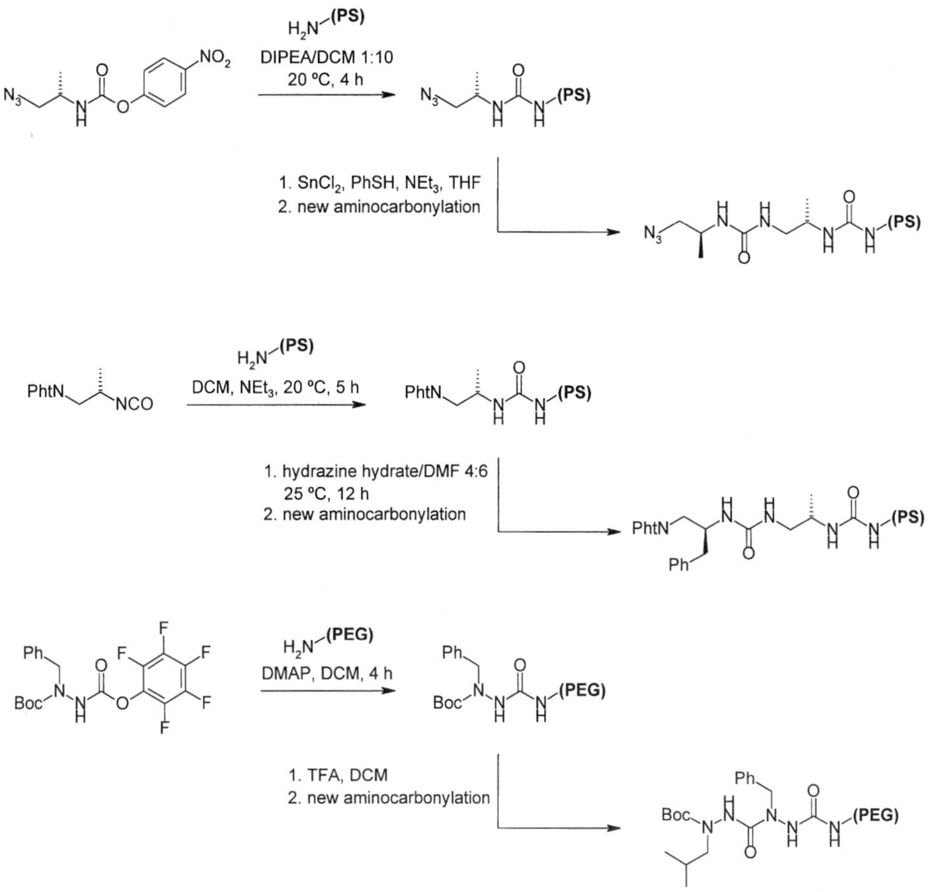

Fig. 16.25. Strategies for the preparation of oligoureas [210–213].

16.4.4 Oligocarbamates

Enantiomerically pure aminoalcohols, which are readily available by reduction of α-amino acids, can be converted into alkoxycarbonylating reagents, suitable for the solid-phase synthesis of oligocarbamates (Figure 16.26). Particularly convenient alkoxycarbonylating reagents are 4-nitrophenyl carbonates, which can be prepared from alcohols and 4-nitrophenyl chloroformate, and which react smoothly with aliphatic primary or secondary amines to yield the corresponding carbamates.

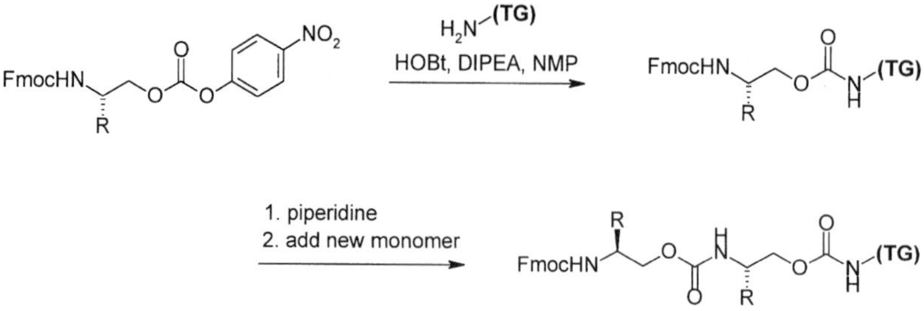

Fig. 16.26. Preparation of oligocarbamates on insoluble supports [215,216].

16.4.5 Oligosulfonamides

Because sulfonamides are generally more stable towards enzymatic degradation than amides, the preparation of oligosulfonamides as peptide mimetics has been thoroughly investigated. The synthesis of α-aminosulfonamides has not yet succeeded, probably because of the instability of this class of compound [217]. For the more stable β-amino sulfonamides and vinylogous α-aminosulfonamides (3-amino-1-propene-1-sulfonamides), however, suitable synthetic routes have been developed. Unfortunately, the preparations of the required monomers are not as easy as those of related, amino acid derived monomers (Figure 16.27).

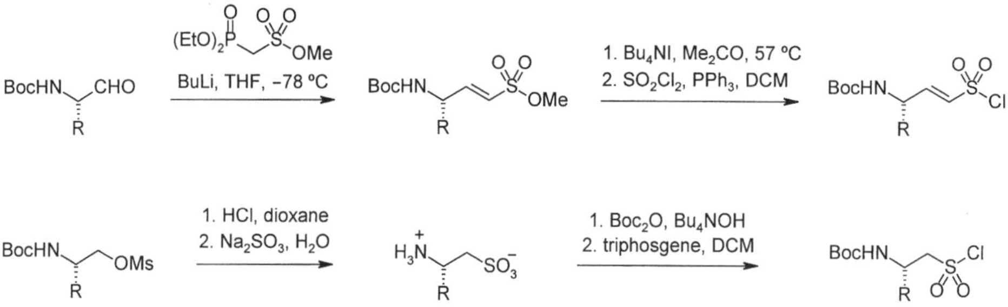

Fig. 16.27. Preparation of amino acid derived, protected aminosulfonyl chlorides suitable for the solid-phase synthesis of oligosulfonamide peptidomimetics [217–219].

The sulfonyl chlorides sketched in Figure 16.27 have been used to prepare peptide mimetics on solid phase, either alone or in combination with protected α-amino acids [219–222] (see Section 8.4).

16.4.6 Oligomeric Phosphoric Acid Derivatives

The synthetic protocols used for the preparation of oligonucleotides on supports can also be used to prepare oligomers from diols other than nucleosides. Symmetric or unsymmetric diols, such as, for instance, N-acylated 4-hydroxyprolinol [223] or cyclopentane-derived diols (carbocyclic deoxyribose analogs [224]), can be selectively mono-tritylated and then converted into a phosphoramidite, suitable for the solid-phase synthesis of oligophosphates. An illustrative synthesis of protected *H*-phosphonates from diols, and their conversion on CPG into oligomeric phosphoramidates, is outlined in Figure 16.28.

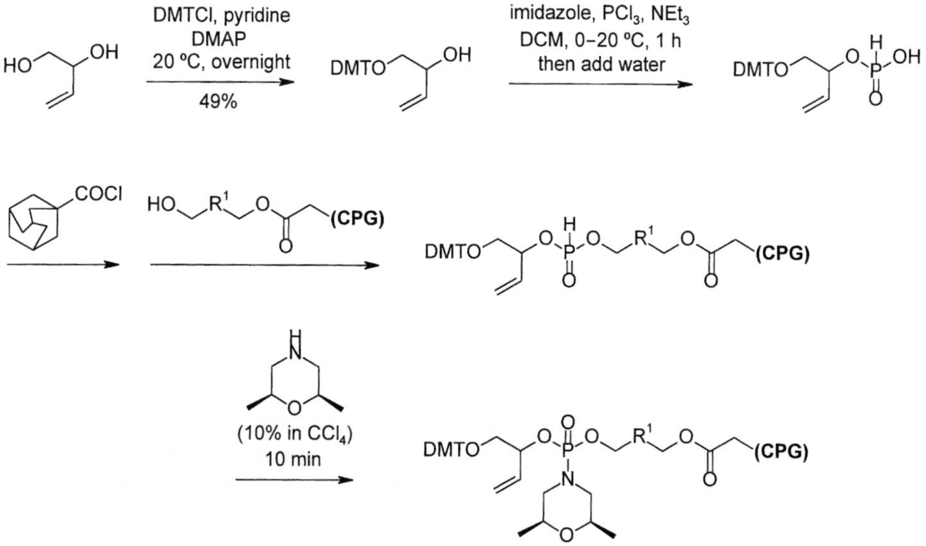

Fig. 16.28. Conversion of diols into protected *H*-phosphonates and the preparation of oligomeric phosphonate diesters and phosphoramidates [118].

16.4.7 Peptide-Derived Oligomeric Compounds

Peptides can be transformed chemically on insoluble supports to yield other types of oligomer. These transformations include N-alkylation and reduction. N-Alkylation of polystyrene-bound peptides enables the preparation of peptide-mimetics with improved enzymatic stability [225]. Polyamines have been prepared by exhaustive reduction of peptides with borane [226] (see Section 10.1.6). N-Alkylated polyamines can be prepared on solid phase by reduction of the corresponding N-alkylated peptides [227].

16.4.8 Oligomers Prepared by C–C Bond Formation

Few examples have been reported of the preparation of oligomers by C–C bond formation. Reggelin described a synthetic protocol which enables the sequential, stereocontrolled assembly of polyketides on cross-linked polystyrene (Figure 16.29; see also [228]). This strategy involves many steps, however, and only short polyketides have so far been prepared. Shorter and more efficient synthetic protocols will probably have to be developed to give access to larger and structurally more complex polyketides.

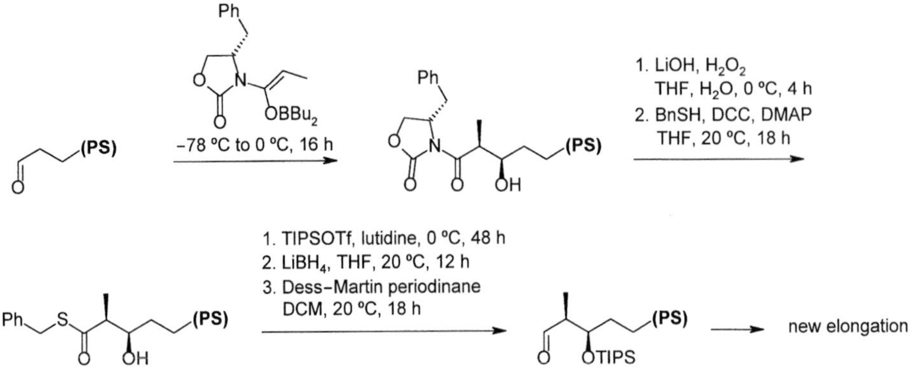

Fig. 16.29. Iterative, stereocontrolled preparation of polyketides on cross-linked polystyrene [229].

Another example of oligomer preparation via C–C bond formation is sketched in Figure 16.30. In this synthesis nitroalkyl phenyl selenides are converted into nitrile oxides in the presence of support-bound terminal alkenes, forming isoxazolines. Oxidative elimination of the selenide yielded a new alkene, which could again be subjected to 1,3-dipolar cycloaddition with a new nitrile oxide. Although this synthesis is short and easy to perform, the cycloadditions proceed with low diastereoselectivity and mixtures of stereoisomers are obtained. Furthermore, these oligomers cannot be readily derivatized with functionalized side chains.

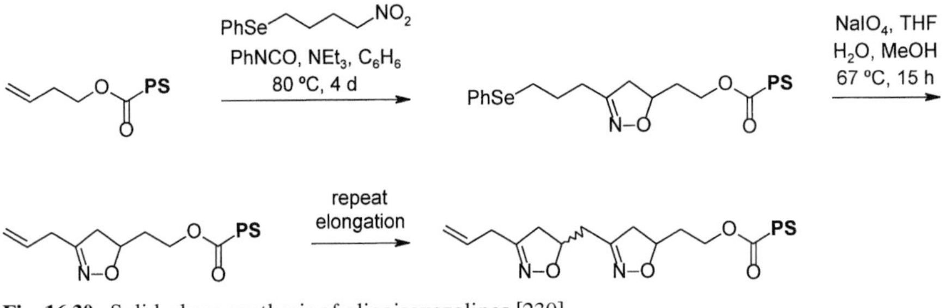

Fig. 16.30. Solid-phase synthesis of oligoisoxazolines [230].

Oligomeric phenylalkynes, which are potentially interesting as liquid crystals [231] or for the construction of molecular electronic devices [232], have been prepared on insoluble supports by the strategy outlined in Figure 16.31. Substituted TMS-protected (ethynyl)iodobenzenes are used as monomers.

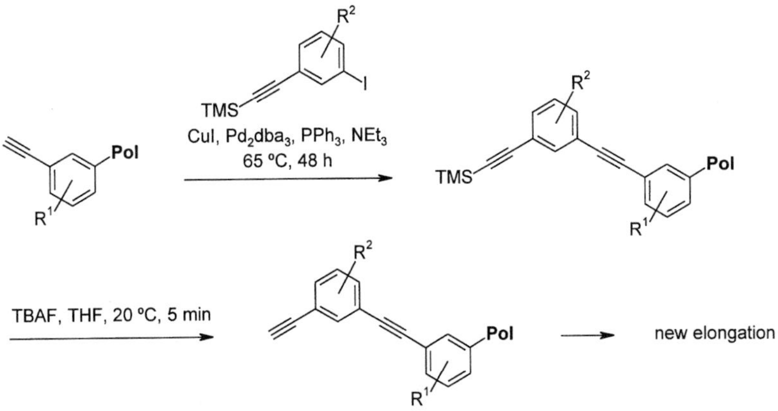

Fig. 16.31. Solid-phase synthesis of oligomeric phenylalkynes [231,232].

16.4.9 Dendrimers

Dendrimers are branched, spherical oligomers which can be prepared both in solution and on supports. Several interesting applications of dendrimers have been proposed including, for instance, their use as carriers for peptides to induce immunogenic response in vivo (multiple antigen peptide system, MAP [16,233]), and as carriers for DNA or other molecules into cells [233]. Dendrimers linked to insoluble supports have also been used to increase the loading of commercially available resins [234]. Some of the common strategies used to prepare dendrimers on supports are outlined in Figure 16.32. Most examples reported are oligoamides of lysine, or oligomers prepared from alkanediamines and acrylic acid ('polyamidoamines'; Figure 16.32). Branched oligonucleotides have also been prepared on insoluble supports [235].

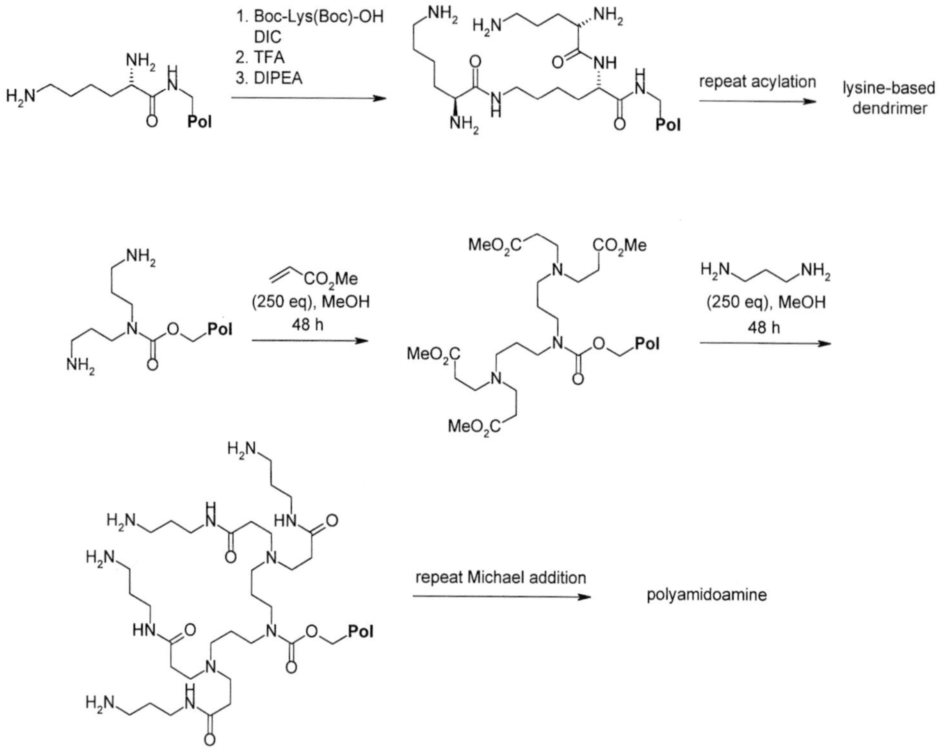

Fig. 16.32. Strategies for the preparation of dendrimers on insoluble supports [16,233,236].

References for Chapter 16

[1] Lam, K. S.; Lebl, M.; Krchnák, V. *Chem. Rev.* **1997**, *97*, 411–448.
[2] Ganesan, A. *Angew. Chem. Int. Ed. Engl.* **1998**, *37*, 2828–2831.
[3] Gordon, E. M.; Gallop, M. A.; Patel, D. V. *Acc. Chem. Res.* **1996**, *29*, 144–154.
[4] Williard, X.; Pop, I.; Bourel, L.; Horváth, D.; Baudelle, R.; Melnyk, P.; Déprez, B.; Tartar, A. *Eur. J. Med. Chem.* **1996**, *31*, 87–98.
[5] Shao, Y. F.; Still, W. C. *J. Org. Chem.* **1996**, *61*, 6086–6087.
[6] Still, W. C. *Acc. Chem. Res.* **1996**, *29*, 155–163.
[7] Cheng, Y. A.; Suenaga, T.; Still, W. C. *J. Am. Chem. Soc.* **1996**, *118*, 1813–1814.
[8] Gennari, C.; Nestler, H. P.; Salom, B.; Still, W. C. *Angew. Chem. Int. Ed. Engl.* **1995**, *34*, 1765–1768.
[9] Yoon, S. S.; Still, W. C. *Tetrahedron* **1995**, *51*, 567–578.
[10] Still, W. C.; Borchardt, A. *J. Am. Chem. Soc.* **1994**, *116*, 373–374.
[11] Fessmann, T.; Kilburn, J. D. *Angew. Chem. Int. Ed. Engl.* **1999**, *38*, 1993–1996.
[12] Merrifield, R. B. *J. Am. Chem. Soc.* **1963**, *85*, 2149–2154.
[13] Letsinger, R. L.; Kornet, M. J. *J. Am. Chem. Soc.* **1963**, *85*, 3045–3046.
[14] Matsueda, R.; Maruyama, H.; Kitazawa, E.; Takahagi, H.; Mukaiyama, T. *J. Am. Chem. Soc.* **1975**, *97*, 2573–2575.
[15] Arshady, R.; Atherton, E.; Clive, D. L. J.; Sheppard, R. C. *J. Chem. Soc. Perkin Trans. 1* **1981**, 529–537.
[16] Novabiochem Catalog and Peptide Synthesis Handbook, Läufelfingen, **1999**.

[17] Advanced Chemtech Handbook of Combinatorial and Solid-Phase Organic Chemistry, Louisville, **1998**.

[18] Mitchell, A. R.; Erickson, B. W.; Ryabtsev, M. N.; Hodges, R. S.; Merrifield, R. B. *J. Am. Chem. Soc.* **1976**, *98*, 7357–7362.

[19] Urban, J.; Vaisar, T.; Shen, R.; Lee, M. S. *Int. J. Pept. Prot. Res.* **1996**, *47*, 182–189.

[20] Beyermann, M.; Bienert, M. *Tetrahedron Lett.* **1992**, *33*, 3745–3748.

[21] Lenard, J. *Chem. Rev.* **1969**, *69*, 625–638.

[22] Jones, J. *The Chemical Synthesis of Peptides*; Oxford University Press: Oxford, **1994**.

[23] Lloyd-Williams, P.; Albericio, F.; Giralt, E. *Tetrahedron* **1993**, *49*, 11065–11133.

[24] Carpino, L. A.; Han, G. Y. *J. Org. Chem.* **1972**, *37*, 3404–3409.

[25] Atherton, E.; Hübscher, W.; Sheppard, R. C.; Woolley, V. *Hoppe-Seyler's Z. Physiol. Chem.* **1981**, *362*, 833–839.

[26] Fields, G. B.; Noble, R. L. *Int. J. Pept. Prot. Res.* **1990**, *35*, 161–214.

[27] Albericio, F.; Pons, M.; Pedroso, E.; Giralt, E. *J. Org. Chem.* **1989**, *54*, 360–366.

[28] Kruijtzer, J. A. W.; Hofmeyer, L. J. F.; Heerma, W.; Versluis, C.; Liskamp, R. M. J. *Chem. Eur. J.* **1998**, *4*, 1570–1580.

[29] Robertson, N.; Ramage, R. *J. Chem. Soc. Perkin Trans. 1* **1999**, 1015–1021.

[30] Raman, P.; Stokes, S. S.; Angell, Y. M.; Flentke, G. R.; Rich, D. H. *J. Org. Chem.* **1998**, *63*, 5734–5735.

[31] Kunz, H. *Pure Appl. Chem.* **1993**, *65*, 1223–1232.

[32] Osborn, H. M. I.; Khan, T. H. *Tetrahedron* **1999**, *55*, 1807–1850.

[33] Nakamura, K.; Ishii, A.; Ito, Y.; Nakahara, Y. *Tetrahedron* **1999**, *55*, 11253–11266.

[34] Habermann, J.; Kunz, H. *Tetrahedron Lett.* **1998**, *39*, 265–268.

[35] Meldal, M.; Auzanneau, F. I.; Hindsgaul, O.; Palcic, M. M. *J. Chem. Soc. Chem. Commun.* **1994**, 1849–1850.

[36] Paulsen, H.; Merz, G.; Peters, S.; Weichert, U. *Liebigs Ann. Chem.* **1990**, 1165–1173.

[37] Christiansen-Brams, I.; Meldal, M.; Bock, K. *J. Chem. Soc. Perkin Trans. 1* **1993**, 1461–1471.

[38] Chadwick, R. J.; Thompson, J. S.; Tomalin, G. *Biochem. Soc. Trans.* **1991**, *19*, 406S.

[39] Hilaire, P. M. S.; Lowary, T. L.; Meldal, M.; Bock, K. *J. Am. Chem. Soc.* **1998**, *120*, 13312–13320.

[40] Atherton, E.; Sheppard, R. C. *Solid Phase Peptide Synthesis; A Practical Approach*; Oxford University Press: Oxford, **1989**.

[41] Simmonds, R. G. *Int. J. Pept. Prot. Res.* **1996**, *47*, 36–41.

[42] Hyde, C.; Johnson, T.; Owen, D.; Quibell, M.; Sheppard, R. C. *Int. J. Pept. Prot. Res.* **1994**, *43*, 431–440.

[43] Johnson, T.; Quibell, M.; Sheppard, R. C. *J. Pept. Sci.* **1995**, *1*, 11–25.

[44] Quibell, M.; Packman, L. C.; Johnson, T. *J. Am. Chem. Soc.* **1995**, *117*, 11656–11668.

[45] Quibell, M.; Turnell, W. G.; Johnson, T. *J. Chem. Soc. Perkin Trans. 1* **1995**, 2019–2024.

[46] Ede, N. J.; Ang, K. H.; James, I. W.; Bray, A. M. *Tetrahedron Lett.* **1996**, *37*, 9097–9100.

[47] McMurray, J. S.; Lewis, C. A. *Tetrahedron Lett.* **1993**, *34*, 8059–8062.

[48] Bourne, G. T.; Meutermans, W. D. F.; Alewood, P. F.; McGeary, R. P.; Scanlon, M.; Watson, A. A.; Smythe, M. L. *J. Org. Chem.* **1999**, *64*, 3095–3101.

[49] Cabrele, C.; Langer, M.; Beck-Sickinger, A. G. *J. Org. Chem.* **1999**, *64*, 4353–4361.

[50] Fridkin, M.; Patchornik, A.; Katchalski, E. *J. Am. Chem. Soc.* **1965**, *87*, 4646–4648.

[51] Flanigan, E.; Marshall, G. R. *Tetrahedron Lett.* **1970**, 2403–2406.

[52] Ehrlich, A.; Heyne, H.-U.; Winter, R.; Beyermann, M.; Haber, H.; Carpino, L. A.; Bienert, M. *J. Org. Chem.* **1996**, *61*, 8831–8838.

[53] Klose, J.; El-Faham, A.; Henklein, P.; Carpino, L. A.; Bienert, M. *Tetrahedron Lett.* **1999**, *40*, 2045–2048.

[54] Alsina, J.; Rabanal, F.; Chiva, C.; Giralt, E.; Albericio, F. *Tetrahedron* **1998**, *54*, 10125–10152.

[55] Spatola, A. F.; Darlak, K.; Romanovskis, P. *Tetrahedron Lett.* **1996**, *37*, 591–594.

[56] Eichler, J.; Lucka, A. W.; Pinilla, C.; Houghten, R. A. *Mol. Diversity* **1996**, *1*, 233–240.

[57] Bloomberg, G. B.; Askin, D.; Gargaro, A. R.; Tanner, M. J. A. *Tetrahedron Lett.* **1993**, *34*, 4709–4712.

[58] Aletras, A.; Barlos, K.; Gatos, D.; Koutsogianni, S.; Mamos, P. *Int. J. Pept. Prot. Res.* **1995**, *45*, 488–496.

[59] Romanovskis, P.; Spatola, A. F. *J. Pept. Res.* **1998**, *52*, 356–374.

[60] Lanter, C. L.; Guiles, J. W.; Rivero, R. A. *Mol. Diversity* **1999**, *4*, 149–153.

[61] Annis, D. A.; Helluin, O.; Jacobsen, E. N. *Angew. Chem. Int. Ed. Engl.* **1998**, *37*, 1907–1909.

[62] Nishino, N.; Xu, M.; Mihara, H.; Fujimoto, T.; Ohba, M.; Ueno, Y.; Kumagai, H. *J. Chem. Soc. Chem. Commun.* **1992**, 180–181.

[63] Ösapay, G.; Profit, A.; Taylor, J. W. *Tetrahedron Lett.* **1990**, *31*, 6121–6124.

[64] Ösapay, G.; Taylor, J. W. *J. Am. Chem. Soc.* **1990**, *112*, 6046–6051.
[65] Mihara, H.; Yamabe, S.; Niidome, T.; Aoyagi, H.; Kumagai, H. *Tetrahedron Lett.* **1995**, *36*, 4837–4840.
[66] Sabatino, G.; Chelli, M.; Mazzucco, S.; Ginanneschi, M.; Papini, A. M. *Tetrahedron Lett.* **1999**, *40*, 809–812.
[67] Crozet, Y.; Wen, J. J.; Loo, R. O.; Andrews, P. C.; Spatola, A. F. *Mol. Diversity* **1998**, *3*, 261–276.
[68] Eichler, J.; Lucka, A. W.; Houghten, R. A. *Pept. Res.* **1994**, *7*, 300–307.
[69] Eckstein, F. *Oligonucleotides and Analogues; A Practical Approach*; Oxford University Press: Oxford, **1991**.
[70] Ghosh, P. K.; Kumar, P.; Gupta, K. C. *J. Indian Chem. Soc.* **1998**, *75*, 206–218.
[71] Engels, J. W.; Uhlmann, E. *Angew. Chem. Int. Ed. Engl.* **1989**, *28*, 716–734.
[72] Trotta, P. P.; Beutel, B. A.; Sherman, M. I. *Med. Res. Rev.* **1995**, *15*, 277–298.
[73] Uhlmann, E.; Peyman, A. *Chem. Rev.* **1990**, *90*, 543–584.
[74] Letsinger, R. L.; Mahadevan, V. *J. Am. Chem. Soc.* **1965**, *87*, 3526–3527.
[75] Hayatsu, H.; Khorana, H. G. *J. Am. Chem. Soc.* **1966**, *88*, 3182–3183.
[76] Cramer, F.; Helbig, R.; Hettler, H.; Scheit, K. H.; Seliger, H. *Angew. Chem.* **1966**, *78*, 640–641.
[77] Hayatsu, H.; Khorana, H. G. *J. Am. Chem. Soc.* **1967**, *89*, 3880–3887.
[78] Köster, H.; Biernat, J.; McManus, J.; Wolter, A.; Stumpe, A.; Narang, C. K.; Sinha, N. D. *Tetrahedron* **1984**, *40*, 103–112.
[79] Kusama, T.; Hayatsu, H. *Chem. Pharm. Bull.* **1970**, *18*, 319–327.
[80] Köster, H.; Cramer, F. *Liebigs Ann. Chem.* **1974**, 946–958.
[81] Köster, H. *Tetrahedron Lett.* **1972**, 1527–1530.
[82] Seliger, H.; Aumann, G. *Tetrahedron Lett.* **1973**, 2911–2914.
[83] Schott, H.; Brandstetter, F.; Bayer, E. *Makromol. Chem.* **1973**, *173*, 247–251.
[84] Chapman, T. M.; Kleid, D. G. *J. Chem. Soc. Chem. Commun.* **1973**, 193–194.
[85] Brandstetter, F.; Schott, H.; Bayer, E. *Tetrahedron Lett.* **1973**, 2997–3000.
[86] Gait, M. J.; Sheppard, R. C. *Nucleic Acids Res.* **1977**, *4*, 1135–1158.
[87] Catlin, J. C.; Cramer, F. *J. Org. Chem.* **1973**, *38*, 245–250.
[88] Itakura, K.; Bahl, C. P.; Katagiri, N.; Michniewicz, J. J.; Wightman, R. H.; Narang, S. A. *Can. J. Chem.* **1973**, *51*, 3649–3651.
[89] Dembek, P.; Miyoshi, K.; Itakura, K. *J. Am. Chem. Soc.* **1981**, *103*, 706–708.
[90] Reese, C. B. *Tetrahedron* **1978**, *34*, 3143–3179.
[91] Ito, H.; Ike, Y.; Ikuta, S.; Itakura, K. *Nucleic Acids Res.* **1982**, *10*, 1755–1769.
[92] Sekine, M.; Hata, T. *Curr. Org. Chem.* **1999**, *3*, 25–66.
[93] Gait, M. J.; Matthes, H. W. D.; Singh, M.; Sproat, B. S.; Titmas, R. C. *Nucleic Acids Res.* **1982**, *10*, 6243–6254.
[94] Witkowski, W.; Birch-Hirschfeld, E.; Weiss, R.; Zarytova, V. F.; Gorn, V. V. *J. Prakt. Chem.* **1984**, *326*, 320–328.
[95] Gough, G. R.; Brunden, M. J.; Gilham, P. T. *Tetrahedron Lett.* **1981**, *22*, 4177–4180.
[96] Letsinger, R. L.; Finnan, J. L.; Heavner, G. A.; Lunsford, W. B. *J. Am. Chem. Soc.* **1975**, *97*, 3278–3279.
[97] Ogilvie, K. K.; Nemer, M. J. *Tetrahedron Lett.* **1980**, *21*, 4159–4162.
[98] Matteucci, M. D.; Caruthers, M. H. *Tetrahedron Lett.* **1980**, *21*, 719–722.
[99] Matteucci, M. D.; Caruthers, M. H. *J. Am. Chem. Soc.* **1981**, *103*, 3185–3191.
[100] Beaucage, S. L.; Caruthers, M. H. *Tetrahedron Lett.* **1981**, *22*, 1859–1862.
[101] Sinha, N. D.; Biernat, J.; McManus, J.; Köster, H. *Nucleic Acids Res.* **1984**, *12*, 4539–4557.
[102] Adams, S. P.; Kavka, K. S.; Wykes, E. J.; Holder, S. B.; Galluppi, G. R. *J. Am. Chem. Soc.* **1983**, *105*, 661–663.
[103] Beier, M.; Pfleiderer, W. *Helv. Chim. Acta* **1999**, *82*, 633–644.
[104] McCollum, C.; Andrus, A. *Tetrahedron Lett.* **1991**, *32*, 4069–4072.
[105] Reddy, M. P.; Michael, M. A.; Farooqui, F.; Girgis, S. *Tetrahedron Lett.* **1994**, *35*, 5771–5774.
[106] Birch-Hirschfeld, E.; Földes-Papp, Z.; Gührs, K. H.; Seliger, H. *Helv. Chim. Acta* **1996**, *79*, 137–150.
[107] Birch-Hirschfeld, E.; Eickhoff, H.; Stelzner, A.; Greulich, K. O.; Földes-Papp, Z.; Seliger, H.; Gührs, K. H. *Collect. Czech. Chem. Commun.* **1996**, *61*, S311–S314.
[108] Micura, R. *Chem. Eur. J.* **1999**, *5*, 2077–2082.
[109] Wright, P.; Lloyd, D.; Rapp, W.; Andrus, A. *Tetrahedron Lett.* **1993**, *34*, 3373–3376.
[110] Bayer, E.; Bleicher, K.; Maier, M. *Z. Naturforschung Sect. B* **1995**, *50*, 1096–1100.
[111] Gao, H.; Gaffney, B. L.; Jones, R. A. *Tetrahedron Lett.* **1991**, *32*, 5477–5480.
[112] Bonora, G. M.; Biancotto, G.; Maffini, M.; Scremin, C. L. *Nucleic Acids Res.* **1993**, *21*, 1213–1217.
[113] Wörl, R.; Köster, H. *Tetrahedron* **1999**, *55*, 2941–2956.

[114] Bonora, G. M.; Baldan, A.; Schiavon, O.; Ferruti, P.; Veronese, F. M. *Tetrahedron Lett.* **1996**, *37*, 4761–4764.
[115] Hayakawa, Y.; Kataoka, M. *J. Am. Chem. Soc.* **1998**, *120*, 12395–12401.
[116] Frieden, M.; Grandas, A.; Pedroso, E. *Chem. Commun.* **1999**, 1593–1594.
[117] Wada, T.; Mochizuki, A.; Sato, Y.; Sekine, M. *Tetrahedron Lett.* **1998**, *39*, 5593–5596.
[118] Fathi, R.; Rudolph, M. J.; Gentles, R. G.; Patel, R.; MacMillan, E. W.; Reitman, M. S.; Pelham, D.; Cook, A. F. *J. Org. Chem.* **1996**, *61*, 5600–5609.
[119] Sergueev, D. S.; Shaw, B. R. *J. Am. Chem. Soc.* **1998**, *120*, 9417–9427.
[120] Pon, R. T.; Ogilvie, K. K. *Tetrahedron Lett.* **1984**, *25*, 713–716.
[121] Stutz, A.; Pitsch, S. *Synlett* **1999**, 930–934.
[122] Ecker, D. J.; Vickers, T. A.; Hanecak, R.; Driver, V.; Anderson, K. *Nucleic Acids Res.* **1993**, *21*, 1853–1856.
[123] Schneider, D. J.; Feigon, J.; Hostomsky, Z.; Gold, L. *Biochemistry* **1995**, *34*, 9599–9610.
[124] Seeberger, P. H.; Caruthers, M. H.; Bankaitis-Davis, D.; Beaton, G. *Tetrahedron* **1999**, *55*, 5759–5772.
[125] Yang, X. B.; Sierzchala, A.; Misiura, K.; Niewiarowski, W.; Sochacki, M.; Stec, W. J.; Wieczorek, M. W. *J. Org. Chem.* **1998**, *63*, 7097–7100.
[126] Zhang, Z.; Nichols, A.; Tang, J. X.; Han, Y.; Tang, J. Y. *Tetrahedron Lett.* **1999**, *40*, 2095–2098.
[127] Guo, M. J.; Yu, D.; Iyer, R. P.; Agrawal, S. *Bioorg. Med. Chem. Lett.* **1998**, *8*, 2539–2544.
[128] Ecker, D. J.; Wyatt, J. R.; Vickers, T.; Buckheit, R.; Roberson, J.; Imbach, J. L. *Nucleosides and Nucleotides* **1995**, *14*, 1117–1127.
[129] Ecker, D. J.; Wyatt, J. R.; Vickers, T.; Buckheit, R.; Roberson, J.; Imbach, J. L. *Nucleosides and Nucleotides* **1995**, *14*, 1117–1127.
[130] Le Bec, C.; Wickstrom, E. *J. Org. Chem.* **1996**, *61*, 510–513.
[131] Fathi, R.; Patel, R.; Cook, A. F. *Mol. Diversity* **1997**, *2*, 125–134.
[132] Lin, J.; Shaw, B. R. *Chem. Commun.* **1999**, 1517–1518.
[133] Linkletter, B. A.; Szabo, I. E.; Bruice, T. C. *J. Am. Chem. Soc.* **1999**, *121*, 3888–3896.
[134] Morvan, F.; Sanghvi, Y. S.; Perbost, M.; Vasseur, J. J.; Bellon, L. *J. Am. Chem. Soc.* **1996**, *118*, 255–256.
[135] Khan, S. I.; Grinstaff, M. W. *J. Am. Chem. Soc.* **1999**, *121*, 4704–4705.
[136] Davis, P. W.; Vickers, T. A.; Wilson-Lingardo, L.; Wyatt, J. R.; Guinosso, C. J.; Sanghvi, Y. S.; De Baets, E. A.; Acevedo, O. L.; Cook, P. D.; Ecker, D. J. *J. Med. Chem.* **1995**, *38*, 4363–4366.
[137] Bergmann, F.; Bannwarth, W. *Tetrahedron Lett.* **1995**, *36*, 1839–1842.
[138] Basu, S.; Wickstrom, E. *Tetrahedron Lett.* **1995**, *36*, 4943–4946.
[139] de la Torre, B. G.; Aviñó, A.; Tarrason, G.; Piulats, J.; Albericio, F.; Eritja, R. *Tetrahedron Lett.* **1994**, *35*, 2733–2736.
[140] Juby, C. D.; Richardson, C. D.; Brousseau, R. *Tetrahedron Lett.* **1991**, *32*, 879–882.
[141] Adinolfi, M.; Barone, G.; De Napoli, L.; Guariniello, L.; Iadonisi, A.; Piccialli, G. *Tetrahedron Lett.* **1999**, *40*, 2607–2610.
[142] van der Laan, A. C.; Meeuwenoord, N. J.; Kuyl-Yeheskiely, E.; Oosting, R. S.; Brands, R.; van Boom, J. H. *Recl. Trav. Chim. Pays-Bas* **1995**, *114*, 295–297.
[143] Bergmann, F.; Bannwarth, W.; Tam, S. *Tetrahedron Lett.* **1995**, *36*, 6823–6826.
[144] McMinn, D. L.; Hirsch, R.; Greenberg, M. M. *Tetrahedron Lett.* **1998**, *39*, 4155–4158.
[145] Peyman, A.; Weiser, C.; Uhlmann, E. *Bioorg. Med. Chem. Lett.* **1995**, *5*, 2469–2472.
[146] Yoo, D. J.; Greenberg, M. M. *J. Org. Chem.* **1995**, *60*, 3358–3364.
[147] McMinn, D. L.; Greenberg, M. M. *Tetrahedron* **1996**, *52*, 3827–3840.
[148] Hovinen, J.; Guzaev, A.; Azhayev, A.; Lönnberg, H. *Tetrahedron* **1994**, *50*, 7203–7218.
[149] Guzaev, A.; Lönnberg, H. *Tetrahedron* **1999**, *55*, 9101–9116.
[150] Frank, R.; Meyerhans, A.; Schwellnus, K.; Blöcker, H. *Methods Enzymol.* **1987**, *154*, 221–249.
[151] Frank, R.; Heikens, W.; Heisterberg-Moutsis, G.; Blöcker, H. *Nucleic Acids Res.* **1983**, *11*, 4365–4377.
[152] Ott, J.; Eckstein, F. *Nucleic Acids Res.* **1984**, *12*, 9137–9142.
[153] Markiewicz, W. T.; Markiewicz, M.; Astriab, A.; Godzina, P. *Collect. Czech. Chem. Commun.* **1996**, *61*, S315–S318.
[154] Dittrich, F.; Tegge, W.; Frank, R. *Bioorg. Med. Chem. Lett.* **1998**, *8*, 2351–2356.
[155] Ito, Y.; Manabe, S. *Curr. Opinion Chem. Biol.* **1998**, *2*, 701–708.
[156] Seeberger, P. H.; Danishefsky, S. J. *Acc. Chem. Res.* **1998**, *31*, 685–695.
[157] Guthrie, R. D.; Jenkins, A. D.; Roberts, G. A. F. *J. Chem. Soc. Perkin Trans. 1* **1973**, 2414–2417.
[158] Excoffier, G.; Gagnaire, D.; Utille, J. P.; Vignon, M. *Tetrahedron Lett.* **1972**, 5065–5068.
[159] Fréchet, J. M.; Schuerch, C. *J. Am. Chem. Soc.* **1971**, *93*, 492–496.
[160] Zehavi, U.; Patchornik, A. *J. Am. Chem. Soc.* **1973**, *95*, 5673–5677.

446 *16 Preparation of Oligomeric Compounds*

[161] Chiu, S. H. L.; Anderson, L. *Carbohyd. Res.* **1976**, *50*, 227–238.
[162] Yan, L.; Taylor, C. M.; Goodnow, R.; Kahne, D. *J. Am. Chem. Soc.* **1994**, *116*, 6953–6954.
[163] Silva, D. J.; Wang, H.; Allanson, N. M.; Jain, R. K.; Sofia, M. J. *J. Org. Chem.* **1999**, *64*, 5926–5929.
[164] Sofia, M. J.; Allanson, N.; Hatzenbuhler, N. T.; Jain, R.; Kakarla, R.; Kogan, N.; Liang, R.; Liu, D.; Silva, D. J.; Wang, H.; Gange, D.; Anderson, J.; Chen, A.; Chi, F.; Dulina, R.; Huang, B.; Kamau, M.; Wang, C.; Baizman, E.; Branstrom, A.; Bristol, N.; Goldman, R.; Han, K.; Longley, C.; Midha, S.; Axelrod, H. R. *J. Med. Chem.* **1999**, *42*, 3193–3198.
[165] Liang, R.; Yan, L.; Loebach, J.; Ge, M.; Uozumi, Y.; Sekanina, K.; Horan, N.; Gildersleeve, J.; Thompson, C.; Smith, A.; Biswas, K.; Still, W. C.; Kahne, D. *Science* **1996**, *274*, 1520–1522.
[166] Wang, Y.; Zhang, H.; Voelter, W. *Chem. Lett.* **1995**, 273–274.
[167] Nicolaou, K. C.; Winssinger, N.; Pastor, J.; DeRoose, F. *J. Am. Chem. Soc.* **1997**, *119*, 449–450.
[168] Doi, T.; Sugiki, M.; Yamada, H.; Takahashi, T.; Porco, J. A. *Tetrahedron Lett.* **1999**, *40*, 2141–2144.
[169] Zheng, C.; Seeberger, P. H.; Danishefsky, S. J. *J. Org. Chem.* **1998**, *63*, 1126–1130.
[170] Zhu, T.; Boons, G. J. *Angew. Chem. Int. Ed. Engl.* **1998**, *37*, 1898–1900.
[171] Ito, Y.; Kanie, O.; Ogawa, T. *Angew. Chem. Int. Ed. Engl.* **1996**, *35*, 2510–2512.
[172] Rodebaugh, R.; Joshi, S.; Fraser-Reid, B.; Geysen, H. M. *J. Org. Chem.* **1997**, *62*, 5660–5661.
[173] Heckel, A.; Mross, E.; Jung, K. H.; Rademann, J.; Schmidt, R. R. *Synlett* **1998**, 171–173.
[174] Rademann, J.; Schmidt, R. R. *J. Org. Chem.* **1997**, *62*, 3650–3653.
[175] Shimizu, H.; Ito, Y.; Kanie, O.; Ogawa, T. *Bioorg. Med. Chem. Lett.* **1996**, *6*, 2841–2846.
[176] Paulsen, H.; Schleyer, A.; Mathieux, N.; Meldal, M.; Bock, K. *J. Chem. Soc. Perkin Trans. 1* **1997**, 281–293.
[177] Adinolfi, M.; Barone, G.; De Napoli, L.; Iadonisi, A.; Piccialli, G. *Tetrahedron Lett.* **1998**, *39*, 1953–1956.
[178] Adinolfi, M.; Barone, G.; De Napoli, L.; Iadonisi, A.; Piccialli, G. *Tetrahedron Lett.* **1996**, *37*, 5007–5010.
[179] Douglas, S. P.; Whitfield, D. M.; Krepinsky, J. J. *J. Am. Chem. Soc.* **1995**, *117*, 2116–2117.
[180] Wang, Z. G.; Douglas, S. P.; Krepinsky, J. J. *Tetrahedron Lett.* **1996**, *37*, 6985–6988.
[181] Hunt, J. A.; Roush, W. R. *J. Am. Chem. Soc.* **1996**, *118*, 9998–9999.
[182] Savin, K. A.; Woo, J. C. G.; Danishefsky, S. J. *J. Org. Chem.* **1999**, *64*, 4183–4186.
[183] Danishefsky, S. J.; McClure, K. F.; Randolph, J. T.; Ruggeri, R. B. *Science* **1993**, *260*, 1307–1309.
[184] Zheng, C. S.; Seeberger, P. H.; Danishefsky, S. J. *Angew. Chem. Int. Ed. Engl.* **1998**, *37*, 786–789.
[185] Schuster, M.; Wang, P.; Paulson, J. C.; Wong, C. H. *J. Am. Chem. Soc.* **1994**, *116*, 1135–1136.
[186] Yamada, K.; Nishimura, S. I. *Tetrahedron Lett.* **1995**, *36*, 9493–9496.
[187] Blixt, O.; Norberg, T. *J. Org. Chem.* **1998**, *63*, 2705–2710.
[188] Moran, E. J.; Wilson, T. E.; Cho, C. Y.; Cherry, S. R.; Schultz, P. G. *Biopolymers* **1995**, *37*, 213–219.
[189] Liskamp, R. M. J. *Angew. Chem. Int. Ed. Engl.* **1994**, *33*, 633–636.
[190] Vázquez, E.; Caamaño, A. M.; Castedo, L.; Mascareñas, J. L. *Tetrahedron Lett.* **1999**, *40*, 3621–3624.
[191] Herman, D. M.; Turner, J. M.; Baird, E. E.; Dervan, P. B. *J. Am. Chem. Soc.* **1999**, *121*, 1121–1129.
[192] Baird, E. E.; Dervan, P. B. *J. Am. Chem. Soc.* **1996**, *118*, 6141–6146.
[193] König, B.; Rödel, M. *Chem. Commun.* **1998**, 605–606.
[194] Rivier, J. E.; Jiang, G. C.; Koerber, S. C.; Porter, J.; Simon, L.; Craig, A. G.; Hoeger, C. A. *Proc. Natl. Acad. Sci. USA* **1996**, *93*, 2031–2036.
[195] Fletcher, M. D.; Campbell, M. M. *Chem. Rev.* **1998**, *98*, 763–795.
[196] Long, D. D.; Smith, M. D.; Marquess, D. G.; Claridge, T. D. W.; Fleet, G. W. J. *Tetrahedron Lett.* **1998**, *39*, 9293–9296.
[197] Müller, C.; Kitas, E.; Wessel, H. P. *J. Chem. Soc. Chem. Commun.* **1995**, 2425–2426.
[198] Byrgesen, E.; Nielsen, J.; Willert, M.; Bols, M. *Tetrahedron Lett.* **1997**, *38*, 5697–5700.
[199] Zuckermann, R. N.; Kerr, J. M.; Kent, S. B. H.; Moos, W. H. *J. Am. Chem. Soc.* **1992**, *114*, 10646–10647.
[200] Hamper, B. C.; Kolodziej, S. A.; Scates, A. M.; Smith, R. G.; Cortez, E. *J. Org. Chem.* **1998**, *63*, 708–718.
[201] Nielsen, P. E.; Egholm, M.; Berg, R. H.; Buchardt, O. *Science* **1991**, *254*, 1497–1500.
[202] Will, D. W.; Langner, D.; Knolle, J.; Uhlmann, E. *Tetrahedron* **1995**, *51*, 12069–12082.
[203] Seitz, O. *Tetrahedron Lett.* **1999**, *40*, 4161–4164.
[204] Egholm, M.; Nielsen, P. E.; Buchardt, O.; Berg, R. H. *J. Am. Chem. Soc.* **1992**, *114*, 9677–9678.
[205] Seitz, O.; Bergmann, F.; Heindl, D. *Angew. Chem. Int. Ed. Engl.* **1999**, *38*, 2203–2206.
[206] Stetsenko, D. A.; Lubyako, E. N.; Potapov, V. K.; Azhikina, T. L.; Sverdlov, E. D. *Tetrahedron Lett.* **1996**, *37*, 3571–3574.
[207] Richter, L. S.; Zuckermann, R. N. *Bioorg. Med. Chem. Lett.* **1995**, *5*, 1159–1162.
[208] Aldrian-Herrada, G.; Rabié, A.; Wintersteiger, R.; Brugidou, J. *J. Pept. Sci.* **1998**, *4*, 266–281.

[209] Kuisle, O.; Quiñoá, E.; Riguera, R. *Tetrahedron Lett.* **1999**, *40*, 1203–1206.
[210] Burgess, K.; Ibarzo, J.; Linthicum, D. S.; Russell, D. H.; Shin, H.; Shitangkoon, A.; Totani, R.; Zhang, A. J. *J. Am. Chem. Soc.* **1997**, *119*, 1556–1564.
[211] Han, H.; Janda, K. D. *J. Am. Chem. Soc.* **1996**, *118*, 2539–2544.
[212] Kim, J. M.; Bi, Y. Z.; Paikoff, S. J.; Schultz, P. G. *Tetrahedron Lett.* **1996**, *37*, 5305–5308.
[213] Boeijen, A.; Liskamp, R. M. J. *Eur. J. Org. Chem.* **1999**, 2127–2135.
[214] Smith, J.; Liras, J. L.; Schneider, S. E.; Anslyn, E. V. *J. Org. Chem.* **1996**, *61*, 8811–8818.
[215] Wendeborn, S.; De Mesmaeker, A.; Brill, W. K. D. *Synlett* **1998**, 865–868.
[216] Cho, C. Y.; Moran, E. J.; Cherry, S. R.; Stephans, J. C.; Fodor, S. P. A.; Adams, C. L.; Sundaram, A.; Jacobs, J. W.; Schultz, P. G. *Science* **1993**, *261*, 1303–1305.
[217] Gennari, C.; Salom, B.; Potenza, D.; Williams, A. *Angew. Chem. Int. Ed. Engl.* **1994**, *33*, 2067–2069.
[218] Gude, M.; Piarulli, U.; Potenza, D.; Salom, B.; Gennari, C. *Tetrahedron Lett.* **1996**, *37*, 8589–8592.
[219] de Bont, D. B. A.; Dijkstra, G. D. H.; den Hartog, J. A. J.; Liskamp, R. M. J. *Bioorg. Med. Chem. Lett.* **1996**, *6*, 3035–3040.
[220] Gennari, C.; Nestler, H. P.; Salom, B.; Still, W. C. *Angew. Chem. Int. Ed. Engl.* **1995**, *34*, 1763–1765.
[221] de Bont, D. B. A.; Moree, W. J.; Liskamp, R. M. J. *Bioorg. Med. Chem.* **1996**, *4*, 667–672.
[222] Gennari, C.; Longari, C.; Ressel, S.; Salom, B.; Piarulli, U.; Ceccarelli, S.; Mielgo, A. *Eur. J. Org. Chem.* **1998**, 2437–2449.
[223] Hébert, N.; Davis, P. W.; De Baets, E. L.; Acevedo, O. L. *Tetrahedron Lett.* **1994**, *35*, 9509–9512.
[224] Domínguez, B. M.; Cullis, P. M. *Tetrahedron Lett.* **1999**, *40*, 5783–5786.
[225] Dörner, B.; Husar, G. M.; Ostresh, J. M.; Houghten, R. A. *Bioorg. Med. Chem.* **1996**, *4*, 709–715.
[226] Ostresh, J. M.; Schoner, C. C.; Hamashin, V. T.; Nefzi, A.; Meyer, J. P.; Houghten, R. A. *J. Org. Chem.* **1998**, *63*, 8622–8623.
[227] Nefzi, A.; Ostresh, J. M.; Houghten, R. A. *Tetrahedron* **1999**, *55*, 335–344.
[228] Gennari, C.; Ceccarelli, S.; Piarulli, U.; Aboutayab, K.; Donghi, M.; Paterson, I. *Tetrahedron* **1998**, *54*, 14999–15016.
[229] Reggelin, M.; Brenig, V.; Welcker, R. *Tetrahedron Lett.* **1998**, *39*, 4801–4804.
[230] Kurth, M. J.; Randall, L. A. A.; Takenouchi, K. *J. Org. Chem.* **1996**, *61*, 8755–8761.
[231] Nelson, J. C.; Young, J. K.; Moore, J. S. *J. Org. Chem.* **1996**, *61*, 8160–8168.
[232] Huang, S.; Tour, J. M. *J. Am. Chem. Soc.* **1999**, *121*, 4908–4909.
[233] Wells, N. J.; Basso, A.; Bradley, M. *Biopolymers* **1998**, *47*, 381–396.
[234] Mahajan, A.; Chhabra, S. R.; Chan, W. C. *Tetrahedron Lett.* **1999**, *40*, 4909–4912.
[235] Hudson, R. H. E.; Robidoux, S.; Damha, M. J. *Tetrahedron Lett.* **1998**, *39*, 1299–1302.
[236] Swali, V.; Wells, N. J.; Langley, G. J.; Bradley, M. *J. Org. Chem.* **1997**, *62*, 4902–4903.

17 Index

F

G

O

Olefin metathesis
 - cross   109, 156, 157, 159
 - ring–closing   108–110, 157, 158, 346, 381, 400, 403
 - as cleavage reaction   108–110, 381, 400, 403
Olefins, see Alkenes
Oligoamides   433–435
Oligocarbamates   437, 438
Oligoesters   436
Oligoglycines   434
Oligoisoxazolines   440
Oligonucleotide analogs   428
Oligonucleotides
 - linkers for   91, 93, 95
 - photocleavable linkers for   95
 - preparation of   422–428
 - protection of   425
Oligophosphates   438, 439
Oligoribonucleotides   427, 428
Oligosaccharides   428–432
Oligosulfonamides   438
Oligothiophenes   360
Oligothioureas   436
Oligoureas   436, 437
On–bead screening   3
On–resin analysis   5–8
Organoaluminum compounds, as catalysts for aminolysis of esters   60
Organoboron compounds, see Boranes
Organocadmium compounds   277
Organocalcium compounds   133–135
Organocopper compounds
 - 1,4–addition of   278, 281, 316, 381
 - alkylation of   146
 - preparation of   133–135
Organogermanium compounds, as linkers   98, 113
Organolithium compounds
 - preparation of   133–136, 191, 192, 215
 - reaction of
 - with alkyl halides   135, 145, 146, 166
 - with amides   276
 - with arenes   133, 134
 - with aryl halides   133–135, 281
 - with carbonyl compounds   187–191
 - with chlorosilanes   137
 - with disulfides   215
 - with imines   244

 - with Merrifield resin   135
 - with oxygen   192
 - with sulfur   209, 210
 - with sulfur dioxide   216
Organomagnesium compounds, see Grignard reagents
Organomanganese compounds   187, 189
Organomercury compounds
 - attachment to supports   136
 - as radical precursor   149, 150
Organometallic compounds
 - acylation of   275–277
 - alkylation of   145–149, 166, 187–192
 - arylation of   162, 163, 166, 168–171
 - hydrolysis of   111–114, 143
 - preparation of   133–139, 191, 192
 - vinylation of   162, 163
Organoselenium compounds   106, 107, 152, 227
Organotin compounds, see Stannanes
Organotin hydrides
 - addition to alkenes and alkynes   137, 138
 - as cleavage reagent   107, 116, 117
Organozinc compounds
 - alkylation of   146, 147, 281
 - arylation of   147, 170, 171, 359, 360, 380
 - preparation of   133, 135
Orthoformates   240
Osmium complexes
 - asymmetric dihydroxylation with   192
 - support–bound   139
Osmium tetroxide   139, 215
Oxadiazoles   376, 377
Oxalic acid derivatives   93, 307, 392
Oxazines   400–402
Oxaziridines   192, 262, 263
Oxazocines   406
Oxazoles   373
Oxazolidines   373
 - hydrolysis of   69, 102, 103
Oxazolones, racemization of   415
Oxidation
 - of alcohols   278–280
 - of aldehydes   303, 305
 - of alkenes   278, 279, 304, 305
 - of aryl ethers   195, 197
 - of benzylamines   72, 385
 - of benzyl cyanides   278, 281
 - of benzyl ethers   86, 88
 - of benzyl halides   279
 - of dihydropyridines   378, 379

– preparation of heterocycles   108–110,
157, 158, 346, 381, 400, 403
Ring–opening cross metathesis   158, 159
Rink acid resin   37, 66, 87, 89, 312
Rink amide linker   56–58, 73, 79
Ruthenium complexes
– as catalysts for olefin metathesis   108,
109, 156–159
– support–bound   139
Ruthenium trichloride   219

S

Safety–catch linkers   43, 44, 57, 58, 61, 64
Samarium iodide   356–358
Saponification
– of carboxylic esters   39–42, 91, 93, 115,
191, 195, 196, 257
– of thiol esters   43, 44, 209, 210
Sasrin   36, 60, 66, 94
Sasrin aldehyde   202
Scandium complexes   139
Scavengers   34
Schiff bases, see Imines
Schotten–Baumann procedure   250, 254
– Experimental Procedure   254
Selenides
– oxidation of   106, 107, 227
– preparation of   227
– reduction of   106, 107, 116, 117, 349
Self metathesis   108, 157
Semicarbazones   101
Sephadex   27
Sepharose   27
Sequential cleavage   34
Serine
– linkers based on   95, 96, 102, 103
– oxidative cleavage of   280
– protection of   194, 195, 419
– racemization of   309
β–Sheet formation   420
Side–chain protection
– of oligonucleotides   425
– of peptides   419, 420
Sieber linker   56, 57, 73
Silanes
– as linkers
– for alcohols   90
– for alkenes   113, 114
– for arenes   111–113, 405, 406
– for carboxylic acids   39, 41

– for halides   98, 99
– preparation of   112, 137
– reaction with alcohols   89, 90
– reaction with alkenes   112, 137, 144
– reaction with carbonyl compounds   89,
90, 187, 189
– reaction with iminium salts   338, 340
– as scavengers   34
Silica   25, 26
Silyl enol ethers
– as linkers   100, 101
– preparation of   100
Silyl ethers
– cleavage of   195, 196
– as linkers   89, 90
– preparation of   89, 90, 195
Silyl ketene acetals   189, 190, 316
Sodium acetoxyborohydride   241–243
Sodium borohydride, reduction
– of aldehydes   186
– of anhydrides   186, 187
– of esters   93, 94
– of imines   240, 241, 243
– of ketones   186
– of nitro compounds   246
– of thiol esters   209, 210
– of tosylhydrazones   116
Sodium cyanoborohydride   240–243
Sodium periodate   104, 152, 219, 278, 280
Soluble supports   13, 14, 22
Solvents, swelling properties   15, 16
Sonogashira coupling   165–168
Spacers   26
Squaric acid derivatives   191, 193
Stannanes
– C–C couplings with   161, 163, 166–171,
276, 355, 360, 380
– as linkers   98, 106, 107
– preparation of   106, 135–138
– transmetallation of   135, 136
Stannous chloride (SnCl$_2$)   246–248
Stereoselective, see Diastereoselective,
Enantioselective
Stilbenes   160
Stille coupling   159, 161, 163, 166–171, 276,
355, 360, 380
Styrenes
– copolymers of   14–22
– polymerization of   14, 15, 19
– preparation of   106, 107, 109, 152–155,
159–163